Signal Processing

Springer
*London
Berlin
Heidelberg
New York
Barcelona
Budapest
Hong Kong
Milan
Paris
Santa Clara
Singapore
Tokyo*

Ezio Biglieri and Marco Luise (Eds)

Signal Processing in Telecommunications

**Proceedings of the 7th International Thyrrhenian Workshop on Digital Communications
Viareggio, Italy, September 10-14, 1995**

With 260 Figures

 Springer

Ezio Biglieri
Politecnico di Torino
Dipartimento di Elettronica, Corso Duca degli Abruzzi 24, 10129 Torino, Italy

Marco Luise
Università degli Studi di Pisa
Dipartimento di Ingegneria dell'Informazione, Via Diotisalvi 2, 56126 Pisa, Italy

Cover design based on: Figure 5, page 325. AfriSpace downlink coverage provided by the DAB system.

ISBN 3-540-76019-9 Springer-Verlag Berlin Heidelberg New York

British Library Cataloguing in Publication Data
Tyrrhenian International Workshop on Digital Communications
(7th : 1995 : Viareggio, Italy)
 Signal processing in telecommunications : proceedings of
the 7th International Thyrrhenian Workshop on Digital
Communications, Viareggio, Italy, September 10-14, 1995
 1.Signal processing 2.Signal theory (Telecommunication)
3.Signal processing - Digital techniques
I.Title II.Biglieri, Ezio III.Luise, Marco
621.3'822
ISBN 3540760199

Library of Congress Cataloging-in-Publication Data
Tyrrhenian International Workshop on Digital Communications (7th : 1995
 : Viareggio, Italy)
 Signal processing in telecommunications : proceedings of the 7th
International Thyrrhenian Workshop on Digital Communications,
Viareggio, Italy, September 10-14, 1995 / Ezio Biglieri and Marco
Luise, eds.
   p.   cm.
 Includes bibliographical references.
 ISBN 3-540-76019-9 (pbk. : alk. paper)
 1. Signal processing - - Digital techniques - - Congresses. 2. Digital
communications - - Congresses.   I. Biglieri, Ezio. II. Luise, Marco.
III. Title.
TK5102.9.T57   1995   95-51546
621.382'2 - - dc20   CIP

Apart from any fair dealing for the purposes of research or private study, or criticism or review, as permitted under the Copyright, Designs and Patents Act 1988, this publication may only be reproduced, stored or transmitted, in any form or by any means, with the prior permission in writing of the publishers, or in the case of reprographic reproduction in accordance with the terms of licences issued by the Copyright Licensing Agency. Enquiries concerning reproduction outside those terms should be sent to the publishers.

© Springer-Verlag London Limited 1996
Printed in Great Britain

The publisher makes no representation, express or implied, with regard to the accuracy of the information contained in this book and cannot accept any legal responsibility or liability for any errors or omissions that may be made.

Typesetting: Camera ready by authors
Printed and bound at the Athenæum Press Ltd, Gateshead
69/3830-543210 Printed on acid-free paper

Preface

It is probably an overstatement to say that the discipline of telecommunication systems is becoming an application of digital signal processing (DSP). However, there is no doubt that by the mid-1980s integrated circuit technology has advanced to such an extent that revolutionary advances in telecommunications are fostered by the introduction of new and powerful DSP algorithms. Actually, DSP has been recently playing a major role in the development of telecommunications systems: to name just one of the most widespread applications where this interaction has been most effective, we may mention the use of intelligent DSP to improve the performance of transmission systems by allowing sophisticated algorithm to be implemented in radio transmitters and receivers for personal communications. Other areas have equally benefited by the latest advances of DSP: speech coding and synthesis, speech recognition and enhancement, radar, sonar, digital audio, and remote sensing, just to cite a few.

With this in mind, when choosing the topic for the 7th Tyrrhenian Workshop on Digital Communications, whose contributions are collected in this book, we aimed at focusing on the state of the art and the perspectives of the interaction between DSP and telecommunications, two disciplines that are becoming increasingly intertwined.

Although by no means exhaustive of all the applications of DSP to telecommunications, we believe that the material presented in this book pinpoints the most interesting among them, and hence it will be considered as a useful tool for investigating this complex and highly challenging field. Particular emphasis is given to the description of advanced techniques and system architectures (such as neural networks, adaptive systems, transform-domain processing and the like) in the fields of signal encoding and transmission for multimedia and personal communications, ranging from modem design to image coding, from on-board processing in satellites networks to array processing in multiple-antenna radio stations.

We take the occasion to express our sincere appreciation to all the authors and organizers who have contributed to the Workshop.

Ezio Biglieri　　　　　　　　　　　　　　　　　　　　　　　　　　Marco Luise

Acknowledgements

The editors are much indebted and wish to express their sincere thanks to the components of the Technical Committee of the 1995 Edition of the International Tyrrhenian Workshop on Digital Communications, namely Riccardo De Gaudenzi from ESA/ESTEC in Noordwijk, The Netherlands, Anibal Figueiras-Vidal from the Polytehcnic University of Madrid, Spain, Umberto Mengali from the University of Pisa, Italy, James Modestino from Rensselaer Polytechnic, Troy NY, USA, Ramjee Prasad from the Technical University of Delft, The Netherlands, and Gregori Vazquez-Grau form the Polytechnic University of Cataluña, Barcelona, Spain, whose precious cooperation was essential to the organization of the Workshop and to the publication of this book.

The Workshop would not have come into being without the support of the components of the Organizing Committee, namely Prof. M. Mancianti and Prof. F. Russo from the University of Pisa, Italy, and Prof. G. Prati from the University of Parma, Italy, and without the sponsorship of the following organizations and institutions, which are gratefully acknowledged hereafter:

Consorzio Nazionale Interuniversitario Telecomunicazioni CNIT

European Space Agency

European Union - Human Capital and Mobility Program

STET S.p.A. - gruppo IRI

University of Pisa

Contents

1 Applications of DSP to Personal Communication Networks 1

BER Comparison of DS-CDMA and MC-CDMA for Frequency Selective Fading Channels 3
S. Hara, T.H. Lee and R. Prasad

Performance Enhancement of Indoor Channels Using Quasi-Coherent Combining of Multiple Antenna Signals 15
G.J.M. Janssen

Applications of Walsh Functions and the FHT in CDMA Technology 27
E. Zehavi

Decision Feedback Multi-User Detection for Multitone CDMA Systems 39
L. Vandendorpe and O. van de Wiel

Blind Adaptive Interference-Cancelling Detectors for DS/SS CDMA: Algorithms and Performance 53
R. De Gaudenzi, F. Giannetti and M. Luise

The Bootstrap Decorrelating Algorithm: A Promising Tool for Adaptive Separation of Multi-User CDMA Signals 68
Y. Bar-Ness

2 Joint Source/Channel Coding for Multimedia Applications 83

Rate-Distortion Performance of Several Joint Source and Channel Coding Schemes for Images 85
M. Ruf and J.W. Modestino

Bounds on the Performance of Vector Quantizers operating under Channel Errors over all Index Assignments 100
G. Ben-David and D. Malah

Adaptive Temporal & Spatial Error Concealment Measures in MPEG-2 Video Decoder with Enhanced Error Detection 112
S. Aign

Source and Channel Coding Issues for ATM Networks 124
V. Parthasarathy, J. Modestino and K. Vastola

Combined Source and Channel Coding for Wireless ATM LANs 135
E. Ayanoglu, P. Pancha, A.R. Reibman and S. Talwar

3 DSP in Channel Estimation/Equalization and Modem Design 151

Block-by-Block Channel and Sequence Estimation for ISI/Fading
Channels 153
K.H. Chang, W.S. Yuan and C.N. Georghiades

Timing Correction by Means of Digital Interpolation 171
M. Moeneclaey and K. Bucket

Digital Demodulator Architectures for BandPass Sampling Receivers ... 183
A.M. Guidi and L.P. Sabel

Quadrature Phase Error and Amplitude Imbalance Effects on Digital
Demodulator Performance 195
J.J. Wojtiuk and M. Rice

Maximum-Likelihood Sequence Estimation for OFDM 207
A. Vahlin and N. Holte

A New Analysis Method for the Performance of BandPass Sampling
Digital Demodulators Considering IF Filter Effects 220
L.P. Sabel

Comparison of Demodulation Techniques for MSK 232
U. Lambrette, R. Mehlan and H. Meyr

Variable Symbol-rate Modem Design for Cable and Satellite TV
Broadcasting 244
G. Karam, K. Maalej, V. Paxal and H. Sari

A New Class of Carrier Frequency and Phase Estimators Suited
for MPSK Modulation 256
I. Mortensen and M.L. Boucheret

4 Signal Processing in Satellite Networks 269

Payload Digital Processor Hardware Demonstration for
Future Mobile and Personal Communication Systems 271
A. D. Craig and F. A. Petz

Frequency Domain Switching: Algorithms, Performances,
Implementation Aspects 283
G. Chiassarini and G. Gallinaro

Design Study for a CDMA Based LEO Satellite Network:
Downlink System Level Parameters 294
S. G. Glisic, J. Talvitie, T. Kumpumäki, M. Latva-aho and J. Iinatti

A Multi-user Approach to Combating Co-channel Interference in Narrow
band Mobile Communications 308
J. Ventura-Traveset, G. Caire, E. Biglieri and G. Taricco

Digital Radio for the World: High-Grade Service Quality Through
On-Board Processing Techniques 321
G. Losquadro

5   Advanced Signal Processing Techniques 335

Digital Equalization using Modular Neural Networks: an Overview 337
J. Cid-Sueiro and A. R. Figueiras-Vidal

Vector Quantization Using Artificial Neural Network Models 346
A.S. Galanopoulos, J.E. Fowler, Jr., and S. C. Ahalt

Neural Networks for Communications and Signal Processing: Overview
and New Results 358
M. Ibnkahla and F. Castanie

Transform-Domain Signal Processing in Digital Communications 374
H. Sari and P.Y. Cochet

Fuzzy-Rule-Based Phase Estimator Suited for Digital Implementation .. 385
F. Daffara

6   Recent Applications of DSP Techniques 395

Analog Envelope Constrained Filters for Channel Equalization 397
A. Cantoni, B. Vo, Z. Zang and K.L. Teo

Design of All-Digital Wireless Spread-Spectrum Modems Using
High-Level Synthesis 409
P. Yeung, R. Subramanian, M. Barberis and M. Paff

Efiicient VHDL Code Generation for Digital Receiver Design 423
T. Grötker, U. Lambrette and H. Meyr

Blind Joint Equalization of Multiple Synchronous Mobile Users
for Spatial Division Multiple Access 435
D.T.M. Slock

Base-Station Antenna Arrays in Mobile Communications 447
B. Ottersten and P. Zetterberg

Part 1

Applications of DSP to Personal Communication Networks

BER Comparison of DS–CDMA and MC–CDMA for Frequency Selective Fading Channels

Shinsuke Hara[†], Tai-Hin Lee[‡] and Ramjee Prasad[‡]

[†]Department of Communication Engineering, Faculty of Engineering,
Osaka University, 2–1, Yamada–oka, Suita–shi, Osaka 565 Japan,
E–MAIL :hara@comm.eng.osaka-u.ac.jp
[‡]Telecommunications and Traffic–Control Systems Group,
Delft University of Technology, P.O.Box 5031, 2600GA, Delft,
The Netherlands

Abstract: This paper presents the advantages and disadvantages of DS–CDMA (Direct Sequence–Code Division Multiple Access) and MC–CDMA (Multi-Carrier-Code Division Multiple Access) systems in synchronous down–link mobile radio communication channels. Furthermore, the bit error rate (BER) performance is analyzed in frequency selective slow Rayleigh fading channels. We theoretically derive the BER lower bound for MC–CDMA system, and propose a simple multi–user detection method. In the BER analysis, we use the same multipath delay profiles for both DS–CDMA and MC–CDMA systems, and discuss the performance theoretically and by computer simulation. Finally, we theoretically prove that the time domain DS–CDMA Rake receiver is equivalent to the frequency domain MC–CDMA Rake receiver for the case of one user.

1 Introduction

Direct Sequence–Code Division Multiple Access (DS–CDMA) technique has been considered to be a candidate to support multi–media services in mobile communications, because it has its own capabilities to cope with asynchronous nature of multi–media data traffic, to provide higher capacity over conventional access techniques such as TDMA and FDMA, and to combat the hostile channel frequency selectivity.

Recently, another CDMA technique based on a combination of the CDMA and the orthogonal frequency division multiplexing (OFDM) signaling has been reported in [1],[2]. This technique is called "Multi-Carrier–CDMA (MC–CDMA) technique", and much attention has been paid to it, because it is potentially robust to the channel frequency selectivity with a good frequency utilization efficiency.

In the bit error rate (BER) analysis of DS–CDMA system, "independent fading characteristic at each received path" is assumed, on the other hand, "independent fading characteristic at each sub–carriers" is considered in the BER analysis of MC–CDMA system[1],[2]. When the multipath channel is a wide sense stationary uncorrelated scattering (WSSUS) one, the assumption for the DS–CDMA system is correct, however, the assumption for the MC–CDMA system is not correct. In the BER analysis of MC–CDMA system, we should take account of frequency correlation function determined by the multipath delay profile of the channel, and in the BER comparison, we should make a fair assumption for both DS–CDMA and MC–CDMA systems using the same channel frequency selectivity, that is, the same multipath delay profile. To the best of the authors' knowledge, no paper has been reported on the fair BER comparison.

In this paper, we discuss the advantages and disadvantages of DS–CDMA and MC–CDMA systems in synchronous down–link mobile radio communication channels, and analyze the bit error rate (BER) performance in frequency selective slow Rayleigh fading channels theoretically and by computer simulation. Section 2 explains the DS–CDMA and MC–CDMA systems and discusses the advantages and disadvantages of MC–CDMA system over DS–CDMA system. Section 3 shows the theoretical derivation of BER lower bound for the MC–CDMA system, and proposes a simple multi–user detection method. Section 4 discusses the BER performance in frequency selective slow Rayleigh fading channels. Section 5 theoretically proves the equivalence of time domain DS–CDMA Rake receiver and frequency domain MC–CDMA Rake receiver for the case of one user. Section 6 draws the conclusions.

2 DS–CDMA and MC–CDMA Systems

2.1 DS–CDMA System

DS–CDMA transmitter spreads the original signal using a given spreading code in the time domain (see Fig.1). The capability of suppressing multi–user interference is determined by the cross–correlation characteristic of the spreading codes. Also, a frequency selective fading channel is characterized by the superimposition of several signals with different delays in the time domain[3]. Therefore, the capability of distinguishing one component from other components in the composite received signal (time resolution) is determined by the auto–correlation characteristic of the spreading codes.

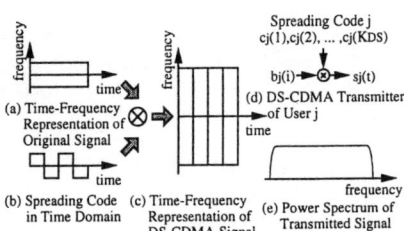

Fig. 1: Concept of DS–CDMA System

Fig.1(d) shows the DS–CDMA transmitter of the j-th user for Binary PSK modulation/coherent demodulation (CBPSK) scheme. The complex equivalent low-pass transmitted signal is written by

$$s_j(t) = \sum_{i=-\infty}^{+\infty} \sum_{k=1}^{K_{DS}} b_j(i) c_j(k) p(t - kT_c - iT_s), \qquad (1)$$

where, $b_j(i)$ and $c_j(k)$ are the i-th information and the k-th bit of the spreading code with length K_{DS} and chip duration T_c, respectively, T_s is the symbol duration, and $p(t)$ is the pulse waveform defined as:

$$p(t) = \begin{cases} 1 & (0 \le t \le T_c) \\ 0 & (otherwise). \end{cases} \qquad (2)$$

The BER is determined by the path diversity strategy in the receiver, and the diversity order depends on how many fingers the Rake receiver employs. Usually, the diversity order of 1 (Non–Rake), 2 or 3 is used depending on hardware limitation. Furthermore, when the Nyquist filters are introduced in the transmitter and receiver for base band pulse shaping, the Rake receiver may wrongly combine paths. This is because noise causing distortion in auto-correlation characteristic often results in wrong correlation. Finally, it is difficult for the DS–CDMA Rake receiver to use all the received signal energy scattered in the time domain.

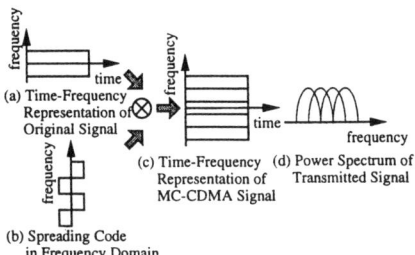

Fig. 2: Concept of MC–CDMA System

2.2 MC–CDMA System

The MC–CDMA transmitter spreads the original signal using a given spreading code in the frequency domain (see Fig.2). It is crucial for multi-carrier transmission to have frequency non-selective fading over

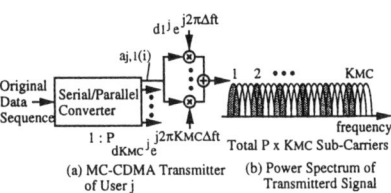

Fig. 3: MC–CDMA Transmitter of User j

each sub-carrier. Therefore, if the original symbol rate is high enough to become subject to frequency selective fading, the signal needs to be first serial-to-parallel converted before spreading over the frequency domain. Also, in a synchronous down-link mobile radio communication channel, we can use the Hadamard Walsh codes as an optimum orthogonal code set, because we do not have to pay attention to the auto-correlation characteristic of the spreading codes.

Fig.3(a) shows the MC–CDMA transmitter of the j-th user for CBPSK scheme, where the input information sequence is converted into P parallel data sequences ($a_{j,1}(i)$, $a_{j,2}(i), \cdots, a_{j,P}(i)$). The complex equivalent low-pass transmitted signal is written by

$$s_j(t) = \sum_{i=-\infty}^{+\infty} \sum_{p=1}^{P} \sum_{m=1}^{K_{MC}} a_{j,p}(i) d_m^j p(t - iT_s) e^{j2\pi \Delta f(m + \frac{p-1}{P})t}, \quad \Delta f = 1/T_s, \qquad (3)$$

where $\{d_1^j, d_2^j, \cdots, d_{K_{MC}}^j\}$ is the Hadamard Walsh code for the j-th user (the length is K_{MC}) and Δf is the sub-carrier separation for $a_{j,p}(i)$. Also, the total number of sub-carriers is $P \times K_{MC}$.

2.3 Advantages and Disadvantages

2.3.1 Advantages of MC–CDMA System over DS–CDMA System

We define the following parameters:

- Transmission Rate : $R \ (= 1/T_s) \ [bits/sec]$,

- Processing Gain : K_{DS} (DS–CDMA), K_{MC} (MC–CDMA),

- The Maximum Number of Users : M_{DS} (DS-CDMA), M_{MC} (MC-CDMA),
- The Number of Sub-Carriers : $N\ (= P \times K_{MC})$.

The required frequency bandwidth (main-lobe) for the DS-CDMA system becomes

$$B_{DS} = 2 \cdot R \cdot K_{DS}, \tag{4}$$

on the other hands, for the MC-CDMA system,

$$B_{MC} = R \cdot K_{MC} \cdot (N+1)/N \approx R \cdot K_{MC}. \tag{5}$$

From Eqs.(4) and (5),

$$K_{MC} = 2 \cdot K_{DS} \quad if \quad B_{MC} = B_{DS}. \tag{6}$$

Eq.(6) shows that the processing gain of MC-CDMA system is twice as large as that of DS-CDMA system for a given frequency bandwidth. Furthermore, the DS-CDMA system cannot accommodate K_{DS} users ($M_{DS} < K_{DS}$) because we need to choose the spreading codes with good auto- and cross-correlation characteristics carefully, while the MC-CDMA system can accommodate K_{MC} users ($M_{MC} = K_{MC}$) using the Walsh Hadamard codes:

$$M_{MC} > 2 \cdot M_{DS} \quad if \quad B_{MC} = B_{DS}. \tag{7}$$

2.3.2 Disadvantages of MC-CDMA System over DS-CDMA System

At the MC-CDMA receiver, we have to make every effort in the FFT window position synchronization, the frequency offset compensation and the coherent detection at each sub-carrier. Also, the MC-CDMA transmitter requires a large input backoff in the amplifier, because it is very sensitive to nonlinear amplification.

It is also pointed out that the BER of multi-carrier modulation itself is inferior to that of single-carrier modulation mainly due to the power loss associated with guard interval insertion[4].

3 BER Analysis

3.1 Frequency Selective Slow Rayleigh Fading Channel Model

We assume a wide sense stationary uncorrelated scattering (WSSUS) channel model[3] with L received paths in the complex equivalent low-pass time-variant impulse response:

$$h(\tau;t) = \sum_{l=1}^{L} g_l(t)\delta(\tau - \tau_l). \tag{8}$$

where t and τ are the time and the delay, respectively, $\delta(t)$ is the Dirac delta function, $g_l(t)$ is the complex envelope of the signal received on the l-th path which is a complex Gaussian random process with zero mean and variance σ_l^2, and τ_l is the propagation delay for the l-th path.

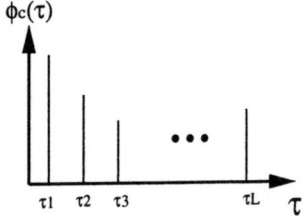

Fig. 4: Multipath Delay Profile

Fig.4 shows the corresponding multipath delay profile given by

$$\phi_c(\tau) = \frac{1}{2} E\left[h^*(\tau;t) \cdot h(\tau;t)\right] = \sum_{l=1}^{L} \sigma_l^2 \delta(\tau - \tau_l), \tag{9}$$

where $E[\cdot]$ is the expectation.

3.2 Communication System Model

We consider a synchronous down-link communication channel, where the signal is transmitted in a burst format with a preamble and a postamble. In this paper, we assume that the receiver can correctly estimate all the channel state information.

3.3 BER of DS-CDMA System

The transmitted signal for total J users is written by

$$s(t) = \sum_{j=1}^{J} s_j(t). \tag{10}$$

The received signal through the frequency selective slow Rayleigh fading channel given by Eq.(8) is written by

$$r(t) = \int_{-\infty}^{+\infty} s(t-\tau) * h(\tau;t) d\tau + n(t) = \sum_{l=1}^{L} r_l(t) + n(t), \quad r_l(t) = s(t-\tau_l) g_l(t), \tag{11}$$

where $n(t)$ is the complex additive Gaussian noise. Defining $\mathbf{r_t}$ as the received signal vector, the time domain covariance matrix $\mathbf{M_t}$ is given by

$$\mathbf{r_t} = [r_1, r_2, \cdots, r_L]^T, \quad \mathbf{M_t} = \frac{1}{2} E\left[\mathbf{r_t} \cdot \mathbf{r_t}^T\right] = \begin{bmatrix} \sigma_1^2 & 0 & \cdots & 0 \\ 0 & \sigma_2^2 & & \vdots \\ \vdots & & \ddots & 0 \\ 0 & \cdots & 0 & \sigma_L^2 \end{bmatrix}, \tag{12}$$

where T is the transpose. In the above equation, we assume a perfect auto-correlation characteristic for the spreading codes.

The BER of time domain I-finger DS-CDMA Rake receiver is uniquely determined by the eigenvalues of $\mathbf{M_t}$ (in this case, the eigenvalues are clearly $\sigma_1^2, \sigma_2^2, \cdots, \sigma_L^2$)[5]. For example, when σ_l^2 ($l = 1, \cdots, L$) are different each other, the BER is expressed as

$$BER = \sum_{l=1}^{I} w_l \cdot \frac{1}{2} \left\{ 1 - \sqrt{\frac{\sigma_l^2/N'}{1 + \sigma_l^2/N'}} \right\}, \quad w_l = \frac{1}{\prod_{\substack{n=1 \\ n \neq l}}^{I} \left(1 - \frac{\sigma_n^2}{\sigma_l^2}\right)}, \tag{13}$$

$$E_b = \sum_{l=1}^{L} \sigma_l^2, \quad N' = N_0 + \frac{2(J-1)}{3K_{DS}} E_b, \tag{14}$$

where E_b and N_0 are the signal energy per bit and the noise spectral density, respectively. Also, when σ_l^2 ($l = 1, \cdots, L$) are all the same ($= \sigma^2$) [3],

$$BER = \left(\frac{1-\mu}{2}\right)^I \sum_{l=0}^{I-1} \binom{I-1+l}{l} \left(\frac{1+\mu}{2}\right)^l, \quad \mu = \sqrt{\frac{\sigma^2/N'}{1 + \sigma^2/N'}}. \tag{15}$$

Eqs.(13) and (15) are based on the Gaussian approximation for multi-user interference[6].

3.4 BER of MC–CDMA System

The transmitted signal for total J users is written by

$$s(t) = \sum_{j=1}^{J} s_j(t). \tag{16}$$

The received signal through the frequency selective slow Rayleigh fading channel is written by

$$r(t) = \sum_{i=-\infty}^{+\infty} \sum_{p=1}^{P} \sum_{m=1}^{K_{MC}} \sum_{j=1}^{J} z_{m,p} a_{j,p}(i) d_m^j p(t - iT_s) e^{j2\pi \Delta f(m + \frac{p-1}{P})t} + n(t), \tag{17}$$

where $z_{m,p}$ is the complex received signal at the $(mP + p - 1)$-th sub-carrier.

Fig.5 shows the MC-CDMA receiver of the j'-th user, where after the serial-to-parallel conversion using the FFT, the m-th sub-carrier component for the received data $a_{j,p}(i)$ is multiplied by the gain G_m and despreading code $d_m^{j'}$ to combine the energy of received signal scattered in the frequency domain. The decision variable is given by (we can omit the subscriptions p and i without loss of generality)

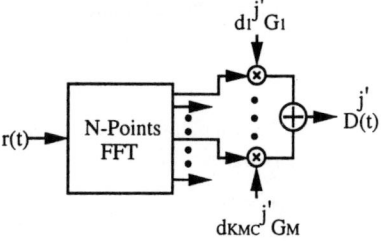

Fig. 5: MC-CDMA Receiver of The j'-th User

$$D^{j'}(t) = \sum_{m=1}^{K_{MC}} \sum_{j=1}^{J} G_m d_m^{j'} \{z_m a_j d_m^j + n_m(t)\}, \tag{18}$$

where $n_m(t)$ is the complex additive Gaussian noise at the m-th sub-carrier.

3.4.1 Orthogonality Restoring Single–User Detection Method

Choosing the gain G_m as

$$G_m = z_m^* / |z_m|^2, \tag{19}$$

the receiver can eliminate the multi–user interference perfectly. However, low–level sub–carriers tend to be multiplied by the high–gains, so the BER degrades due to noise amplification (a weak signal suppression method using a detection threshold is presented in [8]).

3.4.2 Frequency Domain Rake Receiver with No Simultaneous Other User

When there is no simultaneous other user, the frequency domain MC-CDMA Rake receiver based on the maximum ratio combining method in the frequency domain ($G_m = z_m^*$) can achieve the best BER performance (the BER lower bound)[3].

Defining $\mathbf{r_f}$ as the received signal vector, the frequency domain covariance matrix $\mathbf{M_f}$ is given by

$$\mathbf{r_f} = [z_1, z_2, \cdots, z_{K_{MC}}]^T, \quad \mathbf{M_f} = \frac{1}{2} E\left[\mathbf{r_f} \cdot \mathbf{r_f}^T\right] = \{m_f^{a,b}\}, \tag{20}$$

$$m_f^{a,b} = \Phi_C((a-b)\Delta f), \tag{21}$$

where $m_f^{a,b}$ is the $a-b$ element of $\mathbf{M_f}$, and $\Phi_C(\Delta f)$ is the spaced frequency correlation function defined as the Fourier transform of the multipath delay profile given by Eq.(9):

$$\Phi_C(\Delta f) = \int_{-\infty}^{+\infty} \phi_c(\tau) e^{-j2\pi\Delta f\tau} d\tau. \quad (22)$$

Defining $\lambda_1, \lambda_2, \cdots, \lambda_{K_{MC}}$ as the eigenvalues of $\mathbf{M_f}$, the BER is given by a form similar to Eq.(13) or Eq.(15)[5], where we can substitute I, N', L and σ^2 for K_{MC}, N_0, K_{MC} and λ, respectively.

3.4.3 Simple Multi–User Detection Method

In this paper, we propose a simple multi-user detection method. If the preamble contains information on the spreading codes used by the simultaneous users, any user can know it easily (although providing multi-user information in the downlink channel is questionable from the viewpoint of security). In this method, the user first estimates information for simultaneous other $J-1$ users using the orthogonality restoring single–user detection method. After removing the interference component from the received signal, the user detects its own information using the frequency domain maximum ratio combining method. If the decisions for the other users are correct, this detection method can minimize the BER.

4 Numerical Results

We assume the following system parameters to demonstrate the BER performance: • $R = 3.0[Mbits/sec]$, • $K_{DS} = 31(Gold Codes)$, • $K_{MC} = 32$.

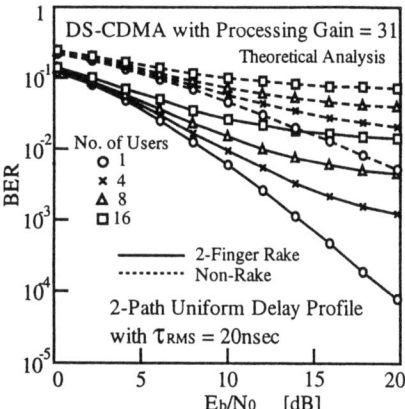

Fig. 6: BER of DS–CDMA System (2–Path Uniform Delay Profile)

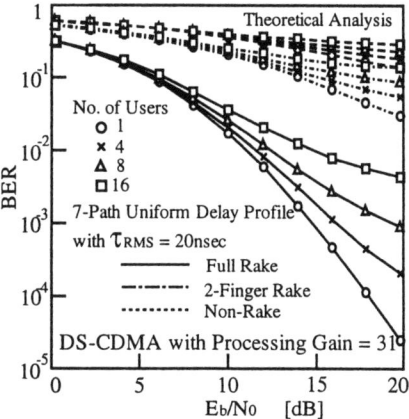

Fig. 7: BER of DS–CDMA System (7–Path Uniform Delay Profile)

4.1 BER Performance of DS–CDMA System

Figs.6, 7 and 8 show the BER performance of DS-CDMA system in frequency selective slow Rayleigh fading channels with 2-path uniform, 7-path uniform and 7-path exponential multipath delay profiles, respectively. All the delay profiles have the same RMS delay spread $\tau_{RMS}=20[nsec]$.

As compared with the full–finger Rake receiver, the performance of Non–Rake receiver is poor. For the delay profiles where several delayed paths have the same average power as the first path, there is a large difference in the attainable BER between the Non–Rake and full–finger Rake receivers. Also, as the number of users increases, the performance gradually degrades. For the 7–path uniform and exponential multipath delay profiles, the full–finger Rake receiver means 7–finger Rake receiver. From the practical point of view, its realization could be difficult.

4.2 Design of MC–CDMA System

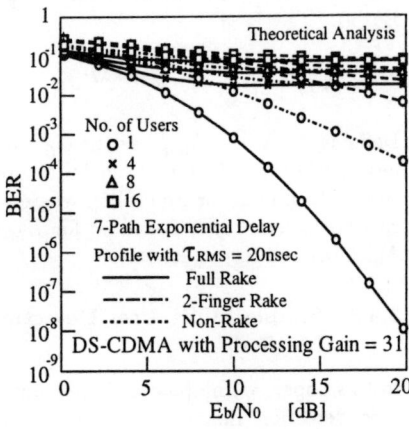

Fig. 8: BER of DS–CDMA System (7–Path Exponential Delay Profile)

As the number of sub–carriers (N) increases, the transmission performance becomes more sensitive to the time selectivity because the wider symbol duration is less robust to the random FM noise. On the other hand, as N decreases, it becomes poor because the wider power spectrum of each sub–carrier is less robust to the frequency selectivity. Therefore, there exists an optimum value in N to minimize the BER[9].

Also, as the guard duration (Δ) increases, the transmission performance becomes poor because the signal transmission in the guard duration introduces the power loss. On the other hand, as Δ decreases, it becomes more sensitive to the frequency–selectivity because the shorter guard duration is less robust to the delay spread. Therefore, there exists an optimum value in Δ to minimize the BER[9].

In [9], it is shown that when the product of the maximum Doppler frequency f_D and the root mean square delay spread τ_{RMS} introduced in the channel satisfies the following condition:

$$\tau_{RMS} \cdot f_D < 1.0 \times 10^{-6}, \tag{23}$$

the multi–carrier modulation scheme can achieve almost the same BER performance as a single–carrier modulation scheme with equalization.

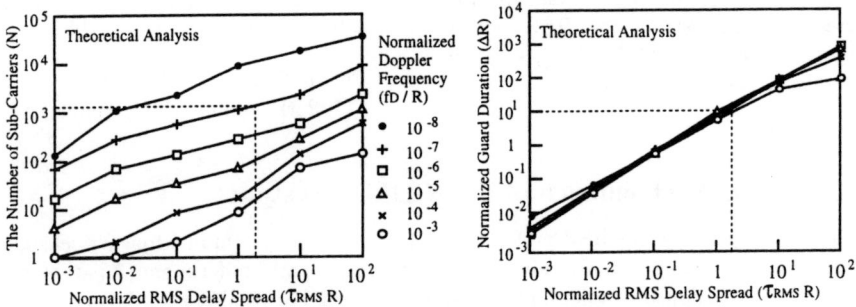

Fig. 9: Design of The Number of Sub–Carriers and Guard Period

Fig.9 shows the optimum values of the number of sub-carriers and guard period for $f_D = 10[Hz]$ and $\tau_{RMS} = 20[nsec]$. The MC-CDMA system with $N = 1024$ and $\Delta = 100[nsec]$ can minimize the BER, where the original information sequence is first converted into 32 parallel sequences ($P=32$), and then each sequence is mapped onto 32 sub-carriers. Also, the power loss associated with guard period insertion is negligible small (the normalized guard duration is about 1%).

4.3 BER Performance of MC-CDMA System

Fig.10 shows the BER performance of MC-CDMA system, where the delay profiles are all the same as those used in the analysis of DS-CDMA system. The frequency domain Rake receiver with no simultaneous other user can achieve the best performance easily, because it can effectively combine the energy of received signal scattered in the frequency domain using a lot of sub-carriers. On the other hand, the performance of orthogonality restoring single-user detection method is poor, although it is insensitive to the number of users. Among three delay profiles, the performance in the 2-path uniform delay profile is slightly better, because there is a less distortion in the frequency domain.

Fig.11 shows the BER performance of the proposed multi-user detection method for the 2-path uniform delay profile. The proposed multi-user detection method is simple and can improve the BER as compared with the single-user detection method. However, the performance gradually degrades as the number of users increases. If more sophisticated (but more complicated) multi-user detection methods such as the Wiener filtering detection[10], the maximum-likelihood detection[7] and the decorrelating interference canceler[11], are employed, the performance can be more improved.

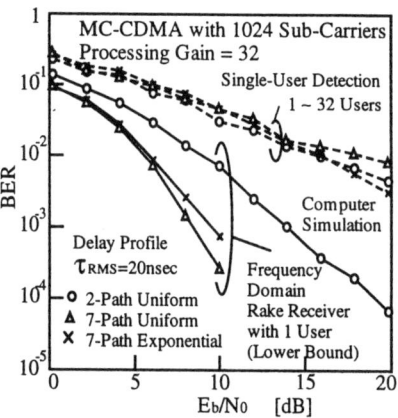

Fig. 10: BER of MC-CDMA System for Single-User Detection and Frequency Domain Rake Methods

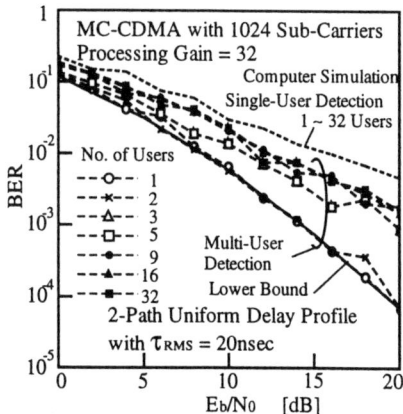

Fig. 11: BER of MC-CDMA System for Multi-User Detection Method (2-Path Uniform Delay Profile)

4.4 BER Comparison of DS–CDMA and MC–CDMA Systems

Fig.12 shows the BER comparison of DS–CDMA and MC–CDMA systems. The best performance of MC–CDMA system agrees well with that of DS–CDMA system. This implies that the frequency domain Rake receiver is equivalent to the time domain Rake receiver for the case of no other user. The next section theoretically proves the equivalence of time domain and frequency doman Rake receivers.

For the performance with 16 users, the MC–CDMA system with the single–user detection method does not work well (but it is insensitive to the number of users). Also, the performance of DS–CDMA system with full-finger Rake receiver is not so good (it becomes worse as the number of users increases).

The performance of DS–CDMA system with non–Rake receiver is much worse than that of MC–CDMA system. Therefore, a Rake receiver could be necessary for the DS–CDMA system.

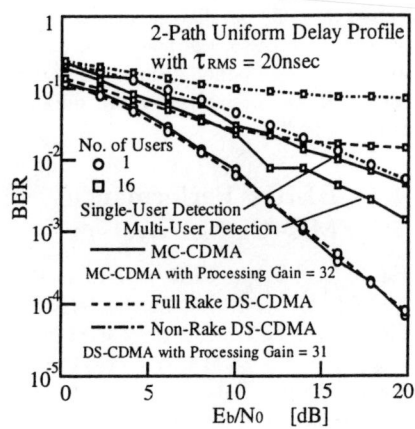

Fig. 12: BER Comparison of DS–CDMA and MC–CDMA Systems (2–Path Uniform Delay Profile)

5 Equivalence of Time Domain DS–CDMA and Frequency Domain MC–CDMA Rake Receivers

As shown in Section 3, the BER of Rake receiver is determined by the eigenvalues of the channel covariance matrix. Therefore, for a multipath delay profile, if the frequency domain covariance matrix has all the same eigenvalues as the time domain covariance matrix, the BER of frequency domain Rake receiver is all the same as that of time domain Rake receiver.

In the FFT, the frequency resolution is determined by the observation period. Therefore, for the multi–carrier modulation with N sub–carriers, N–point DFT is required in the symbol duration T_0. Assume the following $N \times N$ time domain covariance matrix with the time resolution of T_0/N, for example, for the multipath delay profile shown in Fig.13:

$$\mathbf{M}'_t = \begin{bmatrix} \sigma_1^2 & 0 & \cdots & \cdots & \cdots & 0 \\ 0 & 0 & & & & \vdots \\ \vdots & & \sigma_1^2 & & & \vdots \\ \vdots & & & \sigma_3^2 & & \vdots \\ \vdots & & & & \ddots & 0 \\ 0 & \cdots & \cdots & \cdots & 0 & \ddots \end{bmatrix}, \quad (24)$$

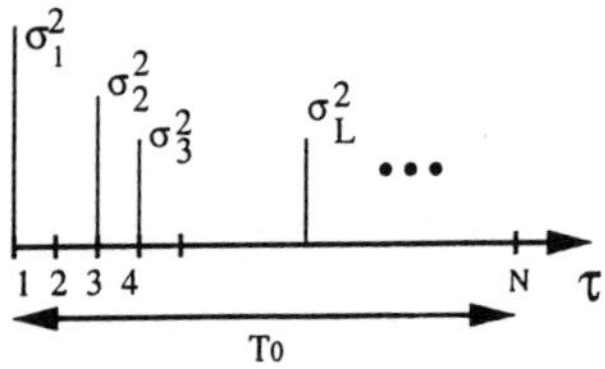

Fig. 13: Multipath Delay Profile

where the non–zero eigenvalues of $\mathbf{M}'_t$ are $\sigma_1^2, \sigma_2^2, \sigma_3^2, \cdots, \sigma_L^2$.

Using $\mathbf{M}'_t$, the frequency domain covariance matrix is written by

$$\mathbf{M}'_f = \mathbf{W}\mathbf{M}'_t\mathbf{W}^*, \tag{25}$$

where $\mathbf{W}$ is the $N \times N$ Discrete Fourier Transform matrix given by

$$\begin{aligned}\mathbf{W} &= \{w^{i,j}\},\\ w^{i,j} &= e^{j2\pi \frac{ij}{N}}.\end{aligned} \tag{26}$$

We define $\mathbf{r}_l$ as the eigenvector corresponding to the eigenvalue σ_l^2:

$$\mathbf{M}'_t\mathbf{r}_l = \sigma_l^2 \mathbf{r}_l \quad (l = 1, 2, \cdots, L). \tag{27}$$

Also, we define $\mathbf{z}_l$ as

$$\mathbf{z}_l = \mathbf{W}\mathbf{r}_l \quad (l = 1, 2, \cdots, L). \tag{28}$$

Now, we can theoretically prove that the frequency domain covariance matrix has all the same eigenvalues as the time domain covariance matrix as follows:

$$\begin{aligned}\mathbf{M}'_f \mathbf{z}_l &= \mathbf{W}\mathbf{M}'_t\mathbf{W}^* \cdot \mathbf{W}\mathbf{r}_l \\ &= \mathbf{W}\mathbf{M}'_t\mathbf{r}_l \\ &= \mathbf{W}\sigma_l^2\mathbf{r}_l \\ &= \sigma_l^2 \mathbf{W}\mathbf{r}_l \\ &= \sigma_l^2 \mathbf{z}_l.\end{aligned} \tag{29}$$

The above equation clearly shows that the eigenvalues of $\mathbf{M}'_f$ are $\sigma_1^2, \sigma_2^2, \sigma_3^2, \cdots, \sigma_L^2$.

Also, we can see that *the assumption of independent fading characteristic at each sub-carrier implies a frequency selective fading at each sub-carrier as long as we employ the OFDM signaling*, because it requires N paths uniformly scattered in the symbol duration.

6 Conclusions

This paper has discussed the advantages and disadvantages of DS-CDMA and MC-CDMA systems, and analyzed the BER performance in given frequency selective Rayleigh fading channels.

The MC-CDMA system can accommodate more users than the DS-CDMA system for a given frequency bandwidth, on the other hand, it has to make an extra effort in the FFT window synchronization at the receiver and the linear amplification at the transmitter.

For the DS-CDMA system, the Rake receiver is necessary to improve the BER in frequency selective fading channels. However, from the hardware limitation, it is difficult to use all the received signal energy scattered in the time domain.

The MC-CDMA system can easily combine the signal energy scattered in the frequency domain. However, a multi-user detection method is also necessary.

In the best BER comparison, there is no difference between the DS-CDMA and MC-CDMA systems. If the DS-CDMA system cannot work well for a given channel condition and a given system condition, in other words, a given frequency selectivity and a given processing gain, the MC-CDMA system can be attractive even at the sacrifice of cost for the sub-carrier synchronization, because it can effectively combine the signal energy scattered in the frequency domain for any frequency selectivity and any processing gain using a number of sub-carriers.

Acknowledgement

The authors wish to thank Prof. Y. Bar-Ness of New Jersey Institute of Technology, Dr. J-P. Linnartz of Philips Research and Assist. Prof. M. Okada of Osaka University for their helpful comments and fruitful discussions.

References

[1] N.Yee, J-P.Linnartz and G.Fettweis : "Multi-Carrier CDMA in Indoor Wireless Radio Networks," Proc. of IEEE PIMRC'93, pp.109-113, Yokohama, Japan, Sept. 1993.

[2] K.Fazel and L.Papke : "On the Performance of Convolutionally-Coded CDMA/OFDM for Mobile Communication System," Proc. of IEEE PIMRC'93, pp.468-472, Yokohama, Japan, Sept. 1993.

[3] J.G.Proakis : "Digital Communications," Mc-Graw Hill, 2nd Ed., 1991.

[4] H.Sari and I.Jeanclaude : "An Analysis of Orthogonal Frequency-Division Multiplexing for Mobile Radio Applications," Proc. of 1994 IEEE VTC'94, pp.1635-1639, Stockholm, Sweden, June 1994.

[5] P.Monsen : "Digital Transmission Performance on Fading Dispersive Diversity Channels," IEEE Trans. Commun., vol.COM-21, pp.33-39, Jan. 1973.

[6] M.B.Pursley : "Performance Evaluation for Phase-Coded Spread-Spectrum Multiple-Access Communications-Part I : System Analysis," IEEE Trans. Commun., vol.COM-25, pp.795-799, Aug. 1977.

[7] K.Fazel : "Performance of CDMA/OFDM for Mobile Communication System," Proc. of IEEE ICUPC'93, pp.975-979, Ottawa, Canada, Oct. 1993.

[8] N.Yee and J-P.Linnartz : "Controlled Equalization of Multi-Carrier CDMA in an Indoor Rician Fading Channel," Proc. of IEEE VTC'94, pp.1665-1669, Stockholm, Sweden, June 1994.

[9] S.Hara, M.Mouri, M.Okada and N.Morinaga : "Transmission Performance Analysis of Multi-Carrier Modulation in Frequency Selective Fast Rayleigh Fading Channel,", accepted for publication in Wireless Personal Communications : An International Journal, Special Issue on Multi-Carrier communications, Kluwer Academic Publishers.

[10] N.Yee and J-P.Linnartz : "Wiener Filtering of Multi-Carrier CDMA in a Rayleigh Fading Channel," Proc. of IEEE PIMRC'94, pp.1344-1347, The Hague, The Netherlands, Sept. 1994.

[11] Y.Bar-Ness, J-P.Linnartz and X.Lin : "Synchronous Multi-User Multi-Carrier CDMA Communication System with Decorrelating Interference Canceler," Proc. of IEEE PIMRC'94, pp.184-188, The Hague, The Netherlands, Sept. 1994.

Performance Enhancement of Indoor Channels Using Quasi-Coherent Combining of Multiple Antenna Signals

Gerard J.M. Janssen
Telecommunications and Traffic Control Systems Group
Delft University of Technology
Delft, The Netherlands
Telephone +31 15 786736; Fax +31 15 781774
E-mail: g.janssen@et.tudelft.nl

Abstract: In this paper, the performance of quasi-coherent combining (QCC) of signals from an antenna array with closely spaced elements, is evaluated for the Rician fading indoor multipath channel at microwave frequencies. The channel model is based on statistical measurement results on channel variation for small antenna displacements at 2.4, 4.75 and 11.5 GHz. Computational BER results for QCC and coherent BPSK modulation are presented and compared to the results for existing diversity techniques.

1. INTRODUCTION

Indoor wireless communication at microwave and millimetre wave frequencies has been a research topic of increasing interest in recent years. The advantages of the higher frequency bands are: *i.* availability of large bandwidths; *ii.* better shielding by concrete walls, which results in smaller cell sizes; and *iii.* small antenna dimensions. The indoor channel, however, has a multipath nature which causes severe signal dispersion, and therefore limits the maximum usable symbol rate.

Different antenna diversity techniques can be applied to combat the adverse multipath effects in the indoor channel. Well known diversity techniques are: *i.* Unity Gain combining (UGC); *ii.* Selection Diversity (SD); *iii.* Maximal-Ratio combining (MRC); and *iv.* Minimum Mean Square Error combining (MMSEC) [1, 2]. MRC and MMSEC perform well in a wideband multipath channel, however, both techniques require a priori knowledge of the channel impulse response at each array element. The terms narrowband and wideband indicate that the occupied bandwidth to transmit information is smaller or larger than the coherence bandwidth of the channel, respectively.

In this paper, a way of combining multiple antenna signals is considered, which does not require a priori knowledge of the channel impulse response. The technique, first presented in [3], is based on quasi-coherent combining of the dominant paths of the signals received by M different closely spaced ($\lambda/2 < d < a\,few\,\lambda$) antenna elements, and is called Quasi-Coherent combining (QCC). This technique exploits the fact that the channel power delay profile

(PDP) is nearly constant for all elements of a small array at microwave frequencies, however, with different relative path phases.

In the following section, the model for the multipath fading indoor channel is reviewed. Statistical results on channel variation for closely spaced elements are presented, based on indoor measurements performed at 2.4, 4.75 and 11.5 GHz. The QCC technique is discussed in section 3. In section 4, antenna diversity techniques as mentioned above are briefly revisited. In section 5, computational bit error rate (BER) results are presented and compared to the results of existing combining schemes for Binary Phase Shift Keying (BPSK) modulation. Finally, conclusions are given in section 6.

2. INDOOR MULTIPATH CHANNEL MODEL

2.1 Channel model

In a multipath channel, multiple replicas of the signal simultaneously arrive at the receiver antenna, each having different amplitude, phase and time delay. This effect causes frequency selectivity and dispersion of the signal. The multipath channel with discrete paths is modelled as [4]:

$$h(t) = \sum_{k=0}^{L} \beta_k \exp(j\theta_k)\delta(t-\tau_k) \qquad (1)$$

where, β_k is the amplitude, θ_k is the phase and τ_k is the time delay of the kth path. $\delta(.)$ is the Dirac delta function. The $L+1$ received path signals consist of one line-of-sight (LOS) signal or otherwise dominant path signal, and L reflected signals. The channel is assumed to be time invariant, so β_k, θ_k and τ_k are taken constant. The power delay profile (PDP) is derived from the impulse response as $PDP(t) = |h(t)|^2$. An important characteristic of the multipath channel is the Root Mean Square (RMS) delay spread τ_{RMS}:

$$\tau_{RMS} = \sqrt{\frac{N_2}{N_0} - \left(\frac{N_1}{N_0}\right)^2} \qquad \text{with } N_n = \sum_{k=0}^{L} \beta_k^2 \tau_k^n \qquad (2)$$

The indoor channel with a dominant path can be modelled as a Rician fading channel where the distribution of the instantaneous received power P_i is

$$f_{Rice}(P_i|K,\overline{P}) = \frac{1+K}{\overline{P}} \exp\left(-\frac{(1+K)P_i + K\overline{P}}{\overline{P}}\right) I_0\left(2\sqrt{\frac{K(1+K)P_i}{\overline{P}}}\right) \qquad (3)$$

Here $\overline{P} = N_0$ is the average received power, and K is the Rice factor which is defined as

$$K = \frac{P_{Dominant}}{P_{Reflected}} = \frac{\beta_0^2}{\sum_{k=1}^{L} \beta_k^2} \qquad (4)$$

where it is assumed that path $k = 0$ is the dominant path. For obstructed (OBS) channels without LOS path often Rayleigh fading is assumed. However, in general, there will be a dominant path. Using the Rician distribution, this would result in a small K-value. For $K \ll 1$ the Rician distribution changes to the Rayleigh distribution.

2.2 Measurement results

A wideband coherent frequency response measurement technique as described in [5], has been applied to measure the complex impulse response of indoor channels. The measurements have been performed in clusters of six positions located on a circle with a diameter of 12.5 cm at three frequencies: 2.4, 4.75 and 11.5 GHz. So the distance between the elements was $\lambda/2 < d < 4\lambda$. Measurement results on path loss, RMS delay spread and coherence bandwidth characteristics are given in [6]. In the following, new statistical results on parameter variations for small antenna displacements as found within the measured clusters, are given.

Correlation factor: The cdf of the correlation factor ρ for the PDP's in a cluster is given in figure 1 for LOS and OBS situations and different frequencies. The influence of frequency on the correlation is small. For LOS the PDP correlation factor is high $\rho_{50\%} = 0.9$. OBS channels have a much lower correlation factor: $\rho_{50\%} \approx 0.3$, and the influence of frequency is larger.

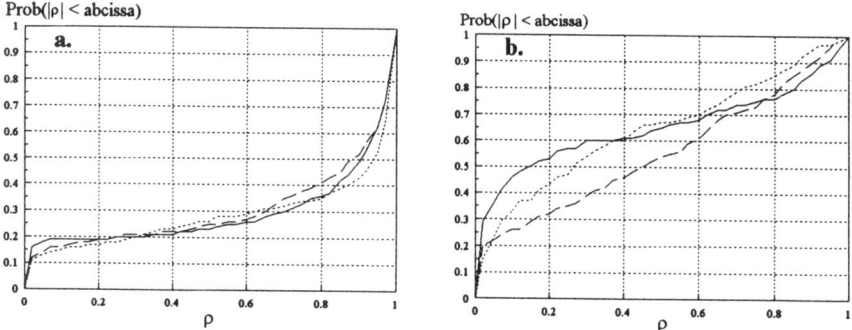

Figure 1: Cumulative distribution of the correlation coefficient ρ of the PDP within a cluster for: a: LOS channel, b: OBS channel and — 2.4 GHz; ---- 4.75 GHz; ······ 11.5 GHz.

Rician K-factor: The cdf of the standard deviation σ_K of the Rician K-factor within a cluster is given in figure 2, for LOS and OBS cases. The frequency influence (not shown) is very small, and also the difference between LOS and OBS is small. $\sigma_{K\_50\%} = 1.7$ dB. The distributions of the K-factor for the measured channels have a median value of +2.5 dB for LOS and -3 dB for OBS channels and show little variation for different frequencies.

RMS delay spread: The cdf of the standard deviation $\sigma_{\tau_{RMS}}$ of τ_{RMS} within a cluster is given in figure 3 for LOS and OBS cases. Again the influence of frequency is small, as is the difference between LOS and OBS, $\sigma_{\tau_{RMS\_50\%}} = 0.85$ ns. The median values of τ_{RMS} are 9 ns for LOS and 15 ns for OBS channels.

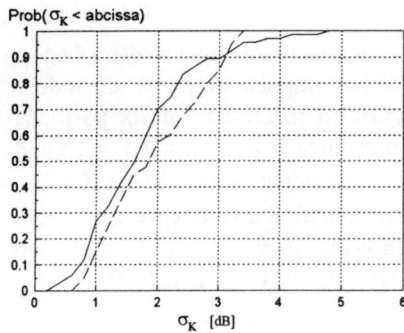
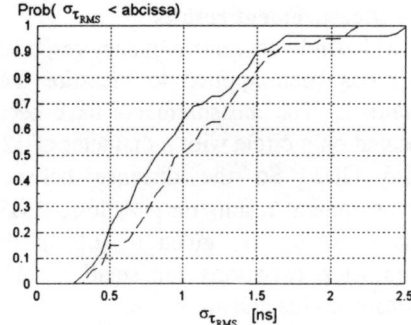

Figure 2: Cumulative distribution of the standard deviation σ_K of the K-factor for: —— LOS and - - - - - OBS channels.

Figure 3: Cumulative distribution of the standard deviation $\sigma_{\tau_{RMS}}$ of τ_{RMS} for: —— LOS and - - - - - OBS channels.

3. QUASI-COHERENT COMBINING

In section 2 and in [8], it is shown that for small displacements of an antenna in a multipath environment, the amplitudes of identical paths in different PDP's are highly correlated, especially for LOS situations. Here, we assume an antenna array with closely spaced elements ($\lambda/2 < d < a\ few\ \lambda$). This means that a path which arrives from a certain direction, is received by all elements of the array with nearly the same amplitude at nearly the same time instant. The aim of the combining technique proposed here, is to sum the dominant paths of M antennae signals coherently. This requires phase shifting of the M signals so that the phases of the dominant path become nearly equal. However, this phase shifting results in uncorrelated phases between the elements for paths which arrive from a different direction as the dominant path. Thus, summation of these signals results only in coherent addition of paths which arrive from the same direction as the dominant path. Signals which arrive from other directions will add incoherently, so the dominant path is strongly enhanced compared to other paths. The resulting signal is given by

$$r(t) = \sum_{m=1}^{M} r_m(t)\exp(-j\theta_m) \qquad (5)$$

with θ_m is the phase of the dominant component of $r_m(t)$. This signal combining technique is equivalent with array beam forming in the direction of arrival of the dominant path. Because the PDP's at the array elements are correlated, the resulting channel after combining is again a Rician fading channel.

3.1 Performance improvement.

In [3, 7], the performance enhancement for ideal coherent combining, with exactly known dominant path phases at each array element, was analyses with the following results.

Rician K-factor: For an M-element array, the resulting channel shows a strongly enhanced dominant path with power $M^2\beta_0^2$. The average increase of power of the uncorrelated path components is $M\beta_k^2$. Therefore, the expected value of the Rician K-factor $\overline{K}_{sum}$ of the combined signals becomes

$$\overline{K}_{sum} = \frac{M^2 \beta_0^2}{M \sum_{k=1}^{L-1} \beta_k^2} = MK \tag{6}$$

Rms delay spread: The expected value of the delay spread $\overline{\tau}_{RMS}$ of the combined signal, can be approximated using (2), by

$$\overline{\tau^2}_{RMS} \approx \frac{1}{M}\frac{\sum_{k=1}^{L} \tau_k^2 \beta_k^2}{\beta_0^2} - \frac{1}{M^2}\left(\frac{\sum_{k=1}^{L} M\tau_k \beta_k^2}{\beta_0^2}\right)^2 \tag{7}$$

where the approximation is valid for large K. The second term of (15) becomes negligible for large M. Then (7) can be further simplified by $\overline{\tau}_{RMS} = \frac{\tau_{RMS}(1)}{\sqrt{M}}$

where, $\tau_{RMS}(1)$ is τ_{RMS} for the signal received by one antenna.

Signal-to-noise ratio: At frequencies > 1 GHz the noise power is mainly generated in the amplifier or mixer after each array element. Thus, M independent noise variables of equal power P_n are summed in the combining process to the final output noise power MP_n. The ratio of $SNR(M)$ after combining, and $SNR(1)$ for a single element, is given by

$$\frac{SNR(M)}{SNR(1)} = \frac{M^2 \beta_0^2 (1+1/MK)/MP_n}{\beta_0^2 (1+1/K)/P_n} \approx M \cdot SNR(1) \tag{8}$$

where the approximation holds for large Rician K-factor.

Coherent combining, which is the optimum case for this scheme, requires exact knowledge of the phases of the dominant path. However, in a practical situation the phases θ_m are unknown and have to be estimated by $\hat{\theta}_m$. Therefore, the technique is called quasi-coherent combining (QCC).

3.2 Practical implementation of Quasi-Coherent combining

The theoretical results for the ideal case of coherent combining of the dominant paths of the profiles require:
i. exact knowledge of the phases of the dominant path at each array element;
ii. identical amplitude of equivalent paths at all antenna elements

$\beta_{m,k}$ (path k of element m) = constant $\forall\ m$.

In a practical situation, however, these requirements are not fulfilled. The phases at the antenna elements, are not known in advance and have to be estimated. As shown in section 2, the amplitudes of equivalent paths are highly correlated from element to element for closely spaced elements in LOS case, but are not constant. In OBS cases the correlation is even quite low. For achieving performance improvements close to the theoretical values, it is crucial that the phases of the dominant path of the array elements are estimated with sufficient accuracy. Therefore, we evaluate the phase variation in a multipath channel modelled by a Rician pdf for an unmodulated carrier. The distribution of the phase difference φ between the resulting phase for the total signal and the phase of the dominant component, can be derived as:

$$p_\varphi(\varphi) = \sqrt{\frac{K}{\pi}} \cos\varphi \exp(-K\sin^2\varphi)\left[1 - Q(\sqrt{2K}\cos\varphi)\right] + \frac{1}{2\pi}\exp(-K) \qquad (9)$$

with $Q(z) = \dfrac{1}{\sqrt{2\pi}} \int\limits_z^\infty e^{-\lambda^2/2}\, d\lambda$.

The cdf of the phase difference φ is given in figure 4, with K as a parameter.

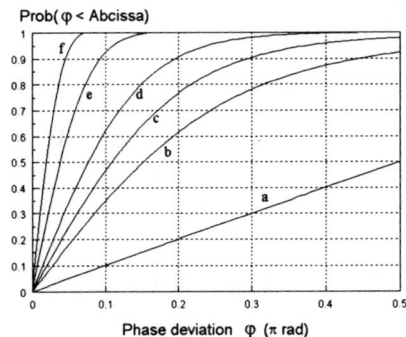

Figure 4: Cumulative distribution function for the phase difference φ in a Rician fading channel for: a.) K = 0 (Rayleigh fading); b.) K = 0 dB; c.) K = 3 dB; d.) K = 6 dB; e.) K = 12 dB; f.) K = 18 dB.

The phase of the total signal is only an accurate estimate of the phase of the dominant signal component for large K-factor, $K > 12$ dB. However, for small values of $K > 0$ dB, which are often found in indoor channels, still large probability is found that $\varphi < \pi/4$, which means a reasonable correlation of the dominant signal components of the array elements after phase shifting.

In the following, BPSK modulation is used to illustrate the phase estimation process and to determine performance of the combining methods. The BPSK modulated signal $x_c(t)$ is given by $x_c(t) = A\, d(t)\, \cos \omega_c t$, where A is the amplitude of the signal, $d(t)$ is the antipodal data signal with bit time T_b and $d(t) \in \{-1,+1\}$, and ω_c is the carrier frequency. The complex baseband received signal at an array element is given by:

$$r_{bb}(t) = \sum_{k=0}^{L} \beta_k d(t - \tau_k) \exp(j\theta_k) \qquad (10)$$

For QCC we need the phase differences between the elements and the reference element. The phase of the reference element #1 is taken $\phi_1 = 0$. We will evaluate and compare two phase estimation techniques:
  i. using signal multiplying, as shown in figure 5,
  ii. using remodulation with decision feedback as shown in figure 6.

i. Phase estimation with signal multiplying. The difference $\Delta \varphi_m$ between the dominant path phases of array element #m and reference element #1, is estimated by:

$$\Delta \varphi_m = \arg\{E[r_m(t) \cdot r_1^*(t)]\} = \arg\left\{ \sum_{k=0}^{L} \sum_{l=0}^{L} \beta_k \beta_l R(\tau_k - \tau_l) \exp(j(\theta_{k,m} - \theta_{l,1})) \right\} \qquad (11)$$

$R(\tau_k - \tau_l)$ is the autocorrelation function which value indicates the fraction of the bit time that two paths carry the same bit during a bit interval, with

$$R(\tau_k - \tau_l) = 1 - \frac{|\tau_k - \tau_l|}{T_b} \quad if \ |\tau_k - \tau_l| \le T_b$$
$$\qquad \qquad = 0 \qquad \qquad if \ |\tau_k - \tau_l| > T_b \qquad (12)$$

The multiplication of the two signals results in an error signal which controls the phase shifter.

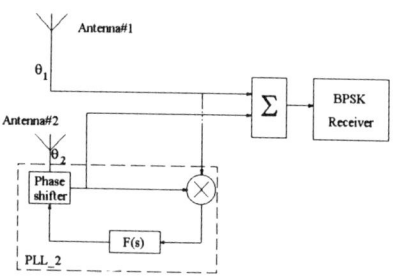
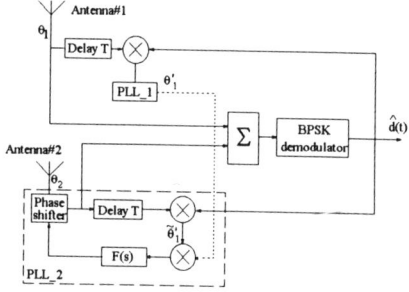

Figure 5: Phase shifting with signal multiplication.

Figure 6: Scheme of the phase estimator using remodulation with decision feedback.

ii. Phase estimation using remodulation with decision feedback. When remodulation with decision feedback is applied for phase estimation, as shown in figure 11, the estimated phase for element #m is given by:

$$\varphi_m = \arg\{E[r_m(t)\hat{d}(t-\tau_s)]\} = \arg\left\{\sum_{k=1}^{L}\sum_{l=1}^{L}\beta_k R(\tau_k - \tau_s)\exp(j(\theta_{k,m})\right\} \quad (13)$$

where $\hat{d}(t-\tau_s)$ is the estimated data sampled at τ_s. The detected desired bit is uncorrelated to the other active bits in the channel. Thus, only the paths which contribute to the desired detected bit influence phase estimation. For increasing bit rate the estimated phase will be less degraded by ISI, because paths with $|\tau_k - \tau_s| > T_b$, do not affect φ_m. The remodulation operation acts as a time filter in the channel impulse response, whereas in method (*i*) all paths contribute to the estimated phase, which results in a larger phase error.

The phase shifting of the signal of antenna#m over $\varphi_1 - \varphi_m$, is performed by a PLL controlled phase shifter. The delay time T_b is added to take into account the time required for optimum detection of $d(t)$.

4. REVIEW OF ANTENNA DIVERSITY SCHEMES

The antenna diversity schemes can be generally modelled as shown in figure 6, where $r_m(t)$ is the received signal of antenna element *m*, and ω_m is the complex weight for signal *m*. The existing combining techniques differ in the criterion to determine the weight factors ω_m. In section 5, QCC will be compared to:

- *Unity Gain combining UGC):* For UGC all weights ω_m are simply set to 1. Though the Rician signals of the array elements have nearly the same PDP, K-factor and τ_{RMS}, the resulting channel is a Rayleigh fading channel due to incoherent summation of the signals. UGC results in a larger spread in BER performance, when compared to the single antenna case, especially for higher K-factor > 3 dB.

- *Selection Diversity (SD):* In SD the best performing array element is selected. SD results in improved channel performance mainly for narrowband signals, because Rician K-factor and τ_{RMS} remain the same.

- *Maximal-ratio combining MRC):* In MRC, the aim of the combining process is to maximise the SNR for the desired bit. No attention is paid to possible ISI.

- *Minimum Mean Square Error combining(MMSEC):* MMSEC aims at joint optimisation of the weights for maximum SNR for the desired bit like in MR-combining, and simultaneous minimisation of the MSE due to ISI by the undesired active bits in the channel, [3].

UGC and SD do not require a priori channel knowledge, whereas MRC and MMSEC require a priori knowledge of the complex channel impulse response at every array element to determine the optimum ω_m.

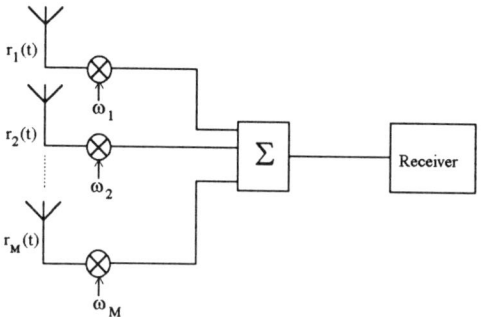

Figure 7: General scheme for antenna diversity.

5. BER PERFORMANCE EVALUATION

The BER performance is calculated using the BER model as presented in [7, 9]. This model is based on the complex channel impulse response and takes into account the effect of ISI. The SNR is accounted for in the model by $SNR(1) = \left(\sum_{k=1}^{L} \beta_k^2\right) / \sigma_n^2 = (1+K)\beta_0^2 / K\sigma_n^2 = 10\ dB$ where $SNR(1)$ is the average SNR for a single antenna and, σ_n^2 is the received noise power. In order to compare the performance for different bitrates R_b in relation to τ_{RMS}, the normalised bitrate defined as $R_{norm} \triangleq R_b \cdot \tau_{RMS} = \tau_{RMS} / T_b$ is used. The BER results are presented as cumulative distribution functions in stead of average BER. This is a more informative way of presenting BER performance results in a fading environment, because the average BER is dominated by the bad results due to the very non-linear behaviour of the BER. The cumulative distributions of the BER, which are each determined from the results of 500 different generated channel impulse responses, have been calculated for different numbers of antennae, different Rice factor, and different R_{norm}.

In figure 8.a - f, the results for QCC are compared to the performance of UGC, SD, MRC and MMSEC for $K = 0$ dB and the number of antennae $M = 1$ and 4. $K = 0$ dB is chosen as the intermediate value for LOS and OBS channels in an indoor environment. QCC shows superior performance enhancement compared to UGC and SDC. UGC performs at $K = 0$ dB about the same as the $M = 1$ case. SD results only in a small performance improvement for low data rate, where the channel with the best narrowband performance is selected. At high bitrate, where ISI dominates, the K-factor and τ_{RMS} are not changed by SD. MRC performs slightly better than QCC at low bitrates because of the amplitude weighting and the exact phase estimates that are used. This results in a slightly better SNR after combining. For high bitrates there is not much difference. MMSEC performs the best because here also ISI is taken into account. A slightly better performance is seen at higher bitrates compared to QCC and MRC, however the improvement is marginal. The results show that the performance of QCC is very close to MRC and MMSEC. However, notice that MRC and MMSEC both require a priori the complex channel impulse response at every array element to determine the combining weights. This information is not required for QCC.

Further simulation results for different K-factor (not shown) result in enhancement for small K due to: *i.* decrease of delay spread and consequently less ISI; and *ii.* increase of *SNR*. At high *K* values, the performance is limited by *SNR*.

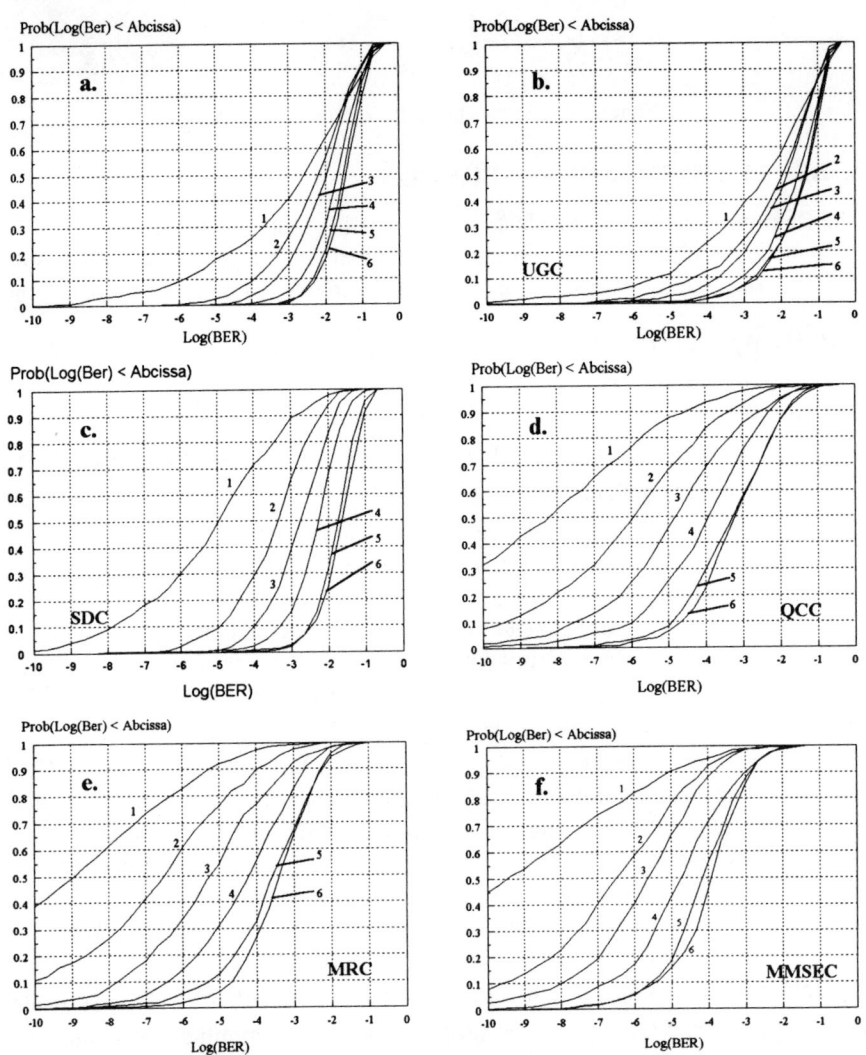

Figure 8: Cumulative distribution function of the BER for Rician channels with K = 0 dB: a.) M = 1; b.) UGC with M = 4; c.) SDC with M = 4; d.) QCC with M = 4; e.) MRC with M = 4; f.) MMSEC with M = 4; and 1.) $R_{norm} = 0.001$; 2.) $R_{norm} = 0.25$; 3.) $R_{norm} = 0.50$; 4.) $R_{norm} = 0.75$; 5.) $R_{norm} = 1.0$; 6.) $R_{norm} = 3.0$;

In figure 9, the difference in performance is given for the two phase estimation methods: the multiplying method and the method using remodulation with decision feedback as discussed in section 3. For low bitrate the difference is hardly noticeable. At high bitrates a clear performance improvement is seen in case ISI is disturbing the phase estimation for the remodulation method. This is the case for low K-factor: K = -3 dB. So we can conclude that the complex phase estimation method using decision feedback only results in improved performance in very bad channels with low K-factor when wideband modulation is used.

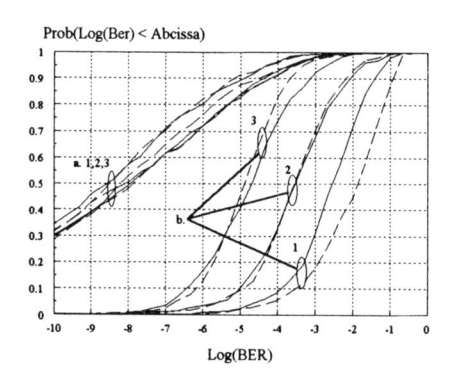

Figure 9: The effect of phase estimation for QCC with M = 4 using the multiplying method (-----) and the method using remodulation with decision feedback (———). With: a.) $R_{norm} = 0.001$; b.) $R_{norm} = 3$ and 1.) K = -3 dB, 2.) K = 0 dB and 3.) K = 3 dB.

Figure 10: Cumulative BER results for QCC with M = 4 with K_{chan} = 3 dB for constant path amplitudes at the elements (———), and for Rician fading path amplitudes with K_{path} = 1 dB (----).
1.) $R_{norm} = 0.001$; 2.) $R_{norm} = 0.25$; 3.) $R_{norm} = 0.50$; 4.) $R_{norm} = 0.75$; 5.) $R_{norm} = 1.0$; 6.) $R_{norm} = 3.0$;

In figure 10, the difference in performance is shown between the models using constant path amplitude at each array element and the more realistic model where the path amplitudes are Rician fading. The Rice factor for the individual paths was set to K_{path} = 1 dB, which gives about the same variation within a cluster as shown in figure 2. The results show hardly any difference between both models.

6. Conclusions

In this paper Quasi-Coherent combining (QCC) of antenna signals of an array with M closely spaced elements is evaluated in the Rician fading frequency selective indoor multipath channel. QCC aims at achieving coherent summation for the dominant path component of the M impulse responses. A receiver model is presented which takes into account the inaccurate phase estimation due to the multipath channel. It is shown that QCC results in a significant improvement of the *BER* due to increase of *SNR* and decrease of ISI after combining. A comparison of QCC with existing diversity techniques shows that it outperforms SD and UGC. Its performance is nearly equal to MRC and shows only a slight degradation to MMSEC, however without the requirement of a priori knowledge of the channel impulse response. QCC is especially attractive at higher frequencies because there the dimensions of the array can be very small. A four element antenna array requires a circle with radius of about $\lambda/2$, which is a relative small area at high frequency.

References

[1] J.H. Winters, "Optimum Combining in Digital Mobile Radio with Cochannel Interference", IEEE Journ. on Sel. Areas in Comm., Vol. SAC-2, No. 4, pp. 528 - 539, July 1984.

[2] M.V. Clark, L.J. Greenstein, W.K. Kennedy, M. Shafi, "MMSE diversity combining for wideband digital cellular radio", IEEE Trans. on Comm., Vol. 40, No. 6, pp. 1128 - 1135, June 1992.

[3] G.J.M. Janssen, W.R. Bredero, R. Prasad, "Performance enhancement for microwave and millimetre wave indoor communication systems by quasi-coherent combining of multiple antenna signals", Proc. PIMRC'94, The Hague, The Netherlands, 18 - 23 September 1994.

[4] A.A.M. Saleh, R.A. Valenzuela: "A statistical model for indoor multipath propagation", IEEE JSAC, Vol. SAC-5, No. 2, pp. 128 - 137, February 1987.

[5] H. Zaghloul, G. Morrison, M. Fattouche, "Frequency response and path-loss measurements of indoor channel", Electronic Letters, Vol. 27, No. 12, pp. 1021 - 1022, June 1991.

[6] G.J.M. Janssen, P.A. Stigter, R. Prasad, "Wideband Indoor Channel Measurements and BER Analysis of Frequency Selective Multipath Channels at 2.4, 4.75 and 11.5 Ghz", To be published in IEEE Transaction on Comunications.

[7] G.J.M. Janssen, B.C. v. Lieshout, R. Prasad, "BER Performance of Millimetre Wave Indoor Communication Sytems Using Multiple Antenna Signals", IEEE Proc. Miniconference, pp. 105 - 109, Globecom'94, San Francisco, December 1994.

[8] H. Hashemi, "Impulse response modelling of indoor radio propagation channels", *IEEE JSAC*, Vol. SAC-11, pp. 967 - 978, Sept. 1993.

[9] G.J.M. Janssen, P.A. Stigter, R. Prasad, "A model for bit error rate evaluation of indoor frequency selective channels using multipath measurement results at 2.4, 4.75 and 11.5 Ghz", *Int. Zürich Seminar on Mobile Comm., Lect. Notes in Comp. Sc. 783*, pp 344 - 355, March 1994.

Applications of Walsh Functions and the FHT in CDMA Technology

Ephraim Zehavi
Qualcomm, Inc.
10555 Sorrento Valley Road
San Diego, CA 92121-1617

Abstract

A unified approach for modulation and demodulation of CDMA signaling based on IS-95 standard [1] is proposed. We propose a simpler forward link design, based on Fast Hadamard Transform (FHT) for generating the composite signal. In some systems, available power is a scarce resource and some power saving can be obtained when the user terminal is able to derive its phase, time and frequency references without the aid of a pilot signal. This necessitates some form of suppressed carrier tracking. A modified receiver using the FHT technique is proposed to implement phase, frequency and time tracking at the user terminal.

1. Introduction

Multiple access communication systems like Cellular CDMA, PCS CDMA, and Globalstar, are based on IS-95 standard [1] and use a combination of frequency division, pseudorandom code division, and orthogonal signaling multiple access techniques. In these systems, Walsh functions and Hadamard matrices play a key role. On the forward link modulation (cell-to-user terminal), Walsh functions are used for channelization. Each user has a dedicated orthogonal channel that is generated by a specific Walsh function. An unmodulated Walsh function is used as a pilot and is shared by all users to extract estimates of time, frequency, phase and received signal power. On the reverse channel the set of M-Walsh functions are used as a M-ary modulation called Walsh Shift Keying which is analog to the traditional MFSK.

In this paper we show that elementary properties of Walsh functions and Fast Hadamard Transforms [2, 3] can be used to simplify and to improve the current implementation of CDMA based on IS-95 technology. In particular we describe a unified approach for modulation and demodulation the IS-95 forward link waveform. This approach allows coherent demodulation with a very weak pilot, and supports non-coherent demodulation using Walsh Shift Keying.

This paper is organized as follows. Section 2 defines the Walsh functions, Fast Hadamard transform and their properties. In Section 3, we introduce a new architecture for the forward link modulator. This design allows multiple forward link CDMA channels to be modulated and combined with a FHT engine. Section 4, describes a scheme by which the receiver uses the transmitted energy (from the same base station) of all the forward link users sharing the same CDMA channel, to estimate the time, frequency and phase of the received signal. This approach was coined Other People's Power (OPP). Section 5 compares the performance of a pilot and OPP based phase tracking systems. Finally, we conclude with some comments and conclusions.

2. Hadamard Matrix and Walsh Functions

A Hadamard matrix of order n is a matrix of +1's and -1's such that, $\mathbf{H}_N \mathbf{H}_N^{tr} = N\mathbf{I}_N$. Multiplying any column by -1 changes $\mathbf{H}_N$ into another Hadamard matrix. A Hadamard matrix of order N contains N sequences (functions), each N chips in length. For example a Hadamard matrix of order $N = 2^k$ can be defined recursively as follows:

$$\mathbf{H}_N = \begin{bmatrix} \mathbf{H}_{N/2} & \mathbf{H}_{N/2} \\ \mathbf{H}_{N/2} & \overline{\mathbf{H}}_{N/2} \end{bmatrix} \text{ and } \overline{\mathbf{H}}_N = -\mathbf{H}_N, \text{ and } \mathbf{H}_1 = 1. \tag{1}$$

This particular construction of a Hadamard matrix is called the Sylvester-type Hadamard matrix. A Walsh function $W_i(N)$ is the i-th row of a Hadamard matrix $\mathbf{H}_N$. We also use the term Walsh function to describe a realization of the i-th row as "squared-wave" function that is span over time interval NT_w, where T_w is the duration of a chip (Walsh chip). Thus, a Hadamard matrix of order 8 can be represented as:

$$\mathbf{H}_{2^3} = \begin{bmatrix} 1 & 1 & 1 & 1 & 1 & 1 & 1 & 1 \\ 1 & -1 & 1 & -1 & 1 & -1 & 1 & -1 \\ 1 & 1 & -1 & -1 & 1 & 1 & -1 & -1 \\ 1 & -1 & -1 & 1 & 1 & -1 & -1 & 1 \\ 1 & 1 & 1 & 1 & -1 & -1 & -1 & -1 \\ 1 & -1 & 1 & -1 & -1 & 1 & -1 & 1 \\ 1 & 1 & -1 & -1 & -1 & -1 & 1 & 1 \\ 1 & -1 & -1 & 1 & -1 & 1 & 1 & -1 \end{bmatrix} \equiv \mathbf{W}_8 = \begin{bmatrix} W_1(8) \\ W_2(8) \\ W_3(8) \\ W_4(8) \\ W_5(8) \\ W_6(8) \\ W_7(8) \\ W_8(8) \end{bmatrix} \tag{2}$$

The set of Walsh functions of order N, has the property that the time-aligned cross correlation between any pair of functions in the set is zero. This can be seen because any Walsh function differs from any other in exactly half of its Walsh chips. Furthermore, there is always one function containing all 1's, and all the other Walsh functions contain exactly equal number of 1's and –1's.

We define an inner product between two vectors, $\mathbf{u} = (u_1,...,u_N)$ and $\mathbf{v} = (v_1,...,v_N)$ of length N, by $(\mathbf{u},\mathbf{v}) = \sum_k u_k v_k$, $u_k, v_k \in \{-1,1\}$. A multiplication operation on the set of vectors is denoted by •, where $\mathbf{u} \bullet \mathbf{v} = \{u_1 v_1, u_2 v_2,...,u_k v_k,...,u_n v_n\}$. A multiplication of a vector by number is denoted by $\alpha \mathbf{v} = \{\alpha v_1,...,\alpha v_n\}$. The following properties[1] of Walsh functions are easily verified [2].

Property 1. The set of Walsh functions of order N and the multiplication operation is a group.

Property 2. $(W_i(N), W_j(N)) = N\delta_{ij}$ where δ_{ij} is the Kronecker delta.

Property 3. $W_i(N) \bullet W_j(N) = W_r(N)$, $r = i \oplus j$. Here $\oplus$ is the module 2 addition operation between two binary vectors representing the indexes i and j.

Property 4. The Hadamard matrix can be decomposed into a product of matrices representing the successive stages of fast transform algorithm. The matrix $\mathbf{W}_{2^k}$ is the product of k matrices $\mathbf{W}_{2^k} \equiv \mathbf{H}_{2^k} = \mathbf{M}_{2^k}^{(1)} \mathbf{M}_{2^k}^{(2)} \mathbf{M}_{2^k}^{(3)} ... \mathbf{M}_{2^k}^{(k)}$, where $\mathbf{M}_{2^k}^{(i)} = \mathbf{I}_{2^{k-i}} \otimes \mathbf{H}_2 \otimes \mathbf{I}_{2^{i-1}}$, $1 \leq i \leq k$, $\otimes$ represents the Kronecker product of two matrices, and I_n is $n \times n$ unit matrix.

Property 5. Let $\mathbf{v}$ and $\mathbf{u}$ be row vectors of length n. Then, the matrix multiplication operation, $\mathbf{u}^T = \mathbf{H}_{2^k} \mathbf{v}^T$, which is also called Hadamard transform can be done by $N \log N$ operations using a fast Hadamard Transform (using Hadamard matrix decomposition of property 4).

Property 6. Let A be a matrix of order $N-1$ whose rows are cyclic shifts of an Maximum Length (ML) sequence of length $N-1$. Then, by adding a row and column of all ones to A a Hadamard matrix is obtained. This form of Hadamard matrix and Sylvester-type Hadamard matrix are permutationally similar [2, 3].

Property 7. Distinct Walsh functions have the distinct spectral properties and correlation characteristics and may response differently to implementation impairments (like timing error or intersymbols interference).

$$\mathbf{M}_8^{(3)} = \mathbf{H}_2 \otimes \mathbf{I}_4 = \begin{bmatrix} 1 & 0 & 0 & 0 & 1 & 0 & 0 & 0 \\ 0 & 1 & 0 & 0 & 0 & 1 & 0 & 0 \\ 0 & 0 & 1 & 0 & 0 & 0 & 1 & 0 \\ 0 & 0 & 0 & 1 & 0 & 0 & 0 & 1 \\ 1 & 0 & 0 & 0 & -1 & 0 & 0 & 0 \\ 0 & 1 & 0 & 0 & 0 & -1 & 0 & 0 \\ 0 & 0 & 1 & 0 & 0 & 0 & -1 & 0 \\ 0 & 0 & 0 & 1 & 0 & 0 & 0 & -1 \end{bmatrix}. \qquad (3)$$

---

[1] Properties 3, 4, and 6 hold (in this form) only for Sylvster type matrix.

Fig. 4. shows a Hadamad transform of order 4 and 8. It can be easily proven that by combining the two transforms of order 4 according to the matrix, $\mathbf{M}_8^{(3)}$, we obtain a Hadamard transform of order 8 where,

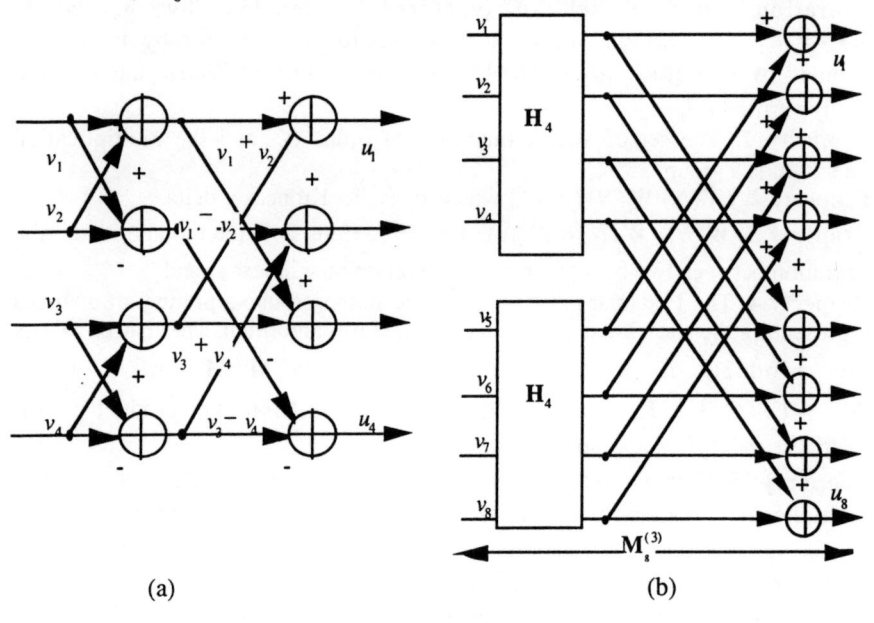

(a)

(b)

Fig. 4. FHT of order 4 (a) and FHT of order 8 (b).

3. New Architecture for Forward Link Modulation

In a CDMA system based on IS-95 [1], the transmitted signal on the forward link for the i-th user is given by

$$S_n(i) = a_n(i) W_i \bullet [PN_I + jPN_Q], \, i = 0,...,63 \qquad (5)$$

where, $S_n(i)$ is the n-th transmitted signal for the i-th user. $a_n(i)$, and W_i are the n-th coded symbol and the Walsh function[2] for the i-th user, respectively. PN_I, and PN_Q are pseudo random binary vectors of length N that are spreading the I and Q channels, respectively, of the n-th symbol (each entries in the vectors can get a value of ±1 with equal probability. The pilot signal is transmitted on the I and Q channels and is unmodulated ($a_n(0) = 1$). Each Walsh function is spread on the I and Q channels by PN_I and PN_Q, respectively. Current spreading

---

[2] Here, and in the sequel we will omit the term N in $W_i(N)$, and in PN_I, and PN_Q in order to simplify the notation.

accommodates up to 63 users each one with 9.6 kbps information rate. Let $\tilde{I}$ and $\tilde{Q}$ denote the combined I and Q components of the transmitted signal. These components are given by the following:

$$\tilde{I}_n = PN_I \bullet \sum_{i=0}^{63} G_i a_n(i) W_i; \quad \tilde{Q}_n = PN_Q \bullet \sum_{i=0}^{63} G_i a_n(i) W_i \qquad (6)$$

Here, G_i is the gain allocated to i-th user. The digital complex baseband signal is shaped digitally by transmission fitter and is converted to analog real signal.

The operations described by equation (3-2) can be performed easily by using a Fast Hadamard Transform (FHT). In the proposed approach, all the coded outputs of the 64 interleavers are fed to the single FHT that generates a composite signal. Fig. 3.1. and Fig. 3.2. illustrate the proposed architecture for forward link. The encoding and interleaving operations can be done, either separately for each CDMA channel, or by a shared hardware and burst operation. The results of all the 64 CDMA interleaved channels are multiplied by the Walsh functions and combined together by using the FHT.

The generating of the composite signal by FHT techniques allows a simple centralized controller for each sector or beam and a flexible structure that enables fast allocation and sharing of traffic channels.

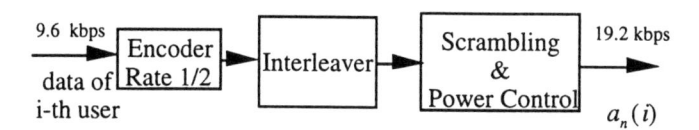

Fig. 7. Block diagarm of the forward link encoder for the i-th User

The proposed architecture supports mixture of QPSK modulation as well M-ary orthogonal modulation (M=2, 4 ...) by allocating to a particular user two or more Walsh functions. These two types of modulation techniques can coexist without mutual interference, since we preserve the orthogonality.

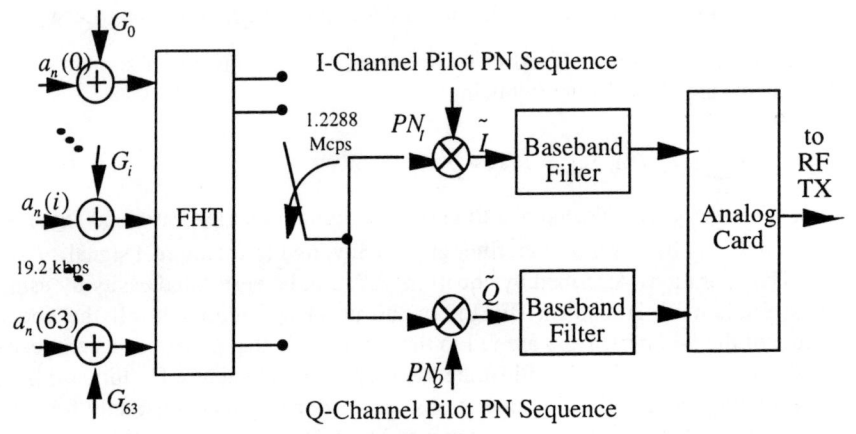

Fig. 8. Combining of all 64 Channels by FHT

4. Forward Link Demodulation with a Very Weak Pilot

In the cellular CDMA system, the user terminal derives its phase, time and frequency references using an unmodulated symbol stream referred to as the pilot. In other systems like GlobalStar, available power is a scarce resource and some power saving can be achieved if the user terminal is able to derive its phase, time and frequency references without the aid of a strong pilot signal. This necessitates some form of suppressed carrier tracking. The bit energy-to-noise density E_b/N_o of a single user link is very low (about 3 dB for AWGN channel). Thus, the accumulated energy over an interval of 10 to 20 symbols is insufficient to form reliable estimates of the phase reference. The OPP (Other People's Power) technique was developed to implement phase, frequency and time tracking at the user terminal with a weak pilot. In the OPP scheme, the receiver uses the received energy of all the active forward link users that are transmitted from the same base station, to estimate the time, frequency and phase of the received signal. In this chapter, we describe how OPP may be utilized for phase tracking. It is emphasized that other tracking schemes, such as frequency and time tracking, follow a similar paradigm.

To simplify notation, only single-path demodulation is described here. Also, most of the analysis assumes Additive White Gaussian Noise, although with minor modifications the analysis is equally valid for fading channels. Assuming that the received signal at the mobile, $R(t)$, has a random phase shift, θ, and a delay, D, with respect to the mobile's internal phase and time references. The received signal is filtered by an IF-matched filter, moved to the baseband, and I and Q

projections are sampled at a rate higher than chip rate. The sampled values R_I and R_Q, after time alignment, can be expressed as follows:

$$R_I(n) = R_I(nT_w) = \tilde{I}(t-D)\cos\theta(t) - \tilde{Q}(t-D)\sin\theta(t) + n_I(t)|_{t=nT_w}$$
$$R_Q(n) = R_Q(nT_w) = \tilde{I}(t-D)\sin\theta(t) + \tilde{Q}(t-D)\cos\theta(t) + n_Q(t)|_{t=nT_w} \quad (9)$$

R_I and R_Q are composed of the original components $\tilde{I}$ and $\tilde{Q}$ and an additive noise n_I, and n_Q with zero mean and variance of σ^2.

A high-level block diagram of the OPP receiver is shown in Fig. 10. Here, the receiver demodulator is composed from the following blocks:
- M-ary Costas loop for frequency and phase tracking. The new version of Costas loop exploits all active user energy for tracking. In addition, this approach provides demodulation as a by-product for all 64 users sharing the same CDMA channel.
- Time-tracking loop for tracking delay change in the received signal
- Energy detection in all Walsh cover functions
- Pilot phase and energy detection
- Deinterleaving and decoding

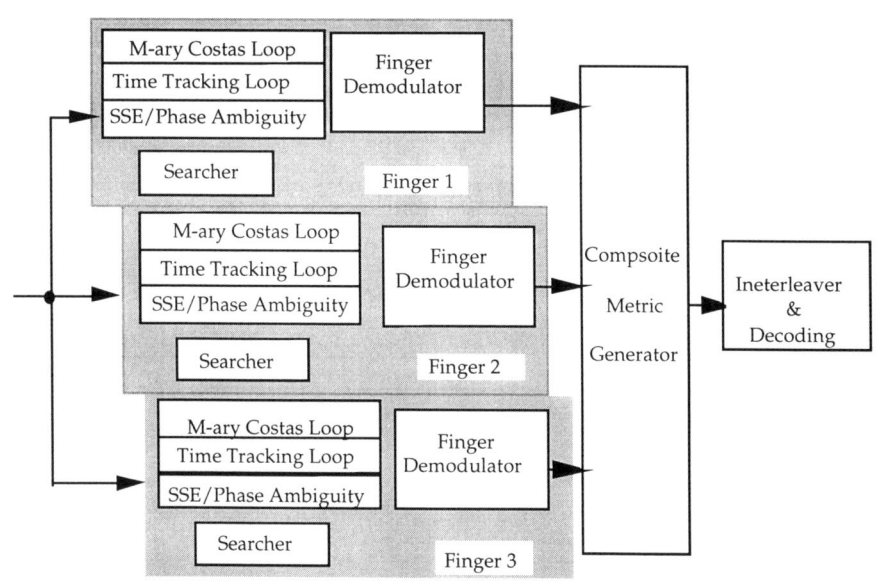

Fig. 10. Forward Receiver using OPP approach

A detailed block diagram of the demodulator structure of a single finger is depicted in Fig. 12.

4.1 Description of the M-ary Costas Loop

A delayed version of the sampled signal, that is associated with the signal coming exactly at Walsh time is despread and rotated by multiplying it by the signal

$$(PN_I(n) - jPN_Q(n)) \bullet \exp(j\phi(n)) \tag{11}$$

The phase $\phi(n)$ is the estimated phase of the incoming signal. The M-ary Costas loop [4-6] changes the phase $\phi(n)$ according to a filtered error signal. If the error signal is zero, then, $\theta(n) = \phi(n)$. The upper arm (in Fig, 10) is called the in-phase arm, and the lower arm is called the quadrature phase arm.

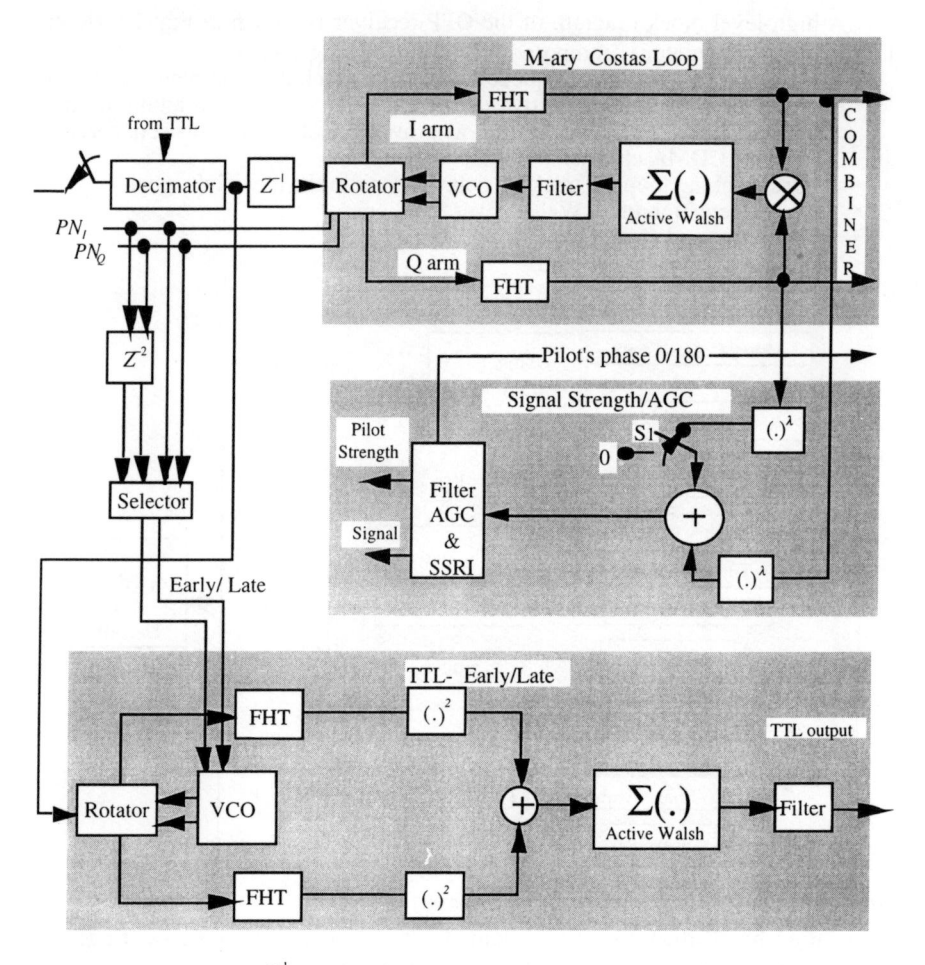

Z^{-1} : Delay of half a chip
S1 : Open for a pilot channel

Fig. 12. Finger Demodulator Structure

These signals can be expressed as follows:

$$I_n = \sum_{i=0}^{63} G_i a_n(i) W_i \cos(\theta(n) - \phi(n)) + N_I; Q_n = \sum_{i=0}^{63} G_i a_n(i) W_i \sin(\theta(n) - \phi(n)) + N_Q \quad (13)$$

Here N_I, and N_Q are the additive noise components. Without noise, I_n and Q_n carry all the modulated information that is transmitted on the forward link for all users. The i-th output of the FHT is given for I and Q as follows:

$$I_n(i) = a_n(i) G_i \cos(\theta(n) - \phi(n)) + N_I(i)$$
$$Q_n(i) = a_n(i) G_i \sin(\theta(n) - \phi(n)) + N_Q(i) \quad (14)$$

Multiplying $I_n(i)$ by $Q_n(i)$ and summing the product over all the active Walsh channels yields the error signal $e(n)$. The $e(n)$ value is a random process with mean:

$$E(e(n)) = \frac{1}{2} \sin(2(\theta(n) - \phi(n))) N_{active} EG_i \quad (15)$$

where E(.) is the expectation of (.), and N_{active} is the number of active users. The $e(n)$ value is passed through a first or second-order loop to correct the estimate of the phase. The actual value of the phase rotator is a sum of two factors: The phase correction from the VCO and phase for despreading operation.

It is preferable to have a weak pilot in order to remove the phase ambiguity. The Costas loop can lock on the correct phase plus 0 degrees or 180 degrees. This phase ambiguity is inherent to all techniques that are used to extract phase without a pilot. We suggest use of a weak pilot to remove phase ambiguity in order to allow the loop to lock onto the correct phase even after a deep fade.

4.2 Demodulation

As a by-product of the FHT, there are N_{active} outputs corresponding to the N_{active} active users (N_{active} <64). For coherent combining of more than one path, the outputs of the I arm are scaled before combining. For non-coherent combining, the outputs of the energy detector of the i-th user are scaled before combining.

In some scenarios, the user terminal receives transmissions from two sectors. One sector of the base station transmits with coherent modulation, while the other transmits with non-coherent modulation. In this case, the combiner combines the results of the two fingers such that the FER is minimal.

4.3 Time-Tracking Loop (TTL)

The time-tracking loop corrects the internal finger timing based on the deviation of the incoming signal from the finger timing. These corrections account for time shifts of the incoming signal due to code Doppler, the changing position of the user terminal compared to cell, or multipath conditions.

The level of time-tracking accuracy of incoming data I and Q chips, is measured by sampling the impulse responses of the incoming data stream at an offset from the nominal chip time. This offset is plus (late), or minus (early), half of a chip period. If the offset data points straddle the nominal despread signal peaks symmetrically, the difference of late, and early sampling values is zero. If the chip clock is not tracking accurately and is fast, then the late-minus-early difference yields a positive value. If the symbol clock is running too slowly, the difference yields a negative value. These correction signals are filtered by a second-order loop, and used in decimator to change the sampling time.

4.4 Signal Strength Estimate (SSE)

The sum of I I^2 and Q^2 of the FHT outputs provides sufficient information for estimating the signal strength in each Walsh function and to determine the active Walsh functions. The signal level estimator of the Walsh function takes a long-term average of the sum of I^2 and Q^2 of every i-th output of the FHT. The final output of this filter is then compared to a known threshold to decide if the channel is active. In addition, a decision on the value of the pilot phase with respect to the phase of the Costas loop is made after averaging the amplitude of the in-phase component. The output of the pilot filter is used for resolving the 180 degree ambiguity of the Costas loop. The time constant of the filter will be fast enough to enable fast phase acquisition after a deep fade. The energy needed to resolve this ambiguity with high reliability is about 8 dB. So, if the pilot strength is above a certain threshold, then the probability of miss detection of a phase jump is low.

5. Comparison between Pilot Only and OPP Approach

In this section, we use the results derived in the previous sections to compare the performance of a pilot and OPP based phase tracking system. The comparison methodology is as follows:
- Select a certain E_b/N_0 and standard deviation of phase error which would generate a fixed value of Frame Error Rate (FER-1 %).
- For the Pilot case, and for a given noise-loop bandwidth, we calculate the relative pilot power α (relative to a user data channel) that is required to maintain the specified standard deviation of phase error.
- For the OPP case, for the same noise-loop bandwidth, we calculate the minimum number of users (at the same E_b/N_0) that are required to maintain the same standard deviation of phase error as the pilot case.
- In the Pilot case, since a portion of the satellite power is used for pilot power, for the same transmit power, a Pilot based system can support fewer number of users than OPP.
- The *maximum* OPP gain is calculated as $\frac{N_\alpha}{N_\alpha - \alpha}$ where N_α is the *minimumum* number of users required to give same performance as pilot with relative power α.

For simplicity, let us consider the special case, where all users have equal power $G_i = G_j$, for all i and j. The signal to noise ratio of the Costas loop output is

$$\left(\frac{S}{N}\right)_{opp} = \frac{E_s}{N_o B_L} G_{OPP}, \qquad (16)$$

where

$$G_{OPP} = \left[\frac{1}{N_\alpha}\left[1 + \frac{1}{2\frac{E_s}{N_0}}\right]\right]^{-1}, \qquad (17)$$

where, E_s is the Energy per coded symbol allocated to a user, B_L is the effective loop bandwidth. Let us compare it to the case where a PLL is used to extract the phase, and the pilot Energy in time period of a coded symbol symbol is αE_s. Then [6],

$$\left(\frac{S}{N}\right)_{PLL} = \frac{\alpha E_s}{N_o B_L}, \qquad (18)$$

Thus, if we assume that the two loops have the same bandwidth, and the same phase noise variance, G_{OPP} has to be equal to α. The gain in capacity due to OPP can be expressed as the extra number of users that the OPP can accommodate relative to OPP, when both have the same phase noise variance. From Eq's (4.6), (4.7), and (w.8) the gain due to the OPP approach is

$$Gain = \frac{N_\alpha}{N_\alpha - \alpha} = 1 + 2\frac{E_s}{N_o} = 1 + 2\frac{E_b \bullet R}{N_o}. \qquad (19)$$

Here R is the code rate and E_b is the nominal value of bit energy -to-noise ratio that is required to provide a given frame error rate performance.

Comments:

In making this comparison we make a few important assumptions.
- The OPP scheme requires a weak pilot for resolving the phase ambiguity. It is assumed that this pilot's strength is insignificant.
- We assume all the users are equal strength users.
- The tracking loop performance is completely captured by the loop SNR. This is a significant assumption, since in reality, the dynamics of the fading process greatly influence the overall performance.
- The equation comparing the OPP and pilot is valid in an AWGN channel only if number of users is equal to the minimum number required to maintain same tracking performance. If the number of users that the system capacity can support is larger than the number N_α, the gain due to OPP decreases.

6. Conclusions

In this paper we show an implementation approach for modulation and demodulation of CDMA based on IS-95 using the FHT engine. In this approach we use a single modulator for all users on the forward link. On the receiver side we described the OPP approach which allows to improve phase and time-tracking algorithms. This also provides the capability of allocating higher data rate channels for some users by simply assigning them more than one traffic channel on a frame-by-frame basis.

The OPP approach is different from the "pilot only" approach that is used currently in cellular systems. All modifications are backward compatible, and guarantee operation with "pilot only" mode. The complexity in the hand set increases due to the need to demodulate (but not decode) many CDMA channels.

References

[1] TIA/EIA/IS-95 Interim Standard, *Mobile Station-Base Station Compatibility Standard for Dual-Mode Wideband Spread Spectrum Cellular System*, Telecommunication Industry Association, July 1993.

[2] F. J. MacWilliams, and N. J. Sloane, *The Theory of Error Correcting Codes*. New York: North-Holland, 1977.

[3] A. Lempel, "Hadamard and M-Sequence transform are permutationally similar," Applied Optics, vol. 18, No. 24, 15 Dec. 1979.

[4] A. J. Viterbi, *Principles of Coherent Communication.*, New York: McGraw-Hill, 1966

[5] W.C. Lindsey and M.K Simon, *Telecommunication Systems Engineering*, New York, Dover Publications, 1973.

[6] R. L. Bogusch, *Digital Communications in Fading Channels: Tracking and Synchronization,* Report WL-TR-90-15 of Mission Research Corporation sponsored by Weapons Laboratory, Kirtland AFB, NM, April 1990.

Decision Feedback Multi-User Detection for Multitone CDMA Systems

Luc Vandendorpe[1] * and O. van de Wiel[1] **

UCL Communications and Remote Sensing Laboratory
2, place du Levant - B 1348 Louvain-la-Neuve - Belgium
Phone : +32 10 47 23 12 - Fax : +32 10 47 20 89
E-Mail : vandendorpe@tele.ucl.ac.be

Abstract

In [9] the combination of multitone modulation with direct sequence spectrum spreading has been introduced. In the present paper we analyze the problem of multi-user detection for such systems in the presence of multipath and multiple access interferences. Two types of decision feedback structures are derived for an MSE criterion. The performance of the two systems is obtained for a constrained comparison and the near-far resistance of the detectors is demonstrated in asynchronous scenarios for two types of channels.

1 Introduction

In [9] a new multiple access system based on the combination of multitone modulation with direct sequence spectrum spreading has been introduced. It has been shown that the larger symbol duration associated with multitone systems was favorable to combat the effect of large delay spreads. The main feature of this MT-CDMA (multitone CDMA) system is that for a constant bandwidth, the ratio between the number of chips per symbol and the number of tones has to be constant. Hence when the number of tones goes up the number of chips per symbol does as well and longer spreading codes can be used. The performance of the system in the presence of multipath propagation and of a multiple access interference has been analyzed in [9]. The receiver was based on matched filters. It has been shown that, for a system constrained by the power, the bandwidth and the bit rate, an increase of the number of tones had a positive effect on the system performance. Nevertheless, in high load conditions, the system performance needs to be improved. It is well-known that communications systems over frequency selective channels can be improved by equalization, and CDMA systems, by interference cancellation or joint detection. Concerning the equalization of multitone spread spectrum signals, the performance of linear devices has been investigated in [10]. The steady-state behavior of MMSE designed equalizers has been analyzed. Adaptive LMS and RLS structures have been put forward

---
* This author would like to thank the Belgian NSF for its financial support.
** This author would like to thank the Belgian FRIA for its financial support.

and their performance has been demonstrated. As regards joint detection, major contributions in the field are due to Lupas and Verdu [6, 7]. Applicable to the same problem is the contribution made by Duel-Hallen in [4] where the problem of equalization for multiple input-multiple outputs devices is analyzed. In [5] she proposed a decorrelating decision-feedback detector for synchronous CDMA which utilizes the decisions of the stronger users when forming decisions for the weaker ones. More recently, a lot of effort has been devoted to this problem. The main challenge is the design of detectors which are near-far resistant without exponential complexity. A very good overview of the recent achievements can be found in [1, 2, 3].

In the present paper, we investigate the performance of decision feedback (DF) multi-user detectors for multitone CDMA. These joint detectors will be referred to as DF joint equalization-interference cancellation (DF-JEIC) devices. As a matter of fact when joint detection is performed one part of the device is associated with equalization (meaning ISI mitigation) while the other one is related to interference cancellation or reduction of the multiple access interference. Linear devices have been proposed recently in [11] for multitone CDMA. In the present paper we investigate the performance of two types of decision feedback systems. The first one is an extension of classical decision feedback equalization : the prediction of a symbol is computed from the present and future matched filter outputs, and past decisions. In the second structure we assume that users received with more power are detected first and the decision made about the current symbols of the stronger users is also used in the prediction of the weakest ones. Both structures follow reception filters matched to the symbol shape and the channel impulse response. Perfect carrier phase and timing recovery is also assumed. In the future fractionally spaced (FS) decision feedback joint detection structures will be investigated.

For the two DF-JEIC structures under consideration in the present paper, the performance will be assessed by means of bit error rates (BERs). The exact BERs will be analytically computed by means of a method based on characteristic functions and proposed in [11]. The steady-state solutions of the DF-JEIC systems will be analyzed, because it gives the potential of the method and the performance an adaptive system would reach if the tracking were perfect. The coefficients of the detectors will be derived in general for MT-CDMA. Then the results will be applied to the case where the different channels are frequency non-selective or multipath.

2 Multitone CDMA systems

The transmitter of an MT-CDMA system is depicted in figure 1.

It can be seen that the transmitted signal associated with a particular user is obtained by multitone modulation first and then spreading of the multitone symbols. As already mentioned in [11] it is expected that the system will benefit from the resistance of multitone signals against large delay spreads while keeping

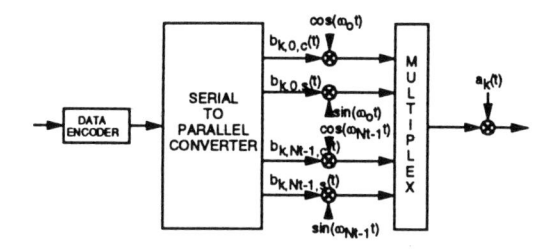

Fig. 1. Transmitter of a multitone CDMA system

the possibility of using the inherent diversity of CDMA signals. The lowpass equivalent signal transmitted by user k is given by

$$x_k(t) = \sqrt{\frac{2P}{N_t}} \sum_{p=0}^{N_t-1} \sum_{n=-\infty}^{\infty} I_{k,p}^n a_k(t) u(t-nT) \exp^{2\pi jpt/T} \quad (1)$$

where $a_k(t)$ is the waveform associated with the periodical pseudo-noise sequence multiplying the multitone signal and associated with user k. The chips are assumed to be rectangular. $I_{k,p}^n$ is the nth symbol conveyed by carrier p of user k and $I_{k,p}^n = b_{k,p,c}^n - jb_{k,p,s}^n$. The symbol shape $u(t)$ is assumed to be rectangular. The RF frequency associated with the pth carrier is given by $f_0 + p/T$ where f_0 is some base frequency and T is the symbol duration. Assuming that the channel between user k and the receiver is a linear channel with equivalent lowpass impulse response $c_k(t)$, the received signal is given by

$$r(t) = \sum_{k=0}^{N_u-1} \sqrt{\frac{2P}{N_t}} \sum_{p=0}^{N_t-1} \sum_{n=-\infty}^{\infty} I_{k,p}^n h_{k,p}(t-nT) + n(t) \quad (2)$$

where N_u is the number of users $u_{k,p}(t) = a_k(t) u(t) \exp^{2\pi jpt/T}$, $h_{k,p}(t) = u_{k,p}(t) \otimes c_k(t)$, and $\otimes$ denotes convolution, $n(t)$ is the AWGN with one-sided power spectral density N_0. Assuming perfect timing and carrier phase recovery, matched filter outputs are computed by

$$y_{k,p}^n = \frac{1}{T\sqrt{2P}} \int_{-\infty}^{\infty} r(t) h_{k,p}^*(t-nT) \, dt \quad (3)$$

It turns out that

$$y_{k,p}^n = \frac{x_{k,k,p,p}^0}{\sqrt{N_t}} \left[I_{k,p}^n + \sum_{m=-\infty}^{\infty} I_{k,p}^{n-m} \frac{x_{k,k,p,p}^m}{x_{k,k,p,p}^0} + \sum_{q=0}^{N_t-1} \sum_{m=-\infty}^{\infty} I_{k,q}^{n-m} \frac{x_{k,k,p,q}^m}{x_{k,k,p,p}^0} \right.$$
$$\left. + \sum_{l=0}^{N_u-1} \sum_{q=0}^{N_t-1} \sum_{m=-\infty}^{\infty} I_{l,q}^{n-m} \frac{x_{k,l,p,q}^m}{x_{k,k,p,p}^0} \right] + \frac{\nu_{k,p}^n}{T\sqrt{2P}} \quad (4)$$

where $\nu_{k,p}^n$ represents samples of the filtered white noise and the normalized channel correlation coefficients are defined as

$$x_{k,k',p,p'}^{n-n'} = \frac{1}{T}\int_{-\infty}^{\infty} h_{k,p}^*(t-nT)\,h_{k',p'}(t-n'T)\,\mathrm{d}t \tag{5}$$

In the present paper we investigate the performance of suboptimal detectors based on the processing of these matched filter outputs which are sufficient statistics. From equation 4, it appears that the overall system between the symbol generation up to the sampling of the matched filter outputs can be modeled by a Multiple Input Multiple Output (MIMO) digital equivalent system. The number of channels is equal to $N_t \times N_u$. Next, it also appears that the samples at the outputs of the matched filters are affected by ISI (inter symbol interference), IBI (inter band interference) and MAI (multiple access interference). This suggests that the DF-JEIC structure also has to be of the MIMO type.

3 MMSE DF-JEIC

In this section we derive the coefficients for the two DF-JEIC devices. This derivation will be made for an MMSE criterion and assuming that the decisions are correct. The overall structure of the system with a DF-JEIC device is shown in figure 2.

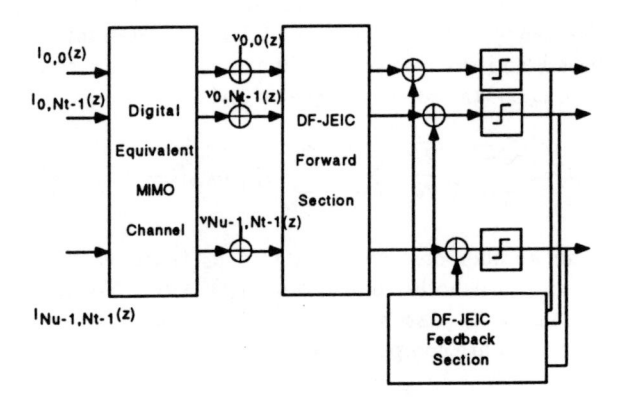

Fig. 2. Overall digital equivalent system including the DF-JEIC MIMO device represented in the z-transform domain.

3.1 DF-JEIC-1

Steady-state solution: The goal of DF-JEIC-1 is to find coefficients $c_{l,k,q,p}^m$ such that the expectation of the squared error between the true symbols $I_{k,p}^n$

and their prediction $\hat{I}_{k,p}^n$ built from the sampled matched filter outputs and previous decisions is minimum. If we assume $2K+1$ coefficients in each branch of the joint device, the prediction is computed as

$$\hat{I}_{l,q}^m = \sum_{k=0}^{N_u-1} \sum_{p=0}^{N_t-1} \sum_{n=-K}^{0} c_{l,k,q,p}^n y_{k,p}^{m-n} + \sum_{k=0}^{N_u-1} \sum_{p=0}^{N_t-1} \sum_{n=1}^{K} c_{l,k,q,p}^n \tilde{I}_{k,p}^{m-n} \qquad (6)$$

where $\tilde{I}_{k,p}^{m-n}$ is the decision about symbol $I_{k,p}^{m-n}$. For mathematical tractability it is assumed that the decisions are always correct, and hence $\tilde{I}_{k,p}^{m-n} = I_{k,p}^{m-n}$. In equation 6, the summation over n is responsible for the equalization effect (suppression of ISI) and the summation over k is related to the mitigation of the multi-user interference. The coefficients have to minimize

$$\frac{1}{N_t N_u} \sum_{l=0}^{N_u-1} \sum_{q=0}^{N_t-1} \mathbf{E}\left[|\epsilon_{l,q}^m|^2\right] = \frac{1}{N_t N_u} \sum_{l=0}^{N_u-1} \sum_{q=0}^{N_t-1} \mathbf{E}\left[|I_{l,q}^m - \hat{I}_{l,q}^m|^2\right] \qquad (7)$$

According to the results reported in [8], we use the orthogonality principle. However we have to handle separately the forward and the feedback sections. For the forward section we require that

$$\mathbf{E}[\epsilon_{l,q}^s (y_{k,p}^{s-m})^*] = 0 \qquad \text{for } -K \leq m \leq 0 \qquad (8)$$

for all combinations of k, l, p, q. This equation leads to a first set of $N_u^2 N_t^2 (K+1)$ equations :

$$\sum_{l'=0}^{N_u-1} \sum_{q'=0}^{N_t-1} \sum_{m'=-K}^{0} c_{l,l',q,q'}^{m'} \times$$

$$\left[\sum_{l''=0}^{N_u-1} \sum_{q''=0}^{N_t-1} \sum_{m''=-\infty}^{\infty} \frac{x_{l',l'',q',q''}^{m''} (x_{k,l'',p,q''}^{m'+m''-m})^*}{N_t} + \frac{x_{l',k,q',p}^{m-m'}}{N_t(E_b/N_0)} \right] +$$

$$\sum_{l'=0}^{N_u-1} \sum_{q'=0}^{N_t-1} \sum_{m'=1}^{K} c_{l,l',q,q'}^{m'} \frac{(x_{k,l'',p,q'}^{m-m'})^*}{\sqrt{N_t}} = \frac{(x_{k,l,p,q}^{-m})^*}{\sqrt{N_t}}$$

(9)

where E_b/N_0 is the energy per bit over white noise ratio; $E_b = PT/N_t$. For the feedback section, we have

$$\mathbf{E}[\epsilon_{l,q}^s (I_{k,p}^{s-m})^*] = 0 \qquad \text{for } 1 \leq m \leq K \qquad (10)$$

for all combinations of k, l, p, q. This equation leads to a set of $N_u^2 N_t^2 K$ equations :

$$\sum_{l'=0}^{N_u-1} \sum_{q'=0}^{N_t-1} \sum_{m'=-K}^{0} c_{l,l',q,q'}^{m'} \frac{(x_{l',k,q',p}^{m-m'})^*}{\sqrt{N_t}} = -c_{l,k,q,p}^m \qquad (11)$$

These two sets of equations can be written in a matrix form. It is interesting to notice that the elements of the matrix affecting the c coefficients do not depend on l nor q. This suggests that the matrix has a block diagonal structure and that the smaller matrix independent from l, q only needs to be inverted once.

Bit error probability: The residual interference after equalization can be computed by

$$\hat{I}_{l,q}^n = \sum_{k=0}^{N_u-1}\sum_{p=0}^{N_t-1}\sum_{m=-\infty}^{\infty} I_{k,p}^{n-m}\, r_{l,k,q,p}^m + \sum_{k=0}^{N_u-1}\sum_{p=0}^{N_t-1}\sum_{m=-K}^{0} c_{l,k,q,p}^m \frac{\nu_{k,p}^{n-m}}{T\sqrt{2P}} \quad (12)$$

where

$$r_{l,k,q,p}^m = \begin{cases} 0 & \text{for } 1 \le m \le K \\ \sum_{k'=0}^{N_u-1}\sum_{p'=0}^{N_t-1}\sum_{m'=-K}^{0} c_{l,k',q,p'}^{m'}\, x_{k',k,p',p}^{m-m'} / \sqrt{N_t} & \text{elsewhere} \end{cases} \quad (13)$$

This is a classical behavior of decision feedback when the decisions are correct : if the feedback section spans K decisions the interference created by the K previous symbols of all tones of all users is perfectly cancelled.

The bit error probability can be computed by means of an approach based on the characteristic functions. It has been explained in [11]. We hereafter summarize the main results. The bit error probability for carrier q of user l is given by :

$$P_{e,l,q} = \frac{1}{2} - \frac{1}{\pi}\int_0^\infty d\omega_1 \frac{\sin\left[\omega_1 \Re(r_{l,l,q,q}^0)\right]}{\omega_1} e^{-\omega_1^2 \sigma_{N,l,q}^2/4}$$

$$\times \prod_{k=0}^{N_u-1}\prod_{p=0}^{N_t-1}\prod_{n=-\infty}^{\infty} \cos\left[\omega_1 \Re(r_{l,k,q,p}^{m-n})\right] \quad (14)$$

In the double product, when $k = l$ and $p = q$, the value of $n = m$ does not have not be taken into account because the value of $I_{l,q}^m$ is assumed. The variance of the noise is given by :

$$\sigma_{N,l,q}^2 = \sum_{k=0}^{N_u-1}\sum_{k'=0}^{N_u-1}\sum_{p=0}^{N_t-1}\sum_{p'=0}^{N_t-1}\sum_{m=-K}^{0}\sum_{m'=-K}^{0} c_{l,k,q,p}^m \left(c_{l,k',q,p'}^{m'}\right)^* \frac{x_{k,k',p,p'}^{m'-m}}{N_t(E_b/N_0)} \quad (15)$$

3.2 DF-JEIC-2

Steady-state solution: In this second structure it is assumed that the users are ordered with decreasing power. For a given user, the prediction of a symbol is built from future matched filter outputs, current matched filter outputs of the reference and weaker users, decisions about the current symbol of stronger users, and decisions made about previous symbols of all users. Hence if we assume that user k is received with a stronger power than user l when $k < l$,

$$\hat{I}_{l,q}^m = \sum_{k=0}^{l-1}\sum_{p=0}^{N_t-1}\left[\sum_{n=-K}^{-1} c_{l,k,q,p}^n y_{k,p}^{m-n} + \sum_{n=0}^{K} c_{l,k,q,p}^n \tilde{I}_{k,p}^{m-n}\right]$$

$$+ \sum_{k=l}^{N_u-1}\sum_{p=0}^{N_t-1}\left[\sum_{n=-K}^{0} c_{l,k,q,p}^n y_{k,p}^{m-n} + \sum_{n=1}^{K} c_{l,k,q,p}^n \tilde{I}_{k,p}^{m-n}\right] \quad (16)$$

where again $\tilde{I}_{k,p}^{m-n}$ is the decision about symbol $I_{k,p}^{m-n}$. For mathematical tractability it is again assumed that the decisions are always correct, and hence $\tilde{I}_{k,p}^{m-n} = I_{k,p}^{m-n}$. The coefficients have to minimize the same objective function as above. Two sets of equations can be found as above, but they are more intricated and hence we omit them for the sake of concision. These two sets of equations can also be written in a matrix form. Compared to the equations associated with DF-JEIC-1, the coefficients depend this time on l. Hence as many matrices as there are users have now to be inverted.

Bit error probability: The residual interference after equalization can be computed as proposed previously. For DF-JEIC-2 we have

$$r_{l,k,q,p}^m = \begin{cases} 0 & \text{for } \begin{cases} k<l \text{ and } 0 \leq m \leq K \\ \text{or} \\ k \geq l \text{ and } 1 \leq m \leq K \end{cases} \\ \sum_{k'=0}^{N_u-1} \sum_{p'=0}^{N_t-1} \sum_{m'=-K}^{0} c_{l,k',q,p'}^{m'} \frac{x_{k',k,p',p}^{m-m'}}{\sqrt{N_t}} & \text{elsewhere} \end{cases}$$

(17)

This is the classical behavior of decision feedback but extended to feedback between users : in addition to the properties about interference cancellation that we already had with DF-JEIC-1, we now have that the interference due to the current symbols of all tones of the stronger users is also perfectly cancelled (as long as there are no decision errors). The bit error probability can still be computed as explained for DF-JEIC-1. The variance of the term associated with the additive noise can again be computed as above but is omitted for the sake of concision.

4 Asynchronous scenarii

4.1 Channel description

The results obtained in the preceding sections have been applied to a multi-user system where the signal associated with each user propagates over a one-path or a two-path channel. However the path delays of the different users may be different and hence an asynchronous scenario is considered. In the most general situation, it is assumed that

$$c_k(t) = \beta_{k1}\delta(t - \tau_{k1}) + \beta_{k2}\delta(t - \tau_{k2}) \qquad (18)$$

We make the following assumptions about the delays : we always have $\tau_{k2} > \tau_{k1}$. Besides, for $l > k$ we have $\tau_{l1} > \tau_{k1}$, $\tau_{l2} > \tau_{k1}$, $\tau_{l2} > \tau_{k2}$. No assumption is made concerning τ_{l1} with respect to τ_{k2}. Concerning the computation of the $x_{k,l,p,q}^n$, the result can be found in [11]. In a general situation, the signal of each user is affected by ISI, IBI and MAI. When there is only one path ($\beta_{k2} = 0$), a given user k does not suffer from self interference, that is there is no interference on carrier

p of user k from neighboring symbols transmitted on the same carrier p, nor from the symbols transmitted on the other carriers q. A very preliminary conclusion would be that there is no need to include in the estimate of $I_{k,p}^n$ matched filter outputs like $y_{k,p}^m$ or $y_{k,q}^m$. However, because of the multiple access interference, the $y_{l,q}^m$ naturally have to be taken into account. It is very important to notice that in the $y_{l,q}^m$ terms, $I_{k,p}^r$ and $I_{k,q}^r$ symbols appear. So, if originally, there is no ISI nor IBI in matched filter outputs, because of the way a prediction is built, ISI and IBI can appear in the symbol estimate. In order to make it possible to compensate these indirect ISI and IBI it makes then sense to use the $y_{k,p}^m$ and $y_{k,q}^m$ as well in the prediction. Thus, for this situation, the prediction have been constructed with all possible degrees of freedom associated with formula 6. It means that we do not force the $c_{k,k,p,p}^{m \neq n}$ nor the $c_{k,k,q,p}^m$ to 0. The performance of the joint structures has been investigated in a power controlled scenario and in the presence of a near-far problem.

4.2 Computational results

Channels with one-path only have first been considered. Simulations results have been derived for a constrained comparison. The comparison has been restricted to numbers of chips of $N_c = 7, 15, 31$, meaning $N_t = 1, 2, 4$, respectively to keep the bandwidth constant. The codes have been generated as Gold sequences. The different user gains β_{k1} have first been set to the same value, 1, meaning that we assume perfect power control. In order to constrain the comparison concerning the channel, the delays of the different paths have been set to $\tau_{11} = 0T$, $\tau_{21} = 0.25T$, and $\tau_{31} = 0.50T$ when $N_c = 7$, and $\tau_{11} = 0T$, $\tau_{21} = 0.125T$, and $\tau_{31} = 0.25T$ when $N_c = 15$ and so on. It is very important to notice that in the comparison, the bandwidth is constrained as well. It means that code lengths of $N_c = 7$ or $N_c = 15$ require identical bandwidths as long as the number of tones N_t is modified accordingly.

With $N_c/N_t = 7/1, 15/2, 31/4$ the bandwidth is roughly constant. Figure 3 gives the bit error rate as a function of the symbol energy over white noise ratio for DF-JEIC-1 and DF-JEIC-2 with $N_c/N_t = 15/2$ and $K = 1$ in both situations. In this present situation, as all gains have the same value, there is no reason to react with the current symbol of particular users on the other ones. However, here user 1 was detected alone, then user 2 with a feedback from user 1 and then user 3 with a feedback from the two other ones. It can be seen that DF-JEIC-2 brings an improvement in the order of half a dB for user 3 in this scenario. In order to prove the concept in high load conditions the system has been tested for $N_u = 8$ users together with the parameters $N_c/N_t = 31/4$. There are therefore 32 BER curves.

Figure 4 shows the result for matched filtering only and DF-JEIC-1 with $K = 1$. This figure clearly shows that the system is still of use even in these load conditions once DF-JEIC-1 is used.

Figure 5 shows the corresponding results for DF-JEIC-2. Again all users have the same path gains. Nevertheless DF-JEIC-2 was applied with detection of user 1, then feedback from 1 to 2, and so on. These results demonstrates the

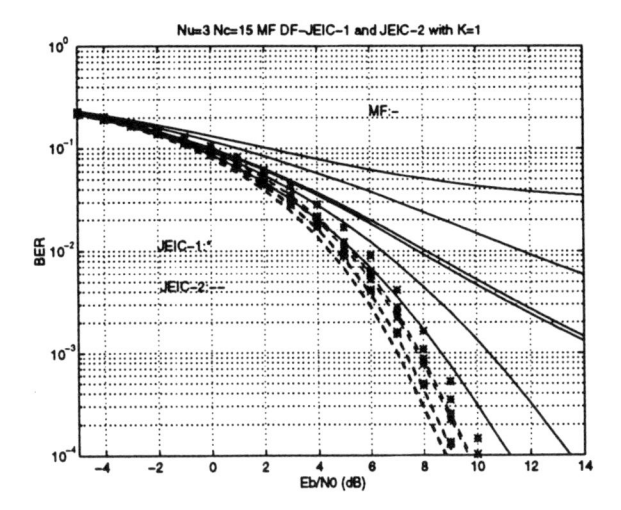

Fig. 3. Bit error rate as a function of E_b/N_0 for $N_c = 15$, $N_t = 2$ and $N_u = 3$. The solid curves are for matched filtering only, the stars are for DF-JEIC-1 with $K = 1$ and the dotted curves are for DF-JEIC-2 with $K = 1$.

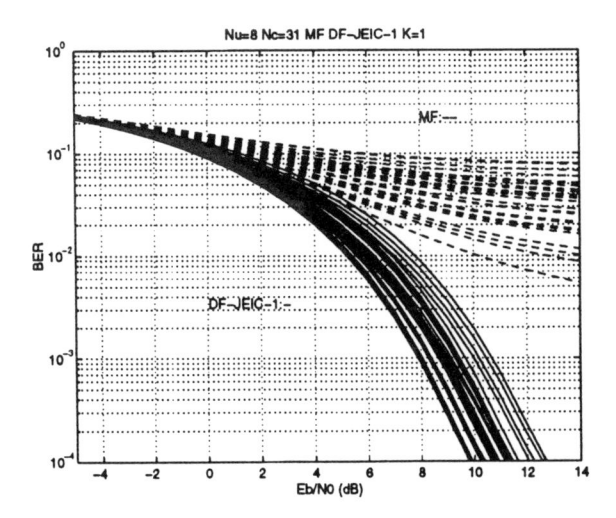

Fig. 4. Bit error rate as a function of E_b/N_0 for $N_c = 31$, $N_t = 4$ and $N_u = 8$. The dotted curves are for matched filtering only, the dotted curves are for DF-JEIC-1 with $K = 1$.

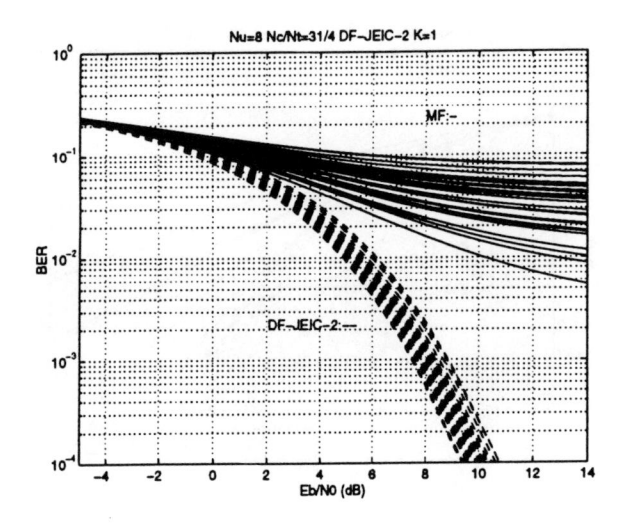

Fig. 5. Bit error rate as a function of E_b/N_0 for $N_c = 31$, $N_t = 4$ and $N_u = 8$. The dotted curves are for matched filtering only, the dotted curves are for DF-JEIC-2 with $K = 1$.

effectiveness of the DF-JEIC devices. Moreover the further improvement brought by using DF-JEIC-2 instead of DF-JEIC-1 is also illustrated. For the two-path scenarii we have arbitrarily set all path gains of all of $N_u = 3$ users to "1". The parameters which have been selected for this second type of channel are : $\beta_{ki} = 1$, $\tau_{11} = 0T_1$, $\tau_{12} = 0.35T_1$, $\tau_{21} = 0.2T_1$, $\tau_{22} = 0.6T_1$, $\tau_{31} = 0.3T_1$, $\tau_{32} = 0.9T_1$.

Figure 6 gives the bit error rate as a function of the symbol energy over white noise ratio for $N_c/N_t = 15/2$, for DF-JEIC-1 and DF-JEIC-2 with $K = 1$ for both situations. Concerning the feedback between users the same procedure as above was followed.

Figure 7 gives similar results for $N_c/N_t = 31/4$. Again, although all users have the same path gains there is a significant advantage brought by DF-JEIC-2. However both detectors behave very efficiently compared to the single matched filter detector.

A very important issue to be addressed is the resistance of the proposed detectors to the near-far effect. Near-far problems arise when a user is received which a larger power than the other users, or more generally, when the powers of the different users significantly differ.

Figures 8 and 9 show the BERs corresponding to DF-JEIC-1 and 2 with $K = 1$, for $N_c = 7$ and $N_c = 15$, respectively for two-paths channels. The amplitudes of the two paths of the third user (reference user) is kept constant, and all amplitudes of the two interfering users vary from $-10\ dB$ up to $10\ dB$ with

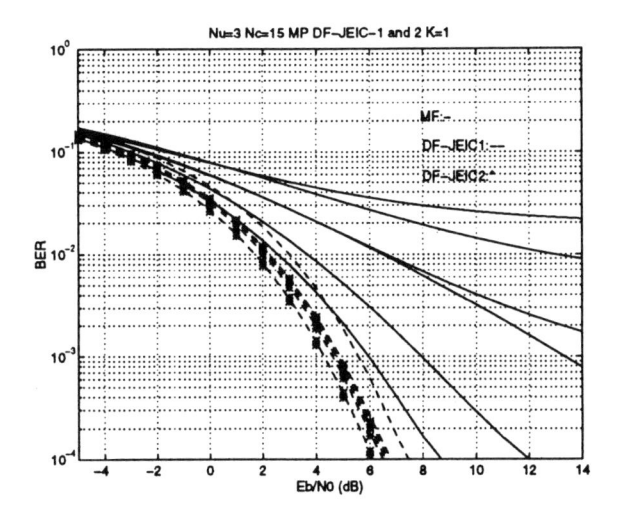

Fig. 6. Bit error rate as a function of E_b/N_0 for $N_c = 15$, $N_t = 2$ and $N_u = 3$ and two-paths channels. The solid lines are for matched filtering only, the dotted lines are for DF-JEIC-1 with $K = 1$ and the stars are for DF-JEIC-2 with $K = 1$.

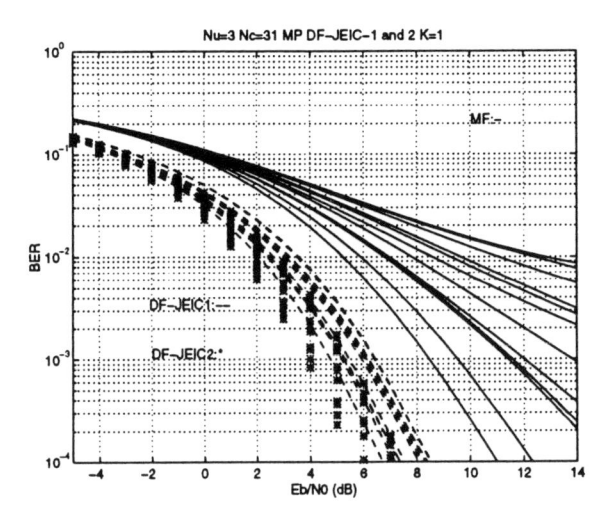

Fig. 7. Bit error rate as a function of E_b/N_0 for $N_c = 31$, $N_t = 4$ and $N_u = 3$ and two-paths channels. The solid lines are for matched filtering only, the dotted lines are for DF-JEIC-1 with $K = 1$ and the stars are for DF-JEIC-2 with $K = 1$.

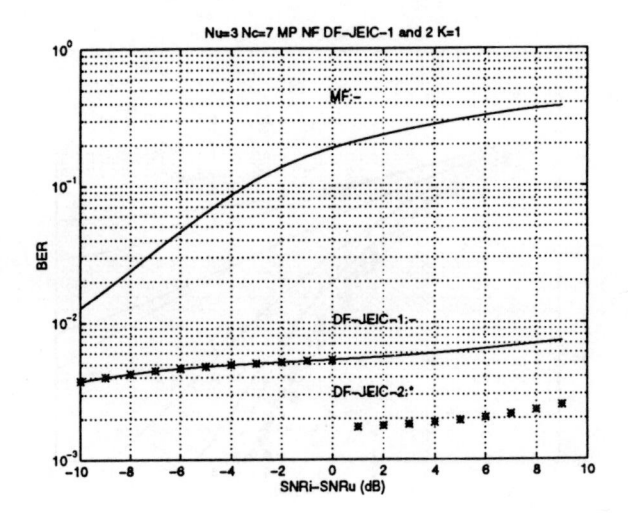

Fig. 8. Illustration of the near-far resistance for $N_c = 7$, $N_t = 1$ and $N_u = 3$ and the two-paths channels. $E_b/N_0 = 7\ dB$. Matched filter detection (solid line), DF-JEIC-1 with $K = 1$ and DF-JEIC-2 with $K = 1$ are compared.

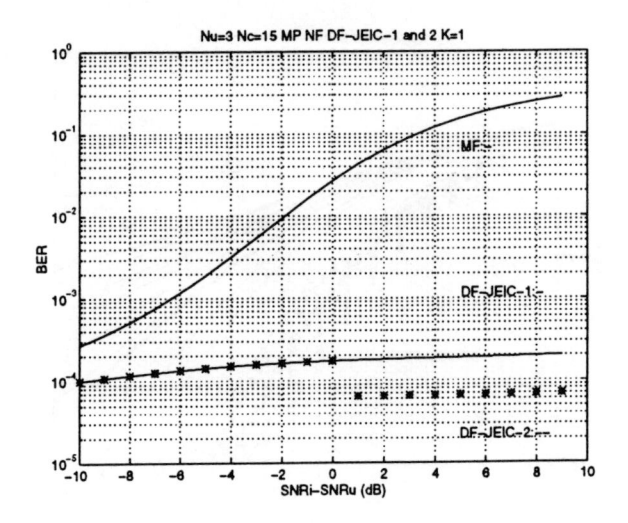

Fig. 9. Illustration of the near-far resistance for $N_c = 15$, $N_t = 2$ and $N_u = 3$ and the two-paths channels. $E_b/N_0 = 7\ dB$. Matched filter detection (solid line), DF-JEIC-1 with $K = 1$ and DF-JEIC-2 with $K = 1$ are compared.

respect to the constant value associated with the reference user. This analysis is made for a constant $E_b/N_0 = 7\ dB$. Here the strategy of DF-JEIC-2 is applied rigorously, meaning that there is no difference between DF-JEIC-1 and 2 when the power of the reference user is the largest. However, when the interfering powers become higher than that of the reference user, DF-JEIC-2 brings an improvement. The evolution type of the BER of the reference user does not differ significantly between DF-JEIC-1 and 2. Nevertheless the gain obtained by using DF-JEIC-2 instead of DF-JEIC-1 is illustrated. It can be concluded that the two types of detectors are near-far resistant for the range of amplitudes and the scenario considered here.

5 Conclusion

In this paper we have derived two structures for decision feedback joint detection of multitone CDMA. The general structures have been derived for an MMSE criterion. The derivation has been made assuming perfect synchronization and detection of the fed back symbols. The system performance has been investigated for asynchronous scenarii and one path or two-paths channels. The computational results have clearly demonstrated that the two structures are effective. It means that using a larger number of tones makes it possible to accommodate more users in a constant bandwidth and this type of receiver makes it possible to detect the different users with a very high efficiency. The near-far resistance of the two detectors has been investigated and demonstrated. Hence multitone CDMA with an appropriate receiver is an attractive candidate for multiuser communications over channels constrained by capacity issues.

References

1. *IEEE Journal on Selected Areas in Communications*, vol. 12, No. 4, May 1994.
2. *IEEE Journal on Selected Areas in Communications*, vol. 12, No. 5, June 1994.
3. *European Transactions on Telecommunications*, vol. 6, No. 1, January-February 1995.
4. A. Duel-Hallen, "Equalizers for multiple input-multiple output channels and PAM systems with cyclostationary input sequences", *IEEE Journal on Selected Areas in Communications*, vol. 10, No. 3, April 1992, pp. 630-639.
5. A. Duel-Hallen, "Decorrelating decision-feedback multiuser detector for synchronous code-division multiple-access channels", *IEEE Transactions on Communications*, February 1993, pp. 285-290.
6. R. Lupas and S. Verdu, "Linear multiuser detectors for synchronous code-division multiple access channels", *IEEE Transactions on Information Theory*, vol. 35, No. 1, January 1989, pp. 123-136.
7. R. Lupas and S. Verdu, "Near-far resistance of multiuser detectors in asynchronous channels", *IEEE Transactions on Communications*, vol. 38, No. 4, April 1990, pp. 496-508.
8. J. G. Proakis, "Digital Communications", New-York, McGraw-Hill, 1989.

9. L. Vandendorpe, "Multitone Spread Spectrum Communications Systems in a Multipath Rician Fading channel", *IEEE Transactions on Vehicular Technology*, Vol. 44, No 2, May 1995, pp. 327-337.
10. L. Vandendorpe and O. van de Wiel, "Performance analysis of linear MIMO equalizers for multitone DS/SS systems in multipath channels", accepted for publication in *Wireless Personal Communications*, December 1994.
11. L. Vandendorpe and O. van de Wiel, "Performance analysis of linear joint equalization and interference cancellation for multitone CDMA", accepted for publication in *Wireless Personal Communications*, September 1995.

This article was processed using the LaTeX macro package with LLNCS style

Blind Adaptive Interference-Cancelling Detectors for DS/SS CDMA: Algorithms and Performance

Riccardo De Gaudenzi[1], Filippo Giannetti[2], and Marco Luise[2]

[1] European Space Agency, European Space Research and Technology Centre, Noordwijk, The Netherlands
[2] University of Pisa, Department of Information Engineering, Pisa, Italy

Abstract

Code Division Multiple Access (CDMA) has emerged as a strong candidate for the air interface of the universal wireless personal communication system planned for the end of the century. In the latest years, a huge research effort has thus been devoted to find new viable techniques to increase the capacity of CDMA radio networks. In particular, different methods to counteract the near-far effect and to diminish the influence of asynchronous Multiple-Access Interference (MAI) have been investigated with the ambition to make CDMA competitive vis-a-vis more conventional access techniques. In this paper, we take into consideration a recently-proposed low-complexity Blind Adaptive Interference-rejection Detector (BAID) scheme that minimizes the detrimental effect of the MAI on the Bit-Error Rate (BER) performance of the data demodulator for Direct-Sequence/Spread-Spectrum (DS/SS) signals. Specifically, we describe a few modifications to the original algorithm that make it more suitable to a practical implementation with QPSK-modulated DS/SS signals and asynchronous multiplexing, and we evaluate the performance of such extended detector in the case of coherent and differential signal detection.

1 Motivation and Outline of the Paper

CDMA with DS/SS signals is known to bear some desirable features insofar as robustness to multipath fading, ease of resources allocation and possibility to exploit diversity reception are concerned. On the other hand, CDMA has also gained a bad reputation as far as the overall capacity (in terms of number of users/bandwidth) is concerned unless very tight power control is achieved jointly with the application of powerful correcting coding. The main factor that was deemed responsible for such throughput limitations is MAI, that limits the number of contemporarily active users operating at a specified system BER.

A step toward receiver robustness is described in Ref. [1] that introduces the *Blind* Adaptive Interference-cancelling Detector (BAID) with good self-adaptation properties and no need of a training sequence. The few simulation results presented therein are relative to the case of *Synchronous* access (S-CDMA) with ideal coherent detection, low Additive White Gaussian Noise (AWGN) level, and do not fully highlight the BAID capabilities. In particular, [1] seems to indicate that the BAID application is restricted to the initial detection stage to bring the equivalent signal-to-noise ratio slightly higher than 0 dB, so that a decision-directed equalizer can take over without the need for a training sequence. Furthermore, in Ref. [1] implementation issues such as application to the more general case of *Asynchronous* CDMA (A-CDMA) and carrier phase recovery are not discussed.

The aim of this paper is therefore to fill such conceptual gaps by introducing a generalized structure of the BAID algorithm presented in [1], that will be shortly addressed in the following as the Extended BAID (E-BAID). The E-BAID turns out to be particularly robust even in the presence of strong A-CDMA interference and we will show how the E-BAID *itself* can attain the MMSE performance over a large range of signal-to-(Gaussian) noise ratios without the need of resorting to any additional equalizer. Consequently, in our opinion, E-BAID concept illustrated in this paper may represent an effective design option for a low-complexity MAI-resilient receiver that must operate in severe conditions of near-far effect and under medium self-noise loading.

One of the main advantages of BAID lies in its *blind* behaviour that produces correct acquisition of the intended user signal with no need of a training sequence and/or knowledge of interfering signals' parameters.

As a result of the topics discussed above, a possible efficient architecture of a CDMA transmission scheme relies on the following techniques:

1. Rough power control based on a mix of open- and closed-loop techniques to avoid "large" signal power unbalance.
2. Pilot-aided downlink carrier frequency estimation [2] and uplink open loop carrier frequency shift correction.
3. Unaided robust chip timing acquisition [3] and tracking [4].
4. Asynchronous-MAI adaptive rejection through the E-BAID.
5. Baseband carrier phase estimation after the E-BAID [5].

While steps 1, 2, 3 and 5 can be achieved by conventional techniques, the E-BAID, that enhances the CDMA network capacity and quality of service relaxing also power control requirements, has still to be devised and will be the subject of this paper.

This Introduction is followed by four more sections. Specifically, in Section 2 we introduce the format of the CDMA signal we are concerned with, and we outline some criteria for the selection of the user signature sequences codebook. Section 3 deals with the extension of the BAID algorithm to the case of A-CDMA with an unknown carrier phase of the intended user signal (non-synchronous detection), while the performance of such a detector is evaluated in Section 4. The relevant numerical results are reported in Section 5, and finally Section 6 ends up the paper with some conclusive remarks.

2 CDMA Signal Format

The DS/SS-CDMA signal format to be presented below is driven by the fundamental requirement that the signalling interval T_s of the un-spread modulated signal be an integer multiple of the period LT_c of the signature code.

Now, assume that user number l intends to transmit the binary data stream $d_k^{(l)} \in \{\pm 1\}$ (with k ticking at the bit-rate $1/T_b$); if we introduce the following mathematical operators

$$|i|_L \triangleq i \bmod L \;,\;\; \lfloor i \rfloor_L \triangleq \text{int}\{i/L\} \qquad (1)$$

the complex envelope of the l-th DS/SS user signal is expressed as follows:

$$s_T^{(l)}(t) = \sqrt{P_S^{(l)}} \sum_{i=-\infty}^{+\infty} \left(c_{p,|i|_L}^{(l)} d_{p,\lfloor i \rfloor_L}^{(l)} + \jmath\, c_{q,|i|_L}^{(l)} d_{q,\lfloor i \rfloor_L}^{(l)} \right) g_T(t - iT_c) \qquad (2)$$

where $P_S^{(l)}$ is the average transmitted power of the l-th user, $c_{p,i}^{(l)}$ and $c_{q,i}^{(l)} \in \{\pm 1\}$ (with i ticking at the chip rate $1/T_c$) are the chips of the composed spreading sequences on the in-phase (P) and quadrature (Q) rails of the l-th transmitter, respectively, $d_{p,h}^{(l)}$ and $d_{q,h}^{(l)} \in \{\pm 1\}$ (with h ticking at the symbol rate $1/T_s$) are the h-th (coded) data symbols on the P and Q rails of the l-th transmitter obtained from the information bits $d_k^{(l)}$ as explained below, L is the spreading code period that, as mentioned above, coincides with the bandwidth spreading factor T_s/T_c, and finally $g_T(t)$ is the Nyquist square-root raised cosine impulse response of the transmit filter. Assuming that $d_k^{(l)}$ is output by a FEC coder with rate r ($r = 1$ in the case of no coding), the chip interval T_c is related to the bit interval T_b through the so-called processing gain $G_p \triangleq T_b/T_c = L/(r \log_2 M)$, M representing the PSK constellation size. As is apparent from equation (2), two independent spreading sequences are used for the P and Q components in order to cope best with transmission impairments (i.e. noisy carrier reference, imperfect carrier synchronization, nonlinear distortions etc. [5]). Such type of system is often addressed as Quadrature Pseudo-Noise (QPN) DS/SS-CDMA. Equation (2) fits the transmitted signal of different QPN DS/SS-CDMA systems currently in use or in advanced testing status either employing QPSK modulation (QPN DS/SS-QPSK, $M=4$, $d_{p,k}^{(l)} = d_{2k}^{(l)}$, $d_{q,k}^{(l)} = d_{2k+1}^{(l)}$, [6]) or with BPSK modulation (QPN DS/SS-BPSK, $M=2$, $d_{p,k}^{(l)} = d_{q,k}^{(l)} = d_k^{(l)}$, [7]). The received signal is made of the multiplexing of N different signals in the form (2), plus the (baseband equivalent of) the AWGN process $\nu(t)$.

Assume we want to demodulate channel # m. In the hypothesis of perfect chip timing recovery, the (ideal) chip sampling instants are $t_h \triangleq hT_c + \tau_m$, and consequently the baseband equivalent of the QPN DS/SS-CDMA signal samples at the output of the Nyquist square-root raised cosine chip matched filter (CMF) $g_R(t)$ of the receiver are as follows:

$$y(h) = y_p(h) + \jmath y_q(h) = e^{\jmath \vartheta_m} \left\{ \sum_{l=1}^{N} \sqrt{P_S^{(l)}} \, e^{\jmath (\Delta\vartheta_l^{(m)} + 2\pi \Delta f_l^{(m)} t_h)} \right.$$

$$\left. \cdot \sum_{i=-\infty}^{\infty} \left(c_{p,\lfloor i \rfloor_L}^{(l)} d_{p,\lfloor i \rfloor_L}^{(l)} + \jmath\, c_{q,\lfloor i \rfloor_L}^{(l)} d_{q,\lfloor i \rfloor_L}^{(l)} \right) g\left[(h-i)T_c + \tau_m - \tau_l \right] + n_h \right\}$$

$$\triangleq e^{\jmath \vartheta_m} \left[r_p(h) + \jmath r_q(h) \right] \tag{3}$$

where ϑ_m is the phase offset of the local oscillator with respect to the m-th transmitted carrier, τ_l is an unknown group delay, $\Delta\vartheta_l^{(m)}$ is the phase shift of the l-th transmitted carrier with respect to the phase of the m-th local oscillator (apparently, $\Delta\vartheta_m^{(m)} = 0$), $\Delta f_l^{(m)}$ is the sum of the relative carrier frequency offset Δf_m of the local oscillator in the m-th user receiver, plus a possible Doppler shift f_{D_l} characteristic of the l-th user, $\Delta f_l^{(m)} = \Delta f_m + f_{D_l}$, $g(t) \triangleq g_T(t) \otimes g_R(t)$ is a unit-energy raised-cosine Nyquist pulse, and $\otimes$ is the time convolution operator. The noise samples $n_h \triangleq n(t_h)$, are taken from the white Gaussian noise process $n(t) \triangleq \nu(t) \otimes g_R(t) = n_p(t) + \jmath n_q(t)$ whose independent real/imaginary components have two-sided power spectral density (PSD) $S_{n_p}(f) = S_{n_q}(f) = N_0 |G_R(f)|^2$, where $G_R(f)$ is the frequency response of the CMF. It is found that $E\{n_i(h)\} = 0$, $\sigma_n^2 \triangleq E\{n_i^2(h)\} = N_0/T_c$, $i = p, q$.

3 Design of the E-BAID Algorithm

3.1 Algorithm Outline

To simplify the description of the operation of the E-BAID, it is expeditious to adopt a vector notation. With reference to a given stream of samples $a(h)$ (with h ticking at the chip rate $1/T_c$), we denote with capital boldface symbols the following $3L$-dimensional, real-valued vector

$$\mathbf{A}(k) \triangleq \begin{bmatrix} \mathbf{a}_{-1}(k) \\ \mathbf{a}_0(k) \\ \mathbf{a}_1(k) \end{bmatrix} \tag{4}$$

where the L-dimensional sub-vectors $\mathbf{a}_w(k)$, denoted by small boldface symbols, are

$$\mathbf{a}_w(k) \triangleq [a((k+w)L), a((k+w)L+1), \ldots,$$
$$a((k+w)L+L-1)]^T \tag{5}$$

$w = -1, 0, 1$, ^{T} denoting vector transposition. Also, we denote with $\cdot$ the standard dot-product between I-dimensional vectors, $\mathbf{a} \cdot \mathbf{b} \triangleq \sum_{i=0}^{I-1} a_i b_i$, where a_i and b_i are the i-th elements of $\mathbf{a}$ and $\mathbf{b}$, respectively, and we also define $\|\mathbf{a}\|^2 \triangleq \mathbf{a} \cdot \mathbf{a}$. According to definitions (4) and (5), $\mathbf{Y}_i(k)$, $i = p, q$, are the $3L$-dimensional vectors containing $3L$ samples of the CMF outputs $y_i(h)$, as in (3). $\mathbf{Y}_i(k)$, span exactly 3 symbol intervals centered on the k-th one; $d_{p,k}^{(m)} + j d_{q,k}^{(m)}$ is thus to be regarded as the *reference symbol*. If we introduce the arrays $\mathbf{R}_p(k)$, $\mathbf{R}_p(k)$ containing the un-phase-shifted samples $r_p(h)$, $r_q(h)$ defined in equation (3), the following relationships hold:

$$\mathbf{Y}_p(k) = \cos\vartheta_m \, \mathbf{R}_p(k) - \sin\vartheta_m \, \mathbf{R}_q(k)$$
$$\mathbf{Y}_q(k) = \sin\vartheta_m \, \mathbf{R}_p(k) + \cos\vartheta_m \, \mathbf{R}_q(k) \tag{6}$$

The approach we pursue here to devise an interference-rejecting detector is akin to that outlined in [1]. In particular, for the sake of generality, as outlined in Fig. 1, we start by considering the more general case of four different linear detectors implemented as (adaptive) Finite Impulse Response (FIR) filters whose impulse responses are stored in opportune $3L$ dimensional vectors denoted as $\mathbf{H}_{i,j}^{(m)}$, $i = p, q$, $j = p, q$. The in-phase detectors (P-P, Q-P), observe the complex-valued signal (3) and output the two variables $z_{p,p}^{(m)}(k)$ and $z_{q,p}^{(m)}(k)$ that undergo phase offset corrections to give the final decision variable $z_p^{(m)}(k)$ relevant to the estimate of the in-phase datum $d_{p,k}^{(m)}$ according to the strategies outlined in Section 4. The quadrature detectors (P-Q, Q-Q) operate analogously on the same observed signal to give the decision variable $z_q^{(m)}(k)$ for the quadrature bit $d_{q,k}^{(m)}$. The four BAID output variables are thus given by

$$z_{i,j}^{(m)}(k) \triangleq \frac{\mathbf{Y}_i(k) \cdot \mathbf{H}_{i,j}^{(m)}(k)}{L} \quad i = p, q, \quad j = p, q \tag{7}$$

The detectors operate at chip rate but provide an output that is decimated down to symbol rate. It is apparent that the detectors replace the correlators with the signature codes (despreader+accumulator) of the conventional single-user detector (SUD) demodulator. The key difference with respect to the original detector introduced in [1] lies in the extension of the length of the impulse response to 3 *symbol intervals*, and the need of *four* detectors to achieve, a phase-invariant adaptation rule (the phase invariance of the four-detector receiver has been verified by means of computer simulations). This

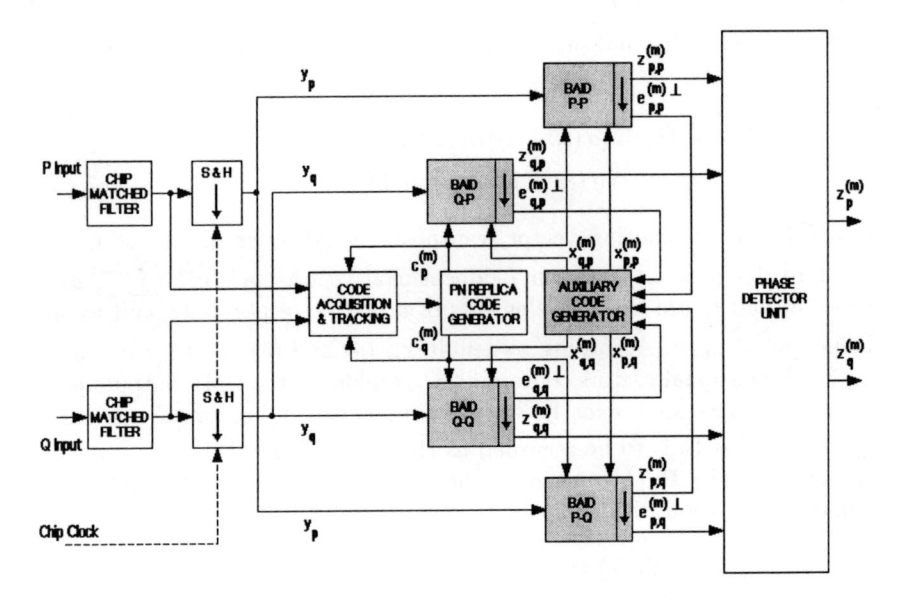

Fig. 1. Overall block diagram of the E-BAID demodulator.

latter feature reveals extremely precious since it allows for a fast convergence of the detector with no need of any chip-time phase offset correction on the received signal. As a consequence, phase error estimation/recovery can be performed at symbol rate on the detector output with a much lower computational effort and in a condition of good Signal-to-Noise Ratio (SNR). Also, lengthening of the impulse response is introduced to enhance the performance of the detector in the presence of *asynchronous* interference. In this case in fact, data transitions in the interfering data signals occur anywhere within the symbol period of the m-th signal. Therefore, the interference-rejection capabilities of the detector can be fully exploited, only if the interferers are observed on a time interval that encompasses the whole duration of the interfering symbols (hence the prudential choice of 3 symbol intervals centered on the reference one).

The impulse response vectors $\mathbf{H}_{i,j}^{(m)}$ are designed so as to minimize the *Mean Square Error* (MSE) between the detector output and the signal sample at the correlator output we would get in the absence of noise and MAI The detector is made adaptive by resorting to a standard stochastic gradient algorithm to find the solution of the minimization problem stated above. In particular, in [1] it is suggested to adopt the following canonical decomposition for the vectors $\mathbf{H}_{i,j}^{(m)}(k)$, $i = p, q$, $j = p, q$:

$$\mathbf{H}_{i,j}^{(m)}(k) \triangleq \mathbf{C}_j^{(m)} + \mathbf{X}_{i,j}^{(m)}(k)$$

$$\mathbf{C}_j^{(m)} \triangleq \begin{bmatrix} \mathbf{0} \\ \mathbf{c}_j^{(m)} \\ \mathbf{0} \end{bmatrix} \;,\; \mathbf{c}_j^{(m)} \triangleq [c_{j,0}^{(m)},\, c_{j,1}^{(m)},\, \ldots\, c_{j,L-1}^{(m)}]^T \qquad (8)$$

where $\mathbf{0}$ is the L-dimensional all-zero vector, with the following auxiliary "chunk" orthogonality condition

$$\mathbf{c}_j^{(m)} \cdot \mathbf{x}_{i,j,w}^{(m)}(k) = 0 \quad w = -1, 0, 1 \qquad (9)$$

As is apparent, the arrays $\mathbf{C}_j^{(m)}$ represent the impulse response we would get in the presence of AWGN only (the optimum L-chip long signature-matched filter), while the (adaptive) interference-rejecting part of the detector resides in the $\mathbf{X}_{i,j}^{(m)}(k)$ term. The goal of the adaptation algorithm is thus to obtain those vector $\overline{\mathbf{X}}_{i,j}^{(m)}$ that yield the *Minimum Mean Square Error* (MMSE), defined as

$$\begin{aligned}
\mathrm{MMSE}_p^{(m)} &\triangleq \min_{\mathbf{X}_{p,p}^{(m)}, \mathbf{X}_{q,p}^{(m)}} \left\{ \mathrm{MSE}_p(\mathbf{X}_{p,p}^{(m)}, \mathbf{X}_{q,p}^{(m)}) \right\} \\
\mathrm{MMSE}_q^{(m)} &\triangleq \min_{\mathbf{X}_{q,q}^{(m)}, \mathbf{X}_{p,q}^{(m)}} \left\{ \mathrm{MSE}_q(\mathbf{X}_{q,q}^{(m)}, \mathbf{X}_{p,q}^{(m)}) \right\}
\end{aligned} \qquad (10)$$

The more general adaptation rule of the E-BAID starting from an arbitrary trial value $\mathbf{X}_{i,j}^{(m)}(0)$ (usually $\mathbf{X}_{i,j}^{(m)}(0)=\mathbf{0}$), is governed by the following set of equations

$$\begin{aligned}
\mathbf{x}_{i,j,w}^{(m)}(k+1) = \mathbf{x}_{i,j,w}^{(m)}(k) - \\
\gamma \, z_{i,j}^{(m)}(k) \left\{ \mathbf{y}_{i,w}(k) - \frac{\mathbf{y}_{i,w}(k) \cdot \mathbf{c}_j^{(m)}}{L} \mathbf{c}_j^{(m)} \right\}
\end{aligned} \qquad (11)$$

where $i = p, q$, $j = p, q$, and γ is the update step, to be set as a compromise between acquisition speed and steady-state BAID performance and $w = -1, 0, 1$. The orthogonality condition (9) is responsible for the second term within braces in (11) and is crucial to get good convergence properties of the adaptation. As noted in [1], such a constraint "anchors" the BAID to minimize the MSE in the space orthogonal to the spreading sequence $\mathbf{c}_j^{(m)}$; this anchoring effect strengthens the convergence of the adaptive detectors towards the MMSE solution. The adaptation rule (11) does not require the use of estimates of the transmitted data, thus ensuring robust blind operation in unfavorable conditions of MAI and AWGN.

As a conventional SUD, the E-BAID performs correlation at chip rate and, due to the quadriphase modulation and spreading adopted, it requires four similar units to recover the full signal power[3] [6] (see Fig. 1).

---

[3] BPSK modulation and quadriphase spreading will only require two units.

4 BER Performance of the E-BAID

In the more general case of QPN DS/SS-CDMA signal, the four outputs of the E-BAIDs can be combined (at symbol time) to remove possible carrier phase offset for coherent data detection, as previously outlined. The result of such an operation is a complex-valued signal $z_p^{(m)}(k) + j z_q^{(m)}(k)$ that is input to a standard QPSK phase detector [5], [6]:

$$z_p^{(m)}(k) \triangleq \Re\left\{[z_{p,p}^{(m)}(k) + j z_{q,p}^{(m)}(k)] \exp(-j\hat{\vartheta}_k)\right\}$$
$$z_q^{(m)}(k) \triangleq \Re\left\{[z_{q,q}^{(m)}(k) + j z_{p,q}^{(m)}(k)] \exp(j\hat{\vartheta}_k)\right\} \qquad (12)$$

where $\hat{\vartheta}_k$ is the k-th estimate of the m-th user carrier phase that can be obtained after standard techniques [2] observing the E-BAID output. Notice that, in the case of RPI, the evolution of the E-BAID is insensitive to phase errors, so that a steady-state condition is attained irrespective of the value of ϑ_m. The carrier synchronization algorithm can thus benefit from the SNR improvement yielded by the E-BAID, and conventional digital carrier recovery techniques can be safely adopted.

In the uplink of a terrestrial/satellite fading channels, due to the absence of a (spread) pilot carrier, coherent detection may result hard to accomplish. Differential detection can be also achieved by combining the four E-BAID outputs as below to yield a differentially-demodulated complex-valued signal sample suited to PSK data decision:

$$z_p^{(m)}(k) \triangleq \Re\left\{\left[z_{p,p}^{(m)}(k) + j z_{q,p}^{(m)}(k)\right]\left[z_{p,p}^{(m)}(k-1) - j z_{q,p}^{(m)}(k-1)\right]\right\}$$
$$z_q^{(m)}(k) \triangleq \Im\left\{\left[z_{p,q}^{(m)}(k) + j z_{q,q}^{(m)}(k)\right]\left[z_{p,q}^{(m)}(k-1) - j z_{q,q}^{(m)}(k-1)\right]\right\} \qquad (13)$$

Coming back to the simpler case of coherent detection of a DS/SS-BPSK signal analyzed in the previous section, it is of interest to evaluate the BER of the E-BAID in such conditions to assess the interference-rejection performance of such a circuit. The k-th decision variable at the E-BAID output is in this hypothesis

$$z^{(m)}(k) = \frac{\mathbf{Y}(k) \cdot \overline{\mathbf{H}}^{(m)}}{L} = \frac{\mathbf{Y}(k) \cdot \left[\mathbf{C}^{(m)} + \overline{\mathbf{X}}^{(m)}\right]}{L} \qquad (14)$$

With some manipulations we get

$$z^{(m)}(k) = \sqrt{E_S^{(m)}}\, d_k^{(m)} + i_k^{(m)} + \nu_k^{(m)} \qquad (15)$$

where $i_k^{(m)}$ is the following MAI term

$$i_k^{(m)} \triangleq \sum_{l=1, l\neq m}^{N} \sqrt{E_S^{(l)}}\, \left\{ d_{k-2}^{(l)}\, \chi_{-1}^{-\,(l)} + d_{k-1}^{(l)}\left[\rho^{-\,(l)} + \chi_{-1}^{+\,(l)} + \chi_0^{-\,(l)}\right] + \right.$$
$$\left. d_k^{(l)}\left[\rho^{+\,(l)} + \chi_0^{+\,(l)} + \chi_1^{-\,(l)}\right] + d_{k+1}^{(l)}\, \chi_1^{+\,(l)} \right\} \qquad (16)$$

and the term $\nu_k^{(m)}$ is a zero-mean Gaussian random variable (RV) with a variance

$$\sigma_\nu^2 = \left[1 + \frac{\|\overline{\mathbf{x}}_{-1}^{(m)}\|^2 + \|\overline{\mathbf{x}}_0^{(m)}\|^2 + \|\overline{\mathbf{x}}_1^{(m)}\|^2}{L}\right] \frac{\sigma_n^2}{L} \qquad (17)$$

We can approximately assume that the MAI term $i_k^{(m)}$ given by (16) is a zero-mean Gaussian random variable with variance

$$\sigma_I^2 \triangleq E\left\{i_k^{(m)\,2}\right\} = \sum_{l=1, l\neq m}^{N} E_S^{(l)} D_2$$

$$\left\{\chi_{-1}^{-\,(l)\,2} + \left[\rho^{-\,(l)} + \chi_{-1}^{+\,(l)} + \chi_0^{-\,(l)}\right]^2 + \right.$$

$$\left. \left[\rho^{+\,(l)} + \chi_0^{+\,(l)} + \chi_1^{-\,(l)}\right]^2 + \chi_1^{+\,(l)\,2}\right\} \qquad (18)$$

so that the total disturbance at the E-BAID output is modeled as a zero-mean Gaussian RV whose variance is $\sigma_{tot}^2 \triangleq \sigma_I^2 + \sigma_\nu^2$. The resulting probability of error is immediately given by

$$P_e \simeq Q\left(\sqrt{E_S^{(m)}}/\sigma_{tot}\right) \qquad (19)$$

The validity of the Gaussian approximation will be demonstrated in Section 5. We observe explicitly that σ_{tot}^2 coincides with the MMSE at the BAID output: $\text{MMSE}^{(m)} = \text{MSE}\left(\overline{\mathbf{X}}^{(m)}\right) = \sigma_{tot}^2$

5 Numerical Results

In the following discussion, we assume an A-CDMA system based upon a QPN DS/SS-QPSK modulation scheme with Nyquist-shaped chip pulses having roll-off 0.4. The users' signatures are preferentially-phased Gold codes with an overlay PN M-sequence, both having period $L = 63$. The first set of results (Fig. 2 through 4) assumes that the interferers carrier phases are all the same as the useful signal. The impact of random interferers carrier phase will be analyzed in Fig. 5-7. Unless otherwise specified, all of the following results are derived for $N=18$ users having delays that are uniformly distributed in the range $(0, T_s/2]$. Since the theoretical analysis dealt with integer delays only, the relevant theoretical results are obtained after rounding the exact values of the users' delays to the nearest integer chip. The simulation results shown hereafter validate such an approximation. The power level of each CDMA interfering channel is related to the desired m-th user by the so-called *useful carrier-to-single interferer power ratio*, $[C/I]_{sc}$.

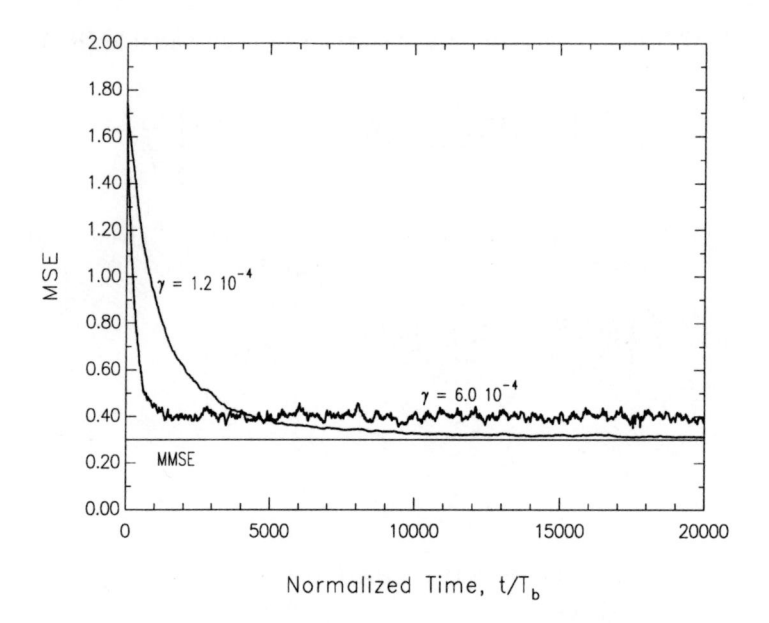

Fig. 2. E-BAID MSE transient sample : $E_b/N_0 = 4$ dB, $L = 63$, $N = 18$, $[C/I]_{sc} = -6$ dB.

An acquisition sample of the E-BAID is shown in Fig. 2 with $[C/I]_{sc} = -6$ dB, $E_b/N_0 = 4$ dB and for two different values of the step-size γ ($1.2 \cdot 10^{-4}$ and $6 \cdot 10^{-4}$). Specifically, it reports the MSE variation with time starting from receiver switch-on in the system configuration described above. The initial setting of $\mathbf{X}^{(m)}(k)$ is just equal to $\mathbf{0}$, i.e. the E-BAID is collapsed to the conventional SUD. In spite of the heavy interference condition, acquisition is steady towards the final value that attains the MMOE. The transient time constant is proportional to $1/\gamma$, and conventional techniques to speed-up the acquisition transient can be envisaged, such as different settings of γ for the acquisition and the steady-state conditions.

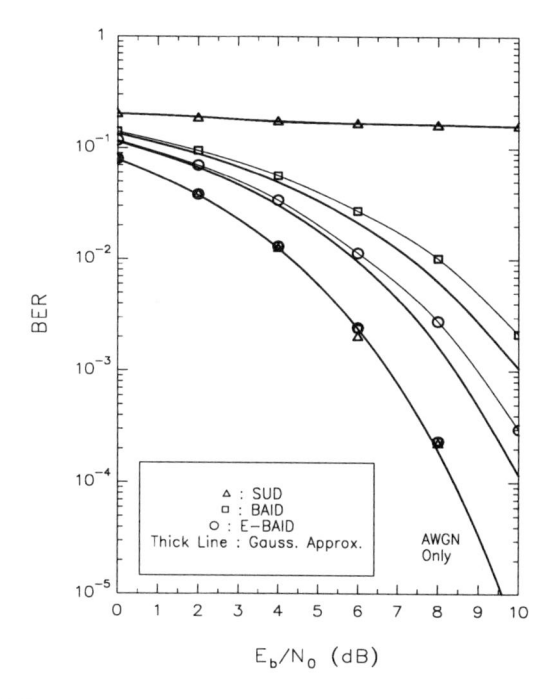

Fig. 3. BER of the coherent E-BAID vs. E_b/N_0 : $L = 63$, $N = 18$. $\gamma = 3 \cdot 10^{-4}$ for BAID, $\gamma = 1.2 \cdot 10^{-4}$ for E-BAID, $[C/I]_{sc} = -6$ dB.

In Fig. 3, the BER of the coherent E-BAID in the steady-state condition is compared to the corresponding results for standard SUD, BAID [1] and the Gaussian approximation (19) for $[C/I]_{sc} = -6$ dB. The step-size γ has been set to $3 \cdot 10^{-4}$ for the BAID and to $1.2 \cdot 10^{-4}$ for the E-BAID. The Gaussian approximation (thick lines) turns out to be remarkably close to Monte Carlo simulations (marks). Also, the E-BAID performance loss in the presence of CDMA interference is about 2 dB with respect to the matched filter bound for the AWGN case. For the sake of completeness, we also derived the performance of the E-BAID in the presence of AWGN only. As it is apparent from Fig. 3, in this case the BAID performs as the optimal SUD, so that the interference cancellation capabilities are not paid by any performance penalty under AWGN operation.

Figure 4, plotted for $E_b/N_0 = 8$ dB, confirms by simulation the claimed near-far resistance of the E-BAID. Simulation results (marks) are compared with the Gaussian approximation (thick lines). Observe that the *individual* worst-case $[C/I]_{sc}$ of -9 dB corresponds to an overall $C/I = [C/I]_{sc}/N = $ -21.3 dB. Again, the step-size γ is $3 \cdot 10^{-4}$ for the BAID and $1.2 \cdot 10^{-4}$ for the E-BAID.

So far, results have been reported for the case of interferers not affected by random phase errors. It can be verified that this additional degree of randomness will further degrade by 3 dB the BAID performance due to the increased dimension of the signal space spanned by the interferers. Fig. 5 shows the theoretical E-BAID performance with interferers having random carrier phase.

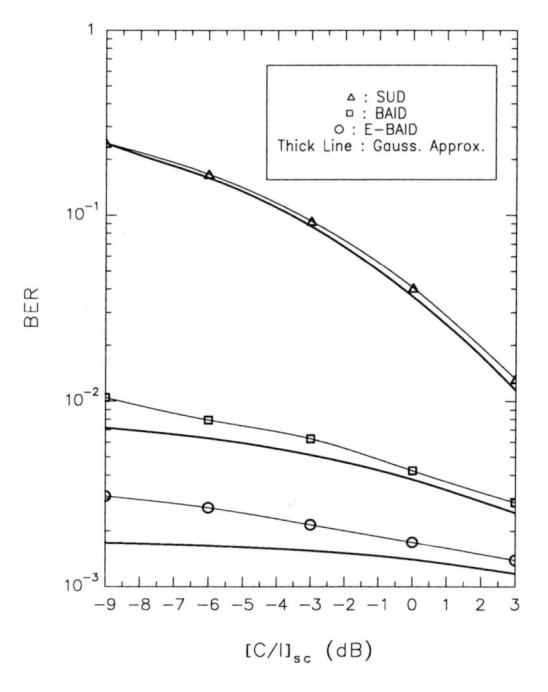

Fig. 4. BER of the coherent E-BAID vs. $[C/I]_{sc}$: $L = 63$, $N = 18$, $E_b/N_0 = 8$ dB, $\gamma = 3 \cdot 10^{-4}$ for BAID, $\gamma = 1.2 \cdot 10^{-4}$ for E-BAID.

It is interesting to analyze the E-BAID behaviour as a function of the number of active carriers each characterized by a different carrier phase and code delay with respect to the useful signal. Figure 6 shows the so called dimensional "clash" phenomenon : when the number of active carriers reaches the space dimension given by the spreading sequence length, the E-BAID performance collapses to the SUD one. Nonetheless, looking at Fig. 6 there is a large loading region where the E-BAID performance is considerably better than the SUD. In addition, the E-BAID performance is maintained also in the case of power-control errors, thus enhancing the quality of service. Finally, the effect of a possible frequency offset between the useful carrier and the MAI (resulting for instance from compensation of the relative Doppler shift of user m with respect to all of the other users) was evaluated by simulation. Results are plotted in Fig. 7 for the worst case wherein all the interfering carriers are affected by the same normalized frequency shift $\Delta f T_s$. It is assumed

$\gamma = 2 \cdot 10^{-4}$, $[C/I]_{sc} = -6$ dB and E_b/N_0 varying from 6 to 10 dB. It is apparent that only a small frequency offset can be tolerated due to strict limitations on the adaptation speed of the detector, related to the particular value of γ. Such a small amount of allowable frequency offset is not of concern for most terrestrial application, whereas is intolerable for non-geostationary satellite personal communication systems. The adaptation speed problem is also of concern when utilizing the E-BAID in a rapidly fading environment. In this case the performance advantages shown above might disappear until the user has reached a quasi-stationary condition.

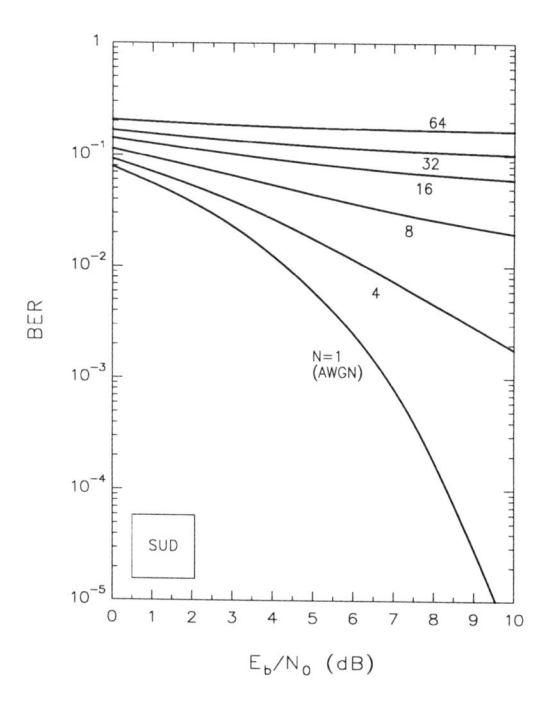

Fig. 5. Theoretical BER of the E-BAID vs. E_b/N_0 : $L = 63$, $N = 18$, $[C/I]_{sc} = 0$ dB, random interferers carrier phase.

6 Conclusions

The paper gives two main original contributions: the first one is the extension of an adaptive blind detector originally conceived for conventional DS/SS BPSK signals with synchronous interference and coherent demodulation to the case of QPN DS/SS-BPSK/QPSK signals with asynchronous MAI and an arbitrary carrier phase offset; the second one is the verification of the remarkably good interference rejection feature of the detector, and the equally good convergence properties of the adaptive filter through a realistic system simulation. The "anchoring" constraint of the filter coefficient vectors is shown to lead to a robust acquisition transient even in conditions of extremely strong interference (down to overall C/I ratios of -20 dB). The phase invariance of the E-BAID has been confirmed by computer simulation. Bit error rate degradation due to the interferers carrier phase randomness has also been pointed out together with the so called *dimensional clash* phenomenon. The E-BAID fragility to interferers carrier frequency offset has been shown together with its potential limitation in a rapidly fading environment due to the limited adaptation speed. The only limiting factor of the analysis shown in the paper is the assumption of perfect signal delay estimation. This still represents an open issue, in that interference-resilient fast code acquisition circuits bearing a reasonable complexity are still to be demonstrated.

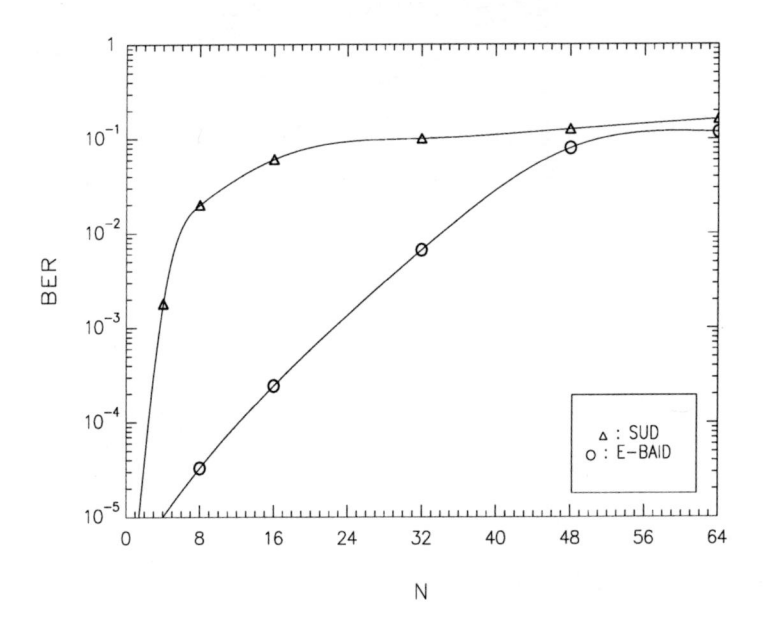

Fig. 6. Theoretical BER of the E-BAID vs. N : $E_b/N_0 = 10$ dB, $L = 63$, $[C/I]_{sc} = 0$ dB, random interferers carrier phase.

References

1. M.L. Honig, U. Madhow, S. Verdú, *"Blind Adaptive Interference Suppression for Near-Far Resistant CDMA"*, In the Proceedings of the IEEE GLOBECOM '94, S. Francisco, California, USA, November-December 1994.

2. F.M. Gardner, *"Demodulator Reference Recovery Techniques Suited for Digital Implementation"*, Final Report, ESTEC Contract No. 6847/86/NL/DG, ESA/ESTEC, Noordwijk, The Netherlands, August 1988.
3. R. De Gaudenzi, F. Giannetti, M. Luise, *"User Recognition and Signature Code Acquisition for BL-DS-CDMA Receivers"*, IEEE ICC '95, Seattle, Washington, June 18-22, 1995.
4. R. De Gaudenzi, M. Luise, R. Viola, *"A Digital Chip Timing Recovery Loop for Band Limited Direct Sequence Spread Spectrum Signals"*, IEEE Transactions on Communications, Vol. 41, No. 11, November 1993.
5. R. De Gaudenzi, F. Giannetti, M. Luise, *"Synchronization for Code-Division Multiple Access Transmission: an Overview"*, IEEE Benelux Workshop on Synchronization and Equalization in Digital Communications, Gent, Belgium, May 11, 1995
6. R. De Gaudenzi, C. Elia, R. Viola, *"Bandlimited Quasi-Synchronous CDMA: A Novel Satellite Access Technique for Mobile and Personal Communication Systems"*, IEEE Journal on Selected Areas in Communications, Vol. 10, No. 2, February 1992, pp. 328-343.
7. *"System and Method for Generating Signal Waveforms in a CDMA Cellular Telephone System"*, Qualcomm Inc., Patent PCT/US91/04400 - WO 92/00639, January 9, 1992.

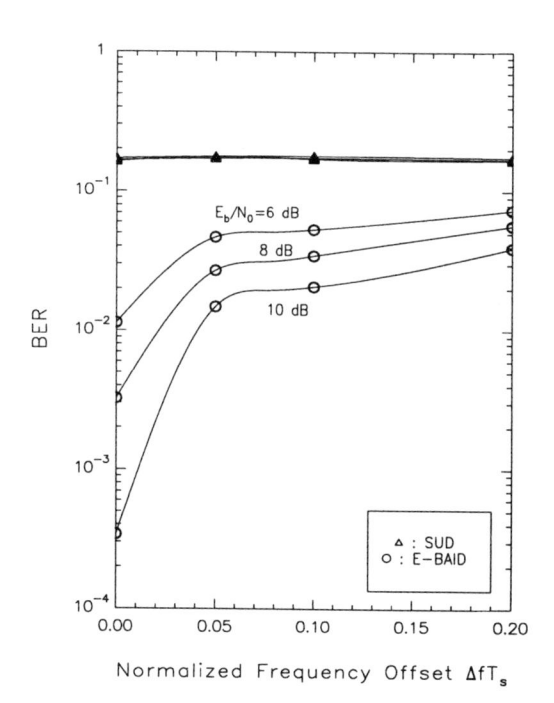

Fig. 7. Simulated BER of the coherent E-BAID vs. $\Delta f T_s$: $[C/I]_{sc} = -6$ dB, $L = 63$, E_b/N_0=6, 8, 10 dB, $\gamma = 2 \cdot 10^{-4}$, $N = 18$.

The Bootstrap Decorrelating Algorithm: A Promising Tool for Adaptive Separation of Multi-User CDMA Signals

by

Y. Bar-Ness

Center for Commun. and Signal Processing Research
NJIT, USA
&
Telecommunications and Traffic Control Systems Group
Delft University, The Netherlands
Telephone +1-201-5963520, Fax: +1-201-5968473
E-mail barness@hertz.njit.edu

Abstract: Decorrelating detectors were proposed as a tool to combat the near-far problem in multiuser CDMA communication systems. They were also used as a first stage of a multistage multiusers signal separator to facilitate tentative decision for the canceler stage. They completely separate signals but with relatively high additive noise. Different separators, which may not totally reject other signals, have been shown to result in better error probability, particularly in regions where interference energies are low. Adaptive structure of this separator termed "Bootstrap" is suggested in this paper and is shown to be suitable in situations where the cross-correlation matrix is unknown, or the codes have been altered by the channel.

1. Introduction

The recent and quickly increasing interest in Vehicular Communication, the need for an increased number of simultaneous users, and the limited availability of channel bandwidth has forced researchers to look into multiple-access schemes with frequency re-use, as in Code Division Multiplexing (CDMA) or time slot sharing, as in time domain multiplexing (TDMA).

The choice of CDMA is attractive because of its potential capacity increases and other technical factors, such as anti-multipath fading capability. For satisfactory performance, however, one must consider the effect of the "near-far" problem resulting from excessive Multiple-Access Interference energy (MAI) from nearby users, compared with the desired user's signal energy. Power control, that is, adjustment of transmitter power, depending on its location and the signal energies of the other users, has been suggested as a solution to this problem. But it requires a significant reduction in the signal energies of the strong users in order for the weaker users to achieve reliable communication. This results in an overall reduction in communication ranges. Commercial digital cellular systems based on CDMA and which use stringent power control are described in [1]. It offers capacity increases over TDMA.

A different approach for combatting the near-far problem was suggested by Verdú [2]. There, a receiver that is optimum in the multiuser interference environment was proposed and shown to eliminate the near-far problem and provide much improved performance. The improvement comes at the expense of high computational complexity. A class of suboptimum receivers that uses decorrelating detectors and which is based on the linear transformation of the sampled matched filters' outputs was considered in [3] and [4]. The decorrelating decision-feedback detector presented in [5] utilizes the difference in received users' energies where the decisions of the stronger users are used to eliminate interference on weaker users.

These decorrelating detectors do not require the knowledge of the receiver amplitude, but need the full information on cross-correlation between all user codes. Particularly in the asynchronous case, one must know accurately the relative delays between users' code sequences so that adequate matched filtering can be implemented and partial code cross-correlation can be calculated and used.

In recent years an adaptive algorithm termed "bootstrap algorithm" has been considered at the Center for Communication and Signal Processing Research, NJIT, for different applications, such as cross-polarization (cross-pol) cancellation with dually polarized co-channel communication. Lately, a version of this algorithm termed "the adaptive bootstrap decorrelator" has been used in multi-user CDMA instead of Verdú's fixed decorrelator, particularly as a first stage of a CDMA receiver. It was shown to perform better to or equal than the latter. Using it in different situations, it proved to be an interesting and promising tool for separation of multi-user CDMA signals. In the next section, we introduce the principle of the bootstrap algorithm. After we define the multiuser CDMA signal model in section 3, we present in section 4 the decorrelating detector of Verdú and another one that has equivalent performance. Next, we present and discuss the adaptive bootstrap detector. Then in section 5 we present two important applications of this algorithm. Conclusions are made in section 6.

2. The Bootstrap Algorithm

One of the most fundamental limitations of the additive interference (noise) canceler is the well known "power inversion" result, which states that cancellation can only be as good as the purity of the interference sample used in the cancellation process. In the case of the cross-pol interference canceler, this means that by using conventional algorithms, the output cross-pol isolation can be no better than the input cross-pol isolation. Beginning with work by Bar-Ness and Rokach [6] in the early 1980's, several closely related algorithms were developed at Bell Laboratories to circumvent this difficulty. These were called bootstrapped algorithms because they provide enhancement for both signal and interference samples using the output of each canceler to improve the other. Recently the same cross-coupled noise cancelers attracted the attention of many European Signal Processing researchers who named them "blind signal separators."

Three structures of bootstrapped algorithms were proposed: the forward-forward, the backward-backward and the forward-backward, as shown in Fig. 1. They are controlled by minimizing the two output powers or the cross-correlation between one output and some transformation of the other. First conceived and thoroughly studied as a two-input/two-output structure to enhance the separation of 16 QAM and 64 QAM dually polarized digital microwave communication [7, 8]. It had been extended to a multi-input/multi-output structure [9, 10] in research for other applications, proving again its effectiveness and robustness.

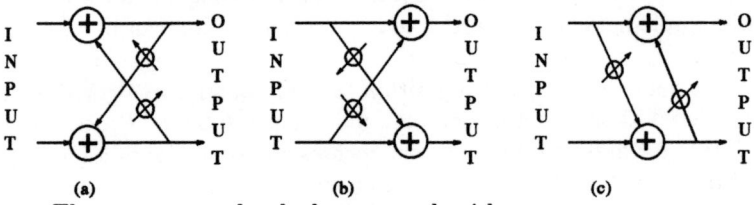

Fig. 1 Three structures for the bootstrap algorithm

It was shown, for example, that the backward/backward structure can achieve total signal separation (with zero additive noise) if the weights are controlled simultaneously to minimize the power at the outputs. Likewise, the forward/for-ward structure can use decorrelation of the outputs as the performance index by which the weights are controlled. In either case, the improved signal-to-interference ratio (SIR) at one output results in further improvement of this ratio at the other. The bootstrapping mechanism acts in such a way that a cleaner output at one of the ports results in better interference cancellation in another port, which in turn results in a still cleaner output of the first one, and so on. The process of bootstrapping will continue to result in a very large SIR in one port and in a very large ISR at the other. These values are ultimately limited by the noise impurities of the system and the control error.

3. Multiuser CDMA Signal Model

For general multi-user CDMA, the equivalent low-pass signal at the input of the matched filter is given by

$$r(t) = \sum_{k=1}^{K} \sum_{i} b_k(i)\sqrt{a_k(i)} s_k(t + iT - \tau_k) + n(t), \qquad (1)$$

where K is the number of users, and a_k, b_k, s_k and τ_k are the signal amplitude, user bit, signature waveform and relative delay of the kth user. $n(t)$ is the zero-mean AWGN, with a variance, σ^2.

For the synchronous channel encountered in down-link communications, the τ_k are zero for all k. In the asynchronous up-link channels the τ_k are not necessarily equal to zero. For the former case the output of the kth matched filter is the composite of bit b_k and all interfering bits given by a linear combination through the cross-correlation factor ρ_{kj}. In matrix notation we can write:

$$x(i) = PAb(i) + n(i), \qquad (2)$$

where P is the cross-correlation matrix of the signature sequence $s_k(t)$, $A = \text{diag}(\sqrt{a_1}, \cdots, \sqrt{a_K})$, $b(i) = [b_1(i), \cdots, b_K(i)]^T$, and $n(i)$ is a zero-mean Gaussian noise vector with covariance $R_n = P\sigma$. For the asynchronous case, we may write (1) as

$$r(t) = \sqrt{a_k}\, s_k(t)\, b_k(0) + \sum_{j \neq k}^{K} \sqrt{a_j\, \epsilon_j}\, \frac{1}{\epsilon_j}\, s_j^{\text{L}}(t) b_j(-1) +$$
$$\sum_{j \neq k}^{K} \sqrt{a_j\, (1 - \epsilon_j)}\, \frac{1}{1 - \epsilon_j}\, s_j^{\text{R}}(t) b_j(0) + n(t), \qquad (3)$$

where $0 = \tau_1 < \tau_2 < \cdots < \tau_K < T$

$$s_j^{\text{L}}(t) = \begin{cases} s_j(t + T - \tau_j) & \text{if } 0 \leq t \leq \tau_j \\ 0 & \text{if } \tau_j \leq t \leq T \end{cases}$$

$$s_j^{\text{R}}(t) = \begin{cases} 0 & \text{if } 0 \leq t \leq \tau_j \\ s_j(t - \tau_j) & \text{if } \tau_j \leq t \leq T \end{cases}$$

$$\epsilon_j = \int_0^{\tau_j} s_j^2(t + T - \tau_j) dt.$$

Then we may use the one-shot matched filter proposed by Verdú [11],(see Fig. 2).

Note from (3) that $r(t)$ contains a linear combination of $b_k(i)$, $b_j(i)$ and $b_j(i-1)$ for $j = 1 \cdots K$ $j \neq k$. Without loss of generality, we will use $k = 1$ (user one). This means that in the asynchronous case P will be replaced by another matrix that we will call the partial cross-correlation matrix. We will assume that P is nonsingular.

$$P^T = \begin{bmatrix} 1 & \rho_{12}^L & \rho_{12}^R & \cdots & \rho_{1K}^L & \rho_{1K}^R \\ \rho_{21}^L & 1 & 0 & & \rho_{2K}^{LL} & \rho_{2K}^{LR} \\ \rho_{21}^R & 0 & 1 & & \rho_{2K}^{RL} & \rho_{2K}^{RR} \\ \vdots & & & \ddots & & \vdots \\ \rho_{K1}^L & \rho_{K2}^{LL} & \rho_{K2}^{LR} & & 1 & 0 \\ \rho_{K1}^R & \rho_{K2}^{RL} & \rho_{K2}^{RR} & \cdots & 0 & 1 \end{bmatrix}. \qquad (4)$$

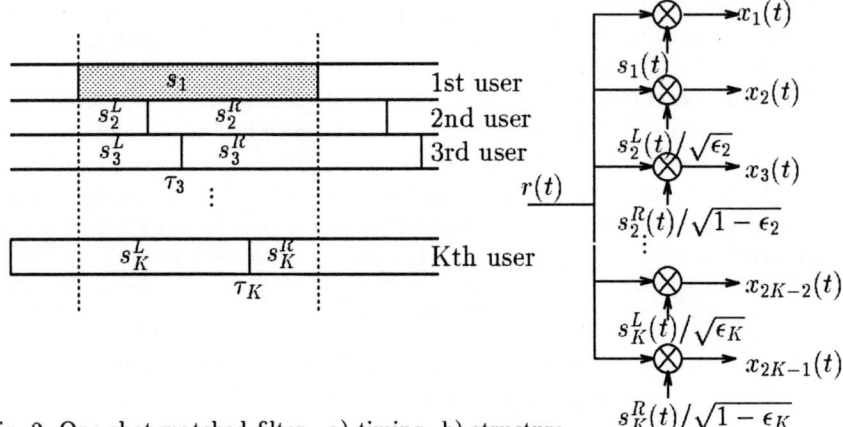

Fig. 2 One-shot matched filter, a) timing b) structure

If the τ_k are known, then all the elements of this matrix are known provided one knows the codes. Therefore, for both synchronous and asynchronous cases the output of the matched filters is given by (2) with the corresponding definition of its components. Also,
$\mathbf{A} = \text{diag}[\alpha_1, \alpha_2, \cdots, \alpha_{2K-1}] = \text{diag}[\sqrt{a_1}, \sqrt{a_2 \epsilon_2}, \sqrt{a_2(1-\epsilon_2)}, \cdots, \sqrt{a_K \epsilon_K},$
$\sqrt{a_K(1-\epsilon_K)}]$ and $\mathbf{b}(0) = [b_1(0), b_2(-1), b_2(0), \cdots, b_K(-1), b_K(0)]^T$. The $\mathbf{n}(t)$ is a zero-mean Gaussian vector with covariance, $\mathbf{P} N_0/2$, $\mathbf{x}(0) = [x_1(0), x_2^L(0), x_2^R(0), \cdots, x_K^L(0), x_K^R(0)]^T$, which corresponds to the output of the filters matched to $s_1(t), s_k^L(t)$ and $s_k^R(t)$, where $k = 2 \cdots K$ respectively. A conventional single-user detector will be using single output.

$$r(t) = \sqrt{a_1} b_1(0) + \sum_{k=2}^{K} \sqrt{a_k \epsilon_k} \rho_{1k}^L b_k(-1) + \sum_{k=2}^{K} \sqrt{a_k(1-\epsilon_k)} \rho_{1k}^L b_k(0) + n_0(t). \quad (5)$$

Clearly, the probability of error will increase when a_k, ρ_{1k}^R and ρ_{1k}^L increase.

4. Multiuser Decorrelating Detector

A. Verdú's Approach

Lupas and Verdú [3] used the inverse of the matrix $\boldsymbol{P}$ to separate the signals. That is:

$$\mathbf{z} = \boldsymbol{P}^{-1} \mathbf{x} = \mathbf{A}\mathbf{b} + \boldsymbol{P}^{-1} \mathbf{n} = \mathbf{A}\mathbf{b} + \xi, \quad (6)$$

where $\mathbf{z} = [z_1, \cdots, z_K]$, and without loss of generality we dropped the dependency on i. It follows that $\mathrm{E}\left[\xi \xi^T\right] = \sigma^2 \boldsymbol{P}$. Apart from noise, note that

$\mathbf{z}$ contains data from only one user. Using a decision on the elements of z_i, we get as output $\hat{\mathbf{b}} = \text{sgn}(\mathbf{z})$. Clearly this detector is near-far resistant. The probability of error for $\hat{\mathbf{b}}_k$ depends on the signal-to-noise ratio

$$\text{SNR}_i = \frac{a_i}{\text{E}[\xi_i^2]} = \frac{a_i |\mathbf{P}|}{\sigma^2 |\mathbf{P}_i|}, \tag{7}$$

where $|\mathbf{P}|$ is the determinant of $\mathbf{P}$ and $|\mathbf{P}_i|$ is the ii-th co-factor of $\mathbf{P}$.

The Verdú separator is only one of many possible separators. In fact any linear transformation that diagonalises matrix $\boldsymbol{P}$ will do. In the next section we present such a transformation and examine its properties and the performance of the corresponding multi-user detector.

B. Equivalent Decorrelating Detector

Let

$$\mathbf{z} = \mathbf{Vx} = \mathbf{VPAb} + \mathbf{Vn} = \mathbf{VPAb} + \zeta, \tag{8}$$

We will try to find $\boldsymbol{V} \neq \boldsymbol{P}^{-1}$ such that $\boldsymbol{VP}$ is a diagonal matrix. We propose to use $\boldsymbol{V} = \boldsymbol{I} - \boldsymbol{W}$, where

$$\boldsymbol{W}^T = \begin{bmatrix} 0 & w_{12} & w_{13} & \cdots & w_{1K} \\ w_{21} & 0 & w_{23} & \cdots & w_{2K} \\ \vdots & & \ddots & & \\ & & & 0 & w_{K-1,K} \\ w_{K1} & \cdots & & w_{K-1,K} & 0 \end{bmatrix}, \tag{9}$$

as shown in Fig. 3.

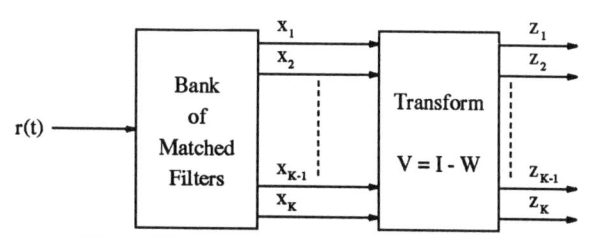

Fig. 3 Transformation of matched filter outputs

Note that $\boldsymbol{W}$ is not necessarily a symmetric matrix. The output of the detector is

$$z_k = x_k - \mathbf{w}_k^T \mathbf{x}_k, \tag{10}$$

where z_k is z without the kth element, $\mathbf{w}_k$ is the kth column of W without w_{kk} and $\mathbf{x}_k$ is the vector $\mathbf{x}$ with x_k taken out. We propose to choose W so that

$$\mathrm{E}\left[z_k b_k\right] = 0, \tag{11}$$

where the kth element is taken out of $\mathbf{b}$ to form $\mathbf{b}_k$. The solution of (11) using (10) is given by [12]

$$w_k = P_k^{-1} \rho_k, \tag{12}$$

where ρ_k is the kth column of P without the kk^{th} element. It is easy to show that

$$VP = \text{diag}(1 - w_i^T \rho_i, \cdots, 1 - w_K^T \rho_K),$$
$$\mathrm{E}[\xi_i^2] = \sigma_{\zeta_k}^2 = (1 - w_k^T \rho_k)\sigma^2$$

and

$$\text{SNR}_k = \frac{\alpha_i^2(1 - \rho_k^T P_k^{-1} \rho_k)}{\sigma^2}. \tag{13}$$

One can also show that $(1 - \rho_k^T P_k^{-1} \rho_k) = \frac{|P|}{|P_k|}$, and therefore when P is non-singular, the performance of this decorrelator is the same as Verdú's decorrelator.

5. The Adaptive Bootstrap Multiuser Detector

We propose using the forward/forward structure of Fig. 1.b principally, which for the multi-dimensional case again involves a linear transformation with $V = I - W$, as in section 4.b. The elements of W, however, will be chosen to satisfy (see Fig. 4)

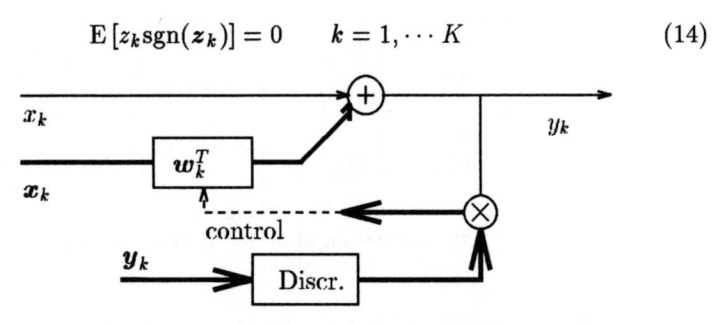

$$\mathrm{E}\left[z_k \text{sgn}(z_k)\right] = 0 \qquad k = 1, \cdots K \tag{14}$$

Fig. 4 Adaptive Bootstrap Multiuser Separator

Only if the signal-to-interference-plus-noise ratio at any of z_j is large, then $\mathrm{E}[z_k \text{sgn}(z_j)] = \mathrm{E}[z_k b_j(1 - P_{e_j})]$, in which case (14) and (11) will give the same result. That is, for high interference-to-noise ratio, the performance is the same as Verdú's. Note that except in these limiting conditions, this linear transformation will not separate the signal totally, but will have some interference residue at the output. The additive noise, however, will be smaller so that the output signal-to-noise-plus-interference ratio will be better (or at least as good) than the total separation signal-to-noise ratio. This gives us improved performance in the low interference region.

Solving (14) analytically without using the high signal-to-noise assumption is rather difficult due to the correlated noise. Using numerical solutions, however, we find that particularly in the region where interference power is low, the output SNR is better and so is the probability of error. In [13], using mathematical derivation and some reasonable approximations, the same conclusion was derived.

As an example we present the analytical result for the two-user case. For a given signal-to-noise ratio of user 1 (a_1/σ^2) the probability of error of this user as a function of SNR_2 (a_2/σ^2)

$$P_{e_1} = \frac{1}{2}\left(Q\left(\frac{\sqrt{a_1}(1-\rho^2+\sqrt{a_2}\delta)}{\sigma\sqrt{1-\rho^2+\delta^2}}\right) + Q\left(\frac{\sqrt{a_1}(1-\rho^2-\sqrt{a_2}\delta)}{\sigma\sqrt{1-\rho^2+\delta^2}}\right)\right), \quad (15)$$

where $w_{12} = \rho + \delta$ and ρ is the cross-correlation between the two codes. For small a_2, $\delta \to -\rho$. Then $P_{e_1} \to Q\sqrt{a_1}\sigma$, the error corresponding to a simple BPSK signal. When a_2 is large, $\delta \to 0$, and $P_{e_1} \to Q\left(\sqrt{a_1}\sqrt{1-\rho^2}/\rho\right)$, the error corresponding to Verdú's decorrelating detector. P_{e_1} is depicted in Fig. 5 as a function of E_b/N_0 of user 2.

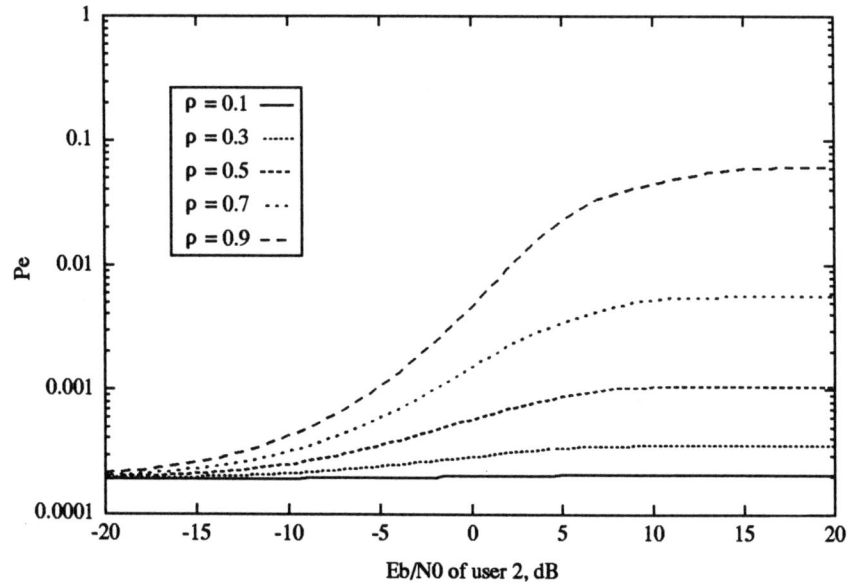

Fig. 5 Theoretical error probability of user 1 as a function of the energy of user 2. $E_b/N_{0_1} = 8dB$.

To obtain an adaptive algorithm for solving (13) we use the following updating formula.

$$w_k(i+1) = w_k(i) + \mu z_k \, \text{sgn}(z_k) \quad k = 1, 2, \ldots, K. \tag{16}$$

This is actually what was previously referred to in the literature as "bootstrap algorithm". Beside its ability to perform better to or equal than Verdú's decorrelating detector, it is suitable for cases where codes have been altered by the channel and hence the exact cross-correlation matrix is not available.

6. Some Applications of the Adaptive Bootstrap Algorithm

Because of its superior performance, the adaptive bootstrap algorithm might be used instead of Verdú's decorrelating detector in a multi-stage CDMA receiver occasionally proposed in the literature (for example [14, 15, 16]). In this section we will present some applications, wherein the ability of the algorithm to handle cases with unknown cross-correlation is demonstrated.

A. Asynchronous Multiuser CDMA Channel with Unknown Relative Delays

Applying a decorrelating detector to an asynchronous channel assumes accurate knowledge of the relative delays between users. In any estimation process, however, errors occur, so that partial codes actually used in Fig. 2 will be based on $\tau_i' \neq \tau_i$. This will lead to cross-correlation matrix $\boldsymbol{P}' \neq \boldsymbol{P}$ (which corresponds to (4) with τ_i'). If we use $\boldsymbol{P}^{-1}$ as a linear transformation, $\boldsymbol{P}^{-1}\boldsymbol{P}'$ is not the identity matrix. With the bootstrap adaptive algorithm, however, $\boldsymbol{V} = \boldsymbol{I} - \boldsymbol{W}$ will diagonalize $\boldsymbol{P}'$, leading to almost similar performance as using $\boldsymbol{P}'^{-1}$ (see [17]).

Example and results (Two-user case):
For $\epsilon_2 = 0.4, \rho_{12}^L = 0.2/\sqrt{0.4} = 0.3101, \rho_{12}^R = 0.6/\sqrt{0.6} = 0.9297$. If $\tau_2' < \tau_2$, then $\epsilon_2' < \epsilon_2$, assumed 0.35 (see partial code sequence timing in Fig. 6)

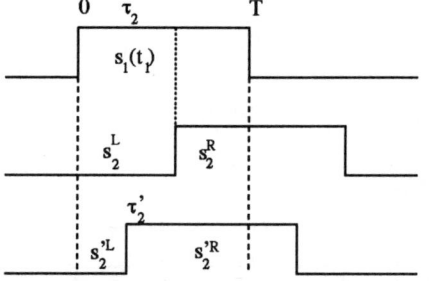

Fig. 6 Partial code sequence timing $\tau_2' < \tau_2$

The partial cross-correlation matrix is then given by

$$\mathbf{P}'^T = \begin{bmatrix} 1 & \rho_{12}^{L'} & \rho_{12}^{R'} \\ \rho'_{12} & \rho_{22}^{L'L} & \rho_{22}^{R'L} \\ \rho'_{12} & \rho_{22}^{L'R} & \rho_{22}^{R'R} \end{bmatrix}.$$

Here $S_2^{R'}$ has larger overlap with S_1 than S_2^R, hence

$\rho_{12}^{L'} = \frac{0.18}{\sqrt{0.35}} = 0.3162 \quad \rho_{12}^{R'} = \frac{0.75}{\sqrt{0.65}} = 0.7746 \quad \rho_{22}^{L'L} = \sqrt{\frac{\epsilon'_2}{\epsilon_2}} = \sqrt{\frac{0.35}{0.4}} = 0.9354$

$\rho_{22}^{R'R} = \sqrt{\frac{(1-\epsilon_2)}{(1-\epsilon'_2)}} = \sqrt{\frac{0.6}{0.65}} = 0.9608 \quad \rho_{22}^{R'L} = \frac{0.05}{\sqrt{(0.65)0.4}} = 0.0981 \quad \rho_{22}^{L'R} = 0$

In Fig. 7 we are showing the probability of error of user 1 as a function of its SNR. For comparison we show the performance of the conventional decorrelator. That is, the one uses the inverse of the cross-correlation matrix which corresponds to the wrong delays (τ'_2).

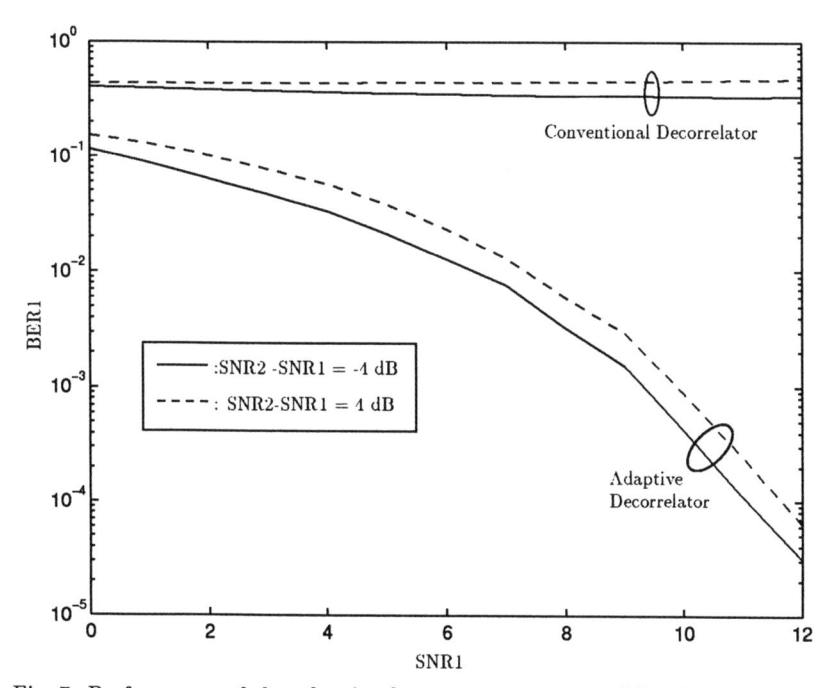

Fig. 7 Performance of the adaptive bootstrap separator with unknown relative delays

B. Multi-Carrier CDMA (MC-CDMA)

Multi-carrier CDMA is a new transmission technique that involves both multi-carrier modulation and spread spectrum [18]. The principle idea is that of transmitting the ith bit of user k, $k = 1, 2, \cdots K$ simultaneously on M orthogonal frequency carriers (OFDM), each BPSK-ed with one chip of a code sequence that corresponds to that user. That is,

$$s_{k,m}(t) = b_k(i)\Pi\left(\frac{t}{T}\right)c_k(m)\cos\left[2\pi\left(f_c + \frac{m-1}{T}F\right)t\right] \qquad (17)$$

$\Pi\left(\frac{t}{T}\right)$ is the bit pulse-shape (assumed rectangular)
$b_k = \pm 1$
$k(m) = \pm\frac{1}{\sqrt{M}}$, $m = 1\cdots M$ is the code sequence of the k-th user, M is the number of subcarriers, chosen such that the frequency spacing is an integer multiple F of the bit rate $\frac{1}{T}$.

If the bit rate is sufficiently low, then fading at each carrier is not frequency selective. Therefore, the channel response can be assumed as $h_{km} = a_{km}e^{j\theta_{km}}$, and hence the received signal at time iT (assumed synchronous)

$$r(t) = \sum_{k=1}^{K} b_k(i)\Pi\left(\frac{t}{T}\right)\sum_{m=1}^{M} c_k(m)a_{km} \qquad (18)$$

$$\cos\left[2\pi\left(f_c + \frac{m-1}{T}F\right)t + \theta_{km}\right] + n(t).$$

Using a coherent detector at each carrier, then multiplying the outputs of these detectors in parallel by code sequence $c_k(m)$, we obtain an output x_k (see Fig. 8).

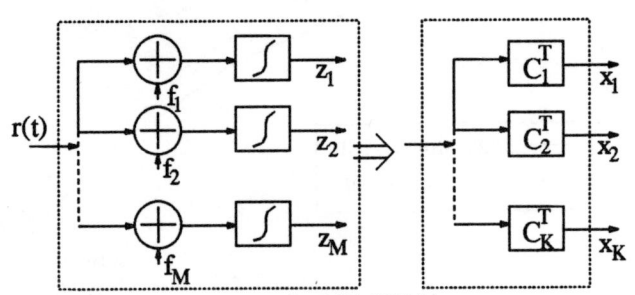

Fig. 8 Receiver structure for MC-CDMA

For all outputs in vector notation we have (see [19] for detail).

$$\mathbf{x}(i) = \mathbf{C}^T\overline{\mathbf{C}}\mathbf{b}(i) + \mathbf{n}(i), \qquad (19)$$

where $\mathbf{x} = [x_1, x_2, \cdots, x_K]^T$, $\mathbf{H}_k = \text{diag}[h_{k1}, h_{k2}, \cdots, h_{kM}]$, and $\mathbf{n} = [n_1, n_2, \cdots, n_K]^T$. The code matrix after the fading channel, $\overline{\mathbf{C}}$ is defined as $\overline{\mathbf{C}} = [\overline{\mathbf{c}}_1, \overline{\mathbf{c}}_2, \cdots, \overline{\mathbf{c}}_K]$ and $\overline{\mathbf{c}}_k = \text{Real}\{\mathbf{H}_k\mathbf{c}_k\}$, $\mathbf{C} = [\mathbf{c}_1, \mathbf{c}_2, \cdots, \mathbf{c}_K]$.

The effect of cross-coupling between users' code sequences and different carrier channels can be reduced with the aid of an adaptive bootstrap separator(see Fig. 9). For the downlink, it is adequate to assume that $\mathbf{H}_k = \mathbf{H}$ for all k, except for a multiplicative constant accounting for different transmit powers. Therefore, equation (19) can be written as:

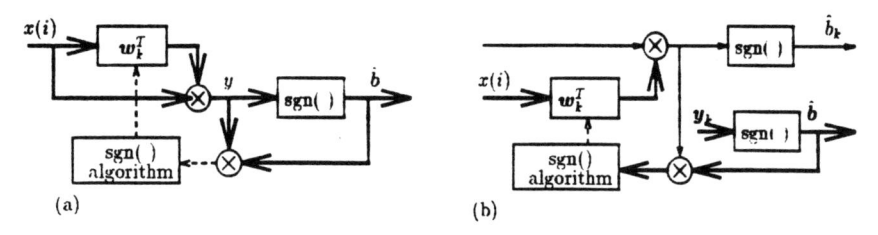

Fig. 9 Bootstrap separator

$$\mathbf{x}(i) = C^T \mathbf{H} C b + \mathbf{n}, \tag{20}$$

where $\mathbf{H} = \mathrm{diag}[h_1, h_2, \cdots, h_M]$, and $h_m = a_m e^{j\theta_m}$. For a multipath fading channel with an obstructed line-of-sight, we may model $[h_1, h_2, \cdots, h_M]$ as complex Gaussian random variables. Then the amplitude a_m has the Rayleigh probability density $f_{a_m} = \frac{a_m}{\sigma^2} e^{-\frac{a_m^2}{2\sigma^2}}$.

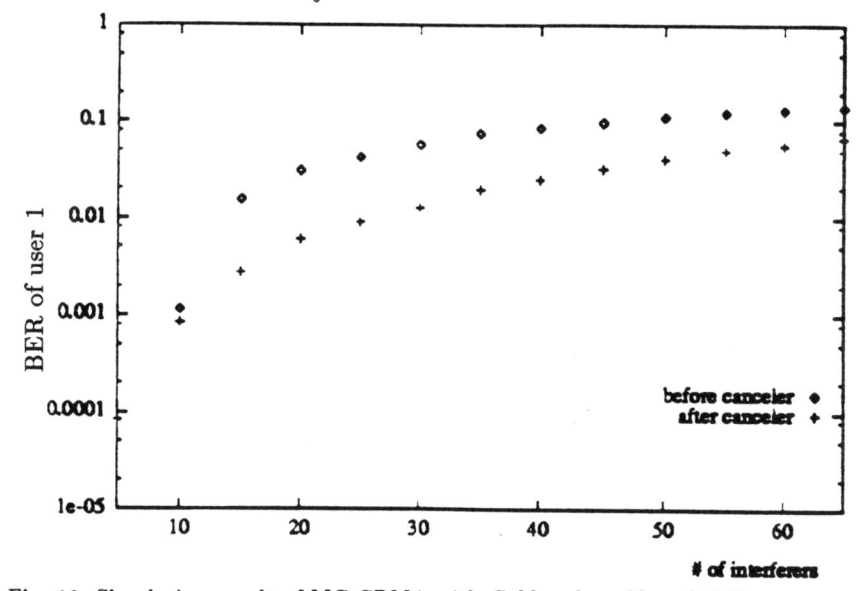

Fig. 10: Simulation result of MC-CDMA with Gold codes of length 127

In our simulation, we limited ourselves to the case of i.i.d-subcarrier amplitudes, i.e., the delay spread T_{rms} is large compared to T, or $F \gg 1$. Fig. 10 presents the BER before and after the adaptive canceler. The spreading codes used were 127 chip Gold codes. The local mean-received signal power-to-noise ratio is 10dB, for the designated user (user 1) and 18dB for other users. We assumed that the channel remains fixed for 1000 symbol-times.

Conclusion

The bootstrap decorrelator performs better than or equal to the conventional (Verdú type) decorrelator. It can be implemented by an adaptive algorithm. For asynchronous multiuser CDMA channels, if the delays are unknown (or known with an estimating error), then we can use a one-shot matched filter with $\tau_k' \neq \tau_k$, the true delay. This results in an unknown partial cross-correlation matrix for the mixture at the output of the matched filters. If one uses the conventional decorrelating detector, i.e., one that uses matrix inversion based on the estimated delay, then the resulting error probability is unsatisfactory. With the adaptive bootstrap algorithm, performance is much better. The bootstrap algorithm has also shown to improve performance of MC-CDMA. Currently it's been examined for other applications where codes have been altered by the channel. As such it is believed to be a promising tool for adaptive sparation of multiuser CDMA signals.

References

1. K. S. Gilhouson, I. M. Jacobs, R. Padovani, A. J. Viterbi, L. A. Weaver and C. Wheatley II , "On the Capacity of Cellular CDMA System," *IEEE Trans. on Vechicular Technology,* vol. VT-40, No. 2, pp. 303-311, May 91.
2. S. Verdú, "Minimum Probability of Error for Asynchronous Gaussian Multiple Access Channels," *IEEE Trans. Inform. Theory,* vol. IT-32, No. 1, pp. 85-96, Jan. 1986.
3. R. Lupas and S. Verdú, "Linear Multiuser Detectors for Synchronous Code Division Multiple Access Channels," *IEEE Trans. Inform. Theory,* vol. IT-35, No. 1, pp. 123-136, Jan. 1989.
4. R. Lupas and S. Verdú, "Near-Far Resistance of Multiuser Detectors for Synchronous Channels," *IEEE Trans. Commun.,* vol. COM-38, No. 4, pp. 496-508, April 1990.
5. A. Duel-Hallen, "Decorrelating Decision-Feedback Multiuser Detector for Synchronous Code Division Multiple Access Channels," *IEEE Trans. Comm.,* vol. COM-41, pp. 285-290, Feb. 1993.
6. Bar-Ness, Y. and Rokach, J., "Cross-Couple Bootstrapped Interference Canceler," *The 1981 AP-S International Symposium, Conference proceedings,* pp. 292-295, Los-Angeles, CA, June 1981.
7. Dinc, A. and Bar-Ness, Y., "Performance Comparison of LMS, Diagonalizer and Bootstrapped Adaptive Cross-Pol Canceler over Non-dispersive Channel," *The 1990 Military Communications Conference,* Monterey, CA, paper 3.7, Sept. 30-Oct. 3 1990.
8. Dinc, A. and Bar-Ness, Y., "Error Probabilities of Bootstrapped Blind Adaptive Cross-Pol Canceler For M-ary QAM Signals over Non-dispersive Fading Channel," *IEEE 1992 International Conference on Communications,* Chicago, IL, paper 353.5, June 15 1992.
9. Dinc, A. and Bar-Ness, Y., "Bootstrap: A fast Unsupervised Learning Algorithm," *IEEE International Conference on Acoustics, Speech and Signal Processing,* ICASSP '92, San Francisco, CA, paper 43.8, March 23 1992.

10. Dinc, A. and Bar-Ness, Y., "Convergence and Performance Comparison of Three Different Structures of Bootstrap Blind Adaptive Algorithm for Multisignal Co-Channel Separation," *MILCOM '92*, San Diego, CA, pp. 913-918, Oct. 12, 1992.
11. S. Verdú, "Recent Progress in Multiuser Detection," *Advances in Communications and Signal Processing*, Springer Verlag, 1989.
12. Y. Bar-Ness and J.B. Punt "An Improved Multiuser CDMA Decorrelating Detector," to be presented at PIMRC '95, Toronto, Canada.
13. Y. Bar-Ness and J.B. Punt "Adaptive Bootstrap CDMA Multiuser Detecter", Accepted to Wireless Personal Comm. :An International Journal, special issue on Signal Separation and Interference Cancelation for PIMRC.
14. M. K. Varanasi and B. Aazhang, "Multistage Detector in Asynchronous Code Division Multiple Access Communications," *IEEE Trans. Commun.*, vol. COM-38, No. 4, pp. 509-519, April 1990.
15. M. K. Varanasi and B. Aazhang, "Near-Optimum Detector in Synchronous Code Division Multiple Access System," *IEEE Trans. Commun.*, vol. COM-39, No. 5, pp. 725-736, May 1991.
16. Z. Siveski, Y. Bar-Ness and D. W. Chen, "Error Performance of Synchronous Multiuser Code Division Multiple Access Detector with Multidimensional Adaptive Canceler," submitted to European Transactions on Telecommunications and Related Technologies, Vol5, no 6, pp 719-724, Nov-Dec 94.
17. Y. Bar-Ness and N. Sezgin, "Adaptive Multiuser Bootstrapped Decorrelating CDMA Detector for One-shot Asynchronous Unknown Channels," ICASSP'95, Detroit, MI, pp. 1733-1736, May 1995.
18. N. Yee, J.P.M.G. Linnartz and G. Fettweis, "Multi-Carrier CDMA Indoor Wireless Radio Networks," IEEE Personal Indoor and Mobile Radio Communications (PIMRC) Int. Conference, Yokohama, Japan, Sept. 1993, pp. 109-113.
19. Y. Bar-Ness, J.P. Linnartz and X. Liu, "Synchronous Multi-User Multi-Carrier CDMA Communication System with Decorrelating Interference Canceler," PIMRC'94 The Hague, Then Netherlands, pp. 184-188, Sept. 1994.

Part 2

Joint Source/Channel Coding for Multimedia Applications

Rate-Distortion Performance of Several Joint Source and Channel Coding Schemes for Images[*]

Michael J. Ruf[1] and James W. Modestino[2]
German Aerospace Research Establishment (DLR), Institute for Communications Technology, D - 82234 Wessling, Germany[1]
ECSE Department, Rensselaer Polytechnic Institute, Troy, NY, 12180, U.S.A.[2]

Abstract

This paper describes a methodology for evaluating the rate-distortion behavior of combined source and channel coding schemes with particular application to images. We demonstrate use of the operational rate-distortion function to obtain the optimum tradeoff between source coding accuracy and channel error protection under the constraint of a fixed transmission bandwidth. Furthermore, we develop information-theoretic bounds on performance and demonstrate that our combined source-channel coding methodology results in rate-distortion performance which closely approaches these theoretical limits. We concentrate specifically on a wavelet-based subband source coding scheme followed by either a scalar quantizer or a product pyramid vector quantizer (PPVQ) and the use of binary rate-compatible punctured convolutional (RCPC) codes for transmission over the additive white Gaussian noise (AWGN) channel.

1 Introduction

Shannon's information theory has established that, in the limit of large block sizes, source and channel coding can be treated separately and if the rate-distortion function of the encoded source is smaller than the channel capacity, theoretically achievable performance is limited solely by source coding errors.

---
[*]This work was partly performed at the ECSE Dept., Rensselaer Polytechnic Institute, Troy, NY 12180 and was supported by the German Educational Exchange Service (DAAD) as part of the HSP II-program, and in part by ARPA under Contract No. F30602-92-C-0030

However, this assumes that there are no constraints on tolerable channel encoding/decoding complexity which is never the case in real-world systems. As a result, some form of combined source and channel coding approach is required to optimize overall performance at reasonable complexity levels. Furthermore, this approach should result in different error protection to individual bits according to their respective effects on the reconstructed image.

This paper is organized as follows. After a brief description of the source coder, the statistical properties of the various data streams and the different quantizers in Section II, we describe in Section III how the quantization errors and the channel error effects contribute to the overall distortion as an explicit function of the number of quantization bits used for the different source data streams and on the specific channel codes employed for operation over an additive white Gaussian noise (AWGN) channel at a specific value of E_s/N_0. In Section IV, we show how to optimize the rate-distortion performance for (real-world) transmission systems by jointly distributing source and channel bits in an optimum way and extend the rate-distortion approach to the development of general information-theoretic bounds and compare different image transmission schemes. Finally, in Section V we provide a summary and conclusions.

2 Source Coder

The image first undergoes a discrete wavelet transform (WVT) [1] to decorrelate the source signal and to make the data more suitable for compression. We make use of the fact that the histograms of the various subimages after the WVT can be modeled very precisely in terms of the generalized Gaussian (GG) distribution [1]

$$p(x) = \frac{\beta}{2\alpha\Gamma(1/\beta)} \cdot \exp\left\{|x/\alpha|^\beta\right\} \;, \tag{1}$$

with α a scale parameter and the parameter β controlling the exponential rate of decay. Results have shown that all AC-subimages match the GG-model very well, while the DC-subimage does not and thus will be modeled by a Gaussian distribution.

2.1 Quantization

In a previous investigation [2] we compared the performance of jointly optimized source and channel coding schemes, using either a uniform threshold (UT) quantizer or an optimum (nonuniform) GG-quantizer. This work showed the much better overall performance of the scalar GG-quantizer. In this work we investigate the improved performance resulting from a jointly optimized approach using a more sophisticated source coding scheme. For this purpose, we compare the performance of the optimum (nonuniform) GG-quantizer to the results of a scheme employing a product pyramid vector quantizer (PPVQ).

For the first scheme, an **optimum (nonuniform) GG-quantizer**, we use the fact that the histograms of the AC-subimages can be modeled as generalized Gaussian (GG) distributions. Then following a procedure described by Roe [3] and applying it to the generalized Gaussian distribution [2], one can calculate the optimum reconstruction levels and quantization thresholds for a particular number of quantization bits n. The expected mse due to quantization can be computed as [4]

$$e_{\text{GG},n} \approx \frac{S_g^3}{3} \cdot \frac{1}{(2(2^n + \gamma_2))^2} , \tag{2}$$

with the correction term γ_2 given by

$$\gamma_2 = \frac{1}{2\alpha} \left(\frac{S_g^3 \Gamma(1/\beta)}{3\Gamma(3/\beta)} \right)^{1/2} - 1 , \tag{3}$$

and

$$S_g = 3^{1/\beta} \cdot \left(\frac{2\alpha \Gamma(1/\beta)}{\beta} \right)^{2/3} . \tag{4}$$

For the second scheme, the **product pyramid vector quantizer** (PPVQ) [5], the coefficients of the subbands are assumed as i.i.d. Laplacian random variables[1] which are quantized by projection onto a lattice structure on the surface of an L-dimensional hyperpyramid $S(L, \rho(\mathbf{X}))$ where the pdf of vectors $\mathbf{X}$ with radius $\rho(\mathbf{X}) = \sum_{i=1}^{L} |x_i|$ formed out of L random variables is constant or sufficiently smooth. The expected distortion due to quantization to one of the lattice points of the pyramid lattice $P(L, K)$

$$P(L, K) = \left\{ \mathbf{X} \in \mathbf{Z}^L; \|\mathbf{X}\|_1 = \sum_{i=1}^{L} |x_i| = K \right\} \quad \text{for} \quad K \in \mathbb{N}_0 , \tag{5}$$

using the PPVQ with $L - K$-thresholding (which further increases the coding efficiency, for details see [5]) and quantizing the radii $\rho(\mathbf{X})$ with a nonuniform 2^n-level Gamma quantizer, can be written as [6]

$$e_{\text{PPVQ},L,K,n} = \begin{cases} \sigma^2(x) ; & \text{for } K = 0 \\ \frac{1}{L} \left[\frac{e_{\text{G},n}}{\zeta} + (\mathbb{E}(\rho(\mathbf{X}))^2 + \sigma^2(\rho(\mathbf{X}))) \cdot \epsilon_{l,k} \right] ; & \text{for } K > 0 , \end{cases} \tag{6}$$

---

[1]Contrary to the generalized Gaussian distribution, the Laplacian distribution results in a mathematically tractable and very efficient hyperpyramidal structure to describe the contours of constant pdf for formed vectors.

with the variance $\sigma^2(\rho(\mathbf{X}))$ and mean $\mathbb{E}(\rho(\mathbf{X}))$ of the radii $\rho(\mathbf{X})$, $\zeta = K$ for $K < L$ and $\zeta = L$ for $K \geq L$ and the second moment $\sigma^2(x) = \frac{1}{s_i}\sum_{i=1}^{s_i} x_i^2$ for a better estimation of the expected error in case of no allocation of quantization bits. Using again Roe's approach, the expected error due to quantization with an optimum nonuniform 2^n-level Gamma quantizer can be written as [7]

$$e_{G,n} \approx \frac{1}{3}\left[2 \cdot \left(\frac{2^n - 1}{S^{3/2}} + \frac{1}{2\beta_g\sqrt{3\alpha_g}}\right)\right]^{-2}, \qquad (7)$$

with

$$S = 3^{\frac{\alpha_g+2}{3}} \cdot \beta_g^{2/3} \cdot \frac{\Gamma\left(\frac{\alpha_g+2}{3}\right)}{[\Gamma(\alpha_g)]^{1/3}}, \qquad (8)$$

and $\beta_g = \sigma^2(\rho(\mathbf{X}))/\mathbb{E}(\rho(\mathbf{X}))$ and $\alpha_g = \mathbb{E}(\rho(\mathbf{X}))/\beta_g$. The error term $\epsilon_{L,K}$ can be written as [6]

$$\epsilon_{L,K} = \begin{cases} \frac{K-1}{12}\left(1+\frac{2}{K}\right)\frac{1}{L^2+L}\left(\frac{1}{K}+Q_{L,L-K}+(1+S_{L,L-K})^2\right)+ \\ \qquad +\frac{1}{L^2+L}\cdot\sum_{i=1}^{L_K}(S_{L,i}^2+Q_{L,i}); & \text{for } 0 < K < L \quad (9) \\ \frac{L-1}{12}\left(1+\frac{2}{L}\right)\frac{1}{K^2}; & \text{for } K \geq L, \end{cases}$$

with $S_{L,i} = \sum_{j=1}^{i}(L-j+1)^{-1}$ and $Q_{L,i} = \sum_{j=1}^{i}(L-j+1)^{-2}$. Finally, the resulting number of quantization bits per pixel (pyramid and radius quantization) can be expressed as

$$n_{\text{PPVQ}} = \frac{1}{L}(\cdot\lceil\log_2(N(L,K))\rceil + n), \qquad (10)$$

where $\lceil\cdot\rceil$ denotes rounding to the nearest higher integer and $N(L,K)$, the number of lattice points on $S(L,\rho(\mathbf{X}))$ will be calculated recursively using [6].

2.2 Coding

Knowing the mean-square distortion e_{i,n_i} for each subband $i = 1, 2, ..., K$ ($i = 1$ denoting the DC-subband and $i = K$ denoting the highest-frequency subband), with n_i the number of quantization bits allocated to the ith subband, one can easily calculate the expected distortion of a compressed image, D_S (measured as the mean-squared error on a per pixel basis), for every possible allocation of different quantizers per subband as $D_S = \sum_{i=1}^{K} D_{S,i} = \sum_{i=1}^{K}(s_i/S) \cdot e_{i,n_i}$, with S the total number of pixels in the original image and $s_i = S/2^{2r_i}$ the number of samples per subband (r_i denoting the resolution of subband i). The corresponding source rate R_S in bits per pixel (bpp) can also be calculated as $R_S = \frac{1}{S}\cdot\sum_{i=1}^{K} s_i \cdot n_i$. The number of quantization bits n_i for a given source rate R_S can then be determined by an optimal bit-allocation algorithm of Westerink et al. [8].

3 Channel Error Effects

For joint source and channel coding, one must consider, in addition to quantization errors, the effects of corrupted source bits on the reconstructed image. In particular, knowing the distribution parameters of the generalized Gaussian distribution associated with a particular subband i, denoted α_i and β_i, or in the case of the PPVQ, the parameters $E(\rho(\mathbf{X}))$ and $\sigma^2(\rho(\mathbf{X}))$, one can then evaluate the contribution to the overall mean-square reconstruction error due to an error in any given bit position. These contributions can be expressed in terms of individual bit-sensitivities to errors, $A_{i,j}$, and lead to a useful and tractable approximation to the combined effect of source and channel coding errors. We used a sign-magnitude representation of quantizer output levels or indices, since it showed a superior performance compared to pure magnitude or to Gray coding [2].

3.1 Derivation of Bit-Sensitivities

For the **optimum (nonuniform) GG-quantizer**, we calculate the bit-sensitivities for the AC-Bands in the following way [2]. Let's denote by $d(x_{l,a}, x_{l,b})$ the distance between the reconstruction levels $x_{l,a}$ and $x_{l,b}$, where, due to symmetry, we refer to the positive range only (i.e., mean=0, $0 \leq x_l \leq \infty$, $a, b \in (0, ..., (2^{n-1} - 1))$). The probability, that a sample to be quantized falls into the interval $[x_{t,k}, x_{t,k+1}]$ with $k \in (0, ..., (2^{n-1} - 1))$ is $\Pr(x_{l,k}) = 1/2 \cdot (P(1/\beta, x_{t,k+1}) - P(1/\beta, x_{t,k}))$, with $P(a, x)$ the incomplete Gamma function. So the sensitivity for the $(n-1)$ magnitude bits can be expressed as [2]

$$A_{\text{GG},\text{m},j} = 2 \cdot \sum_{l=0}^{2^j - 1} \sum_{k=0}^{(2^{n-2-j} - 1)} [\Pr(x_{l,a}) + \Pr(x_{l,b})] \cdot d^2(x_{l,a}, x_{l,b}), \qquad (11)$$

for $j = 0, ..., (n-2)$, with $a = l + k \cdot 2^{(j+1)}$ and $b = l + k \cdot 2^{(j+1)} + 2^j$, where $j = 0$ denotes the LSB and $j = (n-2)$ denotes the MSB. The average sensitivity of the sign bit (SB) can be derived as the summation over all possible corruptions (i.e., all possible magnitude values being corrupted to their negative values, resulting in twice the error of the actual magnitude) and can be written as

$$A_{\text{GG},\text{s}} = 2 \sum_{k=0}^{2^{(n-1)} - 1} \Pr(x_{l,k}) \cdot (2x_{l,k})^2, \qquad (12)$$

with $x_{l,k}$ the quantization level of interval k. For the DC-Band, which is modeled as a Gaussian distribution, the parameters α, β are set to $\alpha = \sqrt{2\sigma^2(x)}$ and $\beta = 2$, respectively, where $\sigma^2(x)$ denotes the variance of the samples x.

To obtain the final effect of a single bit-error on the reconstructed image, one has to normalize the sensitivities $A_{i,j}$ of the samples to the relative number of pixels in each subband by multiplying with the factor s_i/S.

The bit-sensitivities for the **PPVQ** can be expressed in terms of pyramid sensitivities $A_{\text{PPVQ,p}}$ and radius sensitivities $A_{\text{PPVQ,r}}$. As derived in [9], the sensitivity of a pyramid bit (that affects L coefficients) can be upper bounded by

$$A_{\text{PPVQ,p}} \leq \frac{4}{(L+1)} \cdot \left(\sigma^2\left(\rho(\mathbf{X})\right) + E\left(\rho(\mathbf{X})\right)^2\right) \cdot \alpha_p, \tag{13}$$

with $\alpha_p = 1/L$ if $K \geq L$ and $\alpha_p = 1/K$ otherwise.

The radii are coded with a nonuniform scalar 2^n-level quantizer based on the Gamma distribution with the parameters α_g and β_g. The probability of a radius $\rho(\mathbf{X})$ to be quantized to one of the reconstruction levels $x_{l,i}$ ($i = 0,..,(n-1)$) with the quantization thresholds $(x_{t,i}, x_{t,(i+1)})$ is

$$\Pr(x_{l,i}) = P\left(\alpha_g, \frac{x_{t,(i+1)}}{\beta_g}\right) - P\left(\alpha_g, \frac{x_{t,i}}{\beta_g}\right), \tag{14}$$

with $P(a,x)$ again the incomplete Gamma function. So when quantizing with n bits, the sensitivity of the $(n-1)$ amplitude bits (again affecting L coefficients) can be expressed as [7]

$$A_{\text{PPVQ,r,a,i}} = \frac{2\alpha_p}{L+1} \sum_{j=0}^{(2^{(n-2-i)}-1)} \sum_{k=0}^{(2^i-1)} [\ (\Pr(x_{l,a}) + \Pr(x_{l,b})) \cdot d^2(x_{t,a}, x_{t,b}) +$$
$$+ \ (\Pr(x_{l,a+d}) + \Pr(x_{l,b+d})) \cdot d^2(x_{t,a+d}, x_{t,b+d})\ ], \tag{15}$$

for $i = 0, ..., (n-2)$ with $d = 2^{n-1}$ and

$$a = j + k \cdot 2^{(n-1-i)} \quad \text{and} \quad b = j + k \cdot 2^{(n-1-i)} + 2^{(n-2-i)}. \tag{16}$$

Finally, one can write the sensitivity for the sign bit (SB) affecting L coefficients as

$$A_{\text{PPVQ,r,s}} = \frac{2\alpha_p}{L+1} \sum_{i=0}^{(2^{n-i}-1)} \left(\Pr(x_{l,i}) + \Pr(x_{l,(2^n-1-i)})\right) \cdot d^2\left(x_{t,i}, x_{t,(2^n-1-i)}\right). \tag{17}$$

To demonstrate the close correspondence between analytical and simulated bit-sensitivities, we plotted the results for the GG-quantizer in Fig. 1 for a representative bit-allocation at a rate $R_S = 1.0$ bpp for the well-known LENNA image. Results for the UT-quantizer [2] and the PPVQ [9] show a similar correspondence between estimation and simulation results.

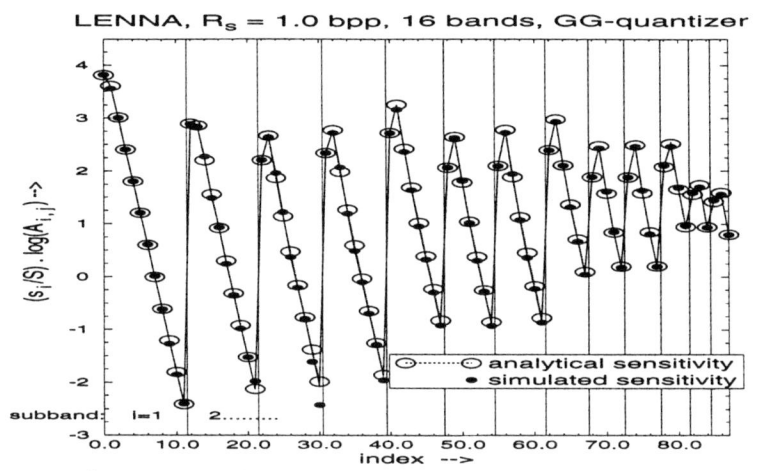

Figure 1: Comparison of analytical and simulated bit-sensitivities for the GG-quantizer, SB, MSB,...,LSB.

3.2 Effects of Channel Errors

It can be shown [2] that for error probabilities $p_{i,j} \ll 1$ for subband i and bit j, the overall distortion due to quantization and channel errors for scalar quantizers can be written as

$$D_{\text{s+c}} = \sum_{i=1}^{K} \frac{s_i}{S} \cdot \left(e_{i,n_i} + \sum_{j=1}^{n_i} A_{i,j} \cdot p_{i,j} \right) . \qquad (18)$$

Considering that one bit-error of a codeword when using the PPVQ affects L of the s_i coefficients within one subband, the joint distortion when using the PPVQ can be written as

$$\begin{aligned} D_{\text{s+c}} &= \sum_{i=1}^{K} \frac{s_i}{S} \cdot \left(e_{i,\text{PPVQ},K,n} + A_{\text{PPVQ},p} \cdot \lceil \log_2\left(N\left(L, K \right) \right) \rceil \cdot p_{p,i} + \right. \\ &\quad + \left. \sum_{j=1}^{n_i} A_{i,\text{PPVQ},r,j} \cdot p_{i,j} \right) , \end{aligned} \qquad (19)$$

where $p_{p,i}$ is the associated bit-error probability for the pyramid bits. The resulting overall rate $R_{\text{s+c}} = R_{\text{S}}/R_{\text{C}}$ (in channel uses per pixel) can be written as

$$R_{\text{s+c}} = \frac{1}{S} \sum_{i=1}^{K} s_i \cdot R_{\text{s+c},i} , \qquad (20)$$

with the joint source-channel rate of subband i for the GG-quantizer as $R_{\text{s+c},i} = \sum_{j=1}^{n_i} 1/R_{i,j}$ and $R_{i,j}$ the allocated channel code-rate to bit j of subband i. For the PPVQ, the joint source-channel rate of subband i can be written as

$$R_{\text{s+c},i} = 1/L \left(\frac{\lceil \log_2(N(L,K)) \rceil}{R_{p,i}} + \sum_{i=0}^{n_i-1} \frac{1}{R_{i,j}} \right), \qquad (21)$$

with $R_{p,i}$ and $R_{i,j}$ the channel code-rates of the pyramid bits and radius bits, respectively.

4 Rate-Distortion Behavior

Similar to the bit-allocation problem for pure source coding, we now have to find the number of quantization bits n_i for every subband i together with the channel code rates $R_{i,j}$ for the different classes of sensitivities to obtain a minimum overall distortion $D_{\text{s+c}}$ under the constraint of a given maximum overall rate $R_{\text{s+c}}$. Therefore, for every subband we calculated the operational rate-distortion function for joint source and channel coding. We allow two different modes: one fixed channel code per subband (**no CA**) with sub-optimum performance, and the more complex but superior case with a code-allocation (**CA**) that assigns different channel codes to the n_i bits with different sensitivities within subband i. The approach to calculate the operational rate-distortion bound for one subband can be seen in Fig. 2. For the GG-quantizer we plotted the joint distortion (in mse) versus the overall rate (in c.u. per pixel) for subband $i = 3$ at an $E_s/N_0 = 0$ dB which corresponds to an uncoded BER of $p = 0.0786$ with available channel code rates of the RCPC-Codes [10] (with memory $\nu = 6$) of 1/1, 8/9, 4/5, 2/3, 4/7, 1/2, 4/9, 4/10, 4/11, 1/3, 4/13, 4/15, 1/4. One can see the decreasing distortion with the increase of quantization bits n_i. The decrease in distortion due to more and more channel coding saturates at a distortion equal to the quantization noise (when quantizing with n_i bits). This is why there is approximately a 6 dB difference of the (horizontal) saturation lines corresponding to each increment in quantization bits. In Fig. 2, we also plotted the final operational joint rate-distortion function as the convex hull for this particular subband.

The calculation of the operational rate-distortion bound in the case of code-allocation (CA) is basically the same, but results in much smoother and steeper behavior of the final operational joint rate-distortion function due to the more optimal assignment of required protection to the differently sensitive bits (see the comparison in Fig. 2).

In the case of the PPVQ, the joint operational rate-distortion function of every subband is calculated in the following way. For every combination of radius bits

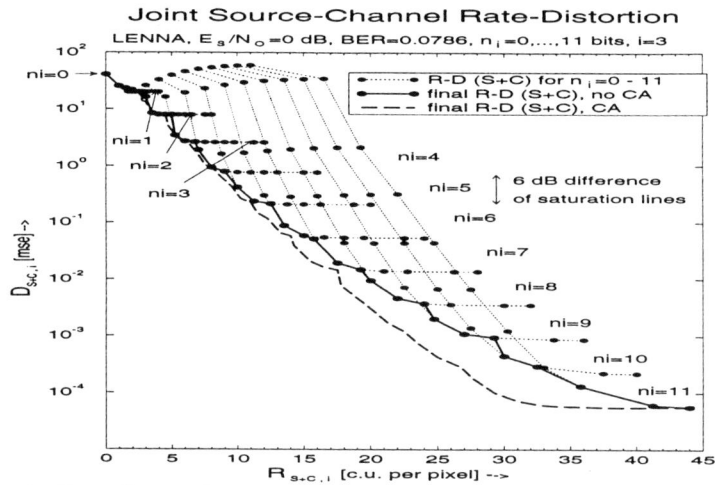

Figure 2: Overall rate-distortion behavior of joint source and channel coding, applied to subband $i=3$, GG-quantizer.

and pyramid bits, a code-allocation will be performed to obtain the minimum overall (source-channel error) for this specific combination. The convex hull of all generated rate-distortion pairs will form the operational rate-distortion bound. To provide a comparison of the R-D behavior of the two different source coding schemes, we plotted the final R-D bound for subband $i = 3$ in Figure 3. The result for the UT-quantizer [2] is plotted also, to relate the performance to this simple scheme. One can see the performance of the GG-quantizer which is always better than that of the UT-quantizer. The characteristics of the PPVQ is somewhat different. For low rates, the distortion is better than both, the GG- and the UT-quantizer, whereas for higher rates, it performs worse. This is due to the higher number of pyramid bits, which are relatively sensitive to errors and thus need a higher amount of redundancy to guarantee sufficient protection. Nevertheless, for the range of actual allocated rates, the PPVQ outperforms the other schemes and thus promises to have a very good overall performance (see Figure 4).

After having calculated the operational rate-distortion bounds of all K subbands, we applied a single bit-allocation procedure to distribute source and channel bits in an optimum way. One can see results with the optimal code-allocation (CA) for transmission over an additive white Gaussian noise (AWGN) channel at E_s/N_0=0 dB in Figure 4 for the (512 by 512) LENNA image. A considerably better performance for the jointly optimized real-world scheme, using additional subband-splitting (1024 field instead of 16 bands) and the much better GG-quantizer instead of the simple UT-quantizer can also be seen. Furthermore, the excellent performance of the PPVQ providing 'near entropy coding' using fixed length coding of vector indices illustrates its suitability for joint source and channel coding. Simulation results in Figure 4 also

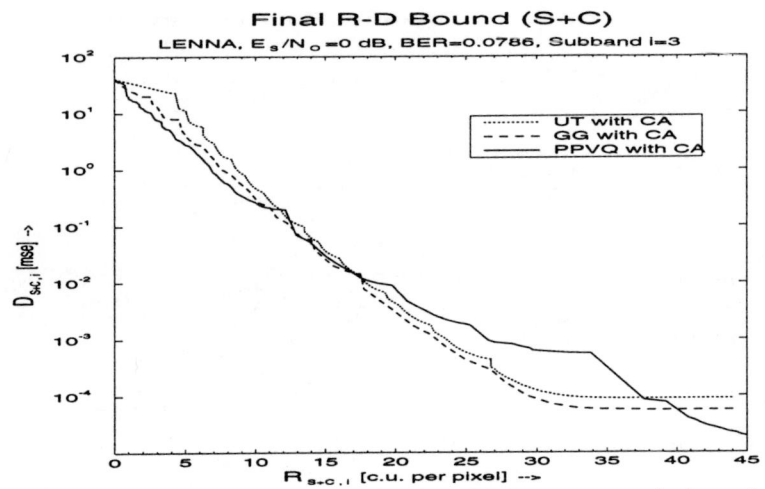

Figure 3: Overall rate-distortion behavior of joint source and channel coding for various compression schemes, applied to subband $i=3$, with CA.

demonstrate the very close correspondence of theory and the actual system performance.

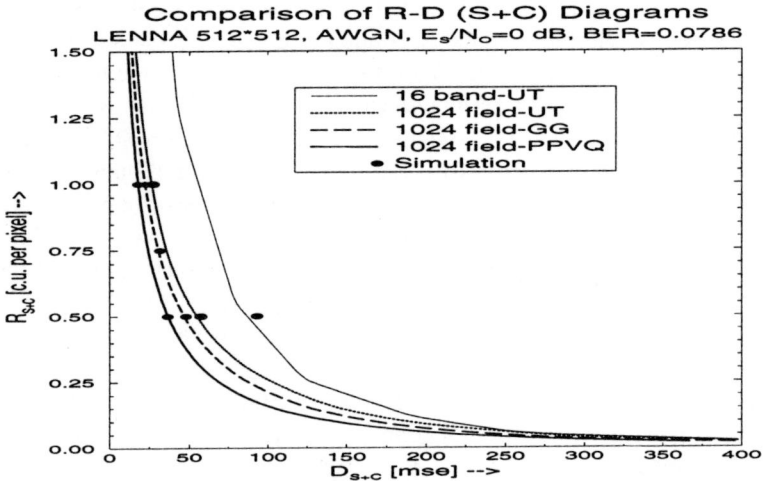

Figure 4: Rate R_{s+c} in c.u. per pixel versus distortion D_{s+c} for various joint S+C schemes (CA).

We also illustrated some representative images in Fig. 5, which were actually transmitted over an AWGN-channel at a signal-to-noise ratio of $E_S/N_0 = 0$ dB and an overall rate of $R_{s+c}=1.0$ bpp. One can get a very good indication of the improvements if one compares the images, beginning from unprotected images with no transmission errors (a: 1024-GG, b: 1024-PPVQ), to unprotected images with channel errors (c: 1024-GG, d: 1024-PPVQ), on to the various schemes with increasing performance (1024-GG (e) and 1024-PPVQ (f)). The comparison of pure source coding performance (a,b) show the better coding efficacy of the PPVQ-scheme. On the other hand, the figures in the case of unprotected transmission (c,d) show the large amount of distortion due to channel errors, where in the case of the PPVQ, the distortions have a much greater impact due to vector quantization. The jointly optimized images of the 1024-GG (e) scheme look satisfactory, but there is some minor distortion noticeable in the background. The best results can be obtained with the properly designed combined source-channel coding scheme if VQ replaces scalar quantization (1024-PPVQ, f), which combines sharpness with less annoying noise. When comparing this quality with the best theoretical quality at reasonable complexity levels, assuming the 1024-PPVQ scheme, but no channel errors when transmitting at the cutoff-rate, we obtain almost no visible improvement (see Table 1). This shows that our best scheme (1024-PPVQ) is operating close to a practical information-theoretic performance limit. The final improvement when assuming transmission at the channel capacity is only about 0.5 dB, which demonstrates the very efficient performance of the jointly optimized PPVQ-scheme. The coding results (PSNR and R_S) of the images are summarized in Table 1.

Joint Source-Channel Coding Scheme	E_S/N_0=0 dB BER=0.079 R_{S+C}=1.0 bpp		E_S/N_0=2 dB BER=0.038 R_{S+C}=0.5 bpp	
	R_S	PSNR	R_S	PSNR
GG: no trans. errors (a)	1.00	38.175	0.50	34.480
PPVQ: no trans. errors (b)	1.00	39.715	0.50	36.102
GG: unprotected (c)	1.00	16.040	0.50	18.224
PPVQ: unprotected (d)	1.00	12.271	0.50	13.461
1024 - GG (e)	0.57	34.594	0.36	32.477
1024 - PPVQ (f)	0.48	35.659	0.32	33.785
PPVQ: Cutoff-rate	0.55	36.569	0.37	34.586
PPVQ: Capacity	0.60	37.052	0.38	34.841

Table 1: Simulation results of the images depicted in Fig. 5.

4.1 Theoretical Performance

We extend our work to obtain theoretical bounds, assuming signaling either at the cutoff-rate or at the channel capacity, and finally, knowing the distribution parameters of each subband we consider source coding at the theoretical source-entropy. In Figure 6, we first compare the overall rate-distortion behavior of joint source and channel coding of the various scalar quantizer schemes. One

a.) GG, source coding only, BER=0.0

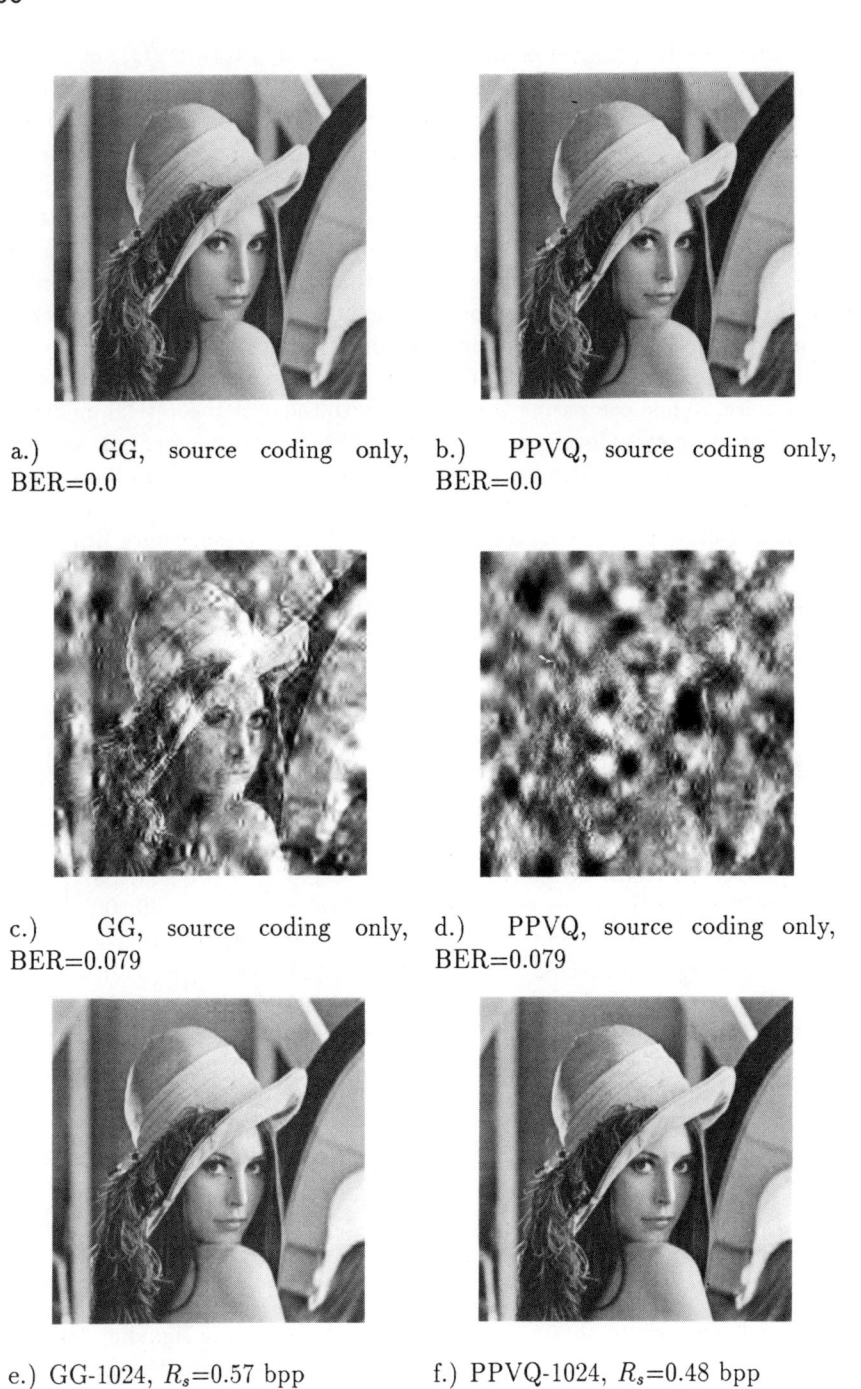

b.) PPVQ, source coding only, BER=0.0

c.) GG, source coding only, BER=0.079

d.) PPVQ, source coding only, BER=0.079

e.) GG-1024, R_s=0.57 bpp

f.) PPVQ-1024, R_s=0.48 bpp

Figure 5: Simulation results for $R_{s+c} = 1.0$ bpp and E_S/N_0=0 dB.

can see the minor improvement, when coding the '1024 field-GG'-scheme at the theoretical cutoff-rate ($R_{c,i} = R_0$), which illustrates the close to optimum performance of our real-world transmission system. A slight improvement is available, assuming the theoretical entropy as the source coding rate and coding at the cutoff-rate (1024 field-GG, $H\text{-}R_0$). This might be achievable, using variable length coding (VLC) and channel coding with high complexity to guarantee an almost errorfree transmission at $R_C = R_0$. The final bound for the wavelet transformed LENNA image when using a scalar quantizer (assuming antipodal signaling over AWGN) is the case when considering the theoretical source entropy and coding at the channel capacity (1024 field-GG, $H\text{-}C$).

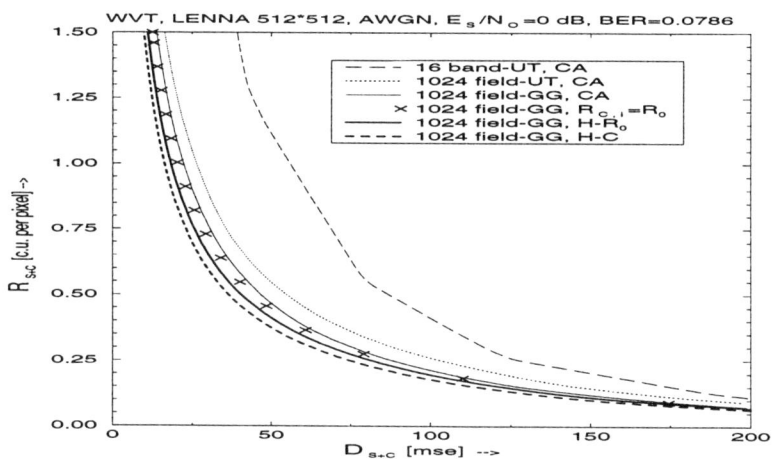

Figure 6: R-D behavior of joint S+C coding (scalar quantizers) for real-world and theoretical systems.

Things look different, when comparing the theoretical bounds for the PPVQ. The theoretical source entropy for radius quantization is calculated assuming an optimum Gamma quantizer, i.e., one knows the probability of a sample (i.e., a radius value) being in one of the possible quantization intervals. For the PPVQ we assume that the probability of mapping the incoming vectors to one of the $N(L,K)$ possible lattice points is $p = 1/N(L,K)$. Figure 7 shows a comparison of the actual performance of the two best real-world systems (1024 - GG, 1024-PPVQ) and their theoretical bounds assuming source coding at the entropy and signaling at either the cutoff-rate or at channel capacity.

First we note the better theoretical bounds of the PPVQ schemes. This is due to the fact, that the lattice points of the PPVQ are adapted to the probability of occurrence ('near-entropy'-coding). We further note that for high compression rates (low R_{s+c}), the actual system performance of the PPVQ is even better than the theoretical bounds of the scalar quantizer, for both signaling at $R_c = R_0$ and $R_c = C$. This demonstrates the superior performance of the PPVQ, enabling optimum combined source/channel coding thru the use

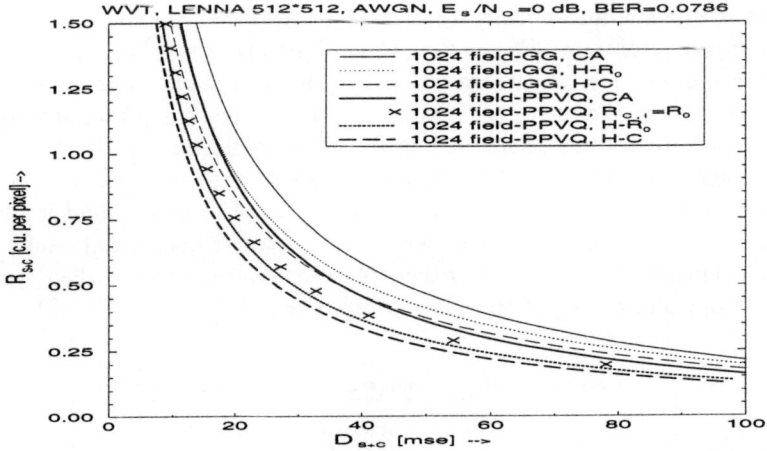

Figure 7: Comparison of R-D behavior of joint S+C coding for real-world (GG and PPVQ) and theoretical systems.

of fixed length coding. Furthermore, the achievable bound ($R_{c,i} = R_0$) is very close to the first theoretical bound (H-R_0). This shows the close to optimum source coding and on the other hand the achievable gain when using more powerful channel coding. This gain is possible since the pyramid bits within the subbands are highly sensitive to channel errors and thus need a very high protection against errors, which could be better achieved with, for example, a more efficient concatenated channel code.

5 Conclusion

In order to derive the dependence of rate and distortion for a combined source and channel coding scheme, in this paper we calculated the effects of channel errors, the bit-error sensitivities to the reconstructed image, for an optimum nonuniform scalar quantizer, based on the generalized Gaussian (GG) distribution, and also for the very efficient product pyramid vector quantizer (PPVQ). We showed how to perform an optimum bit-allocation procedure, based on the knowledge of the operational source-channel rate-distortion function, where the bits were jointly allocated to source and channel coding in one operation. Considerably better performance of the jointly optimized scheme employing the scalar GG-quantizer or the PPVQ (both making use of local adaptivity) under noisy channel conditions (additive white Gaussian noise) was shown. Simulation results demonstrate the very close correspondence of theory and the actual system performance and also show the improvements and the benefits that can be realized by using a combined source and channel coding scheme. Furthermore, we compared the performance to theoretical performance bounds and we showed that the recommended real-world system (PPVQ) closely approaches the rate-distortion behavior of our system operating at the (theoretical) cut-off rate and for high compression rates even operates beyond the theoretical bounds of scalar quantizers, and thus shows the benefits of this approach and the usefulness of combined source and channel coding. It also strongly suggests consideration in addition to current schemes with separate optimization.

References

[1] S. G. Mallat, "A theory for multiresolution signal decomposition: the wavelet representation," *IEEE Trans. Pattern Anal. Machine Intell.*, vol. PAMI-11, pp. 674–693, July 1989.

[2] M. J. Ruf and J. W. Modestino, "Rate-distortion performance for joint source and channel coding of images," *submitted to IEEE Trans. IP*, 1995.

[3] G. M. Roe, "Quantizing for minimum distortion," *IEEE Trans. IT*, vol. 10, pp. 384–385, Oct. 1964.

[4] P. Filip, "Compander quantizer," *unpublished*, 1993.

[5] T. R. Fischer, "A pyramid vector quantizer," *IEEE Trans. Inform. Theory*, vol. IT-32, pp. 568–583, July 1986.

[6] P. Filip and M. J. Ruf, "A fixed-rate product pyramid vector quantzation using a Bayesian model," in *Proc. of IEEE Globecom '92*, (Orlando, Fl.), pp. 240–244, 1992.

[7] M. J. Ruf, *Kombinierte Quell- und Kanalcodierung fuer Festbilder*. Ph. D. Thesis, Technical University Munich, Germany, to appear in 1996.

[8] P. H. Westerink, J. Biemond, and D. E. Boekee, "An optimal bit allocation algorithm for sub-band coding," in *Proc. IEEE ICASSP*, pp. 757–760, 1988.

[9] M. J. Ruf, "A high performance fixed rate compression scheme for still image transmission," in *Proceedings DCC '94, Data Compression Conference*, IEEE Computer Society Press, 1994.

[10] J. Hagenauer, "Rate-compatible punctured convolutional codes (RCPC codes) and their applications," *IEEE Trans. Commun.*, vol. COM-36, pp. 389–400, Apr. 1988.

Bounds on the Performance of Vector-Quantizers operating under Channel Errors over all Index Assignments

Gal Ben-David and David Malah, Technion - Israel Institute of Technology
Department of Electrical Engineering, Haifa 32000, Israel

Abstract

Vector-Quantization (VQ) is an effective and widely implemented method for low-bit-rate communication of speech and image signals. A common assumption in the design of VQ-based communication systems is that the compressed digital information is transmitted through a perfect channel. Under this assumption, quantization distortion is the only factor in output signal fidelity. Moreover, the assignment of channel symbols to the VQ Reconstruction Vectors is of no importance. However, under physical channels, errors may be present, degrading overall system performance. In this case, the effect of channel errors on the VQ system performance depends on the index assignment of the Reconstruction Vectors. For a VQ with N Reconstruction Vectors there are $N!$ possible assignments. Hence, even for relatively small values of N, an exhaustive search over all possible assignments is practically impossible. In this paper, upper and lower bounds on the performance of VQ systems under channel errors over all possible assignments are presented using Linear Programming arguments. These bounds may give the system designer more insight about the gain that could be achieved by improving the index assignment. In numerical examples, the bounds are compared with the performance obtained by using a set of random assignments, as well as with an index assignment obtained by the well-known index switching algorithm.

1. Introduction

Vector Quantization (VQ) is a method for mapping signals into digital sequences [1]. A typical VQ-based communication system is shown in Fig. 1.

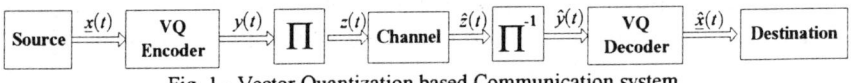

Fig. 1 - Vector Quantization based Communication system

In most Signal Processing applications a discrete-time *Source* emits signal samples over an infinite or a large finite alphabet. These samples should be sent to the *Destination* with the highest possible fidelity. The *VQ Encoder* translates vectors of source samples into *Channel* digital sequences. The task of the *VQ decoder* is to reconstruct source samples from this digital information. Since the analog information cannot be perfectly represented by the digital information some *Quantization Distortion* must be tolerated.

In each channel transmission the VQ encodes a K-dimensional vector of source samples - $\underline{x}(t)$ into a *Reconstruction Vector index* $y(t)$, where the discrete variable t represents the time instant or a channel-use counter. The index is taken from a finite alphabet, $y(t) \in \{0,1,...,N-1\}$, where N is the number of Reconstruction vectors (hence the number of possible channel symbols).

The *Index Assignment* is represented in Fig. 1 by a permutation operator:

$$\Pi : y(t) \in \{0,1,...,N-1\} \rightarrow z(t) \in \{0,1,...,N-1\} \tag{1}$$

where a total of $N!$ possible permutations exist. For just 4-bits quantization there are $16! \approx 2 \cdot 10^{13}$ possible permutations. Examination of all possible permutations is therefore impractical. The channel index $z(t)$ is sent through the channel.

For *Memoryless Channels*, The channel output $\hat{z}(t)$ is a random mapping of its input $z(t)$, characterized by the *Channel Probability Matrix* $\mathbf{Q}$, defined by:

$$\{\mathbf{Q}\}_{ij} = \text{Prob}\{\hat{z}(n) = j | z(n) = i\} \tag{2}$$

Throughout we shall assume that $\mathbf{Q}$ is symmetric (i.e., *Symmetric Memoryless Channels*).

For the special case of the *Binary-Symmetric-Channel* (BSC):

$$\{\mathbf{Q}\}_{ij} = \text{Prob}\{\hat{z}(n) = j | z(n) = i\} = q^{H(i,j)}(1-q)^{L-H(i,j)} \tag{3}$$

where L is the number of bits ($N = 2^L$), q is the *Bit-Error-Rate* (BER) and $H(i,j)$ is the *Hamming Distance* between the binary representations of i and j.

At the receiver, after inverse-permutation, the *VQ Decoder* converts the channel output symbols into one of N possible Reconstruction Vectors. It is desired that the Decoder output $\hat{x}(t)$ be "close" to the original input. The term "close" will be defined by a distortion measure between the input and the output of the VQ system $d(\underline{x}, \underline{\hat{x}})$.

Knowledge of the source statistics $p(\underline{x})$ or a representing *Training Sequence* is assumed. The perfomance of the overall system is measured in terms of the average distortion $E[d(\underline{x}, \underline{\hat{x}})]$.

In "classic" discussions of VQ applications, the channel is assumed to be noiseless ($\mathbf{Q} = \mathbf{I}$, where $\mathbf{I}$ is the unity matrix), [1], so that no errors occur during transmission and $y(t) = \hat{y}(t)$ for every t. This assumption is based upon using a channel encoder-decoder pair to correct channel errors, causing the distortion due to channel-errors to be negligible. The permutation Π has no effect in this case.

Upon knowledge of the source statistics, Lloyd's algorithm [1] may be used to design the VQ. In practice, a training sequence is used and the LBG algorithm [1] is implemented. Both methods are iterative and alternately apply the *Nearest-Neighbor* Condition and the *Centroid* condition.

In some applications, channel-coding is not utilized due to complexity or Bit-Rate constraints. In case of a channel error event, a wrong Reconstruction Vector is selected at the decoder. The distortion due to channel errors is significant and affects the design of the VQ system [2-12].

The Vector Quantization system consists of a partition of the signal space Ω of all possible input vectors - $\underline{x}$. This space is partitioned into N nonoverlapping regions:

$$\cup_i R_i = \Omega; \quad R_i \cap R_j = \emptyset \tag{4}$$

Each partition region R_i has a corresponding *Reconstruction (Representation) Vector* - $\underline{\phi}_i$.

The encoder accumulates a K-dimensional vector of source samples $\underline{x}$. The symbol $y(t) = i$ is emitted if $\underline{x} \in R_i$ and the corresponding channel symbol, $z(t) = \Pi(i)$, is transmitted through the channel. The channel output is a random mapping of this tranmission. Upon receiving the channel symbol $\hat{z}(t) = j$ the decoder emits the Reconstruction Vector - $\underline{\phi}_{\Pi^{-1}(j)}$.

The overall distortion of the VQ-based communication system is:

$$D = E[d(\underline{x}, \underline{\hat{x}})] = \sum_{i=0}^{N-1}\sum_{j=0}^{N-1} \{\pi \cdot \mathbf{Q} \cdot \pi^T\}_{ij} \int_{R_i} d(\underline{x}, \underline{\phi}_j) \cdot p(\underline{x}) \cdot d\underline{x} \tag{5}$$

In (5) the permutation is represented by a permutation matrix - π, whose entries are 0's and 1's and the sum of each of its rows and columns is 1. For the perfect channel, $\mathbf{Q} = \mathbf{I}$, the permutation matrix π is of no importance ($\pi \cdot \pi^T = \mathbf{I}$), and the only factor of the system performance is the *Quantization Distortion*:

$$D_q = E[d(\underline{x}, \underline{\hat{x}})]\big|_{\mathbf{Q}=\mathbf{I}} = \sum_{i=0}^{N-1} \int_{R_i} d(\underline{x}, \underline{\phi}_i) \cdot p(\underline{x}) \cdot d\underline{x} \tag{6}$$

For the region R_i, all vectors $\underline{x} \in R_i$ should be represented by $\underline{\phi}_i$. Yet due to channel errors, other reconstruction vectors may appear at the destination. The probability of receving the channels index corresponding to $\underline{\phi}_j$ given the index corresponding to $\underline{\phi}_i$ was transmitted is $\{\pi \cdot \mathbf{Q} \cdot \pi^T\}_{ij}$.

The *Channel Distortion* is defined by the average distance between the reconstructed vector and the one that would have been reconstructed with no channel errors:

$$D_c = \sum_{i=0}^{N-1} p_i \sum_{j=0}^{N-1} \{\pi \cdot \mathbf{Q} \cdot \pi^T\}_{ij} \cdot d(\underline{\phi}_i, \underline{\phi}_j) = trace\{\mathbf{P} \cdot \pi \cdot \mathbf{Q} \cdot \pi^T \cdot \mathbf{D}\} \tag{7}$$

where p_i is the probability that an input vector $\underline{x}$ belongs to the i-th partition region R_i:

$$p_i = \int_{R_i} p(\underline{x}) \cdot d\underline{x} \tag{8}$$

the diagonal matrix **P** contains these probabilities $\mathbf{P} = diag\{p_0, p_1, \ldots, p_{N-1}\}$, and the entries of the matrix **D** are the distances among all Reconstruction Vectors: $\mathbf{D}_{ij} = d(\underline{\phi}_i, \underline{\phi}_j)$. It is shown in [8],[9] that for the Euclidean distance measure, and Centroid quantizers the overall distortion is the sum of the quantization and channel distortions: $D = D_q + D_c$.

In the literature two main approaches are proposed to improve the performance of Vector Quantizers under channel errors. The first method allows modification of the partition regions and their corresponding codevectors. In the presence of channel errors, and given the transmitted symbol, the received symbol is a random variable. It is suggested to redesign the VQ by modifiying the distortion measure to take all possible output vectors into consideration. This modification results in a *Weighted-Nearest-Neighbor* and *Weighted-Centroid* conditions [7-9]. These conditions are specific to every channel condition. Hence, a VQ designed for a noisy channel should apply a different partition and a different set of codevectors for each possible BER. The main drawbacks of this approach are the large memory consumption and extensive design effort.

The second approach is trying to reduce channel distortion by using a better index assignment. Several suboptimal methods are suggested in the literature. In [7-10] an iterative Index Switching algorithm is proposed. After selecting an initial assignment, the algorithm searchs for a better assignment by exchanging indices of codevectors, and keeping the new assignment if it performs better than its predecessor. This algorithm can only offer a local minima. A more sophisticated algorithm is examined in [7], where Simulated Anealing (SA) is used to search for an optimal index assingnment. The method of SA involves some ad-hoc arguments to define system "temperature" and "cooling" procedures. Moreover, the method of SA has a very slow convergence rate, and cannot assure global optimum during a limited design period.

For the special case of a Uniform Scalar Quantizer with quantization step h, $\phi_i = (i - N/2) \cdot h$; $d(\phi_i, \phi_j) = h^2(i-j)^2$, and a Uniform Source, $p_i = 1/N$, $i = 0, 1, \ldots, N-1$, it is shown in [2],[5],[6] that the Natural Binary Code Assignment, corresponding here to $\pi = \mathbf{I}$, is the optimal assignment.

The remainder of the paper is organized as follows. In section 2, lower and upper bounds on the performance of VQ system over all possible Index Assignments are presented. In section 3 numerical results are shown. Conclusions are given in Section 4.

2. Performance Bounds

In this section we introduce lower and upper bounds on the channel distortion, as defined in (7), under Symmetric Memoryless Channels, over all possible assignments (permutation matrices - π). The bounding technique is based on eigenvalues and Linear Programming arguments. Instead of optimizing over the (discrete) family of matrices covering all possible assignments $\pi Q \pi^T$, we optimize over a wider (continuous) family. A detailed mathematical analysis may be found in [3].

Using the symmetry property of the Channel Matrix, Q, we combine the matrices D and P into a single symmetric matrix $\hat{D} = DP + P^T D^T$. The channel distortion is given then by:

$$D_c = \frac{1}{2} trace\{Q\pi^T \hat{D} \pi\} \tag{9}$$

Recalling that Q represents probabilities, the sum of any of its rows is one, so the vector $\underline{1} = [1 \ 1 \ \cdots \ 1]^T$ is an eigenvector of Q since $Q \cdot \underline{1} = \underline{1}$.

A fundamental step in the bounding technique is that the matrix $\hat{D}$ is replaced by another symmetric matrix $\tilde{D}$, also having $\underline{1}$ as an eigenvector. This replacement changes D_c by a known additive constant. This goal is achieved by adding "Cross Structured" matrices, as we define shortly, to the matrix $\hat{D}$.

We define a "Column Structured" matrix as:

$$C_i = \begin{bmatrix} 0 & & 0 & 1 & 0 & & 0 \\ 0 & \cdots & 0 & 1 & 0 & \cdots & 0 \\ \vdots & & \vdots & \vdots & \vdots & & \vdots \\ 0 & & 0 & 1 & 0 & \cdots & 0 \end{bmatrix}$$
$$\uparrow \ i-\text{th column} \tag{10}$$

Using the property $Q \cdot C_i = C_i$, it is shown in [3] that adding "Cross Structured" matrices $\alpha(C_i + C_i^T)$, where α is a scalar, changes the r.h.s. of (9) just by the addition of the scalar α, for any permutation matrix π:

$$\frac{1}{2} trace\{Q\pi^T [\hat{D} + \alpha(C_i^T + C_i)]\pi\} = \frac{1}{2} trace\{Q\pi^T \hat{D}\pi\} + \alpha \tag{11}$$

In order to achieve the desired property $\tilde{D} \cdot \underline{1} = \omega_0 \underline{1}$, for some ω_0, all rows of $\tilde{D}$ must have the same sum of entries. Let us examine the effect of adding a "Cross Structured" matrix $\alpha(C_i + C_i^T)$ to a general matrix of size $N \times N$. The sum of all rows except for the i-th row is increased by α, while the sum of the i-th row is increased by $(N+1) \cdot \alpha$.

An algorithm for obtaining $\tilde{D}$ having the desired property is shown in Table 1. Throughout the algorithm, a variable S is needed to store the sum of all "α" constants added to the r.h.s. of (9). By adding at most $N-1$ "Cross Structured" matrices we get a symmetric matrix where all rows have the same sum of elements, resulting in the matrix $\tilde{D}$, with the desired property $\tilde{D} \cdot \underline{1} = \omega_0 \underline{1}$. We shall refer to $\tilde{D}$ as the *Weighted Distances Matrix*. The channel distortion can now be written as:

$$D_c = \frac{1}{2} trace\{\mathbf{Q}\pi^T \tilde{\mathbf{D}}\pi\} - S \qquad (12)$$

Initialization: a. Set the matrix: $\tilde{\mathbf{D}} \leftarrow \hat{\mathbf{D}} = \mathbf{DP} + \mathbf{P}^T \mathbf{D}^T$.
b. Clear the sum of additive constants: $S \leftarrow 0$.

Step 1: Calculate the sum of all rows.
Denote the sum of the i-th row by $S_i = \sum_{j=0}^{N-1} (\tilde{\mathbf{D}})_{ij}$.

Step 2: Search all rows for the maximal sum of elements.
Assume that the row with the maximal sum is labeled k.

Step 3: For each row $i \neq k$:
a. Add the *"Cross Structured"* matrix $\tilde{\mathbf{D}} \leftarrow \tilde{\mathbf{D}} + \frac{1}{N}(S_k - S_i)(\mathbf{C}_i + \mathbf{C}_i^T)$.
b. Update $S \leftarrow S + \frac{1}{N}(S_k - S_i)$.

Table 1 - An algorithm for obtaining $\tilde{\mathbf{D}}$ having the property $\tilde{\mathbf{D}} \cdot \underline{1} = \omega_0 \underline{1}$, without affecting the optimization problem

Note that now both the channel matrix $\mathbf{Q}$ and the Weighted Distances Matrix $\tilde{\mathbf{D}}$ are symmetric, have nonnegative entries, and have $\underline{1} = [1 \ 1 \ \cdots \ 1]^T$ as an eigenvector. All eigenvalues of both matrices are real. Next, we use the following Theorem adopted from [14 Section 15.7].

Theorem: The Perron-Frobenius eigenvalue of a nonnegative-entries symmetric matrix $\mathbf{M}$ with the property $\mathbf{M} \cdot \underline{1} = \beta \underline{1}$ is β. This eigenvalue is positive and is the largest in absolute value.

Corollary: The eigenvalue 1 of the matrix $\mathbf{Q}$ and the eigenvalue $\omega_0 > 0$ of the matrix $\tilde{\mathbf{D}}$, both corresponding to the eigenvector $\underline{1}$, are the largest in absolute value for each matrix.

Next, we perform a unitary diagonalization on both matrices:

$$\begin{aligned} \mathbf{Q} = \mathbf{V} \cdot \Lambda \cdot \mathbf{V}^T \quad & \mathbf{V} \cdot \mathbf{V}^T = \mathbf{I} \\ \tilde{\mathbf{D}} = \mathbf{W} \cdot \Omega \cdot \mathbf{W}^T \quad & \mathbf{W} \cdot \mathbf{W}^T = \mathbf{I} \end{aligned} \qquad (13)$$

Without loss of generality, we arrange the eigenvalues (and their corresponding eigenvectors) in Λ and Ω to be in decreasing order. Substituting (13) into (12):

$$D_c = \frac{1}{2} trace\{V \Lambda V^T \pi^T W \Omega W^T \pi\} - S = \frac{1}{2} trace\{\Lambda V^T \pi^T W \Omega W^T \pi V\} - S =$$
$$= \frac{1}{2} trace\{\Lambda \Psi \Omega \Psi^T\} - S = \frac{1}{2} \sum_{i=0}^{N-1} \sum_{j=0}^{N-1} \lambda_i \omega_j \Psi_{ij}^2 - S \qquad (14)$$

where we define $\lambda_i = \Lambda_{ii}$, $\omega_i = \Omega_{ii}$, $i = 0, 1, \ldots, N-1$ and the matrix Ψ is defined as $\Psi = V^T \pi^T W$. The matrix Ψ is also unitary since: $\Psi \Psi^T = V^T \pi^T W W^T \pi V = I$.

Since the first column of both V and W is $\underline{v}_0 = \underline{w}_0 = \frac{1}{\sqrt{N}} \underline{1}$, and the remaining columns are orthogonal to the vector $\underline{1}$, the structure of $\Psi = V^T \pi^T W$ is:

$$\Psi = \begin{bmatrix} 1 & 0 & \cdots & 0 \\ 0 & & & \\ \vdots & & ? & \\ 0 & & & \end{bmatrix} \qquad (15)$$

where the question mark represents unknown entries.

In order to obtain upper and lower bounds over all possible Index Assignments, we relax the constraint that the matrix Ψ equals to $V^T \pi^T W$ for some permutation matrix - π (a discrete set of possible Ψ matrices). Instead, we only require the property that the sum of squares of the elements in each row and column of a unitary matrix (Ψ in this case) is equal to 1 (a wider, continuous set of possible Ψ matrices), and state the following optimization problem:

$$\min_{\Psi} / \max_{\Psi} \left(\sum_{i=1}^{N-1} \sum_{j=1}^{N-1} \lambda_i \omega_j \Psi_{ij}^2 \right)$$
$$s.t. \quad \sum_{i=1}^{N-1} \Psi_{ij}^2 = 1 \quad j = 1, 2, \ldots, N-1$$
$$\sum_{j=1}^{N-1} \Psi_{ij}^2 = 1 \quad i = 1, 2, \ldots, N-1 \qquad (16)$$

Note that the first row and the first column were omitted from the optimization problem. The problem in (16) is a standard Assignment problem in Operations Research, e.g., optimaly assigning N workers to N machines. Using Linear Programming arguments, it is shown in [15] that an optimal solution for the Assignment problem is a permutation matrix Ψ_{opt} that has a single 1 in each row and column, while the remaining elements of the matrix are zero. Nevertheless, Ψ_{opt} does not necessarily correspond to a legal Index Assignment matrix π.

Observing the target function in (16) $\sum_{i=1}^{N-1}\lambda_i\sum_{j=1}^{N-1}\omega_j\Psi_{ij}^2$, we see that the permutation matrix Ψ_{opt} does a one-to-one (permutation) matching between the eigenvalues λ_i and ω_i, $i = 1,2,\ldots,N-1$, while always matching λ_0 and ω_0.

Recalling that both λ_i and ω_i were arranged in decreasing order, it is shown in [3] that the highest (lowest) possible value is obtained by matching the eigenvalues λ_i and ω_i, $i = 1,2,\ldots,N-1$ in the same (reversed) order. The minimum and maximum values of the optimization problem are:

Minimum value : $\sum_{i=1}^{N-1}\lambda_i \cdot \omega_{N-i}$ \quad\quad Maximum value : $\sum_{i=1}^{N-1}\lambda_i \cdot \omega_i$

Corresponding to: \quad\quad\quad\quad\quad\quad\quad Corresponding to:

$$\Psi_{min} = \begin{bmatrix} 1 & 0 & \cdots & 0 & 0 \\ 0 & 0 & & & 1 \\ \vdots & & & 1 & \\ 0 & & \cdot\cdot & & \\ 0 & 1 & & & 0 \end{bmatrix} \quad \Psi_{max} = \begin{bmatrix} 1 & 0 & \cdots & 0 & 0 \\ 0 & 1 & & & 0 \\ \vdots & & \ddots & & \\ 0 & & & 1 & \\ 0 & 0 & & & 1 \end{bmatrix} \quad (17)$$

and the bounds on Channel Distortion over all possible Index Assignments are:

$$\frac{1}{2}\left(\lambda_0\omega_0 + \sum_{i=1}^{N-1}\lambda_i \cdot \omega_{N-i}\right) - S \le D_c \le \frac{1}{2}\left(\lambda_0\omega_0 + \sum_{i=1}^{N-1}\lambda_i \cdot \omega_i\right) - S \quad (18)$$

In conclusion, in order to find the desired bounds one should perform the following steps:

1. Calculate the *Weighted Distances Matrix*, $\tilde{D}$, and the sum of added scalars S, using the algorithm stated in Table 1.
2. Calculate the eigenvalues of the *Channel Matrix* Q (λ_i, $i = 0,1,\ldots,N-1$), and of the *Weighted Distances Matrix* $\tilde{D}$ (ω_i, $i = 0,1,\ldots,N-1$). For the Binary-Symmetric-Channel, λ_i are given in [2], [5].
3. Calculate the upper and lower bounds using (18).

For the special case of an L-bit ($N = 2^L$ levels) Uniform Scalar Quantizer and a Uniform Source operating under the Binary Symmetric Channel with Bit-Error-Rate q, these bounds turn out to be [3]:

$$\frac{2(N-1)(N+1)}{3N^2}2q \le D_c \le \frac{2(N-1)(N+1)}{3N^2}\left[1-(1-2q)^L\right] \quad (19)$$

The lower bound coincides with the performance of the Natural Binary Code, which is the optimal Assignment for this case, as shown in [2], [5], [6]. Note that for small Bit Error Rate values, the ratio between the upper and lower bounds in (19) is equal to the number of bits L.

3. Numerical Examples

In this section numerical examples of the performance bounds are presented. Due to the huge number of possible assignments, the lower and upper bounds are compared with the best and worst of 10,000 random assignments. In some cases the performance due to an index assignment obtained by the Index-Switching algorithm [9] is also shown. Further examples may be found in [3].

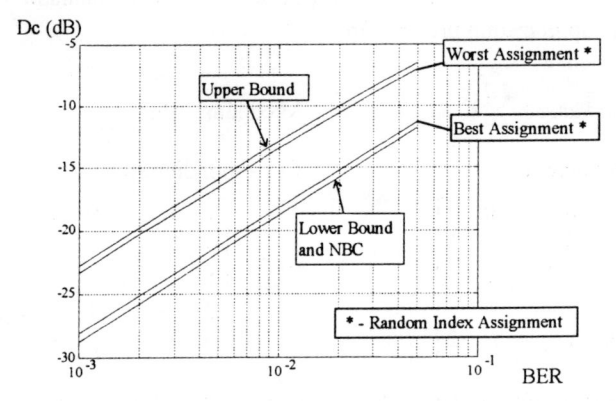

Fig. 2 - Upper and lower bounds over all possible index assignments on the Channel-Distortion of a 4-bit Uniform Scalar Quantizer and a uniform source under the BSC. The bounds are compared with the performance of the best and worst assignments of 10,000 randomly picked assignments. The lower bound coincides with the performance of the optimal assignment (Natural Binary Code)

Example 1: For a 4-bit uniform scalar quantizer, a uniform source and a BSC, bounds were presented in (19). The resulting bounds and the simulation results are shown in Fig. 2. The upper and lower bounds are about 0.5dB away from the best and worst assignments found in the random assignment simulation. As mentioned, The distortion due to the Natural Binary Code (NBC) coincides with the lower bound. The ratio between the upper and lower bound is approximately the number of bits $L=4$, that is 6dB.

Example 2: Consider the 4-bit uniform scalar quantizer, and the uniform source of Example 1. The digital information is assumed to be sent through the BSC utilizing a (7,4) Hamming Error Correcting Code [16]. The channel transition matrix $\mathbf{Q}$ is symmetric, thus enabling us to use the proposed bounds. The eigenvalues and eigenvectors of $\mathbf{Q}$ are different from the BSC case. The resulting bounds and the simulation results are shown in Fig. 3.
It can be seen that the slope of the graphs is 20dB/ Decade, i.e., reducing the Bit Error Rate by a factor of 10, results in a 20dB lower distortion. The channel distortion is approximately proportional to the square of the Bit Error Rate. The upper bound is about 0.6dB away from the worst random assignment, while the lower bound is about 0.1dB from the best random assignment. It is shown in [3], that the NBC is also optimal for this case. The ratio between the upper and lower

bound is approximately 4.5dB. The addition of the channel protection brought the bounds closer together.

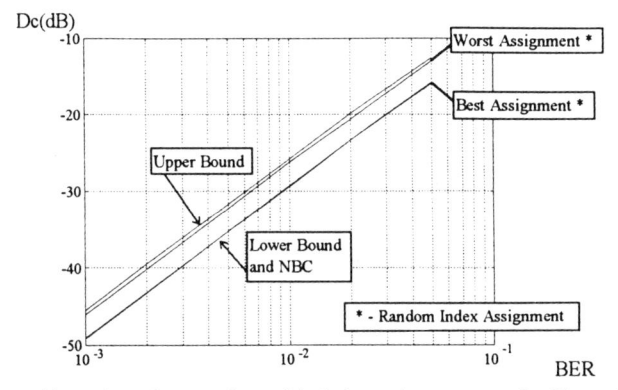

Fig. 3 - Upper and lower bounds over all possible index assignments on the Channel-Distortion of a 4-bit Uniform Scalar Quantizer and a uniform source under the BSC with (7,4) Hamming code. The bounds are compared with the performance of the best and worst assignments of 10,000 randomly picked assignments. The Natural Binary Code coincide with the lower bound.

Example 3: Similar to the first example, we consider now a 4-bit PDF-Optimized Uniform Scalar Quantizer, a Gaussian source and a BSC The resulting bounds and the simulation results are shown in Fig. 4.

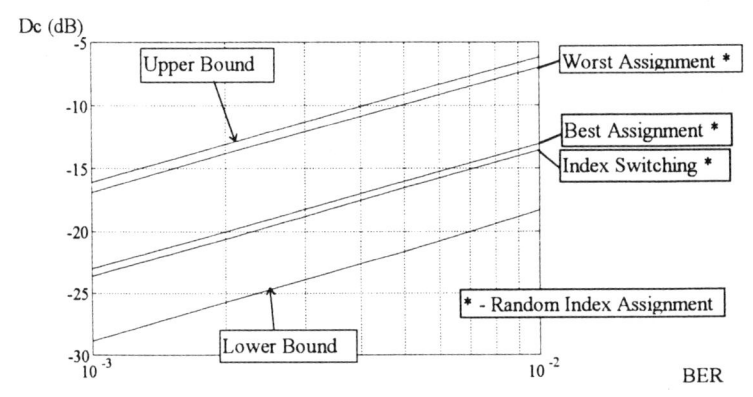

Fig. 4 - Upper and lower bounds over all possible index assignments on the Channel-Distortion of a 4 bit PDF-Optimized Uniform Scalar Quantizer and a Gaussian source under the BSC. The bounds are compared with the performance of the best and worst assignments of 10,000 randomly picked assignments.

The upper bound is about 0.6dB away from the worst assignment found in the random assignment simulation. The lower bound is about 5dB lower than the distortion due to the assignment obtained by the index switching algorithm. As mentioned earlier there are about $2 \cdot 10^{13}$ possible assignments in this example. Since it is not practical to find the best assignment by exhaustive search ,it is not clear at this point how tight the proposed lower bound is. It could well be that the relatively large gap between the lower bound and the performance of the best assignment found in simulations so far, is due to an insuffiecint number of examined assignments (10,000). This issue is presently under investigation.

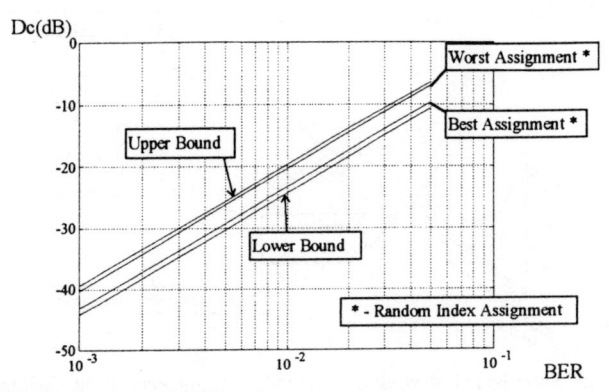

Fig. 5 - Upper and lower bounds over all possible index assignments on the Channel-Distortion of a 4-bit PDF optimized Uniform Scalar Quantizer and a Gaussian source under the Binary Symmetric Channel with (7,4) Hamming code. The bounds are compared with the perforemance of the best and worst assignments of 10,000 randomly picked assignments.

Example 4: Consider the source and quantizer of the previous example. The digital information is sent through a BSC utilizing this time a (7,4) Hamming Error Correcting Code, as in example 2. The resulting bounds and the simulation results are shown in Fig. 5. The upper and lower bounds are about 0.8dB from the best and worst assignment found in the random assignement simulation. As in the case of a uniform source, the addition of channel protection brought the bounds closer together.

As mentioned earlier, further examples may be found in [3]. For a 3-bit PDF-optimized Uniform scalar quantizer, a Gaussian source and the BSC we perform full search over all $8! = 40,320$ possible assignments and the bound appear to be tight. Bounds and simulation result for an 8-bit Vector Quantizer may also be found in [3]

4. Conclusions

In this paper we present upper and lower bounds on the Channel-Distortion for Vector Quantizers operating under channel errors. The bounds were obtained using Linear Programming arguments. Numerical examples are shown for the Binary Symmetric Channel with and without channel Error Correcting Code. For quantizers with 4 bits and more, the bounds are compared with the performance of 10,000 random index assignments. For the Binary Symmetric Channel the upper bounds are close to the performance of the worst assignment found in the random assignment simulation. The lower bounds are sometimes more loose and a significant gap exists between the lower bound and the performance of the assignment obtained by the Index Switching algorithm. This gap may be due to the relatively small number of assignments examined by simulations. This issue is under investigation.

Utilization of an Error Correcting Code decreases the gap between the lower and the upper bounds the gap between the best and the worst assignment found in simulations, and both bounds exhibit a tighter behavior under this conditions.

References

[1] Gersho A. and Gray R.M., *Vector Quantization and Signal Compression*, Kluwer Academic Publishers, Boston 1991
[2] Ben-David G. and Malah D., "Properties of the Natural Binary Code Assignment for Uniform Scalar Quantizers under Channel Errors", Proc. of the ECCTD-93, Elsevier Amsterdam, pp. 773-778, 1993.
[3] Ben-David G., *Vector and Scalar Quantization under Channel Errors*, D.Sc. Thesis, Technion - I.I.T. Haifa, 1995
[4] Ben-David G. and Malah D., "Simple Adaptation of Vector-Quantizers to Combat Channel-Errors", Proc. of the 6th IEEE DSP Workshop Oct. 1994, IEEE NY, pp. 41-44, 1994.
[5] McLaughlin S.W., Neuhoff D.L., Ashley J., "The optimality of the Natural Binary Code", *Proc. of the Joint DIMACS / IEEE Workshop on Coding and Quantization*, Oct. 1992.
[6] Crimmins T.R., Horwitz H.M, Palermo C.J and Palermo R.V., "Minimization of Mean-Square Error for Data Transmitted Via Group Codes", *IEEE Trans. Inform. Theory*, vol. 15, no. 1, pp. 72-78, Jan. 1969.
[7] Farvardin N. "A study of Vector Quantization for Noisy Channels", *IEEE Trans. Inform. Theory*, Vol. 36, No. 4, pp. 799-809, Jul. 1990.
[8] Gersho A. and Zeger K.A., "Zero Redundancy Channel-Coding in Vector Quantization", *Electronics Letters*, Vol. 4, pp. 654-656, June 1987.
[9] Gersho A. and Zeger K.A., "Pseudo-Gray Coding", *IEEE Trans. Comm.*, Vol. 38, No., pp. 2147-2158, 12 Dec. 1990.
[10] DE Marca J.R.B., and Janyant N.S. "An algorithm for Assigning Binary Indices to the Codevectors of a multi-Dimensional Quantizer", *Proc. of IEEE Int. Conf. Comm.*, Seatte WA, pp.1128-1132, June 1987.
[11] Knagenhjelm P., "A recursive Design Method for Robust Vector Quantization", *Proc. Int. Conf. on Signal Processing Applications and Technology*, Boston-92, pp. 948-954, Nov. 1992.
[12] Knagenhjelm P., "How good is your assignment?", *Proc. IEEE ICASSP 93*, Minneapolis, Minnesota, pp. 423-426, Apr. 1993.
[13] Gilmore P.C., "Optimal and Suboptimal Algorithms for the Quadratic Assignment Problem", *J. Soc. Indust. Appl. Math*, Vol. 10, No. 2, pp. 305-313, June 1962.
[14] Lancaster P. and Tismenetsky M., *The Theory of Matrices*, Academic-Press, Orlando, 1985.
[15] Taha H.A, *Operations Research*, Maxwell-MacMillan NY, 1987.
[16] Taub H. and Schilling D.L., *Principles of Communication Systems*, McGraw-Hill, New-York, 1986.
[17] Jayant N.S. and Noll P., *Digital Coding of Waveforms*, Prentice-Hall 1984.

Adaptive Temporal & Spatial Error Concealment Measures in MPEG-2 Video Decoder with Enhanced Error Detection

Susanna Aign [*]
German Aerospace Research Establishment (DLR), Institute for Communications Technology, D-82234 Oberpfaffenhofen, Germany

Abstract

The MPEG-2 source coding algorithm is very sensitive to channel disturbances. A single bit error in the bitstream will cause a high degradation of picture quality due to the use of variable length coding (Huffman code). Therefore, picture replenishment error concealment techniques might be required at the receiver.

The aim of this article is to study error concealment enhancements due to the use of an adaptive error concealment algorithm. The temporal error concealment technique with motion compensation works well in MPEG-2 coded P- and B-pictures. However, for I-pictures in MPEG-2 coded sequences, no motion information exists. A simple temporal error concealment (copying from one previous picture) is only useful in sequences without motion between the pictures. In addition spatial error concealment causes errors in picture domains with high spatial details. The technique considered in this article is to adapt the concealment technique to the spatial and temporal activities.

Reliable error detection techniques are considered for error concealment measures. First, the source decoder can detect errors, however many errors will remain undetected. Therefore, another technique such as the error detection ability provided by channel decoding, is studied. The channel decoder for terrestrial broadcasting consists of an inner convolutional code and an outer RS-code. The outer RS-code can detect packet errors, but the packets are very long and much information will be lost when concealing the entire packet. One

---
[*]The author is with the European dTTb-Race and the German $^H DTV_T$ projects

way to exploit the correct information in one erroneous packet is to use the reliability outputs of an inner SOVA (Soft Output Viterbi Algorithm) for error detection. The error detection technique with the reliability information of the SOVA provides the best results, since the erroneous packet will contain only a few errors.

1 Introduction

The future terrestrial digital TV/HDTV signal will be a highly compressed signal because of the restricted bandwidth of the available UHF and VHF channels. The TV/HDTV source coding algorithms used for this application will be based on MPEG-2 [1], in which motion compensation, hybrid discrete cosine transformation (DCT), variable length coding (VLC) and adaptive quantization are performed.

The MPEG-2 source coding algorithm is very sensitive to channel disturbances. Therefore, for broadcast applications, MPEG-2 hierarchical coding profiles will be needed in order to assume graceful degradation as in analog systems.

The future digital HDTV-system should also support a compatibility with the standard definition TV (SDTV), where the SDTV-image quality will be expected for portable receivers as well. However, HDTV image quality will be supported by stationary receivers with a roof-top antenna. Both system compatibility and some step-wise degradation can be achieved by using the MPEG-2 hierarchical source coding profiles with matched channel coding and modulation techniques [2, 3]. Up to two or three hierarchy layers can be obtained with the so called spatial and/or SNR-scalability specified in the MPEG-2 hierarchical profiles.

In order to achieve graceful degradation, the two (TV1, TV2) or three (TV, HD1, HD2) layers require different protection. This will be done with a combination of Unequal Error Protection (UEP) Codes and Multi-Resolution-Modulation [3]. The base layer (TV or TV1) will be highly protected. However, at very long distances or at the worst reception conditions, i.e. deep fades or impulsive noise, the portable receiver, which suffers from the non-existence of graceful degradation, may risk receiving nothing due to the threshold effect. Even for a fixed receiver, graceful degradation with three different protection levels may not be sufficient. To avoid this problem or to enhance the picture quality, one may apply error concealment techniques at the receiver. For instance, for the most critical layer, error concealment may help to obtain an appropriate picture quality, even if there are many residual errors in the bitstream after channel decoding.

It should be noted that MPEG-2 has its own multiplexing format. The video, audio and auxiliary data after multiplexing and forming the MPEG-2 transport stream, are subjected to channel coding and modulation. The channel coding is based on a concatenated coding scheme with an outer Reed Solomon (RS) code and an inner convolutional code. The inner code is combined with Multi-Resolution Quadrature Amplitude Modulation (MR-QAM) [3]. The outer RS-codewords are adapted to the transport packets of length 188 bytes. An example of a channel coding and modulation scheme for terrestrial hierarchical HDTV-broadcasting is described in [3].

For error concealment measures reliable error detection is needed. One method involves the use of the error detection capability of the source decoder. However, with this technique many errors remain undetected. Therefore, another error detection technique is considered. First, the outer RS-code can give some information about the reliability of each RS-word, but the RS-words are very long. To exploit the correct information in one erroneous packet, the reliability outputs of the inner SOVA may be helpful.

The aim of this article is to study error concealment enhancements in the MPEG-2 Video decoder due to the use of an adaptive error concealment technique with enhanced error detection, taking into account a real transmission medium [3].

The paper is organized as follows: in section 2 the different error detection techniques are briefly described. The different error concealment techniques are analyzed in section 3. Simulation results are given in section 4. Finally, section 5 is devoted to some conclusions.

2 Error Detection

In order to employ error concealment measures, proper error detection is needed. Three methods of error detection are considered: first the source decoder can detect errors [4], but with this method, many errors will still remain undetected and there is the risk of error propagation. Therefore, another way previously mentioned is to use the error detection capability of the channel decoder. The outer RS-decoder can detect incorrect packets with high reliability [5]. Transmission errors in a packet will be indicated by an error flag in the transport header. If the source decoder knows the errors in packets, the slices [1] in which the packet is erroneous, will be concealed until the next synchronization point (the next slice). This error detection technique locates the region in the bitstream where the errors occur. The region is at least 184 bytes long and this information loss will cause the loss of synchronization until the next slice. This detection technique gives no information about the exact location of the errors in the RS-word. In many cases in which the decoder fails, the RS-word has $t_{max} + 1$ errors, where t_{max} is the number of errors that the RS-decoder can handle. If these errors are at the end of the RS-word, the decoding until another detection technique finds errors, could be better because the information, which could be correct in the bitstream, will be used. But this detection technique must be a method with high reliability. One solution for using the correct part of a wrong packet is to use the soft information at the output of the inner channel decoder [6] as a reliability metric for each symbol. The SOVA gives the reliability values for each bit after the inner decoder. From this information one can obtain the reliability for each symbol of the RS-word due to the systematic structure of the RS-code. The correct part of the erroneous packet can therefore be used until the first symbol is incorrect in one packet.

3 Error Concealment

The two main error concealment techniques described here are temporal and spatial error concealment. Temporal error concealment with motion compensation is devoted to inter-coded pictures, since there exists motion information. However, in intra-coded pictures no motion information exists. Therefore, only simple temporal error concealment and spatial error concealment can be applied to intra-coded pictures. In this section different temporal and spatial error concealment techniques and the adaptation of the two algorithms for error concealment in I-pictures are analyzed.

3.1 Temporal Error Concealment

Temporal error concealment techniques benefit from the temporal redundancy in one sequence. In B- and P-pictures the temporal redundancy is used for computing the motion vectors. As these vectors are in the bitstream, it seems worthwhile to use them in inter-coded pictures for error concealment. Temporal error concealment techniques are also easier to implement and the complexity is less than some spatial recovering algorithms.

Different temporal error concealment techniques for inter-coded pictures (P- and B-pictures) are investigated in [4]. The first one is a simple temporal error concealment technique. In this method macroblocks of the previous anchor picture are copied onto the current picture where errors were detected. With this concealment method shifts will be visible, if there is motion.

In addition, the second method uses the motion vectors from one nearest macroblock in the current picture (if motion vectors exist). This method is called motion compensated temporal error concealment. Because a detected error results in one damaged horizontally oriented stripe of the decoded image, the nearest macroblocks are the two macroblocks above and below the current macroblock. Only the first damaged macroblock has three neighboring macroblocks, the left, the top and the bottom macroblock. Therefore, the two or three macroblocks will be investigated first. If a motion vector exists in one of the macroblocks, then the current macroblock will be concealed using this information. If no macroblock with available motion vectors is found in the surrounding macroblocks, then simple temporal error concealment will be applied.

The second method in the most cases is the best one [4], when there is high motion in the sequence. But if the motion from the current macroblock is a motion in a different direction than that of the surrounding macroblocks, this error becomes visible. This occurs also if there are burst errors which cause more than one disturbed slice, in which case the motion vectors are considered to be too far from the current macroblock.

3.2 Spatial Error Concealment

Spatial error concealment techniques are devoted to I-pictures for which no motion information exists. Spatial error concealment techniques make use of the spatial redundancy in a picture.

One method based on simple interpolation is considered in this article. This technique interpolates each pixel of the entire $2N \times 2N$ macroblock with the adjacent pixels of the four neighboring macroblocks. Figure 1 shows the macroblock with the boundary pixels of the neighboring macroblocks ($N = 4$).

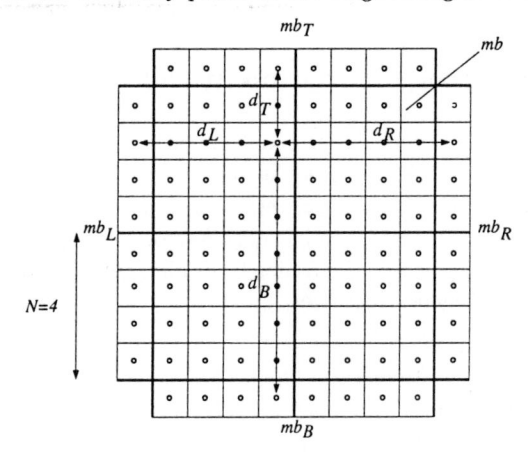

Figure 1: Spatial Interpolation of the entire macroblock

Each pixel of the current macroblock with the size $2N \times 2N$ will be concealed by simple interpolation of the four pixels of the surrounding macroblocks. The equation of this process is

$$mb(i,k) = \frac{1}{d_L + d_R + d_T + d_B} (d_R \cdot mb_L(i, 2N) + d_L \cdot mb_R(i, 1) + d_B \cdot mb_T(2N, k) + d_T \cdot mb_B(1, k)) \quad (1)$$

$$i, k = 1 \ldots 2N \quad ,$$

where mb is the current macroblock, mb_X with $X = R, L, T, B$ are the neighboring macroblocks (Left, Right, Top, Bottom), respectively, and d_X with $X = R, L, T, B$ is the distance from the pixel of the macroblock mb_X to the current pixel $mb(i, k)$, respectively. This concealment technique works better if the surrounding macroblocks exist. In MPEG-2 the synchronization point is at the start of the following slice, therefore, one horizontal stripe of macroblocks must be concealed. The left and the right macroblocks cannot be used for concealment because they do not exist. If some of the macroblocks do not exist, during interpolation the corresponding distance will be set to zero (for instance, if mb_L does not exist d_R will be set to zero).

With only two available macroblocks, mb_T and mb_B, Equation 1 becomes

$$mb(i,k) = \frac{d_B \cdot mb_T(2N, k) + d_T \cdot mb_B(1, k)}{d_T + d_B} \quad i, k = 1 \ldots 2N \quad . \quad (2)$$

This spatial interpolation can be used for MPEG-2 coded pictures where the next synchronization point is at the start of the next slice.

With this technique no abrupt transition occurs in one macroblock (no 'block-effect'). The entire macroblock becomes blurred if there are edges in one macroblock. This then will be a visible 'block- effect' in terms of macroblocks as well. In MPEG-2 coded pictures only the top and the bottom macroblock can be used for interpolation. Therefore, the interpolation is an interpolation in one direction, the vertical direction. Edges in the vertical direction will be interpolated correctly, whereas edges in the horizontal direction will become blurred.

Other possibilities for spatial error concealment are the multi-directional interpolation [7] and the signal loss recovering by imposing smoothing constraints [8]. For MPEG-2 coded pictures multi-directional filtering will be reduced to nearly one-directional filtering, because the left and right macroblocks cannot be used for multi-directional interpolation. Also, for the second method the four surrounding macroblocks and additionally the undamaged DCT-coefficients (which are not given in an entire damaged stripe of macroblocks) are needed, otherwise the complexity will be increased.

For I-pictures, MPEG-2 has the possibility of transmitting additional error concealment motion vectors in the bitstream. This would result in a noticeably better quality and is not as difficult to implement as the spatial concealment algorithms [8]. With the additional transmission of frame motion vectors the overhead will be less than 0.7 % of the total bit-rate [10].

3.3 Adaptive Temporal & Spatial Error Concealment Techniques

The first method for combining the two concealment techniques in a video sequence is to conceal the I-pictures with spatial error concealment techniques (if there are no concealment motion vectors) and to conceal the P- and B-pictures with temporal error concealment techniques with motion compensation.

The second method is based on adaptive spatio-temporal replacement of missing data [9]. The criterion for the decision between spatial and temporal replacement is the measurement of the image activity. If, in I-pictures, the motion between the current picture and the previous picture is higher than the local spatial detail, then spatial error concealment is employed, otherwise simple temporal concealment will be applied. In B- and P-pictures this procedure will only be applied to macroblocks when no neighboring macroblocks with motion vectors exist, because temporal error concealment with motion compensation will in most cases be better than spatial or simple temporal error concealment techniques.

The motion can be measured by computing the mean-square error between two macroblocks of adjacent pictures whereas the spatial activity can be computed by measuring the variance of one macroblock [9]. The motion for the current incorrect macroblock will be computed from the nearest neighboring macroblocks from the current and the previous picture (in MPEG-2 the nearest macroblock will be the top or the bottom macroblock). The mean-square error (motion) will be computed as

$$MSE = \mathrm{E}[(mb_{T/B} - mb'_{T/B})^2] \quad , \qquad (3)$$

where $mb_{T/B}$ is the top or bottom macroblock of the current picture, $mb'_{T/B}$ is the top or bottom macroblock of the previous picture and E[x] is the mean value of x. For computing the spatial detail the nearest neighbored macroblock (top or bottom) of the current picture will be used as well. The spatial detail will be obtained by computing the variance of the neighboring macroblock as

$$VAR = \mathrm{E}[mb^2_{T/B}] - \mathrm{E}[mb_{T/B}]^2 \quad . \qquad (4)$$

One method for determining which concealment technique should be applied, is the following [9]: when the motion is below a certain threshold, the simple temporal error concealment should be applied, and when the motion is above this threshold, the spatial error concealment should be applied only if the motion MSE is greater than the spatial detail VAR. Using this decision metric, there appeared some problems with regions of the picture in which both the motion and the spatial detail is high. In the Flower sequence (Figure 2) the area with the flowers should be concealed with simple temporal error concealment, because the spatial detail is too high to reproduce it using spatial interpolation. But the tree and the surrounding of the tree should be concealed with spatial interpolation, because the motion between the current picture and the previous picture is too high. One I-picture disturbed with $PER = 10^{-1}$ is given in Figure 2. The decisions for the temporal or the spatial concealment using the above mentioned criteria are shown in Figure 6. The dark regions indicate where spatial error concealment will be applied. The bright regions are the regions where simple temporal error concealment will be applied. It is obvious that many macroblocks in the region with the flowers will be concealed with spatial error concealment, where instead the temporal error concealment technique should be applied. In those areas the MSE-value is greater than the VAR-value. It should be noted, however, that both values are very high. To avoid this problem one may introduce a threshold for VAR; if VAR lies above this threshold, temporal error concealment should be applied, otherwise spatial error concealment should be used. With this additional threshold the flowers will be concealed with simple temporal error concealment. But one problem still remains with this technique, the flowers near the tree are concealed with a part of the tree of the previous picture. To avoid this, a third criterion is introduced; the variance of the neighboring macroblock of the previous picture is computed as

$$VAR' = \mathrm{E}[mb'^2_{T/B}] - \mathrm{E}[mb'_{T/B}]^2 \quad . \tag{5}$$

If this variance lies under a threshold, then when VAR is very high, spatial error concealment will be applied. With this third criterion temporal error concealment is avoided even if the motion and the spatial detail are very high. The entire decision metric can be described with the following program-structure:

```
if  (MSE < threshold1) temporal_error_concealment
  else {
    if  (MSE > VAR) {
      if  (VAR > threshold2) {
        if  (VAR' < threshold3) spatial_error_concealment
          else temporal_error_concealment
        }
      else spatial_error_concealment
      }
    else temporal_error_concealment
  }
```

Figure 7 gives the decision regions determined by this algorithm.

With this adaptive spatio-temporal method picture replenishment mainly for I-pictures will be enhanced, because with high motion the simple temporal error concealment will fail and with high spatial detail the spatial error concealment will fail.

4 Simulation Results

Simulation results for different error concealment techniques with different error detection techniques for I-pictures of MPEG-2 coded pictures are given in

Table 1 for the sequences Flower and Calendar, respectively. The different error detection techniques are error detection of the channel decoder without soft information ('channel') and error detection of the channel decoder with soft information ('sova'). Two different error concealment techniques are considered, one is simple temporal error concealment (simply copying from one previous picture, 'simple') and the other is spatial error concealment (interpolation from the pixels of the surrounding macroblocks, 'spatial'). The two error concealment techniques are combined in the adaptive error concealment technique described above ('adaptive').

PER		simple_ channel	simple_ sova	spatial_ channel	spatial_ sova	adaptive_ channel	adaptive_ sova	without errors
				Flower				
10^{-1} :	Y	19.12	19.20	19.90	19.97	19.60	19.68	31.38
	U	27.34	27.42	27.83	27.90	27.53	27.61	33.06
	V	32.21	32.25	32.30	32.35	32.33	32.37	35.11
10^{-2} :	Y	26.94	27.04	27.63	27.72	27.35	27.44	31.38
	U	31.74	31.76	32.03	32.06	31.84	31.86	33.06
	V	34.54	34.56	34.61	34.63	34.59	34.61	35.11
				Calendar				
10^{-1} :	Y	20.34	20.44	20.28	20.44	20.42	20.55	29.38
	U	30.67	30.76	29.14	29.30	30.50	30.62	34.31
	V	31.80	31.93	29.73	29.89	31.56	31.71	35.74
10^{-2} :	Y	26.26	26.29	26.78	26.82	26.39	26.41	29.38
	U	33.50	33.52	33.37	33.41	33.53	33.54	34.31
	V	34.80	34.82	34.36	34.40	34.76	34.77	35.74

Table 1: PSNR values for Flower and Calendar (I-pictures)

The simulations are carried out under the following conditions: the size of the TV picture of 4:2:0 format is 720x576 pixels, one slice consists of 44 macroblocks, and the bit rate is chosen to be 5 Mbit/sec, because it seems to be acceptable for a good TV quality (higher than PAL or SECAM). In the first simulation twenty I-pictures were disturbed with the packet error rate $PER = 10^{-1}$, while in the second they were disturbed with $PER = 10^{-2}$. The mean of the PSNR for the three components is shown in Table 1. The errors were simulated with the channel model [4], which reproduces the errors after the channel decoder (Rayleigh fading channel for a portable receiver or the worst case). The burst errors of the inner Viterbi decoder were simulated as well.

The simulation results show that the PSNR-values for adaptive error concealment lie between the PSNR-values of the simple temporal and the spatial error concealment, but the difference between the PSNR-values of the three techniques is very small. The PSNR-value of the adaptive error concealment technique is not better than the other two techniques, but it will be interesting to investigate the improvements with subjective tests. Figures 3, 4 and 5 show the different techniques. One can see that the tree is concealed with spatial error concealment and the flowers are concealed with temporal error concealment. Because there is motion between the pictures and high spatial detail in the region with flowers, the PSNR-values cannot be significantly improved, if one conceals this region with temporal error concealment. However, subjectively the picture quality is better than the reconstruction with spatial error concealment. The tree is reconstructed better than with simple temporal error concealment, as well.

For the different error detection techniques the simulation results show that the main gain for error detection with the soft output values is only 0.2 dB for the PSNR at $PER = 10^{-1}$ for the luminance component. One reason for this small improvement is that the slices are very long and that the amount of bits in the first part of the RS-word can be very small if one symbol error is at the beginning of one word. Since the synchronization point is at the next start of a slice, only a few macroblocks can be further decoded if using the remaining lossless part at the beginning of the RS-word. One way to obtain improved results is to utilize shorter slices. Table 2 gives the PSNR-values for the disturbed I-pictures, in which the slice length is shorter (22 and 11 macroblocks instead of 44 macroblock per slice).

		\multicolumn{8}{c}{Error Concealment Comparison (PSNR)}
		slice-length=22				slice-length=11		
PER		simple_sova	spatial_sova	adaptive_sova	without errors	simple_sova	spatial_sova	adaptive_sova	without errors
\multicolumn{10}{c}{Flower}
10^{-1} :	Y	20.54	21.68	21.15	31.37	21.82	23.17	22.42	30.98
	U	28.60	29.31	28.85	33.06	29.26	29.94	29.45	32.81
	V	32.91	33.13	33.05	35.11	33.28	33.57	33.39	34.91
\multicolumn{10}{c}{Calendar}
10^{-1} :	Y	21.70	21.68	21.80	29.34	22.49	22.59	22.57	29.22
	U	31.23	29.76	31.09	34.26	31.55	30.49	31.45	34.21
	V	32.30	30.37	32.15	35.69	32.97	31.31	32.89	35.66

Table 2: PSNR values for Flower and Calendar (I-pictures)

The results show that for shorter slices the gain of using additionally soft information is up to 3 dB for the luminance component for the different error concealment techniques in I-pictures at $PER = 10^{-1}$ as compared to the values for a slice-length of 44 macroblocks. The reduction in picture quality due to the use of more synchronization points in a video sequence with the bitrate of 5 Mbit/sec is negligible. In I-pictures high gain can be obtained by improving the error detection techniques, because there are no perfect error concealment techniques available. Therefore, it seems worthwhile to use the correct part of one disturbed RS-word.

5 Conclusions

In this paper different error concealment techniques with different error detection methods for MPEG-2 intra-coded pictures are studied. The first error concealment technique is based on simple temporal error concealment, in which the previous anchor picture is copied onto the current picture, wherever errors are detected. The second technique is spatial error concealment. An algorithm for combining the two techniques in one picture is given in this paper. It has been shown that the adaptation of the two algorithms will result in high subjective quality.

The two detection techniques are error detection in the outer decoder of the channel decoder and the detection of incorrect symbols by additionally using the soft information of the inner SOVA. It is shown that it is worthwhile to use the correct part of the RS-word until a symbol is incorrect even if the concealment technique is not so good. It should be noted that with shorter slices the gain will be high. For the different error concealment techniques the gain for the detection technique with soft information is up to 3 dB for the luminance component at $PER = 10^{-1}$. Therefore, it is worthwhile to use all of the available information in the bitstream.

References

[1] "MPEG-2 video and systems international standard." ISO-IEC/JTC1/SC29/WG11 MPEG 94, Nov. 1994.

[2] G. Schamel, "Graceful degradation and scalability in digital coding for terrestrial transmission," in *Workshop on Digital HDTV*, 1993.

[3] K. Fazel and M. J. Ruf, "A hierarchical digital HDTV transmission scheme for terrestrial broadcasting," in *Proc. Globecom'93*, Dec. 1993.

[4] S. Aign and K. Fazel, "Error detection & concealment measures in MPEG-2 video decoder," in *Proc. of the International Workshop on HDTV'94, Torino*, Oct. 1994.

[5] R. J. McEliece and L. Swanson, "On the decoder error probability for Reed-Solomon codes," *IEEE Transaction on IT*, vol. IT-32, pp. 701–703, Sept. 1986.

[6] J. Hagenauer and P. Hoeher, "A Viterbi algorithm with soft-decision outputs and its application," in *IEEE Global Telecommunications Conference 1990 Conference Record*, vol. 3, pp. 47.1.1–47.1.7, November 1989.

[7] H. Sun and W. Kwok, "Error concealment with directional filtering for block transform coding," in *Proc. Globecom'93*, pp. 1304–1308, Dec. 1993.

[8] Q.-F. Zhu, Y. Wang, and L. Shaw, "Image reconstruction for hybrid video coding systems," in *IEEE Data Compression Conference*, pp. 229–238, March 1992.

[9] H. Sun, J. Zdepski, W. Kwok, and D. Raychaudhuri, "Error concealment algorithms for robust decoding of MPEG compressed video," *submitted to IEEE Trans. on Circuit and System for Video Technology*, 1993.

[10] H. Sun, M. Uz, J. Zdepski, and R. Saint Girons, "A proposal for increased error resilience." ISO-IEC/JTC1/SC29/WG11, MPEG 92, Sept. 30 1992.

123

Figure 2: I-picture, no error concealment, $PER = 10^{-1}$

Figure 3: I-picture, error concealment 'simple'

Figure 4: I-picture, error concealment 'spatial'

Figure 5: I-picture, error concealment 'adaptive'

Figure 6: I-picture, decision regions

Figure 7: I-picture, decision regions

Source and Channel Coding Issues for ATM Networks[†]

V.Parthasarathy, J.W.Modestino and K.S.Vastola
ECSE Department, Rensselaer Polytechnic Institute, Troy, NY 12180, U.S.A
Email: ParthasarathyV@indy.tce.com, {modestin,vastola}@ecse.rpi.edu

Abstract

This paper discusses source and channel coding issues as applicable to ATM networks. Asynchronous Transfer Mode (ATM) has rapidly emerged as the appropriate transport technique for Broadband ISDN. Among the various services offered in future ATM networks, packetized variable bit-rate (VBR) video is likely to be one of the largest users of bandwidth.

However, packet loss is virtually inevitable in such networks for VBR video due to the stochastic nature of traffic. Imperfectly recovered packets lead to error propagation in representative video compression algorithms, particularly those using motion compensation. As a result, it would appear highly beneficial to use some form of *active* recovery scheme, such as forward error-control (FEC) coding, which offers the potential benefit of improved recovery in the event of packet loss and/or errors. This paper discusses different techniques of applying FEC incorporating ideas of *combined* source-channel coding. Furthermore, it introduces a simple code selection strategy which yields codes providing close to optimal performance. Verification of its efficiency is provided by comparing performance of such selected codes to information-theoretic bounds.

1 Introduction

The development of broadband networks has led to the possibility of a wide variety of new and improved service offerings. Packetized video is likely to be one of the most significant high-bandwidth users of such networks. The transmission of variable bit-rate (VBR) video offers the potential promise of constant video quality but is generally accompanied by packet loss which significantly diminishes this potential. In this paper, we study a class of error recovery schemes employing forward error-control (FEC) coding to recover from such losses. In particular, we show that a hybrid error recovery strategy involving the use of

---

[†]This work was performed while the first author was at the ECSE Dept., Rensselaer Polytechnic Institute. He is presently at Thomson Consumer Electronics, Indianapolis, USA. This work was supported in part by ARPA under Contract No. F30602-92-C-0030.

active FEC in tandem with simple passive error concealment schemes offers very robust performance even under high packet losses.

We discuss two different methods of applying FEC to alleviate the problem of packet loss. The conventional method [1]-[4] of applying FEC generally allocates additional bandwidth for channel coding while maintaining a specified average video coding rate. Such an approach suffers performance degradations at high loads since the bandwidth expansion associated with the use of FEC creates additional congestion that negates the potential benefit in using FEC. In contrast, we study a more efficient FEC application technique in our hybrid approach which allocates bandwidth for channel coding by throttling the source coder rate (i.e., performing higher compression) while maintaining a fixed overall transmission rate. More specifically, we consider the performance of the hybrid approach where the bandwidth to accommodate the FEC overhead is made available by throttling the source coder rate sufficiently so that the overall rate after application of FEC is identical to that of the original unprotected system.

Following this we characterize the sensitivity of such a scheme to the choice of the particular code. We devise a simple code selection strategy and demonstrate that it yields codes providing close to optimal performance for a wide range of operating conditions.

2 Preliminaries

Beginning with a broad system framework, we describe in this section the coding details as well as some adaptations to suit network transport. We use an entropy-constrained subband coding scheme (ECSBC) employing pyramid-based hierarchical motion-compensated prediction (HMCP) developed by Kim and Modestino [5, 6] due to its efficient encoding as well as multi-resolution properties. Figure 1 provides a general block diagram of a video coding and prioritization scheme for transmission over a packet-switched network. Although the ideas presented in this paper would be applicable to arbitrary packet-switched networks, this paper focuses on ATM due to its emergence as the appropriate transport scheme for supporting B-ISDN. Fig. 1 diagram is generic in the sense that it is applicable to any chosen source coding scheme (e.g., a subband or a DCT-based system such as MPEG) or transport coding scheme (e.g., single or multiple priorities, with or without FEC).

The output of the video coder, in the form of parallel bit streams, enters the prioritization and transport coder. For example, in the case of a subband-based coding scheme these bit streams might result from coding different subbands. In a DCT-based scheme, they could result from entropy-coding the DC and AC coefficients. These output bit streams can then be packetized individually and classified into separate priority classes by the prioritization and transport

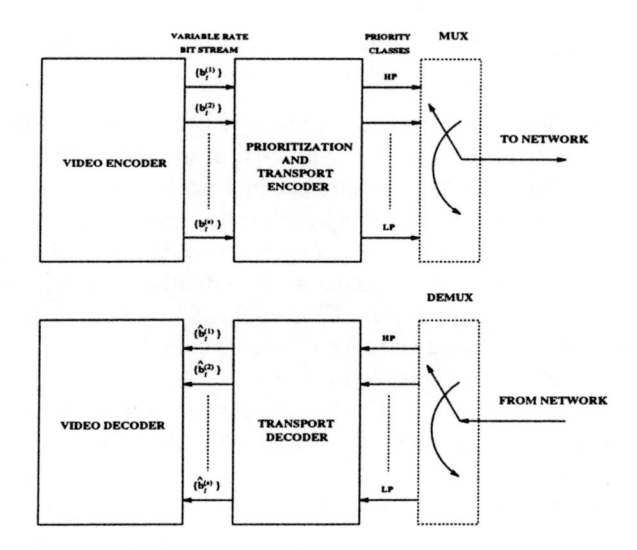

Figure 1: A Generic Block Diagram of a System for Video Transmission over a Packet-Switched Network.

coding block in the figure. The application of FEC would also be performed in this block.

The allocation of priority levels can be performed in a hierarchical multiresolution manner to provide scalable video at different resolutions. As an example, the encoded bits from subbands 1-4 could be allocated the highest-priority (HP) level (priority level 1) and subbands 5-16 allocated the lowest-priority (LP) level (priority level 2). Then two priority packet streams would be the resulting output of the prioritization and transport encoder block. Correct reception of subbands 1-4 would guarantee a low-resolution video sequence while correct reception of all 16 subbands would provide the highest-level video resolution. The backward motion-compensation scheme encodes/decodes the residual frame difference without requiring explicit transport of motion vectors. The frame difference after entropy coding is prioritized, packetized and transported over the network. Further details can be found in [7] and [9].

We employ RS codes for FEC as they make use of the generated overheads efficiently and have attractive minimum distance properties. Accordingly, they can be used effectively for burst erasure recovery which will prove valuable in the face of correlated cell loss. FEC is applied through interlaced coding across packets by grouping the information bits in the packet into q-bit symbols. The technique used here is the same as the approach in [1]-[4]. The packet size we consider to illustrate our approach is 48 bytes which corresponds to a standard ATM cell payload. Details including the delay implications and information theoretic arguments in support of interlacing can be found in [7] and [8].

3 Performance Evaluation

We begin by describing a rule yielding FEC codes which provide excellent performance. Efficiency of this policy has been confirmed in [7] by using more rigorous rate-distortion arguments. This is followed by a brief description of the channel used to model the end-to-end packet loss behavior. Finally, we provide descriptions of the computation of information-theoretic bounds on performance.

3.1 Code selection principle

As we will show, the most efficient FEC application is performed by throttling the coding rate to accommodate the FEC overheads. Consequently, the throttling operation is to be minimized to prevent sacrifice in quality under light loads resulting in small packet loss rates. In other words, a critical parameter is the *code rate* $R = K/N$ which determines the fraction of the overall rate allocated to the source coding operation. Therefore, it is desired to make K/N as close to 1 as possible, which is the ideal code rate under lossless conditions. At the same time, to provide good protection with reasonable delay, it is desired to maintain the FEC coding delay and the decoded loss probability below appropriate thresholds. Therefore, we choose the code which solves the constrained optimization problem:

$$\begin{aligned}&\text{maximize:} \quad R = \tfrac{K}{N} \\ &\text{subject to:} \quad \text{Decoded loss prob} \leq L_{threshold} \\ &\phantom{\text{subject to:}} \quad \text{FEC coding delay} \leq D_{threshold}.\end{aligned} \qquad (1)$$

Here, $L_{threshold}$ and $D_{threshold}$ are the quality-of-service (QOS) constraints which refer to the thresholds below which the decoded loss probability P_{dec} and FEC coding delay are to be maintained. P_{dec} refers to the packet loss probability after the FEC decoding operation is performed. Numerical methods for their computation are provided in [7] and [9]. In such a formulation, we assume the environment to be jitter tolerant. We adopt a brute force method of enumeration in order to find the required code. The maximization of K/N is over all RS codes including the shortened and extended codes. Although it is somewhat tedious, such an enumeration needs to be performed only once for a given set of parameters.

In the next section, we present a Markov model used to capture the behavior of the packet loss process.

3.2 Modeling the end-to-end packet loss behavior

Modeling packet loss in high speed integrated networks is a challenging problem. While a number of sophisticated techniques and models have been developed, our interest in modeling packet loss is only one component in our

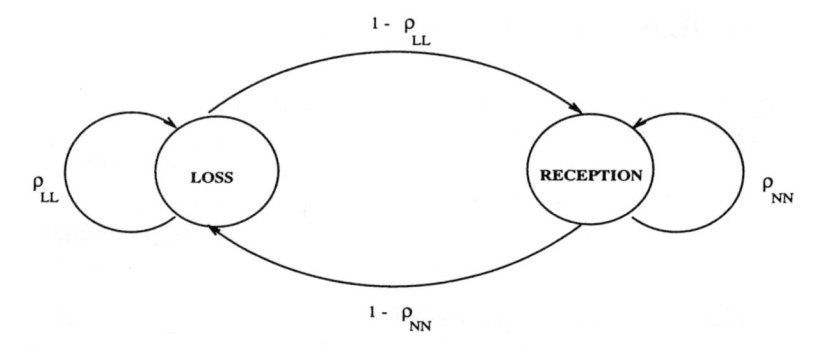

Figure 2: Markov model representation of the packet loss behavior.

end-to-end view of packet video. Thus, it is essential that our model be simple and tractable. The most important characteristic of packet loss in networks which will challenge recovery schemes for video is the correlation between losses caused by buffer overflow and cell dropping by the network (as a form of congestion control). Our Markovian loss representation emulates the basic end-to-end loss characteristics in the entire network without requiring detailed knowledge of the particular topology. In particular, we have modeled the packet loss behavior by a simple two-state Markov chain as illustrated in Fig. 2. Given a desired value of the steady-state loss probability, P_L, once ρ_{LL} is chosen the other parameter ρ_{NN} can be readily calculated.

3.3 Computing an upper bound on achievable performance

In this section, a performance bound is determined for operating at a given average source coding rate over a channel with a certain loss probability. Our approach to calculating the upper bound begins with descriptions of the bit-level and the packet-level behavior of the system. At the bit-level, we use the notion of a *block interference* channel, first developed in [10]. We then model the packet loss behavior as an independent process which typically holds true under ideal interleaving.

3.3.1 Modeling the bit-level transmission as a block interference channel

The channel model we use for this purpose is a special case of a block interference channel introduced by Mcliece and Stark [10]. In [10], successive blocks of length m bits are serially transmitted over one of a finite number of distinct channels with the choice made independently, according to some specified distribution, for each block. We consider only two possible channels; the binary symmetric channel (BSC) and the binary erasure channel (BEC). The particular channel in use for any block is represented by the channel state s, with $s = 0$ corresponding to the BSC while $s = 1$ corresponds to the BEC.

Generally, the probability of selecting either of these channels (the state selection process) would depend on the behavior of the block loss process. For simplicity and tractability in our analysis, we assume in this section that the packet losses occur independently with probability $P_{L,k}$ for the kth priority class (i.e., $\rho_{LL,k} = P_{L,k}$, $\rho_{NN,k} = 1 - P_{L,k}$). As noted previously, this would generally be true under ideal interleaving and provides a theoretical upper bound on performance.

The capacity under perfect CSI of the block interference channel, specialized to represent the packet loss process, can be determined for the kth priority class as [7]

$$\bar{C}_k(m) = \bar{C}_k = (1 - P_{L,k})(1 - H(p)); \text{ bits/channel use}, \quad (2)$$

where $P_{L,k}$ represents the probability of a packet being lost.

3.3.2 Upper bound on the reception quality

Our interest in this section is in calculating an upper bound on the video reception quality for transmission over a *block interference* channel by relating the capacity in (2) to a fidelity metric. Consider the ECSBC scheme where one could distribute the coding rate among the various subbands in many ways. The issue then is to determine the optimal distribution of the coding rate among the various subbands so as to minimize the overall distortion under loss. The minimum distortion, D_{min}, is given by solving the following rate-allocation problem.

$$D_{min} = \min_{R_{s,k}} \sum_{k=1}^{K} D_k(R_{s,k}) \quad (3)$$

subject to

$$R_s = \sum_{k=1}^{K} \frac{R_{s,k}}{\bar{C}_k} \quad (4)$$

In the above equation, $\bar{C}_k$ denotes the capacity computed for the kth priority class, K represents the total number of priority classes and $R_{s,k}$, the source coding rate allocated to the kth priority class. Observe then that $\frac{R_{s,k}}{\bar{C}_k}$ denotes the overall transmission rate. Solving (3) requires the computation of the distortion rate characteristics $D_k(R_{s,k})$ for each of the k priority classes. This distortion-rate characteristics can easily be obtained from the individual distortion-rate characteristics computed for each of the subbands belonging to that priority class by setting up a similar optimization problem as in (3). There are standard methods of solving constrained optimization problems as in (3) (for example, the BFOS algorithm).

4 Results and Discussion

In this section, we use the performance evaluation technique described earlier in our results and discussion. In our FEC-based schemes, for those dropped packets which are not recovered by the RS code, the missing region is then obtained by using a passive error concealment scheme. In our ECSBC coder, the specific passive error concealment scheme employed is that of temporal interpolation.

We now compare the throttled, unthrottled and unprotected scheme performance. To do so we perform the following two-step experiment. We base the choice of the parameters of the two-state Markov loss model on simulations of a multiplexer with the video sources modeled as a discrete-time autoregressive process [11]. The model parameters matched the bit-rate statistics of the coder [11] operating on the *Football* sequence. The parameters of the two-state Markov model, namely P_L and ρ_{LL} were chosen to match those obtained from the simulation. Further details may be obtained from [7] and [8]. Both the throttled and unthrottled scheme employed the RS(15,13) code at an interleaving depth of 1. Results are illustrated in Fig. 3 in the high-load region as the congestion

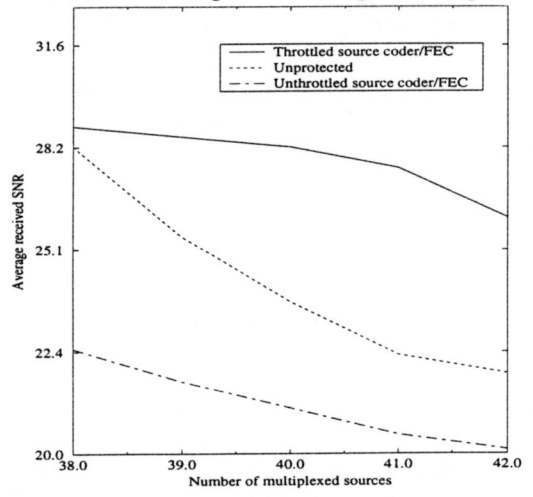

Figure 3: Comparison of the 3 schemes. Multiplexer speed is 100 Mbps and buffer size is 500. Source coding rate = 0.80 bits/pixel for the throttled scheme and 0.93 bits/pixel for the unthrottled, unprotected scheme. RS(15,13) code is used at interleaving depth = 1.

is normally more severe in such cases. We use only a single priority structure in this example as the primary purpose is to demonstrate the superiority of the throttled approach without having to get into the network priority and buffer management issues. Furthermore, as the simulations are time-consuming, they are restricted to a region of maximum interest. The multiplexer operating rate is 100 Mbps (FDDI speed) while the buffer size chosen is 500 packets. As the figure indicates, with increasing load, there is as much as a 6 dB difference between the throttled and unthrottled scheme and up to a 4 dB difference between the throttled and unprotected scheme. This figure also confirms that the unthrottled scheme leads to poor performance under heavy loads due to added congestion. In fact, it behaves much worse than the unprotected system.

We now employ the code selection strategy described earlier and determine its performance. Figure 4 shows the optimized code rates for a particular example of the Markov channel model. In the example, the loss probability P_L is 1×10^{-2} while $D_{threshold} = 5$ msec and $L_{threshold} = 10^{-4}$. As mentioned earlier, we do not include the jitter as a constraint to simplify our evaluation. Although 20 msec typically represents the tolerable end-to-end delay for interactive applications [8], a value of 5 msec is chosen for $D_{threshold}$ to allow for queueing delays which typically dominate end-to-end delay. We also assume in these examples a CCIR 601 video resolution.

Notice that low optimized FEC code rates are obtained when the operating rate is quite small. This means that substantial throttling of the source coding rate, and hence reduced transmission quality, is required to accommodate FEC overheads. Furthermore, observe that the selected code depends on the operating rate. This occurs because the FEC coding delay depends on the operating rate. Hence, a code which satisfies the delay threshold at a given operating rate may not do so at a lower operating rate. Consequently, the philosophy of using only one code (which was originally proposed for AAL 1)ndependent of the operating rate is questionable. A better approach would be to pre-determine a number of codes, one for each operating rate for fixed channel conditions.

Figure 5 demonstrates the performance of different codes at various code rates. In this example, a different code is used for each priority class optimized independently for that class in accordance with the respective values of the loss parameters. The codelengths illustrated ($N = 7, 15, 63$) represent the performance when the corresponding code is applied to both priority classes. The optimized code for the high-priority class is RS(61,56) and for the low-priority class is RS(56,50). Though it seems that the code rate which maximizes the performance for the codelength of 63 does better than the optimizing code, it should be noted that the particular code does not satisfy the delay threshold of 5 msec in this particular example. Notice that the performance of the code of codelength 63 is very close to the optimizing code rate as well as the information-theoretic bound on performance. Consequently, as long as the code rate is properly chosen, a codelength of 63 is sufficient in providing good performance. This is particularly interesting since it indicates that codes of relatively small codelength (the FEC decoding complexity as well as the net FEC delay introduced depend on the codelength) are sufficient to yield good performance. Also shown in the figure is the performance of the RS(128,124) code which has been proposed for the AAL 1 layer [12]. The code optimized according to the selection policy performs much better than the RS(128,124) code.

Though the optimized code was selected here for a particular value of the

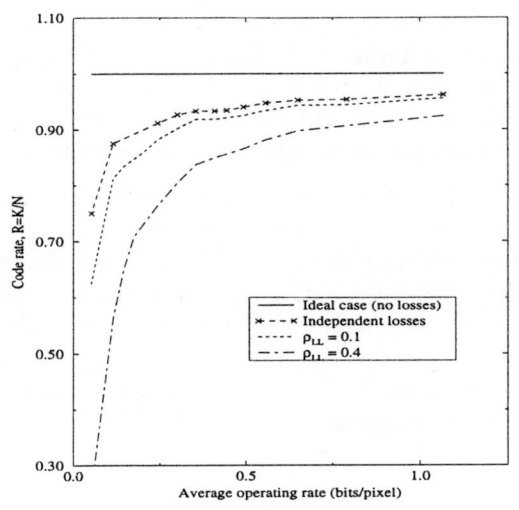

Figure 4: Typical profile of optimized code rate (vs.) transmission rate. In this example, the loss probability $P_L = 1 \times 10^{-2}$ and $D_{threshold} = 5$ msec while $L_{threshold} = 10^{-4}$.

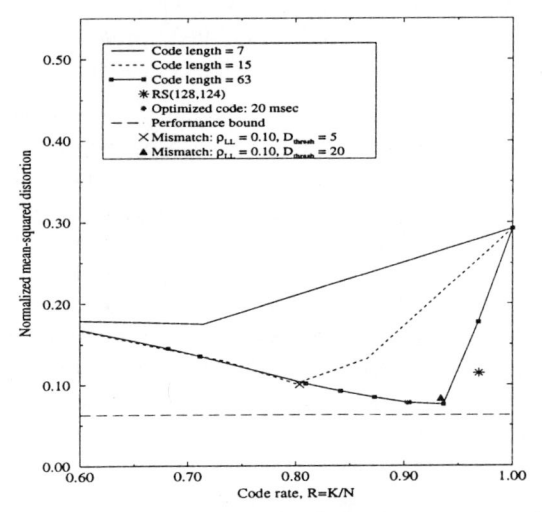

Figure 5: Comparative performance of a hybrid error concealment scheme at different code rates. Loss probability $P_L = 5 \times 10^{-3}$ for the high-priority class, 1×10^{-2} for low-priority class, $\rho_{LL} = 0.40$, sequence is the *Football* sequence, $N_f = 12$ and the overall average transmission rate is 0.85 bits/pixel.

Markov chain parameters, such an accurate description of the loss behavior in the network is seldom available. As a result, it is important to consider the behavior of the optimized codes for slightly different values of the Markov chain parameters so that their behavior can be studied under mismatched conditions. This behavior is illustrated in the figure for two cases, one of which was selected with a relatively strict 5 msec constraint on the delay and the other a more relaxed 20 msec constraint. The 5 msec constraint limits the number of available codes that meet the threshold on the loss. As a result, much lower code rates are required to achieve good performance. Even under mismatched conditions, observe that the optimized codes perform very well indicating the robustness of the code selection strategy.

5 Conclusions

In this paper, a generic transmission scheme for robust transmission of VBR video employing a class of hybrid FEC-based error recovery schemes was studied. Such schemes are particularly useful when there is high motion or rapid scene changes in the encoded video. In such a case, the effect of error propagation due to imperfect packet recovery is greatly reduced by FEC. In using the *active* error concealment technique, the performance of two approaches for applying FEC was studied. In the first approach, additional bandwidth was allocated for the channel coding operation. Such a scheme was shown to suffer performance degradation under higher loads in comparison with an unprotected system. A more judicious approach to applying FEC was investigated in this paper where the bandwidth for FEC application was allocated by throttling the source coder output. Under moderate-to-high packet losses (characteristic of high network loads), employing a throttled source-coder FEC application was shown to offer significantly better performance compared to an unprotected system.

The performance of the scheme is closely related to the code selected. As a result, it is important to devise a clever code selection strategy. A simple strategy of selecting codes based on a constrained optimization technique was outlined and its performance studied. The results indicate that the selection of a single code for all operating rates is questionable. For most cases, codes of small-to-moderate codelength (≤ 63) performed very well as long as the code rates were properly chosen. The code selection strategy provided robust performance even under conditions of mismatch in choice of the channel parameters for which the codes were selected. These results were subsequently evaluated and verified using the MPEG-2 coder. The only noticeable difference was that the gains while employing MPEG-2 were lower [7] due to the relative robustness of the scheme to loss.

References

[1] N. Shachum and P. McKenney, "Packet Recovery in High-Speed Networks using Coding and Buffer Management," *Proc. IEEE INFOCOM*, San Fran., CA, pp. 124–131 , June 1990.

[2] H. Ohta and T. Kitami, "A Cell Loss Recovery Method using FEC in ATM Networks," *IEEE Trans. Commun.*, vol. 39, no. 9, pp. 1471–1483, Dec. 1991.

[3] E.W. Biersack, "Performance Evaluation of Forward Error Correction in ATM Networks," *Proc. ACM SIGCOMM*, Baltimore, MD, pp. 248–257, Aug. 1992.

[4] A.M. McAuley, "Reliable Broadband Communication using a Burst Erasure Correcting Code," *Proc. ACM SIGCOMM* , Philadelphia, PA, pp. 297–306, Sept. 1990.

[5] Y.H.Kim, "Adaptive Entropy Coded Predictive Vector Quantization of Images," Ph.D Dissertation, Dept. of Electrical, Computer and Systems Engg., Rensselaer Polytechnic Institute, Troy, NY, 1990.

[6] Y.H.Kim and J.W.Modestino, "Adaptive Entropy-Coded Subband Coding of Image Sequences," *IEEE Trans. Commun.*, vol. 41, no. 6, pp. 975-987, June 1993.

[7] V.Parthasarathy, "Transport Coding Schemes for Digital Video Transmission over ATM Networks," Ph.D Dissertation, Dept. of Electrical, Computer and Systems Engg., Rensselaer Polytechnic Institute, Troy, NY, June 1995.

[8] V. Parthasarathy, J. W. Modestino and K. S. Vastola, "Reliable Transmission of High-Quality Video over ATM Networks," *submitted to IEEE Trans. Image Processing.*

[9] V.Parthasarathy, J.W.Modestino and K.S.Vastola, "Design of a Transport Coding Scheme for Variable-rate Video Transmission over ATM Networks," *accepted for publication to IEEE Trans. Systems and Circuits for Video Tech.*

[10] R.J.McEliece and W.E.Stark, "Channels with Block Interference," *IEEE Trans. on Inform. Theory*, Vol. IT-30, pp. 44-53, Jan 1984.

[11] B.Maglaris, D.Anastassiou, P.Sen, G.Karlsson, J.D.Robbins, "Performance Models of Statistical Multiplexing in Packet Video Communications," *IEEE Trans. Commun.*, Vol. 36, No. 7, pp.834-844, July 1988.

[12] C.Partridge, *Gigabit Networking*, Addison Wesley Publishing Company, 1993.

Combined Source and Channel Coding for Wireless ATM LANs

Ender Ayanoglu, Pramod Pancha, Amy R. Reibman
AT&T Bell Laboratories
101 Crawfords Corner Road
Holmdel, NJ 07733-3030

Shilpa Talwar[1]
Scientific Computing and Computational Mathematics
Stanford University
Stanford, CA 94305

1 Introduction

It is widely recognized that a viable method to provide "anytime, anywhere" communication needs is by means of wireless technologies. In the indoor wireless environment, the 1-2 Mb/s wireless link technology that is currently available from several vendors (for example, WaveLan) will soon be replaced by higher capacities in the near future. For example, several manufacturers plan to develop radios based on the HiperLan standard that is currently being discussed in ETSI. These radios are intended to support 20 Mb/s wireless links for computer networking. As a result of this increase in physical link capacity, it is now feasible to envision wireless networks that provide similar functionality to current wired Ethernet LANs with the added advantage that the end terminals can be mobile. This in turn makes it feasible for multimedia services, in particular the delivery of video, to be provided over these wireless networks.

In addition to the need for sufficient bandwidth to transport video, it is well-known that guaranteed bandwidth is a basic requirement for video in order to maintain an acceptable quality of service (QoS). The use of ATM transport in the wireless LAN ensures that QoS guarantees for services can be provided and has the additional advantage that it enables seamless internetworking of local area and wide area networks (where ATM is expected to be the primary transport mechanism).

There have been several wireless ATM approaches that have been proposed [1][2]. The work described in this paper is part of an effort to extend ATM from the LAN/WAN infrastructure towards the wireless users that is described in [1], where a design and prototype of a LAN where ATM cells are generated and received by the wireless user terminals is described. In this paper we investigate some of the pertinent video transmission issues for our indoor wireless ATM LAN.

---

1. Work performed while the author was at AT&T Bell Laboratories on a D.O.E. fellowship program

Although the concepts behind the wired and wireless ATM networks are quite similar, there can be significant differences in the characteristics of these networks, as stated previously. While wired ATM networks are usually based on an almost error-free physical medium, wireless links by nature are unreliable at the physical layer. This results in a need for more complex error control algorithms in order to achieve an acceptable cell loss ratio (CLR). However, error control algorithms have the effect of reducing the amount of "useful" bandwidth available on a wireless link and therefore must be used judiciously. The error control policy must therefore be chosen in order to balance the trade-off between these two quantities.

The unreliability of the wireless physical layer combined with the scarcity of bandwidth for video transport also implies that solutions that have been proposed for wired ATM networks may not be applicable in the wireless environment. In wired ATM networks, a basic underlying assumption has been the reliability of end-to-end cell transport. The primary problem for video transport in these networks which must be addressed is performance during congestion. In wireless ATM, reliable end-to-end cell transport is a more difficult proposition. As previously noted, with sufficient error control overhead this can still be achieved. However, for video, this extra bandwidth for overhead may be costlier than higher CLR. The relationship between error control and video transmission must therefore be investigated for this wireless scenario.

In this work, the utility of error control for video transmission in a wireless ATM LAN [1] is quantified. While error control for wireless channels has been studied in the literature in the context of the outdoor cellular channel, the characteristics of indoor high speed wireless LANs are significantly different and will therefore require new error control techniques. We first develop a model for indoor wireless channels which takes into account random losses as well as losses due to multipath fading. By studying the loss characteristics of this model of the indoor wireless channel, it is possible to determine the error control techniques that are likely to be useful in the indoor environment. However, to measure the true effect of error control it is important to determine the impact on service quality. We therefore apply these error control techniques to a video transmission scenario in a wireless ATM LAN link and present the results in terms of video quality. The best combination of video coding algorithm and level of error control are determined from the performance results.

In Section 2 we present the two models for the wireless channel that are used. The loss properties of the realistic multipath loss model are also discussed. In Section 3 we present the error control technique that we propose for the wireless links and discuss its properties. Finally, Section 4 presents the performance results for combined error control and video transmission on the wireless links. Section 6 provides a summary and our conclusions.

2 Wireless Channel Characterization

The error control technique that should be used is dependent on the characteristics of the wireless channel. We investigate two models for the wireless channel, a random loss model and a multipath loss model.

2.1 Random Loss Model

In the random loss model, losses in the wireless channel are characterized as uncorrelated bit errors. These bit errors in turn lead to cell losses on the wireless link. In this model, the main parameter for the wireless channel and which is varied in our simulations is the bit error rate (or equivalently cell loss ratio).

This model characterizes the random noise and co-channel interference components in the wireless channel. Although this model does not capture all the loss effects in the wireless channel, we expect it to give us an insight into the interactions between error control and video coding techniques that will be useful for determining the optimal control policy for a wireless link.

2.2 Multipath Model

In reality, the transmitted signal over a digital mobile channel may undergo severe impairments due to multipath fading, shadowing, and co-channel interference. In this section, we focus on the effects of multipath fading on a signal received at the mobile. We consider time and frequency selective fading channels in an indoor environment (although the model is equally valid for both indoor and outdoor environments). The data rates of interest range from 2 Mbps to 20 Mbps and we consider mobile speeds of 5 m.p.h. which corresponds to a walking pace. We study the bit error performance of a differential detection scheme using $\pi/4$-shifted-DQPSK modulation format for several scenarios.

The digital cellular TDMA system being developed in the U.S. will use $\pi/4$-shifted-DQPSK modulation. Apart from its spectral efficiency, this modulation scheme allows for both coherent and differential detection. Differential encoding is particularly important in fast fading environments, where the channel induces rapid phase fluctuations in the received signal. Furthermore, differential detectors are simpler to implement than coherent ones. A disadvantage of differential encoding is that it may lead to a loss of two symbols when an error occurs. However, this affects the bit error rate only by at most a factor of two, which is insignificant. For this reason, and for simplicity, we study a differential detection scheme in our simulations.

2.2.1 Channel Characterization

The overall model for the wireless channel is shown in Figure 1 and Figure 2. We now consider each block in detail.

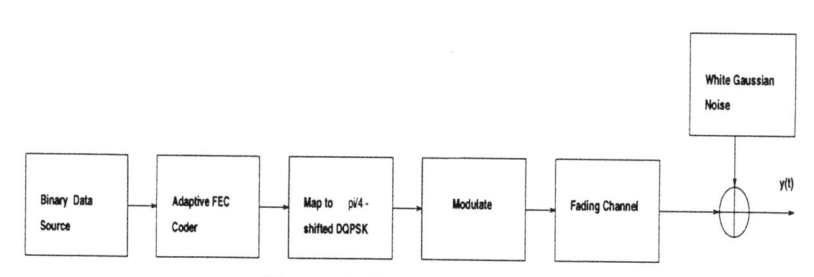

Figure 1. Data Generation

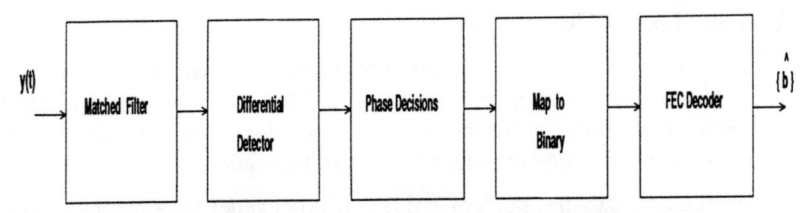

y(t) → Matched Filter → Differential Detector → Phase Decisions → Map to Binary → FEC Decoder → {b}

Figure 2. Demodulation and detection

Modulation: The modulator maps pairs of bits to differentially encoded phases. The Gray-coded mapping of the bits into phase differences is shown below.

Bit Pair	$\Delta\phi_n$
11	$\pi/4$
10	$-\pi/4$
01	$3\pi/4$
00	$-3\pi/4$

Table 1. Mapping of phase differences to bit pairs

Then, the current channel symbol value is calculated as

$$s_n = e^{j(\phi_{n-1} + \Delta\phi_n)}, \phi_0 = 0$$
$$= I_n + jQ_n \quad (1)$$

The channel symbols s_n are then used to modulate a baseband signal $g(t)$ which satisfies the Nyquist condition, (e.g., a rectangular pulse or a raised cosine) so that the modulated baseband waveform becomes

$$s(t) = \sum_n s_n g(t - nT) \quad .$$

Multipath Fading Channel: As it is well-known [3], the multipath fading is due to the reception of multiple paths that arise from scattering of the transmitted wave from walls and structures. Each path is modeled with random amplitude and phase fluctuations. In addition, the mobile causes Doppler shift in each path. The total signal is formed by the sum of the scattered paths

$$r(t) = Re\left[e^{j\omega_c t} \sum_{k=1}^{N} a_k e^{\omega_{dop} t \cos\alpha_k + \phi_k}\right] \quad (2)$$

where ω_c is the carrier radian frequency, ω_{dop} is the maximum Doppler shift frequency; a_k is the random amplitude, α_k is the arrival angle, and ϕ_k is the random phase of the kth path. Due to the central limit theorem, for large N, $r(t)$ becomes an approximately Gaussian complex random process. In addition, if all a_k are about the same strength, and α_k are uniformly distributed, $|r(t)|$ is approximately Rayleigh distributed, and this form of signal reception is known as Rayleigh fading, or diffuse scattering. There are established techniques to simulate (2). In our simulations, we use the method in [3]. In general, the channel can be modeled as the summation of a number of resolvable paths

$$h(t) = \sum_{i=1}^{n_p} r_i(t) \delta(t - \tau_i)$$

where $r_i(t)$ is the impulse response of the ith path, in the form of (2), and τ_i is the delay of the ith path. This is known as the discrete multipath model. In a discrete multipath channel, there exists severe intersymbol interference if a measure of the delay spread τ_i-$min_j\tau_j$ is a significant fraction of the symbol period. A commonly used measure of the delay spread is the rms delay spread, which we will denote by τ in the sequel. It is a general rule of thumb that if the delay spread is more than 10% of the symbol period, the intersymbol interference distortion becomes significant. In this case, the channel can be described by a tapped delay line model.

Demodulation and Detection: After demodulation, the detection process consists of matched filtering to obtain $\bar{I}_n + j\bar{Q}_n$, from which the received phase is estimated as $\bar{\theta}_n = tan^{-1}(\bar{Q}_n / \bar{I}_n)$, and the received phase difference $\Delta\bar{\theta}_n = \bar{\theta}_n - \bar{\theta}_{n-1}$ is computed. The transmitted phase difference $\Delta\bar{\phi}_n$ is then detected to be the closest of $\pi/4$ and $3\pi/4$. Then, $\Delta\bar{\phi}_n$ is mapped to bits using Table 1. As stated above, in a fast fading environment, differential detection is more robust than coherent detection. Although, with differential detection, in the presence of noise, errors occur in pairs which leads to some performance degradation. This is tolerable by virtue of the simplicity of differential detection.

2.2.2 Simulation Results

In the simulations that are presented, the carrier frequency is taken as 850 MHz. The channel model assumes two Rayleigh faded paths of equal power arriving at the receiver with a delay spread which is a multiple of the symbol period. The rms delay spread τ for this model is half the delay interval between the two paths. The transmit pulse waveform used is a raised cosine.

We consider an indoor scenario with the following parameters: mobile velocity is 5 m.p.h., bit rate is 4.5 Mbps, delay spread between two faded paths of 88 nanoseconds (*2T/10*), with a packet size of 54 bytes (53 byte ATM cell + 1 byte sequence number). Figure 3 shows the trace of a typical error event for this case. There are a few bursts of packets in error, and in each burst, the number of bytes in error per packet may be large. The next figure focuses on one of the bursts, showing their duration (about 5 packets), and magnitude. We see that during the bursts, the whole packet is practically lost so interleaving will not help. However, as soon as we determine that a packet is in error, we know that the channel is bad, and the next few packets are likely to be in error. This suggests that per-window, redundant-packet-based FEC can be beneficial to correct these errors. In Figure 4, we observe the error patterns in a similar scenario as before, except that the delay spread between the two paths has now been increased to 176 nanoseconds (*4T/10*). We see that the error bursts that were present previously have increased to about 8 blocks in error, while the duration between error events is relatively unchanged in this example.

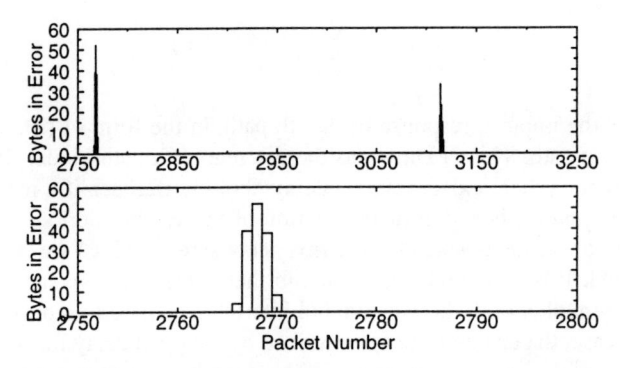

Figure 3. Trace of errored bytes in indoor environment
(Packet Size = 54 bytes, Bit Rate = 4.5 Mbps, v = 5mph, τ = 88ns)

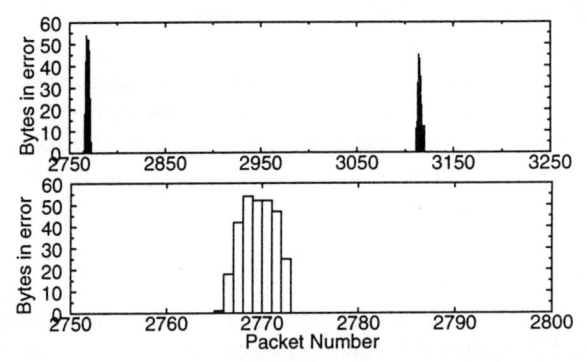

Figure 4. Trace of errored bytes in indoor environment
(Packet Size = 54 bytes, Bit Rate = 4.5 Mbps, v = 5mph, τ = 176ns)

Figure 5 and Figure 6 depict the statistical characteristics of the loss process using the multipath model for the wireless channel. Figure 5 presents the histogram of the number of errored bytes per packet when a loss occurs on the wireless link. For a small delay spread (88 ns), the histogram of errored bytes is a decreasing function while for larger delay spreads (220 ns) there is a peak that occurs at around a loss of 12 bytes per packet. It therefore appears for this case that per-packet error protection may not result in any improvement unless a significant amount of redundant bytes are added to each packet. The histogram for the number of consecutive packets lost in an error burst is shown in Figure 6. From this figure it can be seen that larger delay spreads lead to larger number of packets lost in a burst. It therefore appears that by using error correction on groups of packets it may be possible to compensate for the effect of losses in a multipath environment.

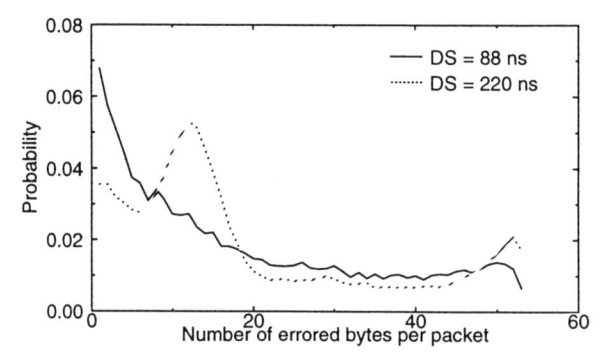

Figure 5. Distribution of bytes in error per packet

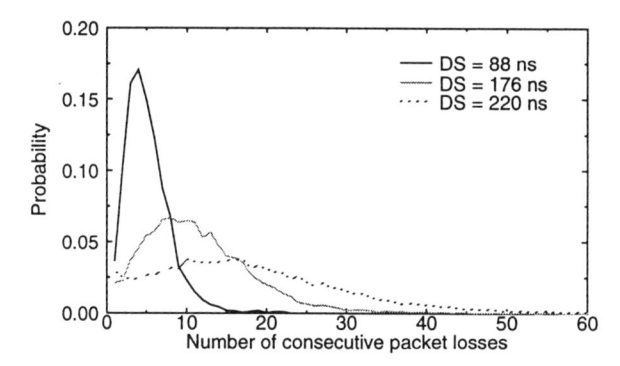

Figure 6. Distribution of packets lost in an error burst

In this section, we observed that when discrete multipath fading occurs, bit error rates higher than 10^{-3} are possible for typical indoor environments. We have also observed that the average bit error rates do not characterize the channel well. In particular, the indoor channel illustrates an on-off behavior: few packets are in error, but those that are have a large number of errored bytes. This suggests redundant-packet, per-window FEC may be useful for the indoor channel.

3 Error Control

Error control in the wireless link is achieved by means of a channel encoder-decoder. In this system Reed-Solomon encoding-decoding is chosen because it leads to minimal overhead bandwidth. Commercially available Reed-Solomon encoder-decoders operate at speeds of up to 80 Mb/s.

For byte-level FEC, the Reed-Solomon (R-S) coder processes a group of N symbols of m bits at a time and generates an overhead of $2t$ symbols. The value t is the number of errored symbols that can be detected and corrected by the R-S coder.

Let p be the bit error rate. Then, the symbol error rate becomes $e=1-(1-p)^m$ and the probability of a block error is

$$P_{L,B} = \sum_{k=t+1}^{N} \binom{N}{k} e^k (1-e)^{N-k} \quad (3)$$

Assuming that there are l blocks in a cell, the CLR is expressed as

$$P_{L,C} = 1 - (1 - P_{L,B})^l \quad (4)$$

For $m=8$ (i.e., a byte), we plot the CLR against bit error rate (BER) in Figure 7 using these R-S codes. For a given channel BER, longer R-S codes lead to lower CLRs after error control has been applied. The R-S code that is utilized will be determined by the requirements of an application since they may have more efficient means to recover from errors.

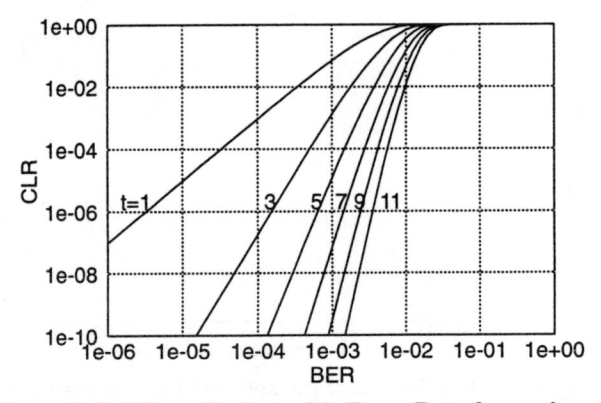

Figure 7. Cell Loss Ratio vs. Bit Error Rate for various t

It is also possible that a fraction of the bit errors are undetected after decoding the FEC codes [4]. A plot of this residual BER which is undetected against the channel BER is shown in Figure 7. It can be seen that this residual BER is very small ($<10^{-8}$) for most combinations of FEC codes and channel BERs. Note that in the region where the residual BER is approximately 10^{-5} ($t=1$ FEC code and channel BER $> 10^{-3}$) the CLR (from Figure 7) is on the order of 10^{-2}. Undetected bit errors are therefore not considered in this work and location of errors are assumed to be known once FEC is applied.

As the wireless channel simulation results indicate, the characteristics of the wireless channel may require error control on blocks of cells in addition to the byte-level FEC described above. If the error locations are known, k R-S symbols are sufficient to recover from k errored symbols. In our scheme, with sequence numbers in cells, it is possible to detect the location of lost cells. Then, to encode blocks of cells, up to s lost cells can be recovered by means of s redundant cells, using a R-S

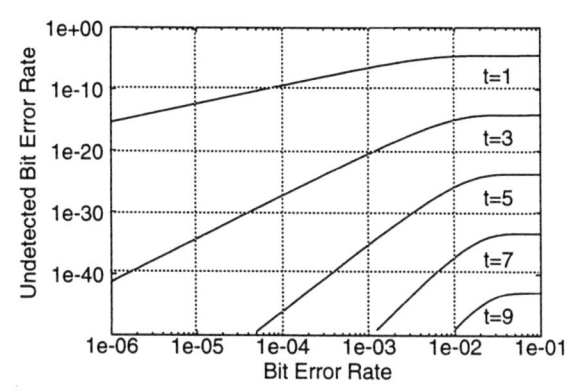

Figure 8. Undetected Bit Error Rate vs. Bit Error Rate for various *t*

coder or another maximum distance separable code, such as a diversity code [5]. In this paper, this method of FEC is called cell-level coding.

4 Video Performance

The error control technique outlined above provides a basis for determining the operational characteristics of the wireless link. As Figure 7 indicates, a longer code results in higher protection for the video sequence (lower CLR for same initial BER). However, the use of error control codes must be consistent with the requirements of the application. For the case of encoded video, we have found that the video quality with a CLR of 10^{-6} is nearly equivalent to a lossless system. Therefore, using long error control codes to achieve a CLR lower than this value is unlikely to improve performance and will lead to unnecessary overhead bandwidth. Judicious use of error control coding can therefore ensure efficient use of bandwidth and optimal video quality.

The primary motivation for this work is to evaluate the performance of video transport schemes over wireless ATM links and, in particular, to assess the effect and impact of error control on video quality. The 150 frame (5 second), MPEG420 format sequence, *Flower Garden*, was utilized in our simulations and a longer 50 second sequence, *You Will*, was used to validate the results. The methodology for the video simulation consisted of packetizing video data into 48 bytes (the user data field in an ATM cell).

It is well-known that encoded video is extremely sensitive to the temporal and spatial location at which a cell loss occurs. To decrease the sensitivity of our results to the location of cell losses, each simulation was run 10 times using a different random number seed for the cell loss process for each run. The average value over all runs for each simulation is presented in the results.

Several video coding techniques are investigated. The 1-layer video coder that is used in the simulations is an MPEG-2 main profile encoder [4] operating at an output bit rate of 4 Mb/s. Performance of 2 different MPEG-2 based 2-layer video coding [7] techniques are also investigated. The first, Data Partitioning, is based on dividing the encoded DCT coefficients into two groups; low frequency components into a base layer and high frequency components into an enhancement layer. The second technique, SNR Scalability, first uses a coarse quantizer to encode DCT coefficients to create a base layer and uses a fine quantizer on the errors to create an enhancement layer. In terms of implementation complexity, Data Partitioning is significantly easier than SNR Scalability. This is primarily because there is a stronger dependency between the base and enhancement layer components in SNR Scalability.

The results for a video sequence are evaluated using the Peak Signal-to-Noise Ratio (PSNR) as an objective measure of video quality for a given error coding scheme as a function of the channel BER. These results, in combination with knowledge of the wireless channel characteristics, provide a good foundation for designing video transmission and error control schemes for wireless ATM networks.

4.1 Random Loss Model

For the random loss model, the main wireless channel parameter that is varied is the BER of the wireless ATM link. For a given BER, the effective CLR can be calculated from Figure 7 as a function of the degree of error control that is applied to the video stream. As noted previously, error control can be applied at several levels. Since byte errors are not correlated in this model, per-packet error coding should efficiently protect the video. Different degrees of per-packet error control are therefore used to investigate video performance for this case.

The trade-off between error control and overhead bandwidth is an important component which determines performance. Table 2 shows the usable video bandwidth for a 1-layer coder for different FEC t values and for a 2-layer coder with FEC of $t=11$ for HP and $t=1$ for LP for different HP, LP traffic fractions. The usable video bit rate is calculated by reducing the raw wireless bandwidth by the FEC overhead. Our simulations were designed for a usable bit rate of 4 Mb/s for the video data when FEC corresponding to $t=1$ is used for the entire bit stream. This value was chosen as it leads to acceptable video quality for the MPEG420 format video that we utilize. This video bit-rate corresponds to a physical layer bit rate for the wireless channel of 4.58 Mb/s. This physical layer bit rate was held constant for all simulation scenarios. For the 2-layer coded sequences, the combination of percentage of data in high and low priority layers combined with the FEC used for each layer determine the overall video rate.

	1-layer encoder	2-layer encoder ($t_{HP}=11$, $t_{LP}=1$)	
t	Video Bandwidth (Mbps)	Bit Rate Ratio (HP, LP - %)	Video Bandwidth (HP, LP - Mbps)
1	4.00	0, 100	0.00, 4.00
3	3.73	25, 75	0.92, 2.75
5	3.49	50, 50	1.69, 1.69
11	2.93	75, 25	2.36, 0.79

Table 2. Usable Video Bandwidths for 1- and 2-layer encoders

4.1.1 Simulation Results

Figure 9 shows the impact of the FEC on a 1-layer MPEG-2 encoding of *Flower Garden*. The graph shows the performance of each FEC for a range of channel BERs before FEC. Note that with the given set of FEC codes, the graphs indicate that it is not possible to obtain good video quality at channel BERs greater than approximately 10^{-2}.

For each FEC case, when the BER increases past a certain point, the PSNR decreases rapidly. This occurs because once losses occur in the headers and motion vectors, there is a dramatic drop in video quality. The knee of the curve occurs at a higher BER for the more powerful FEC codes. However, longer FEC codes result in more overhead which reduces the video bit-rate. Hence the video quality when errors are not present is lower when longer codes are utilized. For each value of BER, an optimal FEC code which maximizes video quality can be determined from Figure 9. For example, if the channel BER=10^{-2}, then optimal error protection would require FEC corresponding to $t=11$, and if BER=10^{-5}, the optimal error protection would be $t=1$.

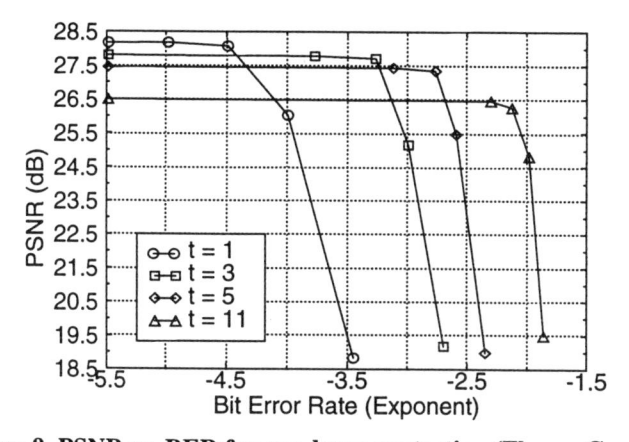

Figure 9. PSNR vs. BER for one-layer protection (Flower Garden)

The graphs of Figure 9 depict the same performance results for a 50 second 1-layer MPEG-2 encoding of *You Will*. The results for this sequence mirror those observed for *Flower Garden*. It is interesting to note that the knee BER, after which PSNR performance degrades rapidly, is very similar for both sequences. This holds even though the PSNR levels for the two sequences are very different. The consistency of these results suggests that the optimal FEC level may be relatively independent of the characteristics of the video sequence.

Unfortunately, in real wireless channels, the instantaneous BER may fluctuate rapidly, and it may be impossible to select a single optimal FEC a priori. An incorrect estimate of the BER could seriously degrade the expected signal quality. For example, if the channel BER is estimated as 10^{-4}, the optimal FEC level would be $t=3$. However, if the actual BER=10^{-2} then this would result in sharply lower video quality. Therefore, we explore the alternative of protecting the high priority video

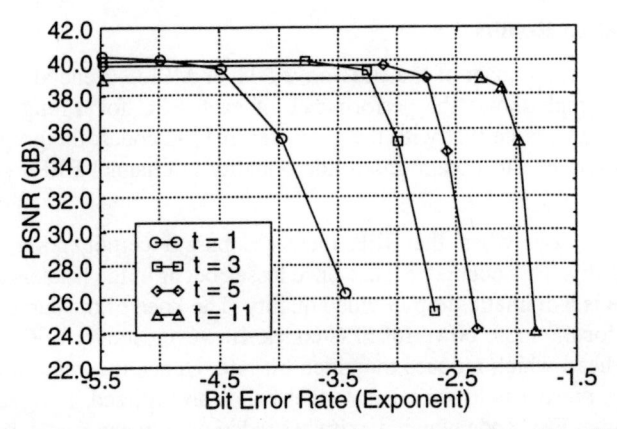

Figure 10. PSNR vs. BER for one-layer protection (You Will)

information with a more powerful FEC code and only protecting the low priority video information with weak FEC ($t=1$). This will have the effect of ensuring that basic video information is always received for a wide range of BER conditions.

Figure 11 shows the performance for *Flower Garden* when MPEG-2 Data Partitioning (DP) is used [7]. In this simulation, the base layer, which contains high priority information, is protected with $t=11$ while the enhancement layer, which contains low-priority, high-frequency information, is protected with $t=1$. Use of this two-layer scheme allows for a higher video bit rate than a one layer $t=11$ scheme because only the high-priority bit stream has a large overhead. This should therefore result in higher video quality when no errors are present.

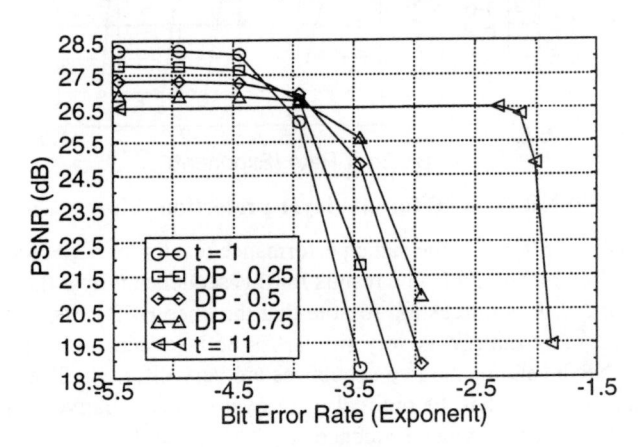

Figure 11. PSNR vs. BER for Data Partitioning (two layer, Flower Garden)

The curves labeled $t=1$ and $t=11$ show the performance when the entire bit stream is protected using these FEC codes (as in Figure 9). The three curves labelled DP show the performance of data partitioning for different percentages of video data in the base layer. For example, DP-0.25 indicates that 25% of the bit stream is designated as high-priority. In all DP cases, since the high-priority data uses a FEC code of $t=11$, there are no cell losses in this layer in our simulations. As explained above, the PSNR for each DP case is greater than for the single layer case with $t=11$ when the channel BER is low. In addition, as the percentage of data in the base layer is increased the video quality in these low BER regions decreases. This is a reflection of the fact that a higher percentage of traffic is protected by a FEC code with larger overhead.

Compared to the $t=1$ coder, the DP cases exhibit a less sharp decrease in performance at the knee and the BER level at which the knee occurs is higher. For example, it can be seen that the DP with 25% high priority has PSNR performance comparable to the single layer $t=1$ case when the BER is between $10^{-4.5}$ and 10^{-6}. For the $t=1$ case a sharp knee occurs at around BER=$10^{-4.5}$. For the DP cases this knee occurs at around the same value but the decrease is more gentle. In general, this type of smooth drop-off in quality with increasing BERs is desirable and is a rationale for using two layer protection.

Performance results for *Flower Garden* using SNR Scalability are shown in Figure 12. As for the DP case, the values that follow the SNR label corresponds to the fraction of the bit stream that is in the base layer. It should be noted that it was not possible to encode the sequence using SNR scalability with a base layer fraction of less than 27.5% because, unlike DP, this technique requires motion vectors to be transmitted in the base layer at all times.

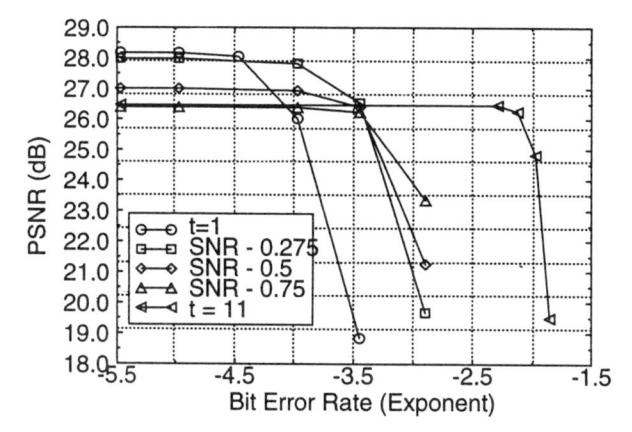

**Figure 12. PSNR vs. BER for SNR Scalability
(two layer protection, Flower Garden)**

The general characteristics of SNR Scalability and DP curves are similar. In both two layer techniques, the knee exhibits a smoother drop-off than for the one layer cases. When comparing the SNR Scalability and DP cases with the same percentage of video data in the base layer, we observe that the SNR Scalability curves consistently are able to tolerate a higher BER without a dramatic loss in quality. This implies that SNR Scalability is a more efficient technique for two layer protection when cells are lost in the low priority bit stream. The cost associated with greater efficiency is the increased complexity of implementing SNR Scalability versus DP.

The SNR Scalability technique also outperforms DP and use of only single layer coding in another sense. The DP curves of Figure 11 are never significantly better than the combination of the $t=1$ and $t=11$ one layer case. However, in the SNR Scalability curves in Figure 12 there is a BER region between $10^{-4.5}$ and $10^{-3.5}$ where SNR Scalability outperforms both the $t=1$ and $t=11$ single layer cases. Therefore SNR Scalability allows good performance over a wider range of BERs than if only one layer schemes are utilized.

4.2 Multipath Model

For the multipath model, the main system parameter that we consider is the delay spread. For each delay spread value, the wireless channel was simulated and the errored bytes were recorded and error correction was then applied. The residual errors which could not be corrected are introduced to the video stream in order to evaluate the effect of these losses. As for the random loss case, video quality is measured in terms of the PSNR.

As noted previously, the statistical characteristics of the loss process for the multipath model seem to indicate that cell-level error control may be most efficient. A number of redundant cells (based on the value of s) containing error control codes for a block of cells are therefore added to the video stream and performance is evaluated as a function of the degree of error protection.

4.2.1 Simulation Results

Figure 13 shows the PSNR ratio performance of a sequence of 50 second video, encoded using one-layer MPEG-2 at a total video bit rate of 4 Mbps. The curve labeled "No Loss" corresponds to video performance when there are no packet losses due to multipath effects. The decrease in this curve with different cell-level error correction codes represents the loss due to overhead bandwidth consumed by the error codes. The curve labeled "DS2" corresponds to a delay spread of 88 ns, "DS4" to 176 ns, and "DS5" to 220 ns. The value of 88 ns corresponds to a large-size office, whereas 176 to 220 ns range corresponds to auditoria size rooms. We conclude from these results that cell-level coding with 10 redundant cells in a block of 256 cells results in acceptable performance for large size offices. Large auditoria require larger overheads, of the size of 50 cells in blocks of 256 cells. Currently, off-the-shelf Reed-Solomon encoder-decoders with parameters $s=10$ exist for block sizes of 255 cells, whereas $s=50$ requires ASICs and is an ambitious goal. For such cases, other alternatives need to be investigated. One such potential technique is to employ two-layer encoding which we plan to investigate.

5 Summary and Conclusions

In this paper, we have shown how to employ forward error correction for protection against errors for MPEG-2 video transmission in a wireless ATM LAN. Errors in an indoor wireless ATM LAN have random and bursty characteristics where bursts

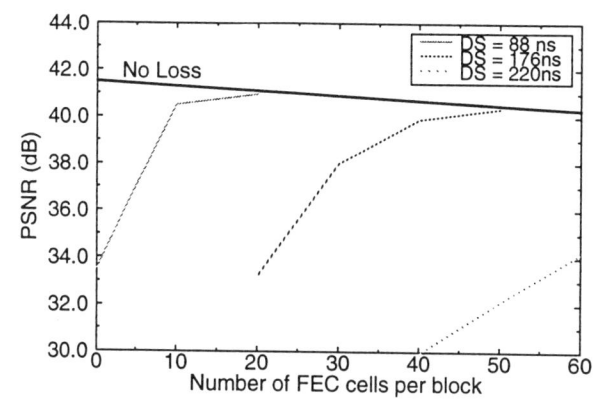

Figure 13. Effect of cell-level coding on video quality

occur due to multipath fading effects. We characterized the error properties of an indoor wireless channel using a channel model which captures the multipath effects and studied the cases of random losses as well as multipath losses. Using the error characteristics of each type of model, we concluded that byte-level coding was appropriate for random losses while cell-level coding provides protection for multipath losses. For random errors, we have shown that 2-layer coding techniques, with each layer coded with different error coding parameters, results in more efficient transmission than single layer MPEG-2. For practical applications, we plan to investigate simple implementations of the techniques described here.

References

[1] K. Y. Eng, et al, "A Wireless Broadband Ad-Hoc ATM Local-Area Network," *Proc. IEEE ICC '95,* Seattle, WA, June 1995.

[2] D. Rayachaudhuri and N. Wilson, "ATM Based Transport Architecture for Multiservices Wireless Personal Communications Networks," *IEEE JSAC*, pp. 1401-1414, Oct. 1994.

[3] W. C. Jakes (Editor), *Microwave Mobile Communications*, Wiley, 1974 (Reissued by IEEE Press, 1994).

[4] J. K. Wolf et al., "Probability of undetected error," *IEEE Trans. Comm.*, Vol. COM-30, pp. 317-324, Feb. 1982.

[5] E. Ayanoglu, C.-L. I, R. D. Gitlin, J. E. Mazo, "Diversity coding for sel-healing and fault tolerant communication networks," *IEEE Trans. Comm.*, Vol. COM-41, pp. 1677-1686, Nov. 1993.

[6] Generic Coding of Moving Pictures and Associated Audio, Recommendation H.262, ISO/IEC 13818-2, March 1994.

[7] R. Aravind, M. R. Civanlar, and A. R. Reibman, "Cell loss resilience of MPEG-2 scalable video coding algorithms", paper I1, 6th International Workshop on Packet Video, Portland, September 1994.

Part 3

DSP in Channel Estimation/Equalization and Modem Design

Block-by-Block Channel and Sequence Estimation for ISI/Fading Channels

Kuor-Hsin Chang*, Warm Shaw Yuan, and C.N. Georghiades
Department of Electrical Engineering, Texas A&M University
College Station, TX 77843, USA
*Wavetek Corporation, CATV and Communications Division
5808 Churchman Bypass, Indianapolis, IN 46203, USA

ABSTRACT

We look at the well studied problems of sequence estimation in unknown or fading channels, and introduce yet another set of algorithms that have reasonable complexity and perform well even in fast changing environments. The algorithms work on a block-by-block basis and utilize periodically, but sparsely inserted known data symbols. In producing their channel estimates, however, the algorithms do make use of the information contained in the received modulated data, thus reducing the rate at which known symbols need to be inserted, and improving performance. Simulation results show good performance for the algorithms.

1 Introduction

The problems of sequence estimation for unknown intersymbol-interference (ISI) and fading channels have been much studied in the literature over the years. The basic problem in both cases is not so much one of obtaining the optimum receiver structure, but one of finding sub-optimal receivers that are easy to implement and yet have a performance close to that of the optimal receiver.

Intersymbol interference (ISI) is one of the primary impediments to reliable data communication, and often occurs when data is sent through unknown, bandlimited channels. In these cases, equalization is used to combat the effects of the channel and to produce reliable data estimates. There are two general categories of equalization algorithms: those that strive to invert the channel, and those that attempt to find the channel and data sequence that most likely will result in the received data. The first category, which invariably results in simpler receivers, works well for channels that are not severely distorting in which case no significant noise enhancement occurs. In particular, channel inversion algorithms, such as those based

---
[1] This work was supported in part by a grant from the STS Group of The Johns Hopkins University Applied Physics Laboratory (APL).

on a peak-distortion or mean-square error criterion, break down for channels that possess spectral nulls. The second category of algorithms, which includes maximum-likelihood sequence estimation (MLSE) algorithms, can perform very well even in severely distorting channels, but the main problem is complexity. Invariably, for these algorithms some complexity-reducing approximation is made to result in an implementable structure.

When data symbols are transmitted over an unknown channel, usually a known training sequence is first sent to initialize the equalizer and aid it in quickly estimating the channel. Equalization techniques (referred to as blind) which attempt to estimate the channel without using a training sequence do exist [1] [2] [3] [4]. Typically, blind equalization algorithms require hundreds to thousands of symbols be processed before convergence to a reliable channel estimate; hence, they generally perform rather poorly in fast-varying ISI channels. Moreover, some of the popular blind equalization algorithms might converge to undesired equilibria whose corresponding equalization parameters cannot eliminate ISI sufficiently [5] [6]. Seshadri [7] [8] has proposed a sub-optimal trellis search algorithm which retains $K \geq 1$ best data sequences into each trellis state. For each of the K best data sequences, the algorithm produces a channel estimate at each state along the trellis; the channel estimate is updated by a least mean square (LMS) algorithm. A step size for the LMS algorithm must be chosen, which depends on the statistics of the channel and affects the performance of the channel estimator and the convergence time. Erkurt and Proakis [9] have proposed a sub-optimal exhaustive search algorithm which expands each node at time t on the tree into Q (assuming Q-ary signaling) nodes at time $t+1$ until $t_f = \mathcal{M} + L - 1$, where $\mathcal{M}$ denotes the level of complexity, and $L-1$ is the channel memory in symbols. For $t \geq t_f$, the algorithm keeps the Q^{t_f} nodes which have the minimum accumulated cost along the trellis. In other words, the algorithm trims the infinite expansion tree into a trellis with Q^{t_f} states after $t \geq t_f$. For each surviving path along the trellis the channel is updated using recursive least square estimation (RLS), or the Baum-Welch algorithm [10]. For binary pulse-amplitude-modulation and a signal-to-noise (SNR) ratio of 10 to 20 dB, the algorithm requires approximately 100 symbols to reach a reliable channel estimate. In other work, Kaleh and Vallet [11] present an iterative algorithm for joint channel parameter estimation and symbol detection using the Baum-Welch algorithm. The algorithm estimates both the channel parameters and the noise variance.

The first part of this paper introduces an iterative joint channel and data estimation algorithm which operates on a block of data at a time without prior knowledge of the channel. The algorithm consists of a maximum-likelihood sequence estimator, which, given a channel estimate produces the most likely sequence, and a maximum-likelihood (ML) channel estimator, which, given a sequence estimate, produces the most likely channel estimate. The algorithm starts with an initial channel estimate, uses it to produce the corresponding sequence estimate, which is then used to produce another channel estimate, and so on until convergence.

The second, but related problem, studied in this paper is that of sequence estimation in the presence of frequency-nonselective fading. The use of a conventional product detector in fast fading channels results in a performance limited by an irreducible error floor. In this case, more complex algorithms must be used for good performance.

Morley and Snyder [12] combined the generalized likelihood formula of [13] and the Viterbi algorithm for ML sequence estimation in randomly dispersive channels. For Gaussian signals with rational spectrum and assuming the fading is sufficiently slow, a discrete-time Kalman filter can be used to estimate the fading process [14]. Although this implementation is optimal, its complexity also grows exponential with increased data. To limit the complexity of the algorithm, Haeb and Meyr [14] restricted the maximum number of possible of sequences to be retained.

To mitigate the problem of phase ambiguity in coherent detection, pilot symbols can be inserted periodically into the data stream. These known symbols correspond to samples of the fading process. If sampling is above the Nyquist rate the receiver is able to unambiguously estimate the fading process. The fading estimates based on these pilot symbols are then interpolated over the data symbols [15].

To further improve the fading estimates, various approaches have been examined in [16], [17], and [18]. Irvine and McLane [16] used a FIR filter to smooth the fading measurements based on the pilot symbols, and used straight line interpolation to estimate the fading over the data symbols. They then update the fading estimate using another FIR filter if the difference between the interpolated fading and the measured fading conditioned on a symbol decision satisfies a confidence threshold. Otherwise, the fading estimate is not updated and the algorithm progresses in the usual manner.

D'Andrea *et al* used the idea of per-survivor processing by incorporating the Viterbi algorithm where the states are points in the signal constellation. The total number of possible sequences is essentially the number of points in the constellation except when the known symbol is processed. For each path, the fading predictions are obtained using a Weiner filter. Paths are allowed to continue to the next data sample if they result in minimum path metric. In their 'modified' algorithm this decision is delayed by an additional sample which effectively increases the number of paths by an order of two. To achieve further improvement, they smooth the tentative fading estimates for each path using a non-causal Weiner filter.

A structured approach was taken by Georghiades and Han in [18] where an EM algorithm was applied to iteratively estimate the fading and the transmitted symbols. They first used an optimal Weiner filter/interpolator as in [15] to provide an initial fading estimate. Based on this fading estimate, tentative symbol decisions are obtained and new fading estimates can be derived.

This iteration continues until the estimates converge.

The estimation/interpolation algorithms in general are relatively simple to implement and have achieved good results. However, the performance of these algorithms is constrained by the Nyquist sampling rate. For example, in a fast fading channel with $B_d T = 0.1$, it is necessary to insert at least one known symbol out of every five symbols transmitted. In Section 3 of this paper we present an algorithm that utilizes the channel information embedded in both the known symbols and the data symbols to estimate the channel response. Combining hypothesis testing and Bayesian estimation, the resulting algorithm has overcome the complexity problem in the optimal implementation and yet it is not constrained by the minimum known symbol insertion rate in the interpolation/estimation algorithms.

Section 2 describes the iterative algorithm for sequence and channel estimation for ISI channels, and presents results for the error-probability, mean-square error, and convergence performance of the algorithm. Section 3 describes the algorithm for sequence estimation in fading channels and its performance, and Section 4 concludes.

2 Iterative Sequence and Channel Estimation

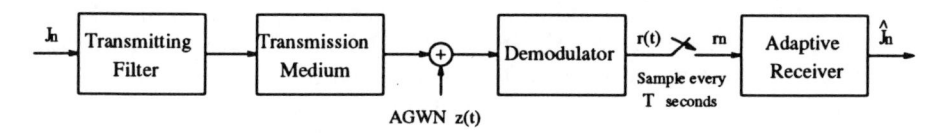

Figure 1: The communication system.

Fig. 1 shows the general system considered, where $J_n \in \{\pm 1, \pm 3, \ldots, \pm(M-1)\}$ is an M-ary transmitted data symbol, $z(t)$ is the usual zero mean, additive white Gaussian noise (AWGN) of power spectral density $N_0/2$, and $r(t)$ is the received baseband signal, which is sampled synchronously every T seconds to yield

$$r_n = \sum_{k=0}^{\infty} J_k h_{n-k} + z_n, \tag{1}$$

where $r_n = r(nT)$, $h_{n-k} = h[(n-k)T]$ and $z_n = z(nT)$. If the channel impulse response has a finite support of $(L-1)T$ seconds, then (1) can be written as

$$r_n = \sum_{k=0}^{L-1} J_{n-k} h_k + z_n. \tag{2}$$

Let N be the length of the data block to be processed, and

$$\mathbf{r} = [r_1, r_2, \ldots, r_N]^T, \quad \mathbf{z} = [z_1, z_2, \ldots, z_N]^T, \quad \mathbf{h} = [h_0, h_1, \ldots, h_{L-1}]^T,$$

$$\mathbf{J}_k = [J_k, J_{k-1}, \ldots, J_{k-L+1}]^T, \quad \mathcal{J} = [\mathbf{J}_1, \mathbf{J}_2, \ldots, \mathbf{J}_N]^T,$$

where superscript T (to be distinguished from symbol duration T) denotes transpose. Then (2) can be written in matrix form as

$$\mathbf{r} = \mathcal{J}\mathbf{h} + \mathbf{z}. \tag{3}$$

2.1 The ML Channel Estimator

Given that $\mathcal{J} = \hat{\mathcal{J}}$, the ML channel estimate is easily obtained as

$$\hat{\mathbf{h}} = \min_{\mathbf{h}} \left[\frac{1}{2} \mathbf{h}^T \hat{\mathcal{J}}^T \hat{\mathcal{J}} \mathbf{h} - \mathbf{r}^T \hat{\mathcal{J}} \mathbf{h} \right]. \tag{4}$$

If $\hat{\mathcal{J}}^T \hat{\mathcal{J}}$ is non-singular, which will be the case in what follows, the ML channel estimate is obtained explicitly as

$$\hat{\mathbf{h}} = \left(\hat{\mathcal{J}}^T \hat{\mathcal{J}} \right)^{-1} \hat{\mathcal{J}}^T \mathbf{r}. \tag{5}$$

When the length of the data block is N, $\mathcal{J}$ has dimension $N \times L$ and $\mathbf{r}$ dimension $N \times 1$. Matrix $\mathcal{J}$ is constructed by data $\{J_{-L+2}, J_{-L+3}, \ldots, J_N\}$ where symbols $\{J_{-L+2}, J_{-L+3}, \ldots, J_0\}$ belong to the previous block of data. To avoid this previous $(L-1)$ data from interfering with the present channel estimation and introduce any propagation error, when using (5) to estimate the channel, we choose $\mathcal{J} = [\mathbf{J}_L, \mathbf{J}_{L+1}, \ldots, \mathbf{J}_N]^T$ and $\mathbf{r} = [r_L, r_{L+1}, \ldots, r_N]^T$, so that the matrix $\mathcal{J}$ only contains data which belongs to the present data block. In (5) the dimension of $\hat{\mathcal{J}}^T \hat{\mathcal{J}}$ is $L \times L$, which shows that the complexity of the channel estimator depends on the length of the channel memory.

2.2 The ML Sequence Estimator

Assuming the channel is known, i.e. $\mathbf{h} = \hat{\mathbf{h}}$, the ML sequence estimate is obtained by performing

$$\hat{\mathcal{J}} = \max_{\mathcal{J}} \left[\mathbf{r}^T \mathcal{J} \hat{\mathbf{h}} - \frac{1}{2} \hat{\mathbf{h}}^T \mathcal{J}^T \mathcal{J} \hat{\mathbf{h}} \right], \tag{6}$$

which can optimally and efficiently be performed using a M^{L-1}-state Viterbi algorithm for M-ary signaling.

2.3 The Overall Receiver

For each received block of data, an initial channel estimate $\hat{\mathbf{h}}^0$ is obtained by processing the known data inserted at its beginning. This initial channel estimate, along with the data $\mathbf{r}$, are then used by the ML sequence estimator to obtain a tentative sequence estimate $\hat{\mathcal{J}}^0$, which is then used by the ML channel estimator with $\mathbf{r}$ to update the estimate of the channel to $\hat{\mathbf{h}}^1$. $\hat{\mathbf{h}}^1$ is then used to refine the sequence estimate to $\hat{\mathcal{J}}^1$, and this iteration process continues until convergence (which occurs when the data sequence does not change from one iteration to the next). The sequence (and channel) estimates are both available at convergence.

In order to improve the performance of the initial channel estimate, the known data sequence is chosen so that it results in minimum variance of the channel estimation error [19], and the resulting matrix $\mathcal{J}^T\mathcal{J}$ is non-singular. Let $\lambda = \{\lambda_1, \lambda_2, \ldots, \lambda_D\}$ be the known (length D) data header, $\lambda_k = [\lambda_k, \lambda_{k-1}, \ldots, \lambda_{k-L+1}]^T$, and $\Lambda = [\lambda_L, \lambda_{L+1}, \ldots, \lambda_D]^T$. Then the covariance matrix of the channel estimation error for $\underline{\lambda}$ is

$$\mathbf{R_h} = E\left[(\hat{\mathbf{h}} - \mathbf{h})(\hat{\mathbf{h}} - \mathbf{h})^T\right] = \frac{N_0}{2}(\Lambda^T\Lambda)^{-1}, \tag{7}$$

and the variance of the channel estimation error is the trace of $\mathbf{R_h}$. Clearly, for $\Lambda^T\Lambda$ to be non-singular with dimension $L \times L$ it is necessary that $D \geq 2L-1$. It is easy to show that, if $\Lambda^T\Lambda$ is non-singular, then so is $\hat{\mathcal{J}}^T\hat{\mathcal{J}}$, which is the condition required in (5).

2.4 Performance Analysis

We evaluate the performances of the sequence and channel estimators through simulation and lower bounds on symbol error probability and mean-square estimation error. The symbol-error rate lower bound is that of Forney [20], assuming perfect knowledge of the channel.

To evaluate the algorithm, 2-PAM signaling and two different channels [21] have been used in the simulations (other channels were also studied with similar results, but are not presented due to space limitations):

$$H_a = [-0.2, -0.5, 0.7, 0.36, 0.2]$$
$$H_b = [0.227, 0.46, 0.688, 0.46, 0.227],$$

where H_a suffers amplitude and phase distortion, and H_b has an in-band spectral null [21].

Figures 2 and 3 plot the lower bound and the simulation results for channels H_a and H_b with a decision delay of $L-1 = 4$ in the Viterbi algorithm.

Because of the decision delay, to detect data sequences with N unknown data in each block, $N+D+L-1$ received data symbols need to be processed with a trellis of length of $N+L-1$. The first N data of the surviving path which has maximum metric are decoded as the transmitted data. This can be done by taking overlapping windows over the received data sequence. By taking overlapping windows, we can introduce a decision delay without losing any received information, and data detection can still be implemented on a block-by-block basis. The figures indicate that the algorithm performs very close to one that knows the channel perfectly, for both channels.

Figures 4 and 5 plot the average number of iterations and channel estimation error vs SNR for the two channels. From the figures, we see that

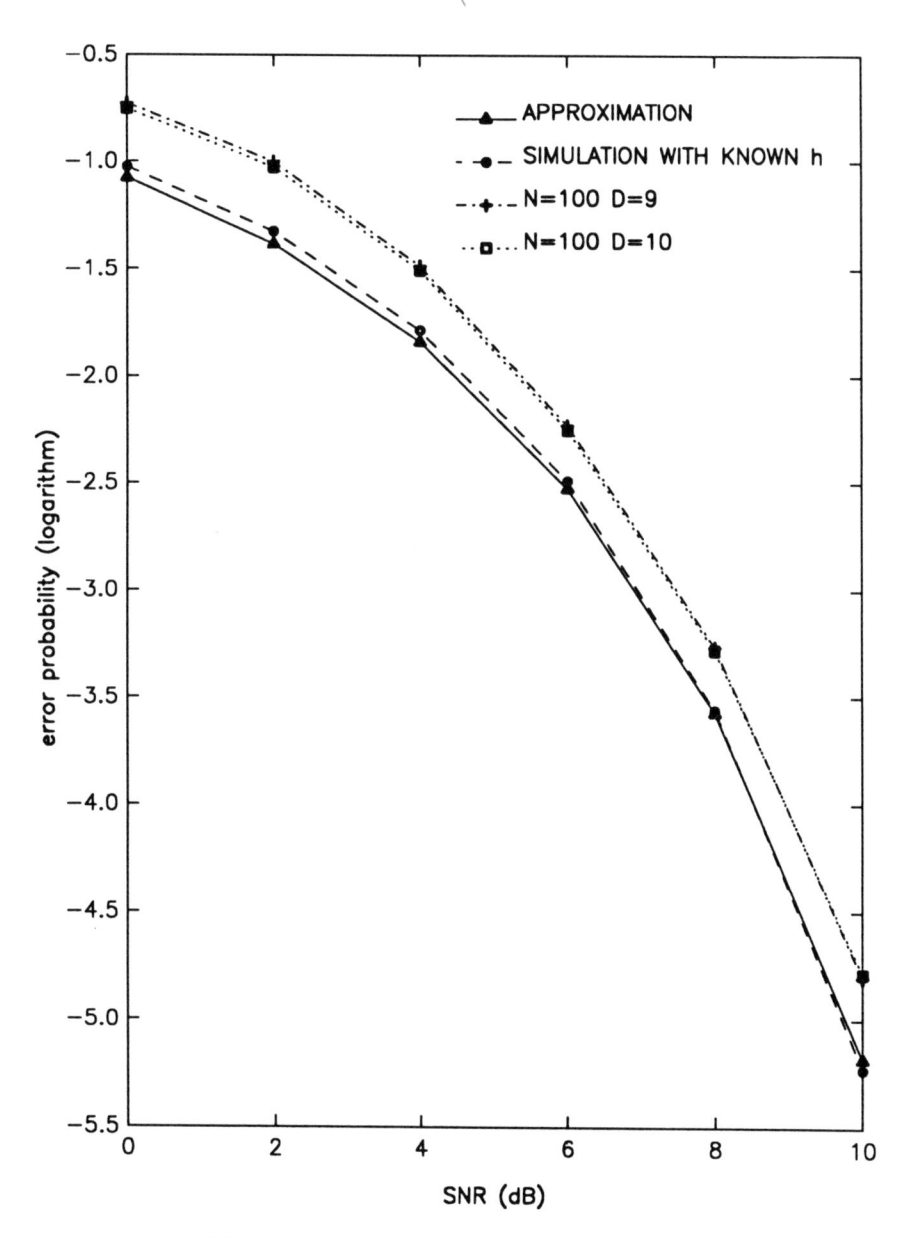

Figure 2: Error probability for channel H_a.

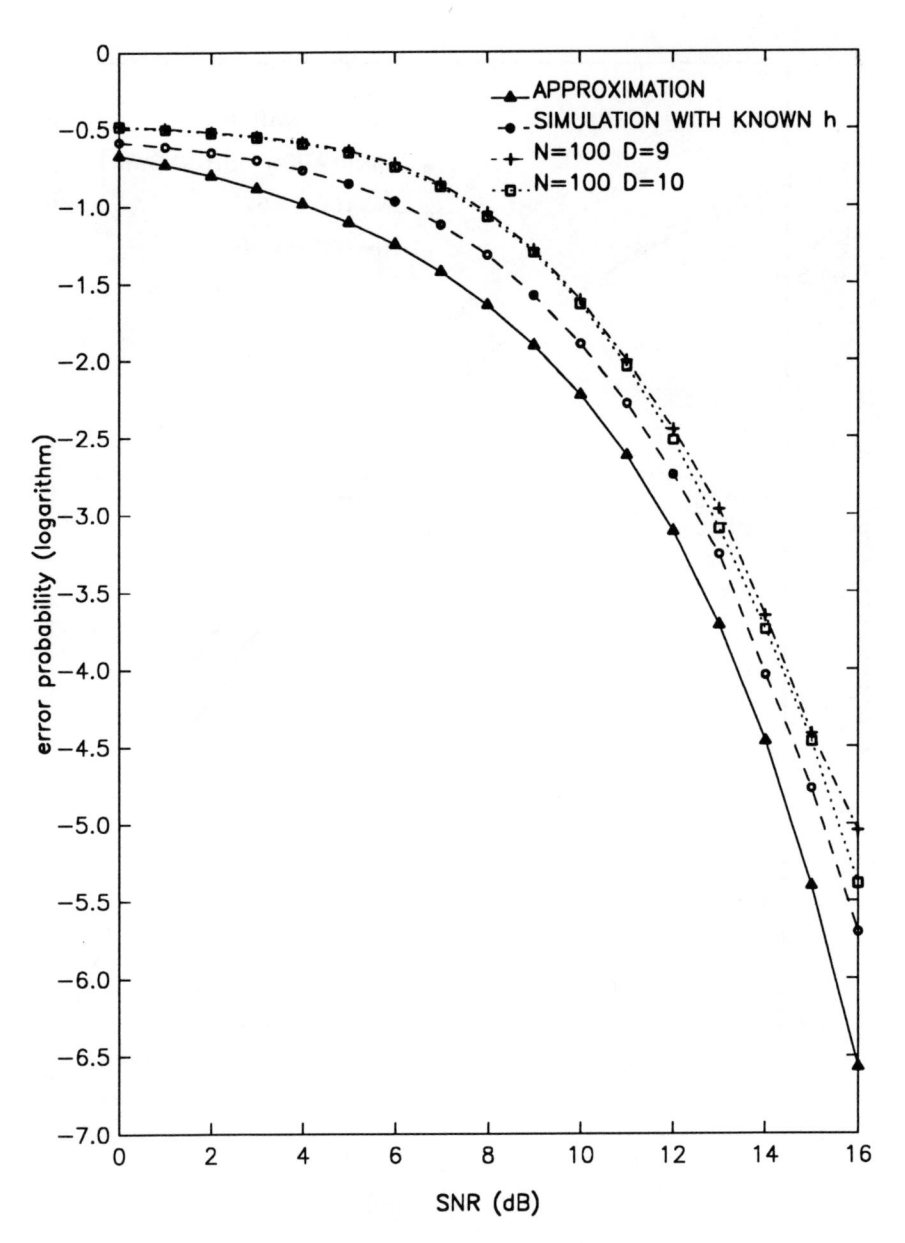

Figure 3: Error probability for channel H_b.

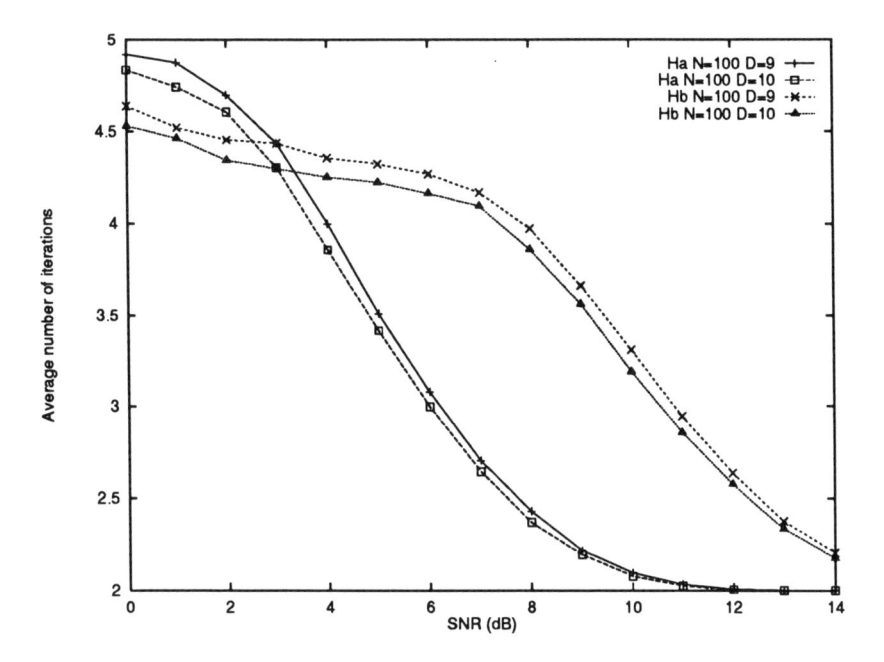

Figure 4: Average number of iterations for convergence vs SNR.

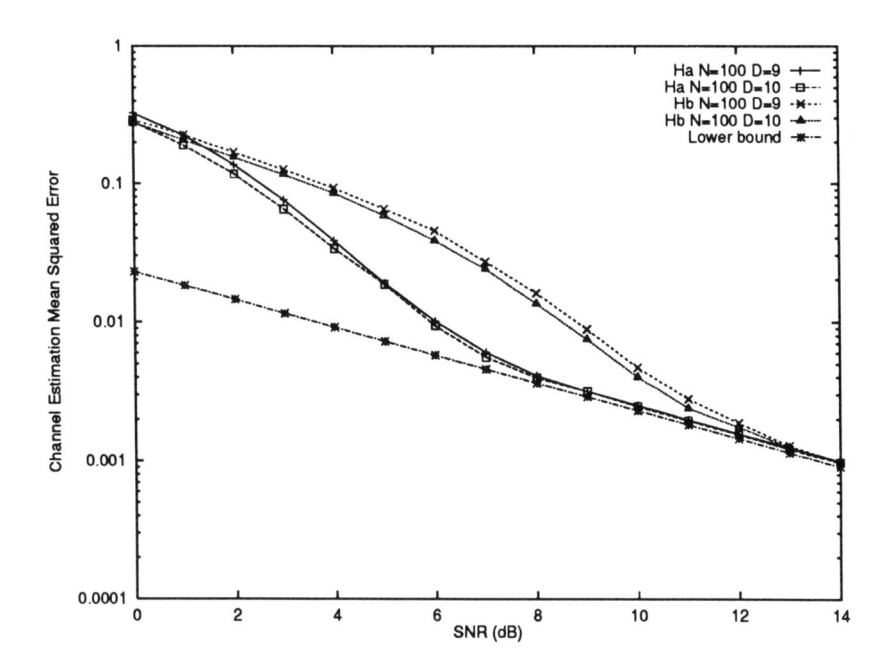

Figure 5: Mean-Square Error vs SNR.

for large SNR, practically only two iterations are needed for convergence, and that the mean-square error approaches closely the Cramer-Rao bound (assuming a known data sequence) for SNR values above about 8 dB.

3 Sequence Estimation for Fading Channels

We assume that fading is sufficiently slow so that it is approximately constant over a symbol interval. The equivalent complex discrete time fading process sampled at the output of the matched filter without intersymbol interference is modeled as

$$\mathbf{x}_k = \mathbf{F}\,\mathbf{x}_{k-1} + \mathbf{v}_k \qquad (8)$$
$$z_k = s_k \mathbf{H}\,\mathbf{x}_k + w_k$$

where s_k is the transmitted symbol; furher, we let $a_k = \mathbf{H}\mathbf{x}_k$ be the complex fading variable with $\mathrm{E}\{a_k^* a_k\} = 1$, and $\sigma_n^2 = \mathrm{E}\{w_k^* w_k\} = N_0/E_b$. For the PSK signaling assumed $|s_k|^2 = 1$. Consequently, for the correct symbol decision, the resulting fading measurement is

$$\hat{s}_k^* z_k = a_k + w_k, \qquad (9)$$

and the optimal minimum mean squared one step fading prediction at time k, $a_{k|k-1} = \mathbf{H}\mathbf{x}_{k|k-1}$, is obtained using a Kalman filter. For a sequence of correct symbol decisions, the error or the innovation is an independent zero mean Gaussian process with covariance in each channel as

$$\frac{1}{2}\mathrm{E}\{|z_k - \hat{s}_k \mathbf{H}\,\mathbf{x}_{k|k-1}|^2\} = \frac{1}{2}\left[\mathbf{H}\mathbf{P}_{k|k-1}\mathbf{H}^T + \sigma_n^2\right] \qquad (10)$$

where $\mathbf{P}_{k|k-1}$ is the covariance of the one step fading state vector prediction $\mathbf{x}_{k|k-1}$.

In this joint sequence estimation problem, it is desired to find a sequence of transmitted symbols, $\hat{\mathbf{s}}^N = \{\hat{s}_i\}_{i=1}^N$, and a sequence of corresponding fading predictions for a given measurement sequence, $\mathbf{z}^N = \{z_i\}_{i=1}^N$, that maximizes the likelihood function, or equivalently minimizes the sum of the magnitude square of the normalized innovation process

$$\ell(\hat{\mathbf{s}}^N|\mathbf{z}^L) = \sum_{k=1}^N \frac{|z_k - \hat{s}_k \mathbf{H}\,\mathbf{x}_{k|k-1}|^2}{\frac{1}{2}\left(\mathbf{H}\mathbf{P}_{k|k-1}\mathbf{H}^T + N_0/E_b\right)}. \qquad (11)$$

Since the fading measurements are functions of the assumed symbols, the fading predictions, $\mathbf{H}\mathbf{x}_{k|k-1}$, are also a function of the assumed symbols. Thus, for each assumed symbol sequence, a corresponding sequence of fading predictions is obtained by the Kalman filter. In (11), the dependence of the fading estimates on the assumed symbol sequence is neglected for notational simplicity.

3.1 The Channel Estimator

For M-PSK, the number of possible sequences of length N that must be considered in (11) is M^N. This exponential growth in the number of sequences as mentioned in [14] is clearly not feasible. The idea here is to eliminate unlikely sequences as quickly as possible so they do not have to be considered further. Given that the state and measurement equations of (8) accurately model the fading process, the innovation is a white Gaussian process. Sequences with incorrectly assumed symbols do not fit this model and would tend to result in larger errors. Using the likelihood ratio test, these unlikely sequences can be eliminated systematically with no noticeable performance degradation. Because the likelihood function of a Gaussian vector is monotone, the uniformly most powerful (UMP) test is the chi-square test with $2N$ degrees of freedom:

$$\sum_{k=1}^{N} \frac{|z_k - \hat{s}_k \mathbf{H} \mathbf{x}_{k|k-1}|^2}{\frac{1}{2}\left(\mathbf{H} \mathbf{P}_{k|k-1} \mathbf{H}^T + N_0/E_b\right)} < \chi^2_{2N}(1-\alpha), \qquad (12)$$

where α is the significance level. Thus by specifying the level of the test, we can control the probability of rejecting the correct sequence while eliminating unlikely ones using only the likelihood that was computed in (11).

Following the pilot symbol insertion technique, the most accurate decoding decision can be made at the end of each decoding frame when the known symbol is processed. Forcing a decision at this time will on the average penalize sequences with incorrect fading estimates more than the correct sequence. If each decoding frame is initiated with one fading estimate, then at the end of each decoding frame we would at most have M^{L-1} possible sequences and corresponding fading estimates. Although we would simply decode the symbols associated with the most likely sequence, the choice of channel estimates to initiate the next decoding frame is not obvious. An incorrect choice of fading estimate would result in error-propagation. Here we propose to use a channel estimate based on all valid sequences formed up to the time of the known symbol. In the absence of modulation, it is well known that the optimal channel estimate is the conditional mean estimate given the measurement sequence

$$\hat{\mathbf{x}}_{k|k} = \int \mathbf{x}_{k|k} p(\mathbf{x}_{k|k}|\mathbf{z}^L) \, d\mathbf{x}_{k|k}. \qquad (13)$$

Because the receiver does not know which sequence was transmitted, each remaining sequence, $m \leq M^{L-1}$, is considered equally likely. Thus, in the presence of the unknown data sequence the optimal fading estimate is

$$\hat{\mathbf{x}}_{k|k} = \int \mathbf{x}_{k|k} \sum_{i=1}^{m} p(\mathbf{x}_{k|k}, \hat{\mathbf{s}}_i^L|\mathbf{z}^L) \, d\mathbf{x}_{k|k} = \sum_{i=1}^{m} \mathbf{x}_{k|k}(\hat{\mathbf{s}}_i^L) p(\hat{\mathbf{s}}_i^L|\mathbf{z}^L). \qquad (14)$$

Here $\mathbf{x}_{k|k}(\hat{\mathbf{s}}_i^L)$ denotes the conditional mean estimate of $\mathbf{x}_k$ given an assumed sequence $\hat{\mathbf{s}}_i^L$. After some manipulation, we obtain the following expression for the fading estimate as a function of the likelihood functions already computed

$$\hat{\mathbf{x}}_{k|k} = \sum_{i=1}^{m} \mathbf{x}_{k|k}(\hat{\mathbf{s}}_i^L) \frac{e^{-\frac{1}{2}\ell(\hat{\mathbf{s}}_i^L|\mathbf{z}^L)}}{\sum_{j=1}^{m} e^{-\frac{1}{2}\ell(\hat{\mathbf{s}}_j^L|\mathbf{z}^L)}}. \tag{15}$$

Thus, without additional complexity we can test sequence decisions using the UMP test, and also compute the conditional mean estimate of the channel at the time of the known symbol. Furthermore, the overall complexity of the algorithm is limited to a maximum of M^{L-1} sequences.

3.2 Simulation Results

Performance results presented in this section are obtained by computer simulation using a fourth order Butterworth fading spectrum. The fading process was generated using the discrete time model of (8). Figures 6 and 7 show the performance of the algorithm at various B_dT values. The insertion rates are 1 : 10 for 2-PSK and 1 : 7 for 4-PSK. Also, for 2-PSK in Figure 6, the performance difference between insertion rates of 1:10 and 1:5 can be seen. In this case, higher insertion rate actually performs worse at high SNR. The results shown here are comparable to those of interpolation/estimation algorithms but the performance does not depend critically on the insertion rate. The SNR used in the simulations was adjusted by a factor of $\frac{L-1}{L}$ to compensate for the pilot symbols.

Figures 8 and 9 show the performance of 2-PSK as a function of the significance level for fixed SNR, and the corresponding average number of sequences at the end of each decoding frame. The results shown here present a significant complexity reduction especially at high SNRs compared to the maximum of 512 possible sequences. Additionally, the standard deviation on the number of sequences has been seen to be small for high SNRs. Thus with the proper cutoff threshold, the performance degradation is negligible and the sequence reduction is significant. In an actual implementation, provisions must be made to handle the situation when all the sequences are cutoff. For the simulations we retain at least one most likely path in each decoding block.

4 Conclusion

We have looked at algorithms that process data on a block-by-block basis and use periodically inserted known data to estimate the channel and data sequences. For the ISI channel, an iterative algorithm was introduced and seen to converge quickly and to perform well even for severely distorting channels. The algorithm could be used in situations where fast convergence is necessary and the channel may be fast changing. For the fading channel, the algorithm introduced makes use of both the known data periodically

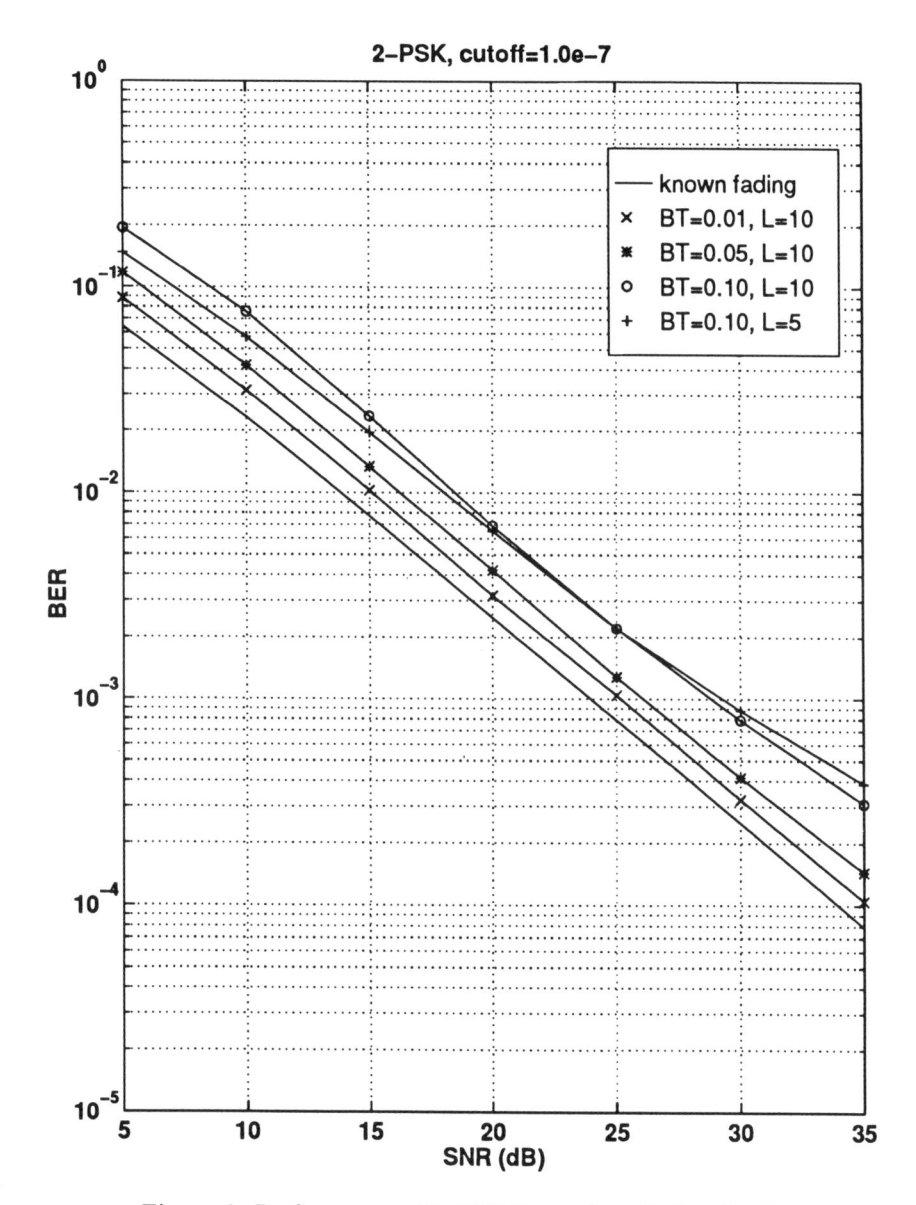

Figure 6: Performance of 2-PSK for various fading levels.

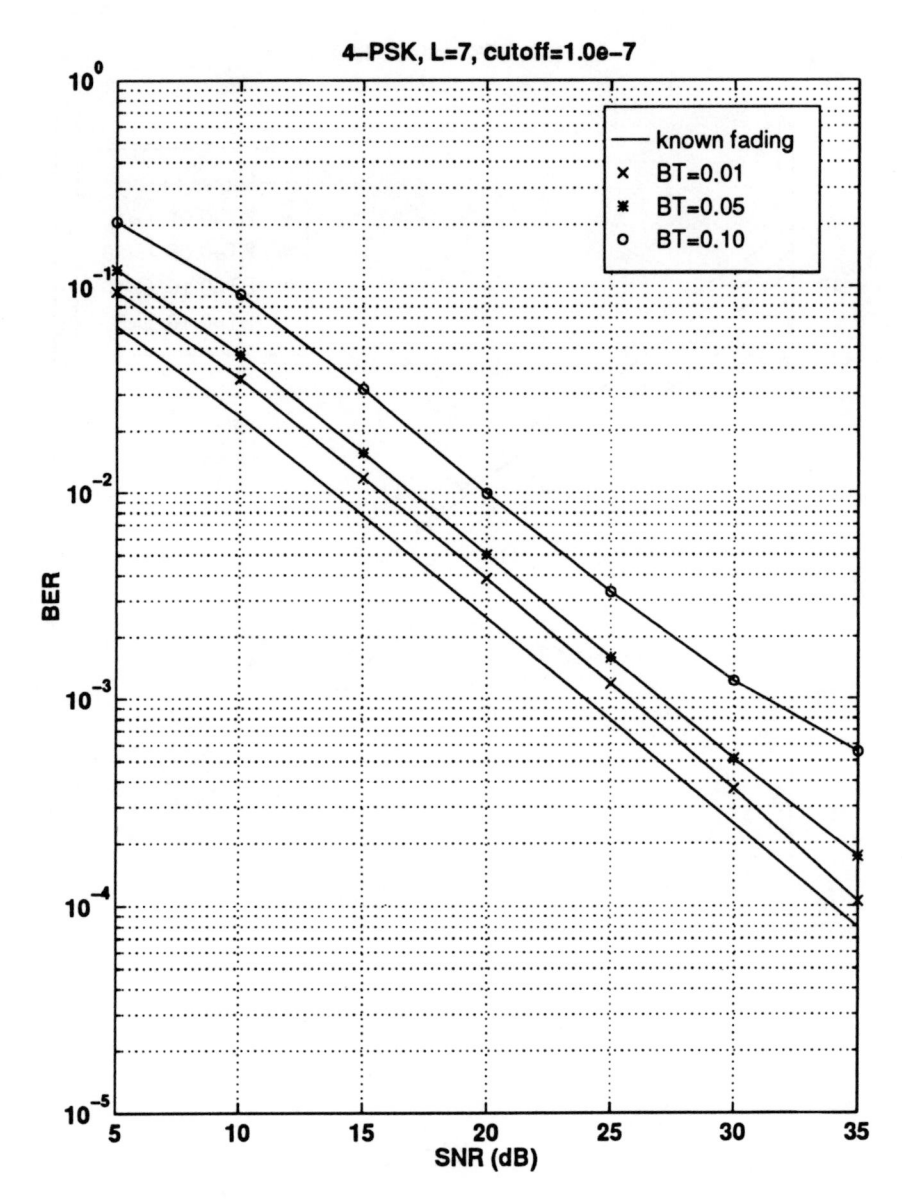

Figure 7: Performance of 4-PSK for various fading levels.

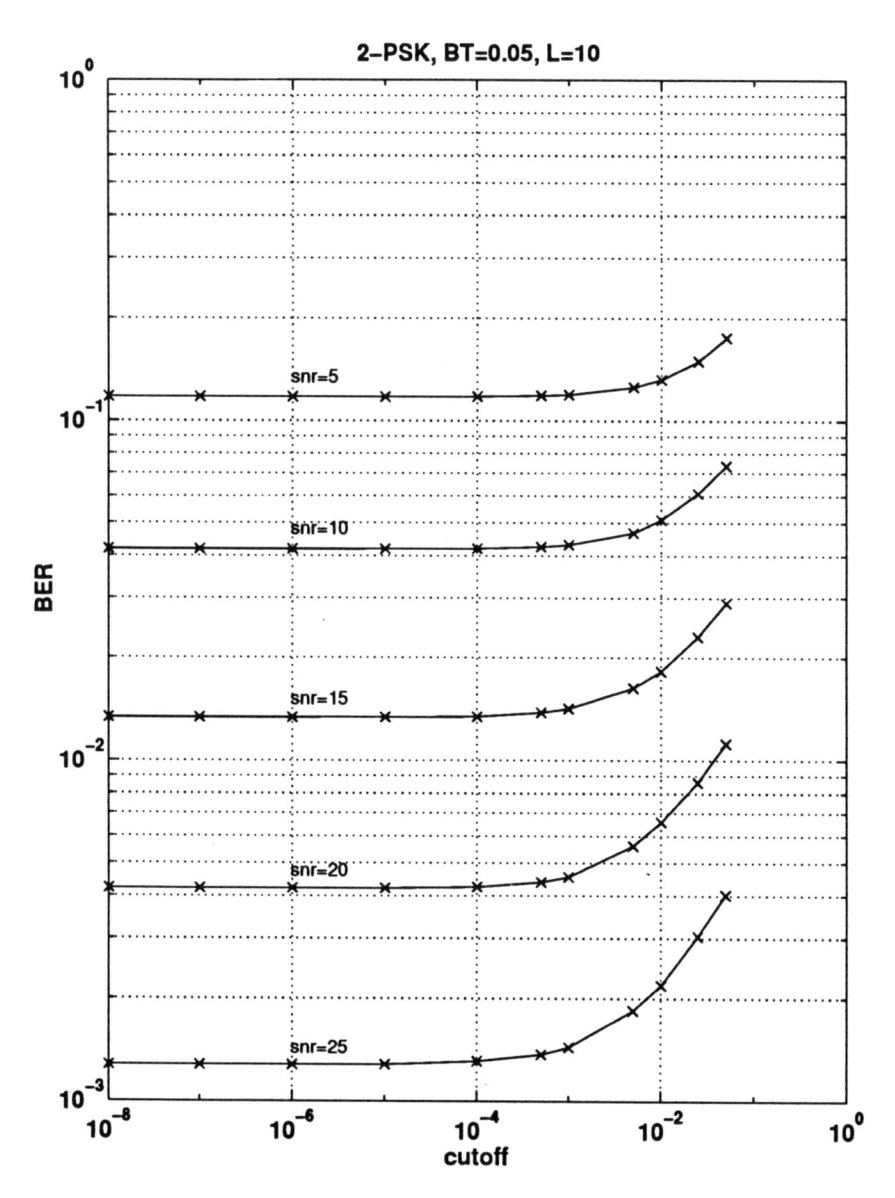

Figure 8: Error probability versus significance level for 2-PSK.

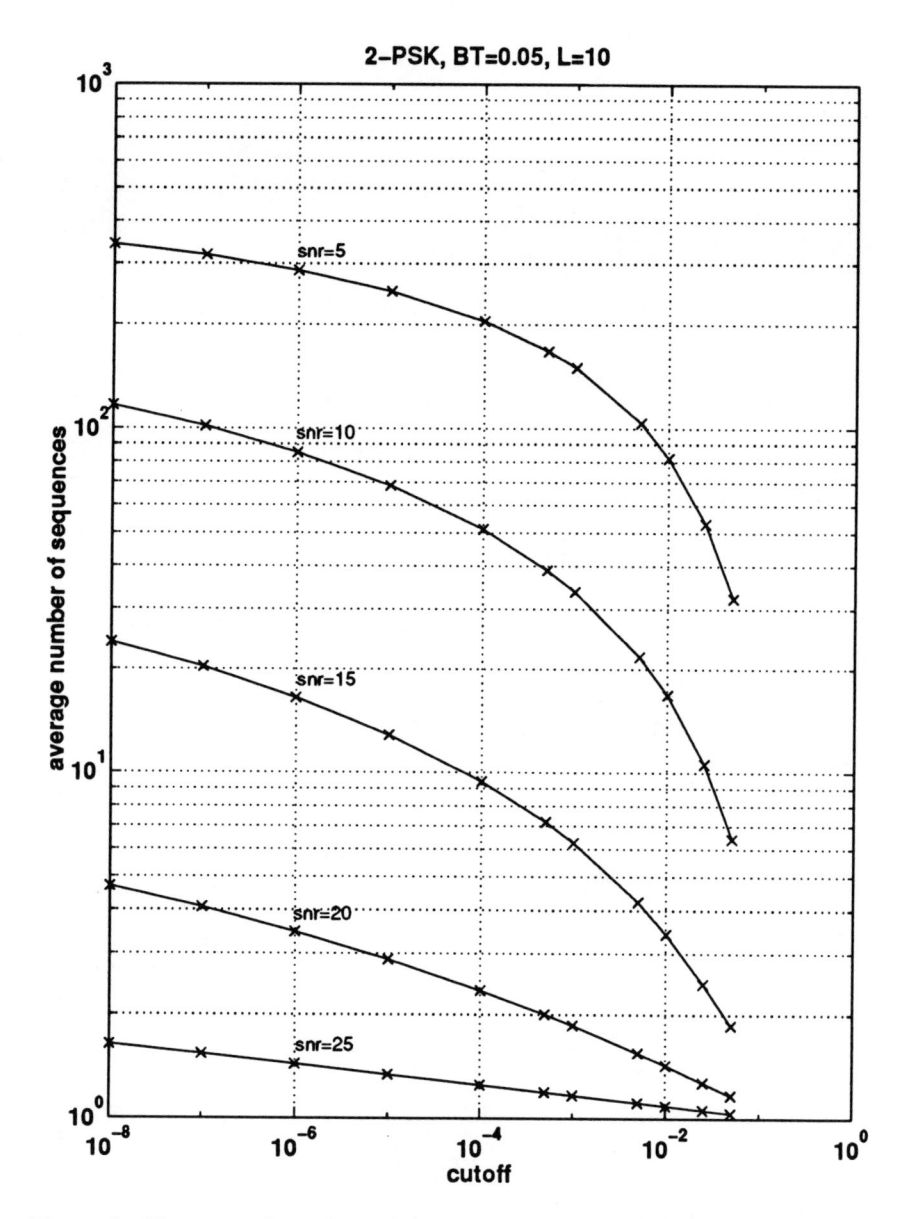

Figure 9: Mean number of surviving sequences versus significance level for 2-PSK.

inserted, as well as the randomly modulated data to improve channel estimation performance. To limit the complexity of the algorithm, a chi-square test is used to screen out unlikely sequences. The algorithm is seen to perform well even for fast fading channels.

References

[1] Y. Sato, "A method of self-recovering equalization for multilevel amplitude modulation," *IEEE Trans. on Communications*, vol. COM-23, pp. 679-682, June 1975.

[2] D. N. Godard, "Self-recovering equalization and carrier tracking in two-dimensional data communication systems," *IEEE Trans. on Communications*, vol. COM-28, pp. 1867-1875, Nov. 1980.

[3] A. Benveniste and M. Goursat, "Blind equalizers," *IEEE Trans. Commun.*, vol. COM-32, pp. 871-883, August 1984.

[4] G. Picchi and G. Prati, "Blind Equalization and Carrier Recovery Using a Stop-and-Go Decision Directed Algorithm," *IEEE Trans. on Communications*, Vol. 35, pp. 877-887, September 1987.

[5] Z. Ding, R. A. Kennedy, B. D. O. Anderson, and C. R. Johnson, Jr., "Ill-convergence of Godard blind equalizers in data communications", *IEEE Trans. on Communications*, vol. 39, pp. 1313-1328, September 1991.

[6] Z. Ding, "Blind equalization based on joint minimum MSE criterion", *IEEE Trans. on Communications*, vol. 42, pp. 648-654, February/March/April 1994.

[7] N. Seshadri, "Joint data and channel estimation using fast blind trellis search techniques," *IEEE Globecom '92* Conf. Rec. 807.1, pp. 1659-1663, Dec. 1990.

[8] N. Seshadri, "Joint data and channel estimation using blind trellis search techniques," *IEEE Trans. on Communications*, vol. 42, pp. 1000-1011, February/March/April 1994.

[9] M. Erkurt and J. G. Proakis, "Joint data detection and channel estimation for rapidly fading channels," *IEEE Globecom '92* Conf. Rec. pp. 910-914, Dec. 1992.

[10] L. E. Baum, T. Petrie, G. Soules and N. Weiss, "A maximization technique occurring in the statistical analysis of probabilistic functions of markov chains," *The Annals of Mathematical Statistics*, vol. 41, pp. 164-171, 1970.

[11] G. K. Kaleh and R. Vallet, "Joint parameter estimation and symbol detection for linear or nonlinear unknown channels," *IEEE Trans. on Communications*, vol. 42, pp. 2406-2413, July 1994.

[12] R. E. Morley Jr and D. L. Snyder, "Maximum Likelihood Sequence Estimation for Randomly Dispersive Channels", *IEEE Trans. on Communications*, vol. 27, No. 6, pp. 833-839, June 1979.

[13] T. Kailath, "A General Likelihood-Ratio Formula for Random Signals in Gaussian Noise,,, *IEEE Trans. on Information Theory*, vol. 15, No. 3, pp. 350-361, May 1969.

[14] R. Haeb and H. Meyr, "A Systematic Approach to Carrier Recovery and Detection of Digitally Phase Modulated Signals on Fading Channels,,, *IEEE Trans. on Communications*, vol. 37, No. 7, pp. 748-754, July 1989.

[15] J. K. Cavers, "An Analysis of Pilot Symbol Assisted Modulation for Rayleigh Fading Channels", *IEEE Trans. on Vehicular Technology*, vol. 40, No. 4, pp. 686-693, Nov. 1991.

[16] G. T. Irvine and P. J. McLane, "Symbol-Aided Plus Decision-Directed Reception for PSK/TCM Modulated on Shadowed Mobile Satellite Fading Channels", *IEEE Journal on Selected Areas in Communications*, vol. 10, no. 8, pp. 1289-1299, Oct. 1992.

[17] A. N. D'Andrea and A. Diglio and U. Mengali, "Symbol-Aided Channel Estimation with Nonselective Rayleigh Fading Channels", *IEEE Trans. on Vehicular Technology*, vol. 44, pp. 41-49, Feb. 1995.

[18] C. N. Georghiades and J. C. Han, "On the Application of the EM Algorithm to Sequence Estimation for Degraded Channels", Proceedings of the Allerton Conference, Urbana, Illinois, September 1994.

[19] J. Salz, "On the start-up problem in digital echo cancelers," *The Bell System Technical Journal*, vol. 62, pp. 1353-1364, 1983.

[20] G. D. Forney, "Maximum-likelihood sequence estimation of digital sequences in the presence of intersymbol interference," *IEEE Trans. on Information Theory*, vol. IT-18, pp. 363-378, May 1972.

[21] J. G. Proakis, *Digital Communications*, McGraw-Hill, 1989.

[22] H. L. Van Trees, *Detection, Estimation, and Modulation Theory, Part I*, New York: Wiley, 1968.

[23] P. A. Frost and T. Kailath, "An Innovations Approach to Least-Squares Estimation – Part III: Nonlinear Estimation in White Gaussian Noise", *IEEE Trans. on Automatic Control*, vol. 16, no. 3, pp. 217-226, June 1971.

Timing Correction by Means of Digital Interpolation [1]

M. Moeneclaey [2] and K. Bucket

Communication Engineering Laboratory, University of Ghent
Sint-Pietersnieuwstraat 41, B-9000 Gent, BELGIUM.

Phone : + 32-9-264 34 13
Fax : + 32-9-264 42 95
E-mail : Marc.Moeneclaey@lci.rug.ac.be

Abstract

This contribution deals with some performance aspects of fully digitally implemented receivers, operating on either a narrowband M-PSK signal or a bandlimited direct-sequence spread-spectrum (DS/SS) M-PSK signal. The considered digital receivers operate on (quantized) samples of the received noisy signal, taken by a fixed clock which is not synchronized to the transmitter clock. The synchronized samples needed for the detection of the information sequence and for the timing synchronization algorithms are computed by interpolating between the available non-synchronized samples. Because of finite memory, interpolation is non-ideal; hence, some amount of distortion is introduced, which affects the performance of the receiver. By means of theoretical analysis, we demonstrate that even simple interpolators, operating at only a few samples per symbol (narrowband M-PSK) or per chip (spread-spectrum M-PSK), yield small BER degradations. Furthermore, we investigate the tracking performance of some specific timing synchronizers. We show that non-ideal interpolation gives rise to a loop noise spectrum containing spectral lines, that mainly occur near f=0 when the sampling frequency is very close to an integer multiple of the symbol rate (narrowband M-PSK) or the chip rate (spread-spectrum M-PSK). Unless a sufficiently small loop bandwidth is chosen, the contribution of these lines could dominate the tracking variance, which then becomes much larger than for synchronized sampling.

---

[1] This work has been funded by the European Human Capital and Mobility Project No. CHRX-CT93-0405.
[2] This author is supported by the Belgian National Fund for Scientific Research (NFWO).

1. Introduction

A fully digitally implemented receiver does not operate on the continuous-time received noisy signal, but rather on a sampled version of it, where the samples are taken by a fixed clock which is not synchronized to the transmitter clock. If the sampling frequency $1/T_s$ is larger than twice the bandwidth of the signal, these samples contain the same information as the continuous-time signal. From these non-synchronized signal samples, the digital receiver must derive the synchronized signal samples needed for data detection and synchronization.

Theoretically speaking, these signal samples can be perfectly reconstructed by interpolating between the available samples [1]. Indeed, the signal value x(t) at any instant t is obtained from the signal samples $x(kT_s)$ by using the following interpolation formula:

$$x(t) = \sum_{k=-\infty}^{+\infty} x(kT_s) p(t - kT_s) \qquad (1)$$

where the ideal interpolation pulse p(t) equals $\sin(\pi t/T_s)/(\pi t/T_s)$.

The signal value x(t) in (1) is obtained from an infinite number of signal samples $x(kT_s)$. In a practical implementation however, only a finite number of signal samples can be used for interpolation. Hence, rather than providing the exact signal value, practical interpolation introduces some amount of distortion, which will affect the receiver performance.

In this contribution, we investigate the effect of this distortion on the BER performance and the timing synchronizer performance, in both cases of a narrowband communication system and a bandlimited direct-sequence spread-spectrum system.

2. System Description

The complex envelope r(t) of the received noisy M-PSK signal is given by

$$r(t) = \sum_k a(k) h(t - kT - \tau) e^{j\theta} + n(t) \qquad (2)$$

where $\{a(k)\}$ is a sequence of independent equiprobable M-PSK symbols taking values from the set $\{ e^{j2\pi k/M} \mid k=0,...,M-1 \}$, T denotes the symbol interval, τ and θ are the unknown time delay and carrier phase, and n(t) is complex-valued white Gaussian noise, whose real and imaginary part are statistically independent and have power spectral density of $N_o/(2E_b)$. The baseband pulse h(t) has unit-energy, and is selected such that h(t) and h*(t-mT) are orthogonal; consequently, the bandwidth of h(t) is at least $1/(2T)$.

In the case of narrowband M-PSK, h(t) is real-valued and given by

$$h(t) = h_{nb}(t) \qquad \text{narrowband M-PSK} \qquad (3)$$

where the one-sided bandwidth of $h_{nb}(t)$ does not exceed the symbol rate $1/T$; in the sequel, it will be assumed that $h_{nb}(t)$ is a unit-energy square-root cosine rolloff pulse with bandwidth $(1+\beta)/(2T)$, where β denotes the rolloff factor.

In the case of bandlimited DS/SS M-PSK with N chips per symbol, $h(t)$ is complex-valued and given by

$$h(t) = \frac{1}{\sqrt{N}} \sum_{n=0}^{N-1} \alpha(n) h_{ss}(t - nT_c) \qquad \text{bandlimited DS/SS M-PSK} \qquad (4)$$

where the chip interval T_c equals T/N and { $\alpha(n)$ | $n=0,...,N-1$ } is a random sequence of N complex-valued statistically independent chips. The unit-energy real-valued chip pulse $h_{ss}(t)$ is orthogonal to $h_{ss}(t-nT_c)$ for $n \neq 0$, which implies that the one-sided bandwidth of $h_{ss}(t)$ is at least $1/(2T_c)$. For bandlimited DS/SS M-PSK, the one-sided bandwidth of $h_{ss}(t)$ does not exceed $1/T_c$; in the sequel it will be assumed that $h_{ss}(t)$ is a square-root cosine rolloff pulse with bandwidth $(1+\beta)/(2T_c)$, where β denotes the rolloff factor.

A fully digital implementation approximating the ML decision as well as containing the timing synchronization is shown in Fig. 1. The received complex envelope $r(t)$ enters an analog anti-aliasing filter, which we assume not to distort the signal component of $x(t)$, and whose output is sampled at a rate $1/T_s$. These samples enter a digital filter, matched to the pulse $h_{nb}(t)$ (narrowband communication) or $h_{ss}(t)$ (spread-spectrum communication). The matched filter output samples are rotated over an angle equal to the carrier phase estimate; in this contribution we assume perfect carrier phase estimation. The resulting non-synchronized samples $x(kT_s)$ then enter an non-ideal interpolator which provides estimates of the synchronized rotated matched filter output samples.

In the case of a narrowband communication system these estimates are used further for decision and symbol synchronization. However, in the case of spread-spectrum communication, the N interpolator output samples corresponding to the same M-PSK symbol are first correlated with the chip sequence, and it is the correlator output which is then further needed for decision and chip synchronization.

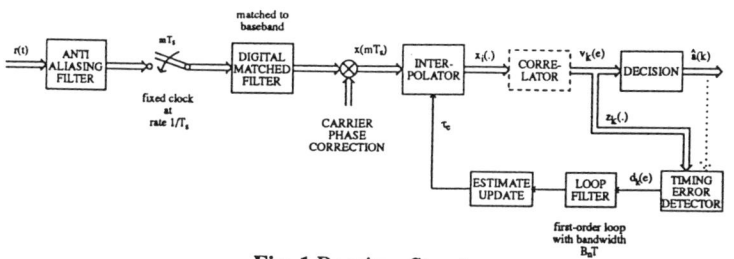

Fig. 1 Receiver Structure

Suppose we need an estimate $x_i(t)$ of the rotated matched filter output $x(t)$ at some instant t. We introduce the quantities n (integer) and u (0≤u<1) defined by $t = nT_s+uT_s$, which indicates that nT_s is the non-synchronized sampling instant occurring immediately before or at the instant t. Note that n and u are related to the instant t by $n = \text{int}(t/T_s)$ and $u = \text{rem}(t/T_s)$, where $\text{rem}(x) = x-\text{int}(x)$ and $\text{int}(x)$ is the largest integer not exceeding x. The interpolator is a linear filter [1],[2], which produces $x_i(t) = x_i(nT_s+uT_s)$ according to

$$x_i(nT_s+uT_s) = \sum_{l=-N_1}^{N_2} h_i(l;u) x(nT_s-lT_s) \quad (5)$$

where $\{ h_i(l;u) \mid l=-N_1,-N_1+1,...,N_2 \}$ represent the N_1+N_2+1 tap gains of the interpolator. Note that the interpolator is a time-variant filter whose impulse response $h_i(l;u)$ is a function of the fraction u only. In the following, we will investigate zeroth-order, first-order and second-order interpolators. The corresponding tap gains are given in Table 1, where U(.) denotes the unit-step function, and γ is a design parameter of the second-order interpolator, which can be selected to minimize the BER degradation [3]. Unless u=0, every interpolator in general yields $x_i(t) \ne x(t)$. Hence, some amount of distortion is introduced at the input of the decision device as well as at the input of the timing error detector, which will affect its performance in terms of bit error rate and tracking error variance, respectively.

Table 1

	Zeroth-order	First-order	Second-order
N_1	1	1	2
N_2	0	0	1
$h_i(-2;u)$	-	-	$-\gamma u(1-u)$
$h_i(-1;u)$	$U(u-1/2)$	u	$u(1+\gamma(1-u))$
$h_i(0;u)$	$U(-u+1/2)$	1-u	$(1-u)(1+\gamma u)$
$h_i(1;u)$	-	-	$-\gamma u(1-u)$

3. BER Performance

In this section we determine the degradation of the BER, caused by non-ideal interpolation; ideal timing is assumed. This degradation L is defined as the reduction of the signal-to-noise ratio at the input ot the decision device, as compared to the case of synchronized sampling. We will restrict our attention to BPSK modulation (i.e. a(k)=±1); the result for QPSK is the same as for BPSK, whereas the result for M-PSK (M>2) can be derived in a similar way.

3.1 Narrowband Communication

In the case of narrowband communication, the detection of the bit a(0) is based on the polarity of the signal v(0) at the input of the decision device. This signal v(0) can be written as

$$v(0) = a(0)S(u_0) + ISI(u_0) + \sqrt{\frac{N_0}{2E_b}}\sqrt{V(u_0)}w_0 \qquad (6)$$

where w_o is a zero-mean unit-variance Gaussian random variable. The first term in (6) is the useful component, the second is zero-mean intersymbol interference and the third is additive noise. Note that all three terms depend on the quantity u_o, which is uniformly distributed over (0,1). In the case of synchronized sampling, (6) is valid with $S(u_o)=V(u_o)=1$ and $ISI(u_o)=0$.

Defining

$$f(u) = \bigl(S(u) + ISI(u)\bigr) / \sqrt{V(u)} \qquad (7)$$

the BER degradation L is given by [4]

$$L = -10 \log \left(\frac{(E[f(u)])^2}{1 + \frac{2E_b}{N_0} Var[f(u)]} \right) \quad [dB] \qquad (8)$$

where E[.] and Var[.] involve averaging over the data symbols that contribute to the ISI term and over the random variable u. The former is done analytically, the latter numerically.

3.2 Spread-Spectrum Communication

In the case of spread-spectrum communication, the signal v(0) at the input of the decision device can be decomposed in a similar way as (6):

$$v(0) = a(0)S(\underline{u},\underline{\alpha}) + ISI(\underline{u},\underline{\alpha}) + \sqrt{\frac{N_0}{2E_b}}\sqrt{V(\underline{u},\underline{\alpha})}w_o \qquad (9)$$

In the above, $\underline{\alpha}$ denotes the random chip sequence, and $\underline{u}$ is the vector of quantities u_k, k=0, ..., N-1, where the index k refers to the k-th interpolator output value that contributes to v(0). Note that the quantities u_k are related by u_{k+1}=rem(u_k+r), with r=rem(T_c/T_s). For moderate and large values of the chip sequence length N, it can be verified that the random nature of $\underline{\alpha}$ yields a negligible fluctuation of $S(\underline{u},\underline{\alpha})$, $ISI(\underline{u},\underline{\alpha})$ and $V(\underline{u},\underline{\alpha})$ with respect to their mean over $\underline{\alpha}$. Hence, we approximate $S(\underline{u},\underline{\alpha})$ and $V(\underline{u},\underline{\alpha})$ by their mean values $S(\underline{u})=E_{\underline{\alpha}}[S(\underline{u},\underline{\alpha})]$ and $V(\underline{u})=E_{\underline{\alpha}}[V(\underline{u},\underline{\alpha})]$, and ignore $ISI(\underline{u},\underline{\alpha})$. The resulting BER degradation L is given by [5]

$$L = -10 \log \left(\frac{(E[f(\underline{u})])^2}{1 + \frac{2E_b}{N_o} Var[f(\underline{u})]} \right) \quad [dB] \tag{10}$$

where $f(\underline{u}) = S(\underline{u})/\sqrt{V(\underline{u})}$.

3.3 BER Degradation Results

Fig. 2 shows the BER degradation in dB as a function of the number of samples per symbol (narrowband communication) or per chip (spread-spectrum communication) at a BER of 10^{-6} and for a rolloff factor β of 50 %. Note that the BER degradation for spread-spectrum communication is smaller than for narrowband communication because in the former case the intersymbol interference can be ignored. In both cases, the degradations decrease with increasing sampling rate and increasing order of interpolation, because then interpolation becomes more accurate.

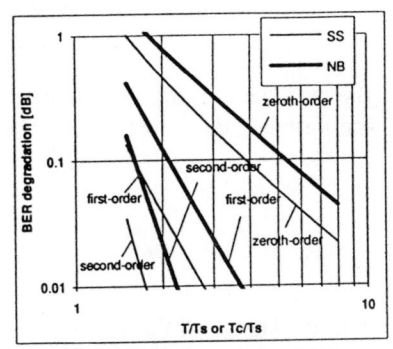

Fig. 2 BER degradation [dB]

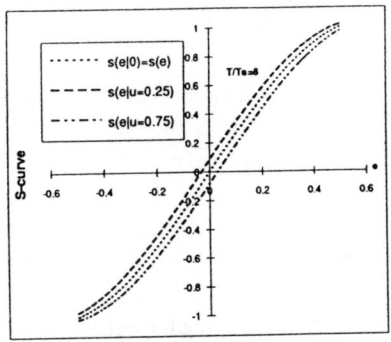

Fig. 3. S-curve for zeroth-order interpolation (NB)

4. Synchronizer Performance

4.1 Narrowband Communication

In the case of a narrowband communication system, we consider a decision-directed feedback symbol synchronizer operating at two synchronized samples per symbol : it needs (an estimate of) the rotated matched filter output samples at the decision instants (from which the receiver's decisions are obtained) and halfway between decision instants; these two synchronized samples per symbol are obtained by interpolating between the available (at rate $1/T_s$) non-synchronized samples. The timing error detector output corresponding to the k-th symbol and a timing error $e_k = (\tau_e - \tau)/T$ is given by

$$d_k(e_k) = \text{Re}\left[a^*(k)\left(x_i(kT + T/2 + \tau_e) - x_i(kT - T/2 + \tau_e)\right)\right] \tag{11}$$

This timing error detector corresponds to the DDEL ($\lambda=1/2$) algorithm, which has been considered in [6] for synchronized sampling (in which case the phase of the sampling clock is controlled by the symbol synchronizer).

As the receiver's decisions are assumed to be correct, the timing error detector output $d_k(e_k)$ depends on e_k only through the instants halfway between decision instants. We introduce the following notation

$$kT - T/2 + \tau_e = (n_k + u_k)T_s \tag{12}$$

with $0 \leq u_k < 1$ and n_k integer. Substitution of (12) in (11) shows that $d_k(e_k)$ depends on u_k and u_{k+1}. For a fixed value of e_k, i.e. $e_k = e$ for all k, u_k and u_{k+m} are related by

$$u_{k+m} = rem(u_k + mr) \tag{13}$$

where $r = rem(T/T_s)$. Hence, when the value of a single component of the sequence $\{u_k\}$ is known, the values of all other components is determined solely by r. When T/T_s is integer, then r=0 and all u_k take on the same value. In practice, however, a small frequency offset between receiver and transmitter yields r≠0, so that the values of u_k change slowly with k.

We will show results only for zeroth-order interpolation; the rolloff factor β is 50 % and the sampling rate T/T_s equals 8. The considered interpolator yields a BER degradation of about 0.05 dB at BER=10^{-6} when perfect symbol synchronization is assumed. The presented results apply only for BPSK. However, for a given E_s/N_o, the timing error variance for BPSK and M-PSK (M≠2) compare as follows : the self-noise contribution of M-PSK is half the one of BPSK, but the additive noise contribution remains the same [6].

- *The s-curve*

For a fixed timing error e, the s-curve s(e) is defined as $E[d_k(e)]$, where E[.] denotes averaging over data symbols, noise and u_k. This averaging is performed in two steps. First we compute the conditional s-curve $s(e|u_k)$ by taking a fixed value of u_k and averaging over noise and data symbols. Fig. 3 shows $s(e|u_k)$ for r=0 and various values of u. Note that the conditional average timing error detector output $s(e|u_k)$ is in general nonzero for e=0. Secondly, $s(e|u_k)$ is averaged over u_k, assuming a uniform distribution in the interval (0,1). This averaging is performed numerically, and the resulting s-curve s(e) is also shown in Fig. 3. Also note that s(0)=0, so that e=0 is a stable tracking point. In the case of synchronized sampling (or, equivalently, ideal interpolation) it can be verified that the corresponding s-curve equals s(e|0), i.e. the average timing error detector output corresponding to u=0.

- *The loop noise spectrum*

The loop noise at the stable tracking point e=0 is given by the zero-mean quantity $d_k(0)$, which can be decomposed into two uncorrelated terms :

$$d_k(0) = d_{c,k}(0) + s(0|u_k) \qquad (14)$$

The first term depends on the data symbols and noise, whereas the second does not. The first term is very close to the loop noise that occurs in the case of synchronized sampling, and gives rise to a continuous loop noise spectrum. The second term is caused only by non-ideal interpolation, and can be written as a Fourier series expansion [7]:

$$s(0|u_k) = \sum_{m=-\infty}^{+\infty} c_m(r) e^{j2\pi m u_k} \qquad , 0 \le u_k < 1 \qquad (15)$$

The spectrum $S_s(fT)$ of the sequence $\{s(0|u_k)\}$ is given by

$$S_s(fT) = \sum_{m=-\infty}^{+\infty} |c_m|^2 \frac{1}{T} \sum_{n=-\infty}^{+\infty} \delta(f - \frac{mr}{T} - \frac{n}{T}) \qquad (16)$$

Hence, this shows that the spectrum $S_s(fT)$ of the sequence $\{s(0|u_k)\}$ consists of spectral lines, that occur at frequencies f_n, determined by

$$f_n T = rem(nr + 1/2) - 1/2 \qquad (17)$$

and their magnitude is denoted $D(f_n T)$ (which equals $|c_n|^2$). When T/T_s is integer, then r=0 so that $s(0|u_k)$ does not depend on k; consequently, the spectrum $S_s(fT)$ consists of a single spectral line at f=0. In practice, there is a small frequency offset between receiver and transmitter, which makes $s(0|u_k)$ slowly varying with k; the corresponding spectrum $S_s(fT)$ contains spectral lines which occur mainly near f=0, as can be seen from Fig. 4.

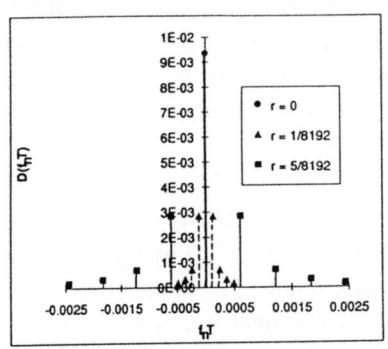

Fig. 4 Dominant lines in loop noise for zeroth-order interpolation (NB)

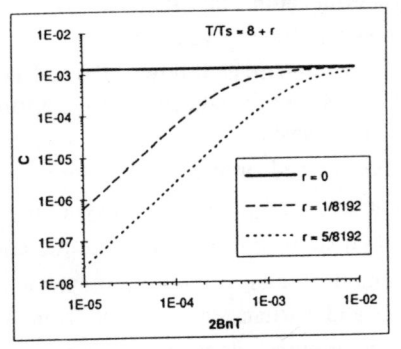

Fig. 5 Term C as a function of B_n for zeroth-order interpolation (NB)

- *The timing error variance*

The first term $d_{c,k}(0)$ of (14) gives rise to a timing error variance that is nearly the same as for synchronized sampling. The second term $s(0|u_k)$ yields an additional contribution C to the timing error variance.

Assuming a first-order loop with a one-sided noise bandwitdth B_n, Fig. 5 shows the behavior of the term C as a function of the loop bandwidth in the case of a zeroth-order interpolator. For r=0, $S_s(fT)$ consists of a single spectral line at f=0 (see (17)), whose effect on the timing error variance cannot be reduced by decreasing the synchronizer bandwidth. A small frequency offset between receiver and transmitter yields a small nonzero value of r. In this case, $S_s(fT)$ consists of spectral lines that occur mainly near f=0 (see (17)). Hence, the effect of these lines on the timing error variance can be reduced only by taking a sufficiently small loop bandwidth. The value r=1/8192 corresponds to a relative clock frequency offset between receiver and transmitter equal to $1.5 \; 10^{-5}$ for zeroth-order interpolation; for r=5/8192, the frequency offset is five times as large. The total timing error variance is shown in Fig. 6, and compared with the timing error variance for synchronized sampling (or, equivalently, ideal interpolation). For practical values of B_nT, the timing error variance for non-synchronized sampling is larger than for synchronized sampling and mainly because of the spectral line contribution C.

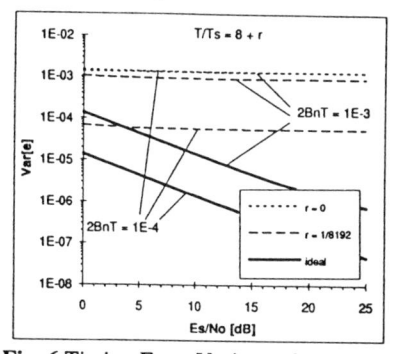

Fig. 6 Timing Error Variance for zeroth-order interpolation (NB)

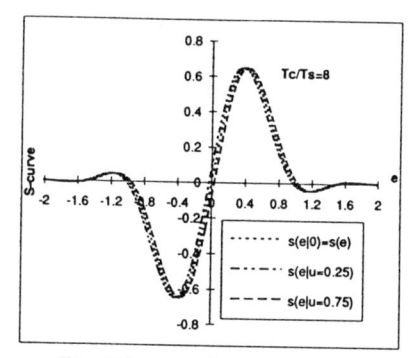

Fig. 7 S-curve for zeroth-order interpolation (SS)

4.2 Spread-Spectrum Communication

In the case of spread-spectrum communication, we consider a nondecision-aided early-late chip synchronizer, requiring (an estimate of) the correlator output samples at the decision instants and instants that are advanced/delayed by half a chip with respect to the decision instants. The chip synchronizer forms a signal $d_k(e)$ which gives an indication about the estimation error $\tau_e - \tau$ during the k-th symbol, and updates the timing estimate e accordingly. We consider the following chip synchronizer, where the timing error detector output corresponding to the k-th symbol and a timing error $e = (\tau_e - \tau)/T$ is defined as

$$d_k(e) = \text{Re}\left[z_k^*(\tau_e)\left(z_k(\tau_e + T_c/2) - z_k(\tau_e - T_c/2)\right)\right] \quad (18)$$

The timing error detector presented above corresponds to the NDANCEL

($\lambda=1/2$) algorithm, which has been considered in [8] for synchronized sampling. It is a noncoherent algorithm, operating independently of the carrier phase estimate.

We introduce the following notation
$$kT + (m-1)T_c/2 + \tau_e = \left(n(2kN,m) + u(2kN,m)\right)T_s \tag{19}$$
with $0 \le u(2kN,m) < 1$ and $n(2kN,m)$ integer. Substitution of (19) in (18) shows that $d_k(e)$ depends on $\underline{u}_k = (u(2kN,0),...,u(2kN,2N))$. For a fixed value of e, $u(2kN,m)$ and $u(2kN,m+1)$ are related by
$$u(2kN,m+1) = rem(u(2kN,m) + \frac{T_c/2}{T_s}) \tag{20}$$

We will present results for a rolloff factor β of 50 % and a sampling rate of 8 samples per chip for a zeroth-interpolator, which yields a BER degradation of about 0.03 dB at BER=10^{-6} when perfect chip synchronization is assumed [4].

- *The s-curve*

In the case of small interpolation errors, it can be verified that the conditional s-curve at e=0, $s(0|\underline{u}_k)$, can be expanded as [9]
$$s(0|\underline{u}_k) = E_s \sum_{n=-\infty}^{+\infty} c_n(r) F_n(N,r) e^{j\pi n(N-1)r} e^{j2\pi n u(2kN,0)} \tag{21}$$
where
$$F_n(N,r) = \frac{1}{N}\frac{\sin(\pi nNr)}{\sin(\pi nr)} \tag{22}$$
and r= rem(T_c/T_s). The value s(0) of the s-curve for zero timing error is obtained by averaging (21) over a uniformly distributed $u(2kN,0)$, yielding $s(0)=c_0(r)$. It can be derived that $c_0(r)=0$; this implies $s(0)=0$, so that e=0 is a stable tracking point.

From (21) we observe that $s(0|\underline{u}_k)=0$ when $F_n(N,r)=0$ for $n \ne 0$, i.e. when Nr is a nonzero integer; in this case, the components of $\underline{u}_k$ are evenly spread over the interval (0,1). It turns out that a nonzero integer value of Nr yields a conditional s-curve $s(e|\underline{u}_k)$ which does not depend on the specific value of $\underline{u}_k$; hence $s(e|\underline{u}_k)=s(e)$, with $s(0)=0$.

When $|F_n(N,r)| \ll 1$ for $n \ne 0$, $s(0|\underline{u}_k)$ (and also $s(e|\underline{u}_k)$) depend very little on $\underline{u}_k$: $s(0|\underline{u}_k) \approx 0$; the condition $|F_n(N,r)| \ll 1$ is satisfied for Nr$\gg$1, in which case the components of $\underline{u}_k$ are nearly evenly spread over (0,1).

The dependence of $s(0|\underline{u}_k)$ on $\underline{u}_k$ (and also of $s(e|\underline{u}_k)$ on $\underline{u}_k$) is largest when $F_n(N,r) \approx 1$, i.e. when Nr is very close to zero (or to N); this implies that r is very close to zero (or to 1), meaning that the number of chips per symbol is very close to an integer. Assuming r=0, so that all components of $\underline{u}_k$ take the same value u, Fig. 7 shows the conditional s-curve $s(e|u)=s(e|\underline{u}_k)$ for various values of u, along with the average s-curve s(e), in the case of a zeroth-order interpolator. The conditional s-curve s(e|0) equals the s-curve for synchronized sampling. Note that s(0|u)≠0 for u≠0, and s(0)=0.

- The loop noise spectrum

As for narrowband communication, the loop noise $d_k(0)$ can be decomposed as
$$d_k(0) = d_{c,k}(0) + s(0|\underline{u}_k) \quad (23)$$
where the first term $d_{c,k}(0)$ is nearly the same as for synchronized sampling, and $s(0|\underline{u}_k)$ is the contribution caused by non-ideal interpolation.

Making use of (21), it follows that the spectrum $S_s(fT)$ of the sequence $\{s(0|\underline{u}_k)\}$ consists of spectral lines, that occur at frequencies f_n, determined by
$$f_n T = rem(nNr + 1/2) - 1/2 \quad (24)$$
Notice from (24) that the position of the spectral lines depends not only on r, but also on the length N of the chip sequence.

-The tracking error variance

The first term of $d_{c,k}(0)$ of (23) gives rise to a timing error variance that is nearly the same as for synchronized sampling. The second term $s(0|\underline{u}_k)$ yields an additional contribution C to the timing error variance. From (21) and (24), it can be derived that

$$C = \frac{1}{\left(s'(0)^2\right)} \sum_{n=-\infty}^{+\infty} |c_n(r)|^2 |F_n(N,r)|^2 |H(e^{j2\pi nNr})|^2 \quad (25)$$

When Nr is a nonzero integer, we obtain $F_n(N,r)=0$: there is no line spectrum present in the loop noise spectrum S(fT), which results into C=0. The tracking performance is very close to the one obtained when performing synchronized sampling. A similar performance is obtained for Nr>>1, because then $|F_n(N,r)|<<1$ so that the contribution from the spectral lines can be ignored.

For Nr<<1, Fig. 8 shows the behavior of the term C as a function of the normalized loop bandwidth $B_n T$, assuming a first-order loop. Similar remarks can be made as in the case of narrowband communication. However, note that for given $B_n T$, the effect of the line contribution can also be reduced now by using a longer chip sequence which moves the spectral lines to higher frequencies (see (24)).

The total tracking error variance is shown in Fig. 9, and compared with the tracking error variance for synchronized sampling (or, equivalently, ideal interpolation). Note that, in the case of synchronized sampling, the coefficient C is equal to zero.

For non-synchronized sampling with Nr<<1, the contribution C, in general, cannot be ignored, and may even dominate the tracking error variance. For r=0, C does not depend on $B_n T$, so that for moderate and large E_s/N_o the tracking error variance is essentially independent of $B_n T$, and much larger than for synchronized sampling. For r≠0, and small enough $B_n T$, C is proportional to $(B_n T)^2$. As the timing error variance for synchronized sampling is essentially proportional to B_nT, the tracking error variance for non-synchronized sampling will approach that of synchronized sampling when B_nT is sufficiently small and r≠0. However, for practical values of B_nT and Nr<<1, the tracking error variance for non-synchronized sampling is mainly dominated by the spectral line contribution C.

References

[1] F.M. Gardner, "Interpolation in Digital Modems. Part I : Fundamentals", IEEE Trans. Comm, vol. 41, pp. 501-507, March 1993.

[2] F.M. Gardner, "Timing adjustment via interpolation in digital demodulators', Final report (Part I) to ESA contract No. 8022/88/NL/DG, June 1990.

[3] K. Bucket and M. Moeneclaey, "Optimization of a second-order interpolator for bandlimited DS/SS communication", Elect. Letters, No. 28, pp. 1029-1031, 1992.

[4] K. Bucket and M. Moeneclaey, "The effect of interpolation on the BER performance of narrowband BPSK and (O)QPSK on Rician fading channels", IEEE Trans. Comm., vol. 42, No. 11, November 1994.

[5] K. Bucket and M. Moeneclaey, "Bit error rate degradation caused by non-ideal interpolation of bandlimited direct-sequence spread-spectrum signals", AEU, Vol. 48, No. 5, pp. 231-236, 1994.

[6] T. Jesupret, M. Moeneclaey and G. Ascheid, "Digital demodulator synchronization", Final Report to ESTEC contract No. 8437/89/NL/RE, June 1991.

[7] K. Bucket and M. Moeneclaey, "The effect of non-ideal interpolation on symbol synchronizer performance", ETT, Vol. 6, No. 6, pp. 627-632, November-December 1995.

[8] M. Moeneclaey and G. Dejonghe, "Tracking performance of digital chip synchronization algorithms for bandlimited DS/SS communications", Elec. Letters, Vol. 27, No. 13, pp.1147-1148, June 1991.

[9] K. Bucket and M. Moeneclaey, "On the influence of non-ideal interpolation on the chip synchronization performance of bandlimited direct-sequence spread-spectrum signals", Proc. IEEE ICC '95, Seattle, pp. 1658-1661, June 1995.

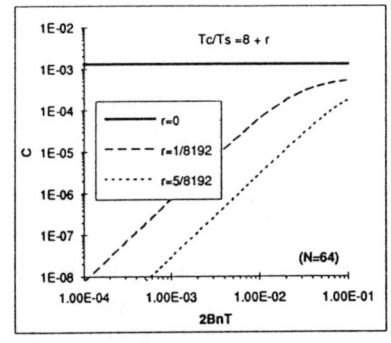

Fig. 8 Term C as a function of B_n for zeroth-order interpolation (SS)

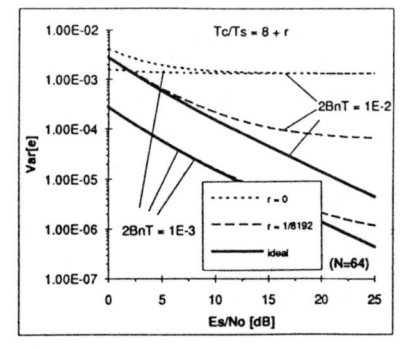

Fig. 9 Tracking Error Variance for zeroth-order interpolation(SS)

Digital Demodulator Architectures for BandPass Sampling Receivers

A. M. Guidi and L. P. Sabel

Institute for Telecommunications Research,
SPRI Building, University of South Australia, The Levels, SA 5095, Australia

Abstract

Implementations of digital demodulator architectures based on bandpass sampling schemes are presented. The architectures proposed achieve robust front–end digital receiver designs by eliminating typical problems associated with conventional analog implementations. Comparisons are made between two bandpass sampled receiver architectures and practical implementation results are presented.

1 Introduction

This paper presents performance results for front–end digital architectures suitable for implementing BandPass Sampled (BPS) multirate receivers. A comparison is made between conventional analog front–end demodulator architectures and BPS digital architectures. The introduction of a digital IF processing stage can improve modem performance since it eliminates typical problems associated with analog implementations including I/Q amplitude and phase imbalances and carrier feedthrough. Higher level constellations such as 8PSK and 16QAM are very sensitive to such imbalances while BPSK and QPSK are more tolerant [1]. A companion paper [2] presents the results of a recent study into the effects of quadrature phase, amplitude imbalance and carrier feedthrough on the Bit Error Rate (BER) of a digital demodulator. The push towards highly spectrally efficient data links using multilevel constellations further emphasises this factor and it stresses the importance of robust receiver architectures such as those demonstrated in this paper. Recent advances in DSP technology has enabled the development of BPS communication receivers for relatively high data rates. The development of commercially available Application Specific Integrated Circuits (ASICs) and Field Programmable Gate Array (FPGA) architectures along with improvements in A/D technology now allow the integration of fully digital BPS receivers capable of supporting data rates up to tens of Mbit/sec. We will discuss the advantages gained from a bandpass sampling scheme with respect to demodulator complexity and will show that careful selection of IF and sampling frequencies can greatly reduce the processing complexity of further DSP within the demodulator. The effect of a multirate design on DSP hardware implementation will also be discussed.

The use of bandpass sampling schemes in communication receivers has been documented in [3] for sonar applications, in [4] for HF radio, in [5], [6] and [7] for satellite communications, while [8] details the theory for bandpass sampling and a accompanying paper [9] discusses further theoretical results on bandpass sampling related to the estimation of BER considering aliased noise and IF filter distortion (ISI) effects. This paper presents recent practical results for two BPS receivers. We note that while theoretical approaches to BPS receiver design are topical, there have been few practical results published.

To compare the proposed bandpass sampling architectures with a conventional analog architecture, consider the block diagram of a traditional lowpass sampling demodulator shown in Fig. 1a. The transmitted signal corrupted by noise is downconverted to a suitable IF frequency f_I (typically 70MHz) where it is bandpass filtered to remove unwanted signal components. The wanted signal is then further downconverted to a low–IF frequency, f_d, and lowpass filtered to remove high frequency images. An analog quadrature downconverter is then used to mix the signal down to its Inphase (I) and Quadrature (Q) baseband components. The resultant signal is then filtered to remove high frequency images produced by downconversion before being sampled. The sampled outputs are matched filtered in order to reduce inter–symbol interference (ISI) which are then used to drive the synchronisation functions within the demodulator. For higher rate applications (typically > 30Msym/sec), the matched filtering is performed in the analog domain.

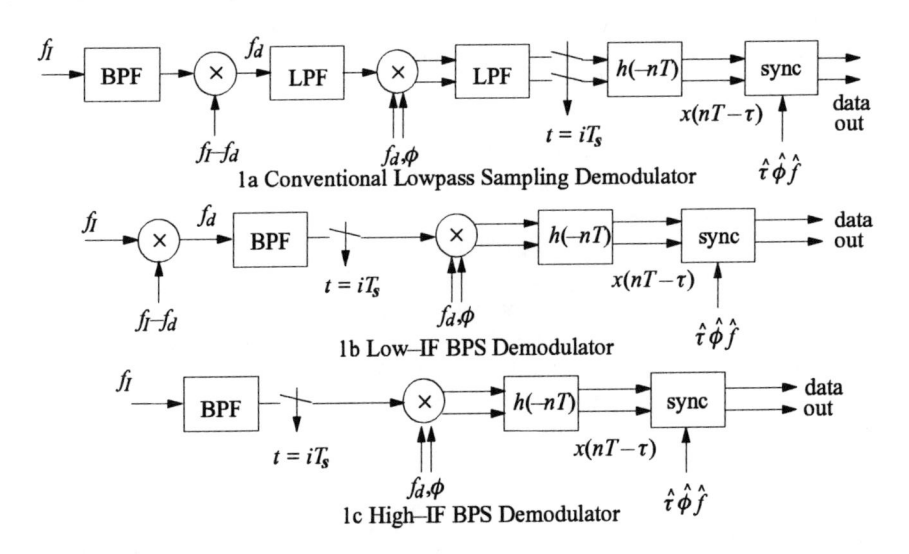

Fig. 1 Demodulator Front–End Architectures

We propose two architectures for implementing BPS digital receivers, Fig. 1b gives the block diagram for the architecture we will term a "low–IF" BPS receiver. The major difference between this architecture with that shown in Figure 1a is that the final downconversion process is performed after sampling. Figure 1c gives the block diagram for the architecture we will term a "high–IF" BPS receiver. The major difference between this architecture and the one shown in Fig. 1b is that there is one less analog downconversion stage.

The synchronisation block shown in all the architectures in Fig. 1 is responsible for performing the functions of estimating and removing timing phase offsets and both carrier phase and frequency offsets. These functions are not discussed in this paper and such effects will be ignored in the analysis. The soft decision outputs from the synchronisation block are then passed to a decoder.

In this paper, Section 2 presents the general theory of bandpass sampling schemes with emphasis on frequency selection. Section 3 describes the implementation of a low–IF sampling scheme which is currently under development. Section 4 discusses a high–IF sampling case. Results from practical systems measurements are presented in both Sections 3 and 4. Section 5 draws conclusions.

2 Background Theory

In this Section we discuss a number of issues which must be considered when developing a bandpass sampling digital receiver. Consider the general signal model for a BPS receiver shown in Fig. 2 which can be used to analyse both the low–IF and high–IF architectures presented in Fig. 1. As shown in Fig. 2 there are four critical parameters to be considered, namely f_I, f_d, f_s and B. The problem involves the selection of suitable values for these parameters which minimise the complexity, and hence the cost, of the receiver without compromising performance. The problem is exacerbated when the receiver must be capable of flexible operation, particularly in multirate receiver applications.

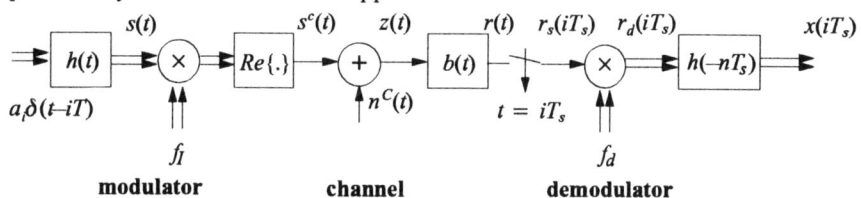

Fig. 2 Complex bandpass sampling signal model.

Observing the model presented in Fig. 2 it can be shown that the spectrum of the signal immediately after sampling is described by

$$R_s(f) = \frac{\sqrt{2}}{2} \sum_{k=-\infty}^{\infty} \left(H_b(f-f_I + kf_s) + H_b^*(-f-f_I + kf_s) \right) \quad (1)$$

where $H_b(f)=H(f)B_L(f)$. $H(f)$ is the transmitted signal spectrum, $H^*(f)$ its complex conjugate and $B_L(f)$ is the complex lowpass filter equivalent spectra of the IF bandpass filter, p. 406 [10]. As $H_b(f)$ is a complex lowpass equivalent spectrum, $H_b(0)$ corresponds to evaluating the frequency response at the centre of the passband, that is DC.

The spectrum of the inphase (I) component of the sampled bandpass signal after downconversion may be described as

$$R_d^I(f) = \frac{1}{2}\begin{cases}\sum_{k=-\infty}^{\infty}((H_b(f-f_d-f_I + kf_s) + H_b^*(-f-f_d-f_I + kf_s) + \\ (H_b(f + f_d-f_I + kf_s) + H_b^*(-f + f_d-f_I + kf_s))\end{cases} \quad (2)$$

A similar result may be obtained for the Q channel. There are some essential criteria when determining the values of f_s, f_d and f_I. These are:

1. To ensure that no signal aliasing occurs during the sampling process. The sampling rate must conform to, using the notation of [8],

$$\frac{2f_u}{N} \leq f_s \leq \frac{2f_L}{N-1} \quad (3)$$

where $f_u = f_I + B/2$, $f_l = f_I - B/2$ and N is an integer given by

$$1 \leq N \leq \left\lfloor \frac{f_u}{B} \right\rfloor \quad (4)$$

$\lfloor x \rfloor$ is the floor function, i.e. the largest integer within x. It is clear from the discussion in [8] that there are bands of allowed and disallowed sampling frequencies which are totally dependent on the signal centre frequency and its bandwidth (assuming that the signal has a symmetrical frequency response).

2. There must be a case where two conjugate images coincide at DC in $R_d^I(f)$.
3. The wanted signal images must be positioned so that they do not overlap and cause aliasing immediately after the sampling process, i.e. in $r_s(iT_s)$. While some aliasing of noise may be allowed as it can be filtered out later, the signal or noise must not be allowed to alias into another signal image as this will cause irreversible signal distortion and hence performance degradation.

We may define the sampling rate $f_s = nr_s$ where $n = T/T_s$ to be always a nominal integer multiple of the symbol rate. Note that this is not essential but is desirable as it eliminates the need for a multirate matched filter, the output rate of which is an integer multiple of the symbol rate. Clearly criterion 1 is the Nyquist sampling criterion and will be used to eliminate low sampling frequency options if they arise. Criterion 2 is adhered to if

$$f-f_d-f_I + kf_s = 0 \text{ and } -f-f_d-f_I + kf_s = 0 \quad (5)$$

or

$$f + f_d-f_I + kf_s = 0 \text{ and } -f + f_d-f_I + kf_s = 0. \quad (6)$$

The solution to the simultaneous equations resulting from (5) and (6) are

$$kf_s = f_I + f_d. \quad (7)$$

and

$$kf_s = f_I-f_d. \quad (8)$$

respectively. Adherence to these relationships will ensure that a complex conjugate image pair is located at DC with the other images being located at integer multiples of $f_s/2$.

Given the use of a root raised cosine transmit spectrum, which is assumed to be ideal at this stage, i.e. at baseband the signal energy is zero for $|f| > (1+a)r_s/2$, the following criteria may also be stated,

1. The sampling frequency is related to the symbol rate by
$$f_s > 2(1 + a)r_s = 2B \tag{9}$$
2. The downconversion frequency is related to the sampling frequency by $f_d < f_s/2$. It is preferable that f_d be an integer sub–multiple of the sampling rate.

As the oscillators used to generate the signals at each of the f_s, f_d and f_I frequencies will not be ideal, there will be a residual frequency offset, and phase offset. The issue of frequency offsets is covered to some extent in [4]. In digital receivers, phase and frequency offsets must be estimated in order to recover the transmitted data. In subsampling receivers the issue is exacerbated by the "magnification" of frequency offsets at IF relative to the frequency offsets at baseband. For example, rearranging (7) to give $f_d = kf_s - f_I$ shows that for large values of k any frequency offset or phase noise in the sampling waveform will be magnified and result in a waveform which is not centred on f_d as required. The existence of frequency offsets may be dealt with to some degree in the dimensioning of the IF BPF, however any frequency offset at baseband will result in a degradation of BER due to matched filter frequency response mismatch.

3 Low–IF demodulator

3.1 Frequency selection

In this case we let $f_d = f_I$. Substituting into (2) we obtain

$$R_d^I(f) = \frac{1}{2}\left\{\sum_{k=-\infty}^{\infty}(H_b(f-2f_d + kf_s) + H_b^*(-f-2f_d + kf_s) + H_b(f + kf_s) + H_b^*(-f + kf_s))\right\} \tag{10}$$

where the image at DC is supplied by the latter two terms. The receiver architecture, for this example, is described above and shown in Fig. 1b. This option gives the lowest, or near lowest, sampling rate as $4r_s$. However it does require the low–IF frequency to be equal to the symbol rate and hence requires an analog down conversion stage from the high–IF. Also, if the received signal has multiple symbol rates, either a variable frequency analog downconversion stage or a multirate digital filter will be required to ensure that the final matched filtered signal has a sample rate which is an integer multiple of the symbol rate.

Further details of a low–IF architecture currently under development are now discussed. The system is based upon the principle outlined in Section 2, where $f_d = r_s$, which as stated results in a near lowest integer sampling rate of $4r_s$. This architecture requires an analog conversion stage to mix the signal down from the high–IF to the low–IF. The low–IF spectrum after sampling is shown in Fig. 3.

The bandpass filter must have suitable stopband attenuation in order to filter out spurious signals. Due to the choice of an oversampling factor of 4 and setting our low–IF frequency f_d equal to our symbol rate r_s, the digital downconversion waveform, $e^{-j(\pi i/2)}$, simplifies the downconversion process to multiplying the sampled bandpass signal by the sequences 1, 0, −1, 0... and 0, 1, 0, −1... to obtain the I and Q baseband channels respectively. The process of downconverting the signal to its baseband components becomes a trivial task and is easily implemented in hardware. For high data rate applications, this function can be performed inside a FPGA as described in Section 3.2.

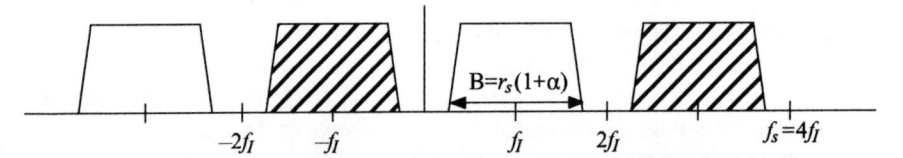

Fig. 3 Spectrum for low–IF receiver after sampling

The front–end architecture of a low–IF digital receiver can also be modified for use in a multirate communication system supporting data rates of similar orders of magnitude. The same bandpass filter, ADC, and digital quadrature downconverter can be used, while the operation of the digital matched filters requires some minor modifications. A cost effective low–IF design of a multirate system is based upon keeping the low–IF frequency "relatively" fixed and independent of lower values of r_s. The low–IF frequency, f_d, is initially determined by the highest symbol rate to be supported by the receiver. As r_s changes, it is undesirable to change f_d accordingly (setting it equal to r_s) as this would require different filtering and frequency synthesis stages in the analog downconverter. With advances in DSP technology, the most cost effective and practical implementation of a multirate system is to keep all IF and RF stages fixed and vary the DSP processing in the demodulator.

For a multirate design, in order to keep the low–IF frequency fixed, the oversampling factor doubles for every halving of the symbol rate. If the symbol rate is not exactly halved, then the low–IF frequency is slightly adjusted such that the oversampling factor is exactly doubled. The bandpass filter is required to have a suitable passband bandwidth such that slight adjustments to the low–IF frequency, f_d, does not result in the wanted bandpass signal being attenuated due to the transition and stopband response of this filter. The I and Q digital receiver matched filters $h(-nT)$ are configured as real decimators in order to reduce the larger oversampling factor existing for lower values of r_s. In this current development, decimation is however also used for the highest symbol rates since the chosen DSP algorithms used to determine both symbol timing offsets and carrier phase offsets require only one sample per symbol.

3.2 Implementation

The main components in the digital front–end design of a low–IF BPS demodulator include a suitable ADC, a digital downconverter (implemented within an FPGA) and a pair of multirate digital filters for matched filtering the I and Q channels. The selection of a suitable ADC requires the investigation of dynamic performance including Signal to Noise ratio for input frequencies up to $f_{IN} = f_d + B/2$, the second harmonic distortion and also the third harmonic distortion. Switching performance, measured in terms of aperture delay and jitter, also needs to be considered. The ADC must also provide a high enough analog bandwidth to accommodate the bandwidth of the input signal. The final factor to be considered is the number of bits and in particular the effective number of bits (ENOB) for input signals centred at f_d. Note that although a bandpass sampled scheme results in the need for only one ADC as compared to two in a baseband design (Fig. 1a), the device and hence initial cost saving may be offset by the need for a superior ADC with greater input analog bandwidth. With the low–IF design, there is a further component saving resulting from no analog 90° phase splitter requirement. There is also one less analog mixer requirement as only one mixer is required to shift the spectrum to a suitable low–IF instead of the two mixers required to perform frequency conversion to baseband.

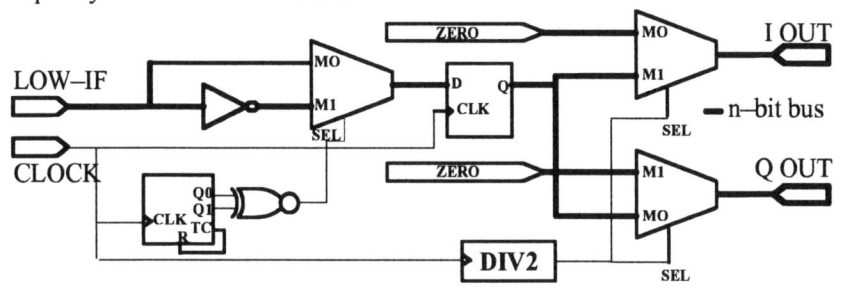

Fig. 4 Logic for Implementing Digital Downconversion

The output of the digital downconversion process consists of the baseband I and Q unfiltered signal components. The symbol rate determines the oversampling factor for the I and Q data streams with the highest symbol rate giving the lowest oversampling factor.

The digital matched filtering in this implementation is performed using commercially available FIR ASICs with a flexible architecture providing up to 208 coefficient latches and the capability of performing decimation. The ASIC allows the highest data rate to be filtered through a 26 tap filter with the number of coefficient registers accessible increasing by a factor of 4 to 104 registers when the device is configured to decimate by a factor of 4. In this particular case the output I and Q signals are given by

$$x^{I,Q}(iT) = \sum_{n=0}^{103} h(nT) r_d((4i-n)T) \tag{11}$$

to give the desired one sample per symbol.

3.3 Results

Initial results obtained from a system based on the low–IF scheme outlined in Section 3.1 are presented. Three spectral plots for the same input to the bandpass sampling analog to digital converter are shown below. The output of the digital matched filters are converted to analog signals using D/A converters and analog reconstruction filters and then fed to a spectrum analyser. The input used is an unmodulated carrier of frequency 5.417MHz and the low–IF frequency is set to 4.827MHz. The output of the signal generator used to generate the ADC input is non-ideal resulting in harmonics of the input occurring at integer multiples of 5.417MHz. As a result these signals are also involved in the mixing process. In the first two plots shown, the Nyquist frequency f_{sNI} = 9.654MHz. Figure 5 gives the output after digital downconversion of the 5.417MHz signal to baseband. As expected there exists a wanted signal component at 590kHz with the other spurious signals being the result of harmonic distortion and aliasing of such components about $f_s/2$. The peak on the right hand side of Fig. 5 is located at 4.827MHz. This peak arises due to some introduced dc offset at the front–end which gets mixed up to 4.827MHz. The second peak from the left is due to aliasing of the mixing image while the next two peaks located at 1.8MHz and 3.7MHz are due to aliasing of the mixed signal generated from the second and first harmonics of the input signal respectively.

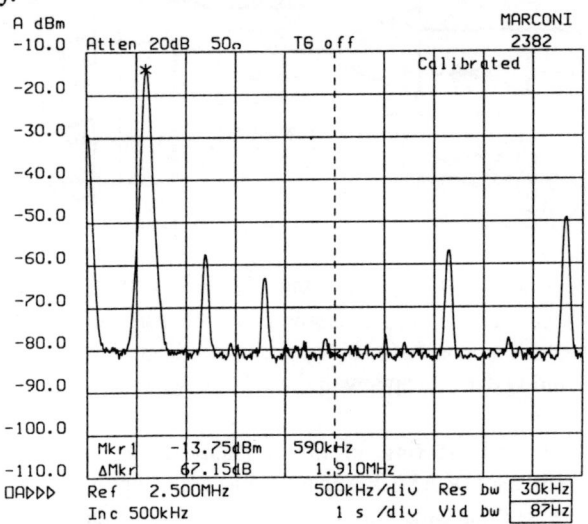

Fig. 5 Spectral plot after digital downconversion (I Channel with 10bit ADC quantisation, receiver filtering and DAC)

The second spectral plot shown in Fig. 6 gives the output after digital matched filtering. As expected all components greater than $r_s/2$ are significantly attenuated due to the characteristic of a root–Nyquist filter with roll–off factor α = 0.4. Note that the design can tolerate DC offsets in the analog signal prior to sampling since after downconversion, the dc offsets are mixed to 4.827MHz and subsequently filtered out by the receiver matched filter. The mixing component due to the first harmonic is still present since it falls in the digital filter bandwidth symmetrical about f_s. The third spectral plot shown in Fig. 7, shows the output after down conversion, digital matched filtering and decimation to 1 sample/symbol. The wanted signal component is located at 590kHz with the other component being the image of the 590kHz component located at 4.237MHz. The process of matched filtering and decimation causes a reduction in the Nyquist frequency from f_{sNI} to f_{sN2} = 2.4135MHz. The image located at 4.237MHz is attenuated due to the effect of the output DAC sinc weighting and the passband response in the analog reconstruction filter.

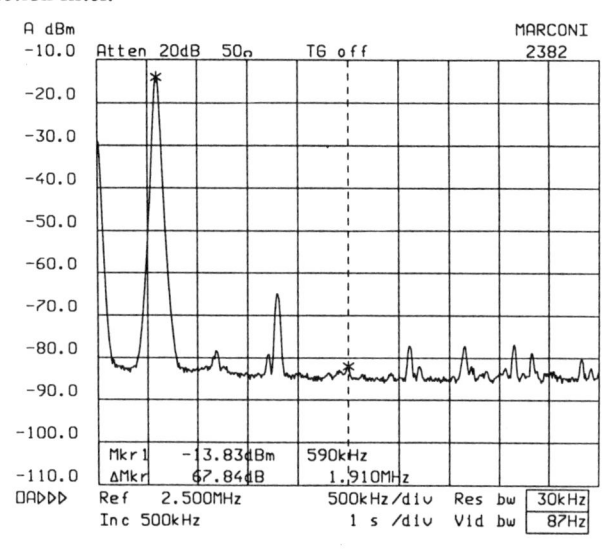

Fig. 6 Spectral plot after digital downconversion and receiver matched filtering (I Channel with 10bit ADC quantisation, receiver filtering and DAC)

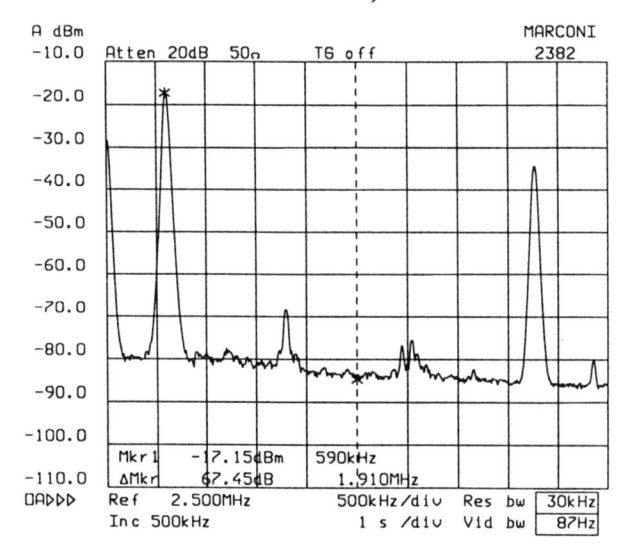

Fig. 7 Spectral plot after digital downconversion, receiver matched filtering and decimation (I Channel with 10bit ADC quantisation, receiver filtering and DAC)

The digital logic to perform downconversion using the simple signal sequences discussed is trivial with an example implementation is shown in Fig. 4.

4 High–IF demodulator

4.1 Frequency selection

In this case, we still require f_d and f_s to be integer multiples of r_s, consequently let $f_d = f_s/m$ and $f_s = nr_s$ and hence (7) and (8) can be written as

$$knr_s = f_I + \frac{nr_s}{m} \Rightarrow f_I^- = nr_s\left(k - \frac{1}{m}\right) \tag{12}$$

and

$$knr_s = f_I - \frac{nr_s}{m} \Rightarrow f_I^+ = nr_s\left(k + \frac{1}{m}\right) \tag{13}$$

respectively. As $m>2$ and preferably $m \geq 4$ the IF frequency is closely related to nkr_s with nr_s/m being an offset. Also, in the general case, $n \geq 4$ as required by (9).

Consider now an example using $r_s=1.024$Msym/sec. The use of $n=8$, $m=4$ and $k=9$ gives an IF frequency of 71.680MHz. Note that (3) and (4) define the allowable sampling frequencies at which aliasing does not occur. Substituting the values, $f_L=70.963$MHz, $f_u=72.397$MHz and $B=1.434$MHz ($\alpha=0.4$) gives

$$\frac{144.794}{N} \leq f_s \leq \frac{141.926}{N-1} \tag{14}$$

where N is an integer given by

$$1 \leq N \leq I_g\left[\frac{72.397}{1.434}\right] = 50 \tag{15}$$

hence the minimum sampling rate is 2.895880<f_s<2.896449MHz a range of only 569Hz! We have chosen $f_s=8r_s=8.192$Msamples/sec as it simplifies later processing. This choice corresponds to the allowed sampling frequencies of 8.044111<f_s<8.348588 where $N=18$ and hence a range of 304.477kHz. Such a range will provide sufficient guard bands in the event of oscillator frequency drift.

4.2 Results

In this particular experiment we perform the bandpass sampling in real–time but perform the post–sampling processes in non–realtime using MATLAB. Fig. 8 shows the spectrum of the sampled signal prior to down conversion. The wanted image is clearly visible at $f=f_d=2r_s=2.048$MHz.

Figure 9 shows an eye diagram composed of 128 symbols of the baseband signal after downconversion, matched filtering and timing and phase synchronisation have been performed. No noise has been added to the signal at IF to allow the eye to be clearly seen. We note that the eye diagram is nearly perfect. The slight blurring of the eye at the symbol centre is due to a combination of distortion caused by the BPF (predominantly amplitude 'ripple'), some uncompensated frequency offset and distortion introduced by the transmit hardware.

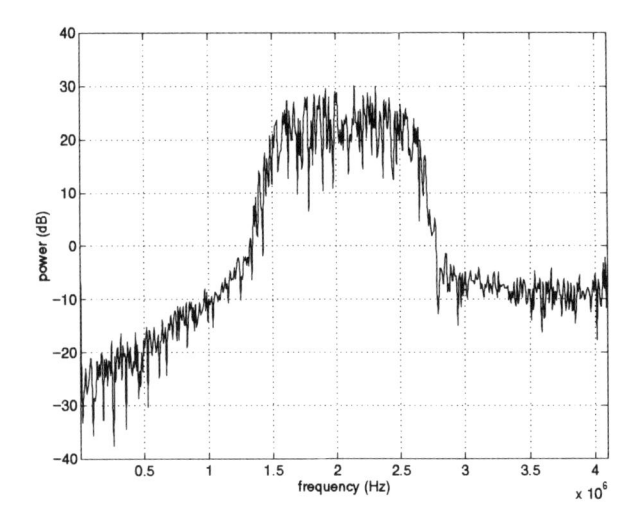

Fig. 8 Spectral plot of the sampled high–IF showing the signal being located at f_d=2.048MHz.

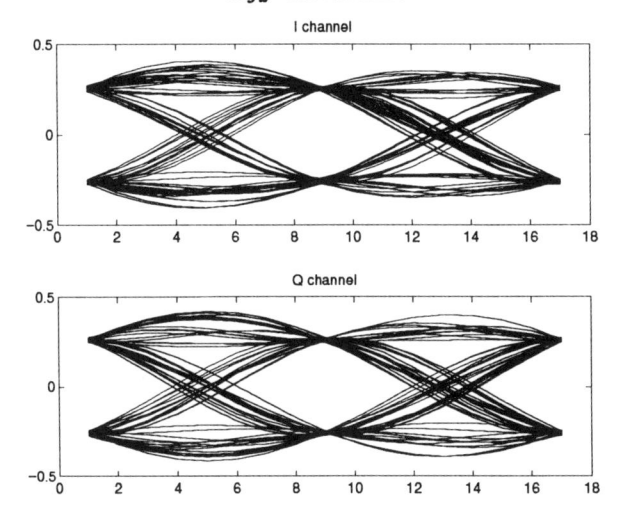

Fig. 9 Eye diagram of the resultant QPSK baseband signal.

The example shown is in fact that of a *sub–sampling* receiver, that is the sampling frequency is less than twice the maximum signal frequency. This particular implementation has the advantages of using a sampling rate which is an integer multiple of the symbol rate and a downconversion simplification as described in the low–IF implementation. The disadvantage is that the high–IF frequency is not standard, i.e. not equal to 70MHz and that the sampling rate is 8 times the symbol rate. This high value will result in additional computation in the matched filter. The use of decimation within the matched filter, however, can greatly reduce the complexity.

5 Discussion and Conclusions

Various design issues to be considered when developing BPS communication receivers have been discussed in this paper. The initial results presented here show that high performance bandpass sampling modem front ends may be implemented using simple digital structures. The development of commercially viable of high–IF implementations will require a more detailed analysis of the trade–offs between savings due to the elimination of an analog downconversion stage and the need for higher performance S/H circuitry. Trends in S/H performance and cost indicate the high–IF implementation will become increasingly viable.

6 Acknowledgments

The authors thank Duong Tran, Gerald Bolding, Jeff Wojtiuk and Philip McIllree for their assistance. They also acknowledge the Australian Research Council (ARC) and Australian Space Office (ASO) collaborative grants.

7 References

[1] J. Wojtiuk, "Analysis of Frequency Conversion for M–QAM and M–PSK Modems", M.Eng Thesis, University of South Australia, Feb. 1995.

[2] J. Wojtiuk and M. Rice, "Quadrature Phase Error and Amplitude Imbalance Effects on Digital Demodulator Performance", *7th Tyrrhenian International Workshop on DSP for Telecommunications,* Italy, Sept. 1995.

[3] W. M. Waters and B. R. Jarrett, "Bandpass Signal Sampling and Coherent Detection", *IEEE Trans. Aerospace and Electronic Systems*, Vol.AES–18. No.4, Nov. 1982.

[4] B. Haller and H. Kaufmann, "Digital DPSK Demodulator Employing Bandpass Subsampling", *I.E.E. Electronic Letters*, Vol. 26 No. 18, 30th Aug. 1990.

[5] R. D. Allan, J. R. Bramwell and D. A. Saunders, "A High Performance Satellite Data Modem using Real–Time Digital Signal Processing Techniques" *Journal of the Institution of Electronic and Radio Engineers*, Vol.58. No.3. May 1988.

[6] L. P. Sabel, "A DSP Implementation of a Robust Flexible Receiver/Demultiplexer for Broadcast Data Satellite Communications", *Proc. Communications 90*, Melbourne, Australia, pp. 218–223, Oct. 1990.

[7] J. K. Cavers and S. P. Stapleton, "A DSP–Based Alternative to Direct Conversion Receivers for Digital Mobile Communications", *Proc. IEEE GLOBECOM*, San Diego, USA, pp. 2024–2029, Dec. 1990.

[8] R. G. Vaughan, N. L. Scott and D. R. White, "The Theory of Bandpass Sampling", *IEEE Trans. Signal Proc.*, Vol. 39, No.9, pp. 1973–1984, Sept. 1991.

[9] L. P. Sabel, "A New Analysis Method for the Performance of BandPass Sampling Digital Demodulators Considering IF Filter Effects", *7th Tyrrhenian International Workshop on DSP for Telecommunications*, Italy, Sept. 1995.

[10] J. G. Proakis and D. G. Manolakis, "Digital Signal Processing : Principles, Algorithms and Applications", second edition, MacMillian, 1992.

Quadrature Phase Error and Amplitude Imbalance Effects on Digital Demodulator Performance

J.J. Wojtiuk, M.Rice
Institute for Telecommunications Research, SPRI Building, University of South Australia, The Levels, SA 5095, Australia

Abstract

This paper analyses the performance of M–QAM and M–PSK modulation schemes after frequency conversion using analogue quadrature hybrid circuits. In particular, the effects of carrier feed–through, quadrature phase error, amplitude imbalance, DC offset in the I and Q channels of the quadrature hybrid circuit are investigated.

1 Introduction

The performance of a digital radio system can be degraded by imperfections that result from the use of analogue circuits in the RF subsystem of a radio modem. This paper is concerned with the analogue implementation of the quadrature hybrid circuit that is commonly used in frequency conversion.

A model will be presented that can analyse the effect of quadrature hybrid circuit imperfections on bit error rate (BER) for a variety of M–PSK and M–QAM schemes. Section 2 gives some background theory and describes the quadrature hybrid circuit and some of the imperfections that can occur. A system of equations are shown that take into account the imperfections, and in section 3 a matrix model is developed from these equations, treating the effect of the imperfections as linear transformations of the transmitted symbol positions in a given constellation. The effect of the down converter on additive white Gaussian noise (AWGN) is considered in section 4, and results are presented in section 5 which show the effect on BER performance.

2 Characterisation of a Modulated Carrier

The complex baseband output of a digital modulator can be given by

$$\tilde{c}(t) = I(t) + jQ(t) \tag{1}$$

where $I(t)$ and $Q(t)$ are the 'in–phase' and 'quadrature' outputs. Note that these outputs may have already been filtered to reduce the effects of intersymbol interference. For wireless transmission the modulating signal must be transferred to a carrier with frequency ω_c. This involves a multiplication operation of the complex baseband signal $\tilde{c}(t)$ with the complex exponential $e^{j\omega_c t}$, to give complex output $\tilde{y}(t)$ as shown below.

$$\tilde{y}(t) = \tilde{c}(t)e^{j\omega_c t}$$
$$= \text{Re}\{\tilde{y}(t)\} + j\,\text{Im}\{\tilde{y}(t)\} \quad (2)$$

The real part of $\tilde{y}(t)$ contains all of the information contained in the modulating signal $\tilde{c}(t)$ and is therefore used for transmission. The modulated signal is given by
$$y_m(t) = I(t)\cos\omega_c t - Q(t)\sin\omega_c t \quad (3)$$

2.1 The Quadrature Hybrid Circuit

Inspection of equation (3) shows the frequency conversion process consists of two real multiplication operations. The baseband signals, $I(t)$ and $Q(t)$, are mixed with a carrier signal. However, the carrier is split into two components that are in phase quadrature. The outputs are then combined to form the modulated carrier. This forms the basis of the quadrature hybrid circuit, which is given in figure 1.

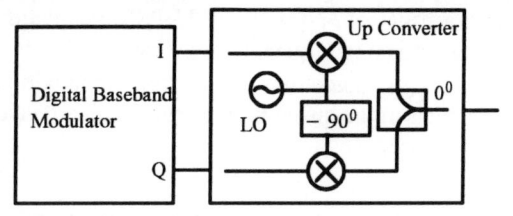

Figure 1 A Quadrature Hybrid Up Converter

Analogue implementations of Quadrature hybrid circuits can have imperfections that can contribute to distortion of the transmitted signal. There are four main sources of imperfection. These are quadrature phase error between the I and Q channels, amplitude imbalance between the I and Q channels, carrier feed–through, and DC offsets in the I and Q channels.

2.2 Up Conversion

A model of a direct conversion up converter with imperfections described in Section 2.1 is shown in figure 2.

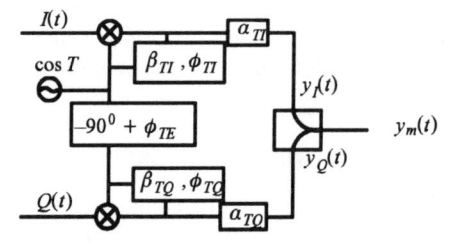

Figure 2 Direct Conversion Up Converter with Imperfections

The digital modulator produces a series of 2 dimensional symbols s_i using the I and Q channels in the same form as equation (1). This is shown below in (4) where s_{Ii} and s_{Qi} belong to the set of real numbers that define the symbol's position in the constellation.
$$s_i = s_{Ii} + js_{Qi} \quad (4)$$

The up converter has gain terms, a_{TI} and a_{TQ}, resulting in an amplitude imbalance between the channels of $\varkappa_T$ as given below.

$$\varkappa_T = \frac{a_{TQ}}{a_{TI}} \tag{5}$$

Two gain terms are used to facilitate the normalisation of the total power to a constant value. The overall power from both channels, λ, is given by (6). The relationship between the channel gains and the amount of imbalance is shown in (7) and (8).

$$a_{TI}^2 + a_{TQ}^2 = \lambda \tag{6}$$

$$a_{TI} = \sqrt{\frac{\lambda}{1+\varkappa_T^2}} \tag{7}$$

$$a_{TQ} = \varkappa_T \sqrt{\frac{\lambda}{1+\varkappa_T^2}} \tag{8}$$

The modulated output signal, $y_m(t)$ from the up converter in Figure 2 is given by:

$$y_m(t) = \mathrm{Re}\{\tilde{f}(t)e^{jT}\} \tag{9}$$

which is in the same form as (3), and where

$$T = \omega_T t + \theta_T \tag{10}$$

$$\tilde{f}(t) = f_I(t) + jf_Q(t) \tag{11}$$

$$f_I(t) = a_{TI} S_I(t) - a_{TQ} \sin\phi_{TE} S_Q(t) + a_{TI} \beta_{TI} \cos\phi_{TI} - a_{TQ} \beta_{TQ} \cos\phi_{TQ} \tag{12}$$

$$f_Q(t) = a_{TQ} \cos\phi_{TE} S_Q(t) + a_{TI} \beta_{TI} \sin\phi_{TI} + a_{TQ} \beta_{TQ} \sin\phi_{TQ} \tag{13}$$

A full derivation is given in [1].

2.3 Down Conversion

The up converter model from figure 2 can also be used for down conversion, with the subscript R replacing T. The input to the down converter corresponds, in the absence of channel imperfections, to the output from the up converter, $y_m(t)$ represented by (9). As with the up converter, the I and Q channels of the down converter have gain terms, a_{RI} and a_{RQ}, the ratio of these representing amplitude imbalance between the channels. An expression for the output signal from the down converter has been derived in [1] and [2] and is shown below.

$$\tilde{r}(t) = r_I(t) + \beta_{DCI} + j(r_Q(t) + \beta_{DCQ})$$

$$= \mathrm{Re}\left\{\frac{a_{RI}\tilde{f}(t)}{2} e^{j(\Delta\omega t + \Delta\phi)}\right\} + j\mathrm{Im}\left\{\frac{a_{RQ} e^{-j\phi_{RE}} \tilde{f}(t)}{2} e^{j(\Delta\omega t + \Delta\phi)}\right\}$$

$$+ \mathrm{Re}\left\{\frac{a_{RI}\beta_{RI}}{2} e^{j\phi_{RI}}\right\} + j\mathrm{Im}\left\{\frac{a_{RQ}\beta_{RQ}}{2} e^{j(\phi_{RQ}-\phi_{RE})}\right\} + \beta_{DCI} + j\beta_{DCQ} \tag{14}$$

The net frequency and phase offset between the up converter and down converter LO signals is given by $\Delta\omega$ and $\Delta\phi$ respectively. The complex baseband signal may be simplified further, assuming $\Delta\omega = 0$ and $\Delta\phi = 0$. The real part, or I channel output is given by:

$$r_I(t) + \beta_{DCI} = a_{RI} f_I(t) + a_{RI} \beta_{RI} \cos\phi_{RI} + \beta_{DCI} \qquad (15)$$

and the imaginary part, or Q channel output is given by:

$$r_Q(t) + \beta_{DCQ} = a_{RQ}\big(f_Q(t)\cos\phi_{RE} - f_I(t)\sin\phi_{RE}\big)$$
$$+ a_{RQ} \beta_{RQ} \sin(\phi_{RQ} - \phi_{RE}) + \beta_{DCQ} \qquad (16)$$

3 Linear Transformation of Transmitted Symbols Due to Frequency Conversion

For ideal signal transmission, the received signal constellation is of the same form as the transmitted one. For signal transmission with imperfect up and down converters the received signal constellation is distorted by the various imperfections in the up and down converters as each transmitted symbol undergoes a transformation, or mapping onto the received signal constellation. The i-th transmitted symbol (before up conversion) may be represented by the 1 x 2 column vector s_i.

$$s_i = \begin{bmatrix} s_{Ii} \\ s_{Qi} \end{bmatrix} \qquad (17)$$

Similarly, the i-th received symbol (after down conversion) may be represented by the 1 x 2 column vector r_i. The difference between each transmitted and received symbol position defines an error vector. If it can be assumed that the random process associated with the channel itself is stationary, then it can be inferred that each symbol is acted upon by the same translation rule. An error vector e_i may be represented in terms of a 1 x 2 column vector.

$$e_i = r_i - s_i \qquad (18)$$

The error vector is shown for a symbol in a constellation in figure 3.

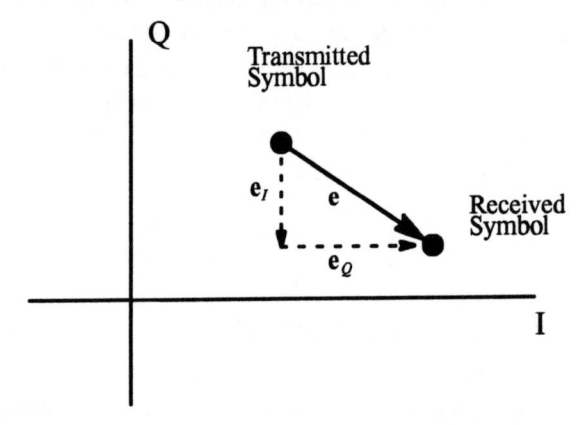

Figure 3 The Error Vector Translating the Transmitted Symbol

3.1 Development of a Matrix Model

Equations developed in sections 2.2 and 2.3 that take into account imperfections in the up and down converters describe a linear transformation of the up and down converter imperfections on the signal constellation. This can be written in matrix form [3]. A model of the communication system used by the analysis is shown in Figure 4.

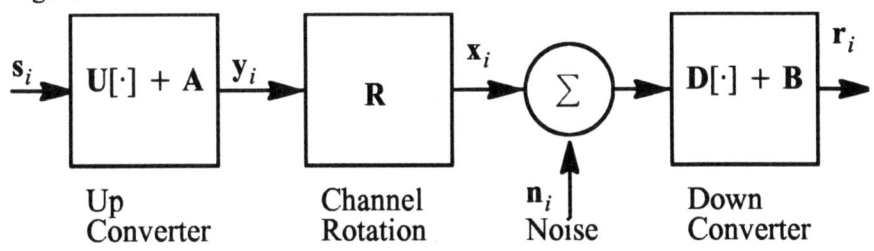

```
s_i → [ U[·] + A ] y_i → [ R ] x_i → (Σ) → [ D[·] + B ] → r_i
       Up            Channel      n_i        Down
       Converter     Rotation     Noise      Converter
```

Figure 4 Model Used for the Analysis

The **U** and **A** matrices are formed from equations (12) and (13).

$$U = \begin{bmatrix} a_{TI} & -a_{TQ}\sin\phi_{TE} \\ 0 & a_{TQ}\cos\phi_{TE} \end{bmatrix} \quad (19)$$

$$A = \begin{bmatrix} a_{TI}\beta_{TI}\cos\phi_{TI} - a_{TQ}\beta_{TQ}\cos\phi_{TQ} \\ a_{TQ}\beta_{TQ}\sin\phi_{TQ} + a_{TI}\beta_{TI}\sin\phi_{TI} \end{bmatrix} \quad (20)$$

A channel rotation transformation can be included to model the effect of phase differences between the up converter and down converter local oscillators as shown below, where **R** is the general rotation matrix.

$$x_i = Ry_i \quad (21)$$

The **D** and **B** matrices are formed from equations (15) and (16).

$$D = \begin{bmatrix} a_{RI} & 0 \\ -a_{RQ}\sin\phi_{RE} & a_{RQ}\cos\phi_{RE} \end{bmatrix} \quad (22)$$

$$B = \begin{bmatrix} a_{RI}\beta_{RI}\cos\phi_{RI} + \beta_{DCI} \\ a_{RQ}\beta_{RQ}\sin(\phi_{RQ} - \phi_{RE}) + \beta_{DCQ} \end{bmatrix} \quad (23)$$

The position of the i-th received symbol after down conversion, r_i, is represented as:

$$r_i = Fs_i + G \quad (24)$$

where

$$F = DRU \quad (25)$$
$$G = DRA + B \quad (26)$$

The error vector for the i-th symbol is shown below, where **I** is the identity matrix.

$$e_i = (F - I)s_i + G \quad (27)$$

4 Transformation of the Input Noise by the Down Converter

The input noise to the down converter is given by **n**. This is also a 2 x 1 column vector representing the I and Q channel noise components:

$$\mathbf{n} = \begin{bmatrix} n_I(t) \\ n_Q(t) \end{bmatrix} \tag{28}$$

where n_I and n_Q are random variables associated with the I and Q channel noise components. The general multi–variate Gaussian Probability Distribution Function (PDF) is given by equation (29) with mean **m** and covariance matrix **C** [4].

$$p_n(\mathbf{n}) = \frac{1}{2\pi\sqrt{\det(\mathbf{C})}} \exp\left[-\frac{1}{2}(\mathbf{n}-\mathbf{m})^T \mathbf{C}^{-1}(\mathbf{n}-\mathbf{m}) \right] \tag{29}$$

The down converter affects the noise in two ways. Firstly the covariance matrix is transformed.

$$\mathbf{W} = \mathbf{DCD}^T \tag{30}$$

By substituting equation (22) for **D**, and $\sigma^2 \mathbf{I}$ for the uncorrelated covariance matrix **C** in (30), the transformed covariance matrix **W** is:

$$\mathbf{W} = \begin{bmatrix} \sigma^2 \alpha_{RI}^2 & -\sigma^2 \alpha_{RI} \alpha_{RQ} \sin\phi_{RE} \\ -\sigma^2 \alpha_{RI} \alpha_{RQ} \sin\phi_{RE} & \sigma^2 \alpha_{RQ}^2 \end{bmatrix} \tag{31}$$

Secondly, the mean of the noise distribution after down conversion, $\mathbf{m}_z$, is changed according to:

$$\mathbf{m}_z = \mathbf{Dm} + \mathbf{B}$$

$$= \begin{bmatrix} \alpha_{RI} \beta_{RI} \cos(\phi_{RI}) + \beta_{DCI} \\ \alpha_{RQ} \beta_{RQ} \sin(\phi_{RQ} - \phi_{RE}) + \beta_{DCQ} \end{bmatrix} \tag{32}$$

It can be seen that the down converter quadrature phase error can transform the input I and Q noise components to introduce correlation between them. The correlation coefficient, ϱ, is related to the amount of quadrature phase error between the I and Q channels.

$$\varrho = -\sin(\phi_{RE}) \tag{33}$$

At the same time the I and Q channel noise variances are scaled by the channel gains α_{RI} and α_{RQ}, which are ultimately related to the degree of amplitude imbalance between the I and Q channels.

$$\sigma_I^2 = \alpha_{RI}^2 \sigma^2 \tag{34}$$
$$\sigma_Q^2 = \alpha_{RQ}^2 \sigma^2 \tag{35}$$

The mean $\mathbf{m}_z$ can be seen to be related to the amount of carrier feed–through at IF in the down converter and also DC offset on the I and Q channels.

4.1 Diagonalization of the Noise Covariance Matrix

Analytical expressions for the BER performance of various modulation schemes are greatly simplified if the I and Q channel noise components are uncorrelated. Therefore, a further transformation is needed to make the covariance matrix, **W**, a diagonal matrix. The required transformation can be determined by selecting a transform matrix, **Y**, to be an orthogonal matrix such that $\mathbf{Y}^T = \mathbf{Y}^{-1}$. This will have its columns equal to the Eigenvectors of **W** and the resulting covariance matrix **Q** will be a diagonal matrix with elements equal to the Eigenvalues of **W** [4].

The transformation is also applied to the signal constellation. However, this does not have a net effect as the transformation is rotational [5] and so the I and Q axes remain orthogonal with the decision regions for the signal constellation remaining unaffected. Thus the error vectors for each symbol are not changed relative to both the received symbol position and the I and Q axes.

5 BER Expressions for *M*–QAM and *M*–PSK

An expression for BER, $P_B(E)$, is derived by adding the symbol error probabilities for every symbol and dividing by the number of symbols M in the constellation and the number of bits per symbol μ. The expression for BER is given as:

$$P_B(E) = \frac{1}{\mu M} \sum_{i=1}^{M} P(E)_i \tag{36}$$

where $P(E)_i$ is the probability of error for the i–th symbol in the constellation. It is assumed that Gray coding is used so that only single bit symbol errors occur. The probability of error for each symbol can be found by integrating the Bivariate Gaussian PDF over the shape of the decision region. If the I and Q noise components are uncorrelated, this can be simplified by integrating the one dimensional Gaussian PDF for the I and Q components separately. This allows the use of well defined functions such as the Q–function.

$$Q(x) = \frac{1}{\sqrt{2\pi}} \int_x^\infty \exp\left(-\frac{1}{2}z^2\right) dz \tag{37}$$

For QAM decision regions the final form of error probability for a symbol will vary depending on the region having two, three, or four finite boundaries. Note that another type of QAM decision region exits in the corners of the 32–QAM constellation. An example expression given below is for a QAM decision region with two finite boundaries. Note also the approximation is due to taking the union bound. This region is shown in figure 5.

$$P_i(E)_i \approx Q\left(\frac{A}{v_I \sigma}(1 + e_{Ii})\right) + Q\left(\frac{A}{v_Q \sigma}(1 + e_{Qi})\right) \tag{38}$$

The signal power is given by A, the down converter input noise variance is given by σ, and the transformed I and Q noise variance components relative to σ are given by v.

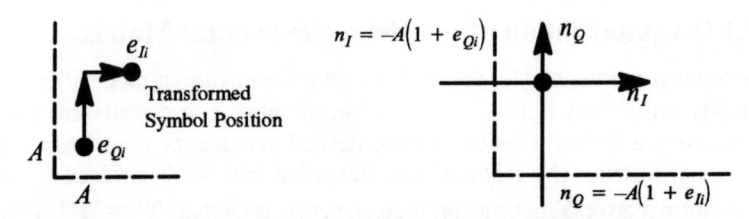

Figure 5 QAM Decision Region with 2 Finite Boundaries

Unlike QAM constellations, the optimum decision regions for the M–PSK constellations form sectors of a circle about each symbol in the constellation and therefore, only one type of decision region exists. The general PSK decision region is shown in figure 6.

$$\phi_1 = \frac{2\pi}{M}(i - 1)$$

$$\phi_2 = \frac{\pi}{M}(2i - 1)$$

$$\phi_3 = \frac{\pi}{M}(2i - 1) + e_{\phi i}$$

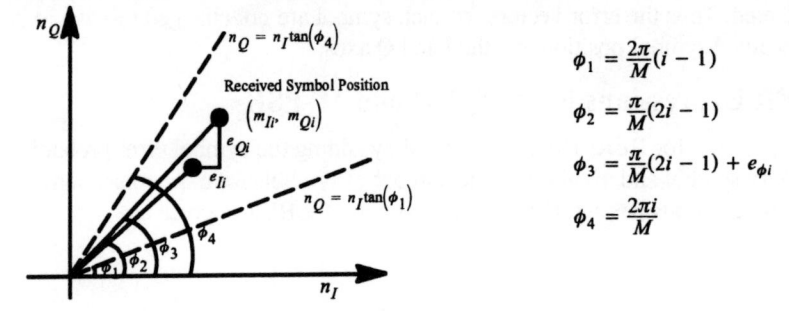

$$\phi_4 = \frac{2\pi i}{M}$$

Figure 6 PSK Decision Region

The probability of a symbol error is given by:

$$P(E)_i \approx Q\{b\cos[L_1] - a\sin[L_1]\} + Q\{a\sin[L_2] - b\cos[L_2]\} \tag{39}$$

where

$$L_1 = \arctan\left\{\left(\frac{v_I}{v_Q}\right)\tan\left[\frac{2\pi}{M}(i-1)\right]\right\} \tag{40}$$

$$L_2 = \arctan\left\{\left(\frac{v_I}{v_Q}\right)\tan\left[\frac{2\pi i}{M}\right]\right\} \tag{41}$$

$$a = \frac{A}{v_I\sigma}e_{Ai}\cos\left[\frac{\pi}{M}(2i-1) + e_{\phi i}\right] \tag{42}$$

$$b = \frac{A}{v_Q\sigma}e_{Ai}\sin\left[\frac{\pi}{M}(2i-1) + e_{\phi i}\right] \tag{43}$$

$$e_{Ai} = \sqrt{(\cos(\phi_2) + e_{Ii})^2 + (\sin(\phi_2) + e_{Qi})^2} \tag{44}$$

$$e_{\phi i} = \arctan\left[\frac{\sin(\phi_2) + e_{Qi}}{\cos(\phi_2) + e_{Ii}}\right] - \phi_2 \tag{45}$$

6 Implementation Loss Results

Graphs of implementation loss against error parameters are presented in the following pages for various PSK and QAM schemes. The error parameters investigated are quadrature phase error and amplitude imbalance on the up converter and the down converter, DC offset on the down converter, phase offset between the up converter and down converter LO's (Local Oscillators). In the case of amplitude imbalance and quadrature phase error, different results were observed for up converter and down converter. These are distinguished by '_UC' to signify an up converter parameter and '_DC' to signify a down converter parameter. In all cases the implementation loss is for a BER of 10^{-6}.

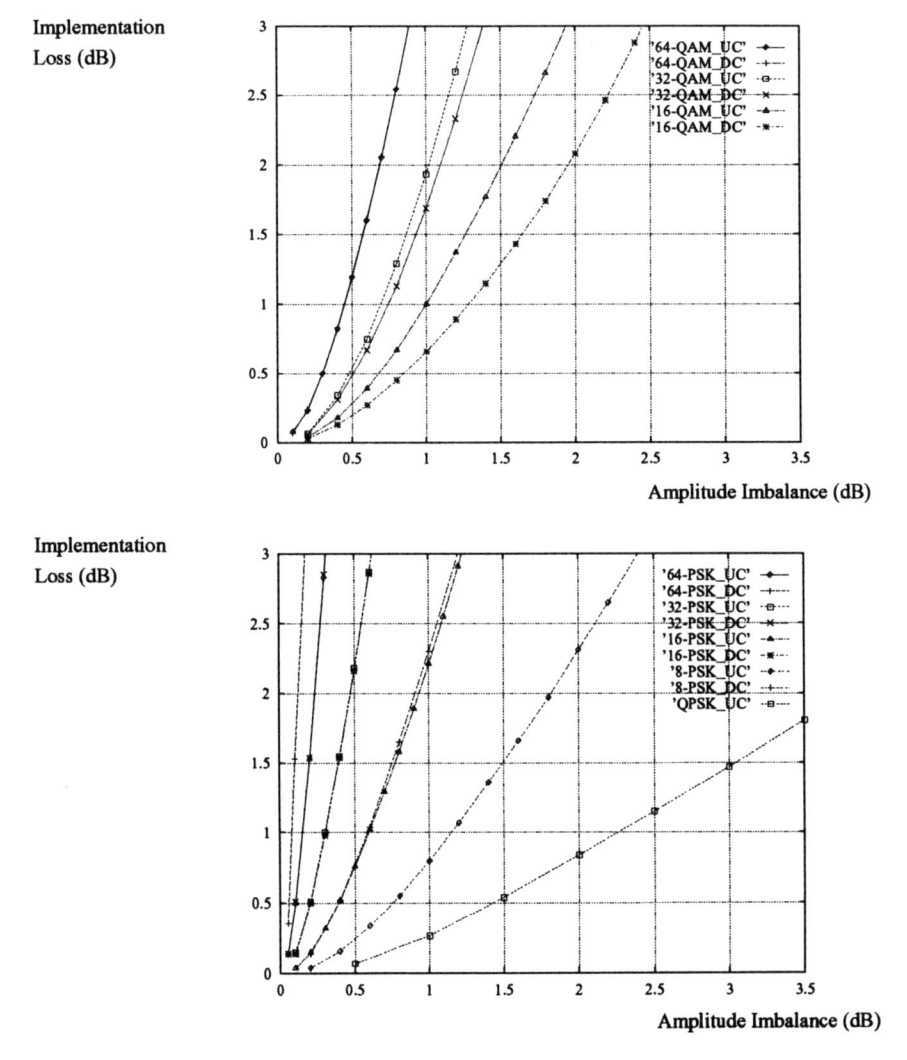

Figure 7 Implementation Loss vs Amplitude Imbalance

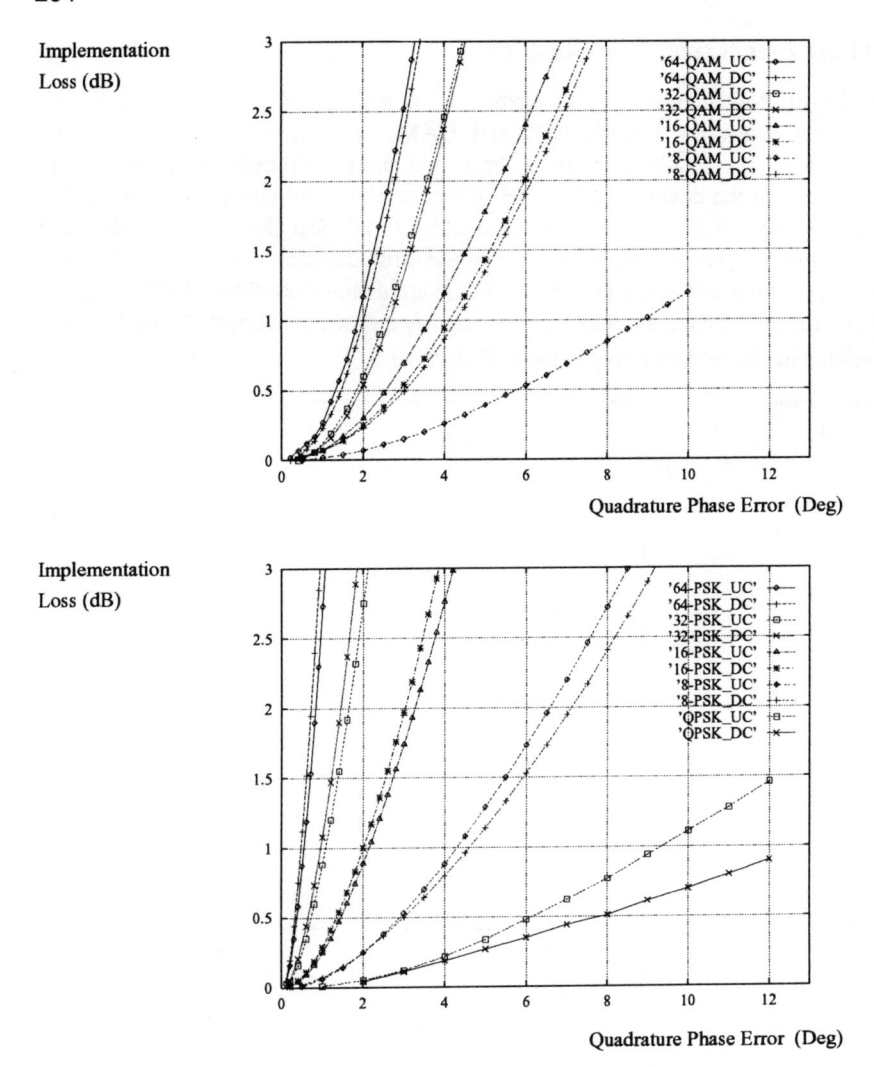

Figure 8 Implementation Loss vs Quadrature Phase Error

In general, the QAM schemes have more implementation loss for up converter error parameters while the PSK modulation schemes have greater implementation loss for down converter error parameters. However, there are exceptions that require further discussion. Of the QAM schemes, 8–QAM was the only one that showed a higher level of implementation loss for quadrature phase error on the down converter. For the PSK schemes, QPSK and 8–PSK had a higher level of implementation loss for quadrature phase error on the up converter. An example of combined quadrature phase error and amplitude imbalance effects are presented in figure 9 for 64–QAM. The parameters are given in the table below for four different cases.

Case	Up Converter			Down Converter		
	Quadrature Error (Deg)	Amplitude Imbalance (dB)	Carrier Suppression (dBc)	Quadrature Error (Deg)	Amplitude Imbalance (dB)	Carrier Suppression (dBc)
a	0.5	0.1	30.0	0.5	0.1	30.0
b	0.5	0.1	Infinite	0.5	0.1	Infinite
c	1.0	0.3	Infinite	1.0	0.3	Infinite
d	3.0	0.5	Infinite	2.0	0.5	Infinite

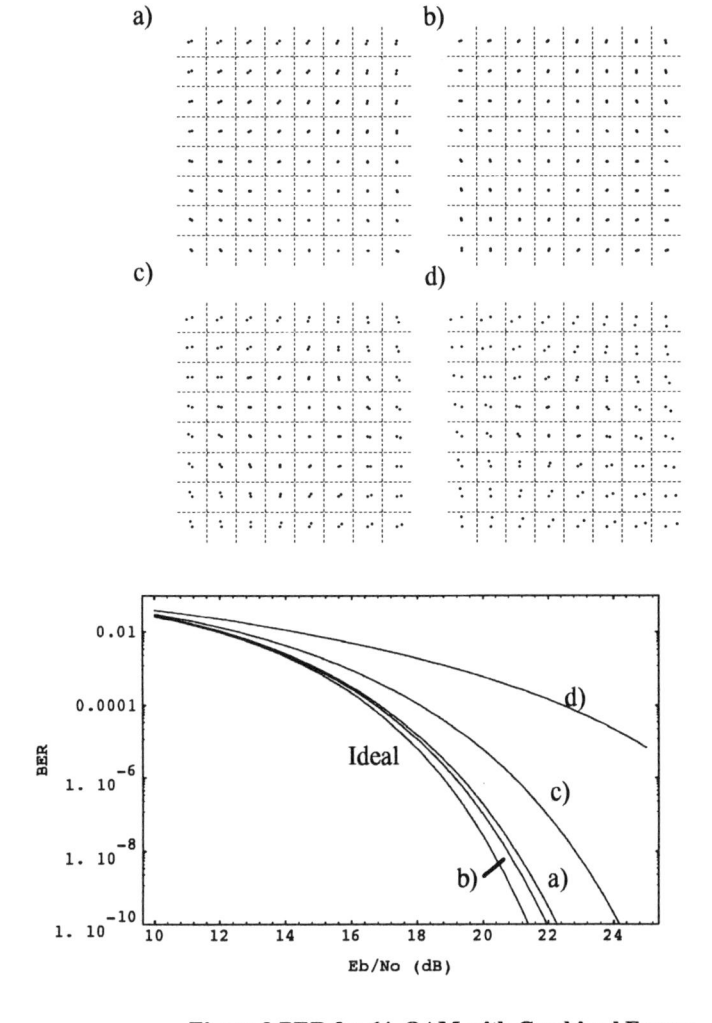

Figure 9 BER for 64–QAM with Combined Errors

7 Conclusions

The results presented in this paper have served to quantify some of the effects of non–ideal frequency conversion on various PSK and QAM modulation schemes. The results generally show that low order schemes such as QPSK can be used in a communications system with up converters and down converters built to physically realisable performance specifications without appreciable BER degradation. The design specifications become tighter for 8–PSK and 16–QAM if negligible implementation loss is required. It is likely that practical frequency converter designs for 64–QAM and higher order QAM schemes system would introduce significant errors. Analog frequency converter specifications for 32–PSK and higher order PSK schemes would not be physically realisable unless some implementation loss could be accepted. For high order modulation schemes these results present a good case for the use of digital radio architectures that have the analog/digital interface already at an intermediate frequency, with the quadrature hybrid frequency conversion to baseband implemented digitally.

8 References

[1]   J. Wojtiuk, "Analysis of Frequency Conversion for M–QAM and M–PSK Modems", M.Eng Thesis, University of South Australia, February 1995.

[2]   Roome, S. J., 'Analysis of Quadrature Detectors using Complex Envelope Notation', IEE Proceedings Vol.136, part F, No.2, April 1989.

[3]   Cavers, J. K., Liao, M., 'Adaptive Compensation for Imbalance and Offset Losses in Direct Conversion Tranceivers'   41st IEEE Vehicular Technology Conference, 1991.

[4]   J. G. Proakis, 'Digital Communications', McGraw Hill

[5]   Wozencraft, J.M., Jacobs, I.M.,   'Principles of Communication Engineering', Wiley.

Maximum-Likelihood Sequence Estimation for OFDM

Anders Vahlin and Nils Holte

Department of Telecommunications
The Norwegian Institute of Technology
N-7034 Trondheim, Norway.

Abstract

In this paper, Maximum-Likelihood Sequence Estimation (MLSE) for OFDM is derived together with a corresponding upper bound on probability of error. The error bound of the MLSE is used to assess the performance of an OFDM system with a two-path channel and a linear zero-forcing equaliser. The results show that the equaliser has close to optimum performance for this channel. Further, the OFDM system with MLSE is compared to a single carrier system which also uses MLSE. The single carrier system gives a lower BER when a two-path channel is used.

1 Introduction

There has recently been an increasing interest in the concept of Orthogonal Frequency Division Multiplex (OFDM) [1]. The technique has been considered for applications such as broadcasting [2] and digital subscriber lines [3].

In single carrier modulation, optimum receiver structures for channels with intersymbol interference has received considerable attention. Maximum-Likelihood Sequence Estimation (MLSE) was developed more than 20 years ago [4]. For OFDM, optimum receivers has not been extensively investigated in the literature.

This paper presents derivation of MLSE for OFDM and a corresponding upper bound on probability of error. The resulting algorithms can be seen as a generalisation of the MLSE algorithm for single carrier modulation presented in [5]. The MLSE algorithms for OFDM will in many applications be too complex to implement with today's technology. They can however, serve as a starting point for derivation of suboptimal algorithms for these applications. The MLSE minimises the probability of sequence error which in most applications also gives minimum probability of symbol error. This probability can therefore be used as a bound on attainable performance when evaluating other types of receivers.

Two examples of performance evaluation based on the MLSE are also given in this paper. In the first example, a linear zero-forcing equaliser is compared to the MLSE for an OFDM system with a two-path channel and white Gaussian noise. This shows how close to optimum the linear zero-forcing equaliser performs in this application. The second example compares the OFDM system to a single carrier system, both of which employ MLSE. This gives the attainable performance of the two systems for this channel.

2 The OFDM System

In this paper, an OFDM system based on N QAM channels is considered. The block diagram of the baseband model of the transmitter and channel is shown in Fig. 1. The complex symbols $c_{k,n}$ are filtered by the pulse shaping filter $h(t)$ and the resulting signal is modulated by multiplication by the complex exponential. The signals from the N channels are added to establish the transmitted signal $x(t)$. Hence, the transmitted OFDM signal can be written

$$x(t) = \sum_k \sum_n c_{k,n} h(t - kT) e^{j\frac{2\pi}{T}nt} \qquad (1)$$

where T is the symbol interval in each QAM channel.

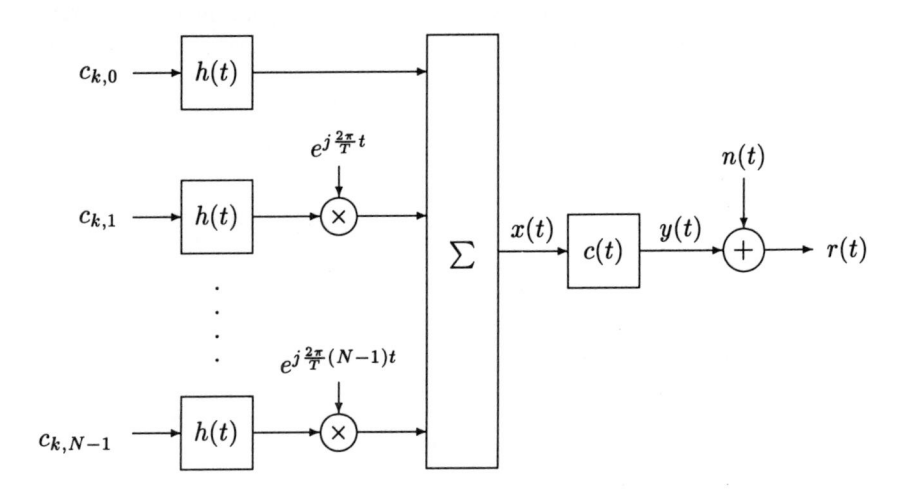

Fig. 1. Block diagram of an N-channel OFDM transmitter and a multipath channel.

The transmitted signal is passed through the channel with impulse response $c(t)$. White Gaussian noise, $n(t)$, is added to the signal before it reaches the receiver. The received signal $r(t)$ then becomes

$$r(t) = x(t) * c(t) + n(t) \qquad (2)$$

where $*$ denotes convolution. This is the observed signal that the receiver should process to retrieve the transmitted symbols $c_{k,n}$.

By substituting (1) in (2) and interchanging the order of summation and convolution, the received signal can be written

$$r(t) = \sum_k \sum_n c_{k,n} h(t - kT) e^{j\frac{2\pi}{T} nt} * c(t) + n(t) \qquad (3)$$

Now define a set of N functions as

$$\psi_n(t) = h(t) e^{j\frac{2\pi}{T} nt} * c(t) \quad n = 0 \ldots N-1 \qquad (4)$$

The received signal r(t) can then be written

$$r(t) = \sum_k \sum_n c_{k,n} \psi_n(t - kT) + n(t) = y(t) + n(t) \qquad (5)$$

This form for the received signal is used when deriving the maximum-likelihood receiver for OFDM. A block diagram corresponding to this equation is shown in Fig. 2. This block diagram is functionally equivalent to that in Fig. 1.

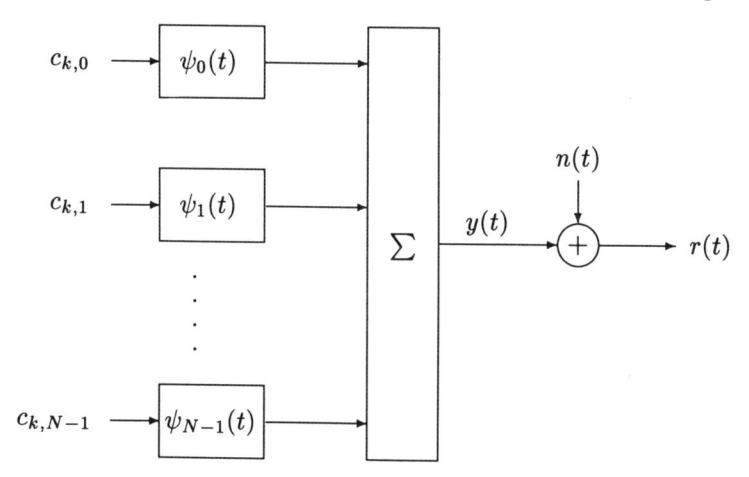

Fig. 2. Functionally equivalent block diagram of OFDM transmitter and multipath channel.

3 Maximum-Likelihood Sequence Estimation

In the previous section, a model of the received signal in an OFDM system was established. The task of the receiver is to estimate, based on the received signal, which symbols were transmitted. A maximum-likelihood sequence estimator is a receiver that chooses the symbol sequence such that the probability of sequence error is minimised [6]. In this section, a maximum-likelihood sequence estimator for OFDM is derived.

3.1 The Likelihood Function for OFDM

The maximum-likelihood receiver chooses, among all possible sequences, the one for which the likelihood function, or equivalently the log-likelihood function has the largest value. The log-likelihood function for the hypothesis that the transmitted symbols are $c'_{k,n}$, when $r(t)$ is received, is [6]

$$\Lambda = -\int_I |r(t) - y'(t)|^2 \, dt \tag{6}$$

where the interval I covers the sequence to be estimated and $y'(t)$ is the hypothesised received waveform in absence of noise, defined as

$$y'(t) = \sum_k \sum_n c'_{k,n} \psi_n(t - kT) \tag{7}$$

By expanding the integrand in (6), the log-likelihood function becomes

$$\Lambda = -\int_I |r(t)|^2 - 2Re\{r(t)y'^*(t)\} + |y'(t)|^2 \, dt \tag{8}$$

where $()^*$ denotes complex conjugate. By substituting (7) for $y'(t)$ and interchanging the order of integration and summation, (8) can be written

$$\Lambda = -\int_I |r(t)|^2 \, dt + 2Re\{\sum_k \sum_n c'^*_{k,n} \int_I r(t)\psi_n^*(t - kT)\} \, dt$$
$$- \sum_k \sum_n \sum_{k'} \sum_{n'} c'_{k,n} c'^*_{k',n'} \int_I \psi_n(t - kT)\psi_{n'}^*(t - k'T) \, dt \tag{9}$$

The first term in this equation is independent of the hypothesis and need therefore not be considered in maximisation of the likelihood. The integral in the second term can be interpreted as samples of a bank of filters matched to the transmitted waveforms $\psi_n(t)$. A block diagram of a corresponding receiver structure is shown in Fig. 3. The input to the filters is the received signal $r(t)$ and samples are taken at $t = kT$. Define new variables for these samples as

$$\hat{r}_{k,n} = \int_I r(t)\psi_n^*(t - kT) \, dt \tag{10}$$

The integral in the last term of (9) represents samples of the cross-correlations of the transmitted waveforms. Define variables for these samples as

$$\rho_{n',n}(k) = \int_I \psi_n(t)\psi_{n'}^*(t - kT)\, dt \tag{11}$$

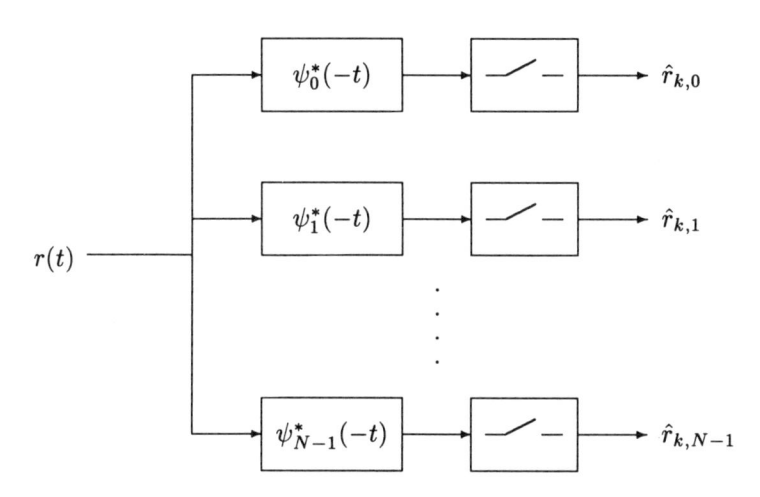

Fig. 3. Block diagram of front end of receiver for maximum-likelihood sequence estimation.

To simplify the notation, the likelihood function can be expressed in vector form. To do so, define vectors of hypothesised transmitted symbols

$$c'(k) = [\, c'_{k,0}, c'_{k,1}, \ldots c'_{k,N-1} \,]^T \tag{12}$$

and vectors of samples of the bank of matched filters

$$\hat{r}(k) = [\, \hat{r}_{k,0}, \hat{r}_{k,1}, \ldots \hat{r}_{k,N-1} \,]^T \tag{13}$$

Also, define a matrix $R(k)$ of cross-correlations where the element on row i and column j is

$$R_{i,j}(k) = \rho_{i,j}(k) \tag{14}$$

The likelihood function in (9) can now be written in vector form as

$$\Lambda = 2Re\{\sum_k c'^H(k)\hat{r}(k)\} - \sum_k \sum_{k'} c'^H(k')R(k'-k)c'(k) \tag{15}$$

The maximum-likelihood receiver uses the sequence of vectors $\{\hat{r}(k)\}$ and chooses, as estimate of the transmitted sequence, the sequence $\{c'(k)\}$ for which the likelihood in (15) has the largest value.

3.2 Recursive Computation of the Likelihood Function

As in single carrier modulation, the likelihood function can be computed recursively using the Viterbi algorithm. To see that, the likelihood function in (15) can be rewritten by breaking the sums. For a sequence of length K and intersymbol interference of length L the likelihood function becomes

$$\Lambda_K = \Lambda_{K-1}\hat{r}(K-1)\} - 2Re\{ \sum_{k=K-1-L}^{K-2} c'^H(K-1)R(K-1-k)c'(k)\}$$
$$+ 2Re\{c'^H(K-1) - c'^H(K-1)R(0)c'(K-1)\} \quad (16)$$

Here, it has been used that $R(-k) = R^H(k)$ and $R(k) = 0$ for $k > L$.

The equation has the same form as for single carrier modulation but there are vectors and matrices instead of scalars. The number of states in the Viterbi is M^{NL} where M is the size of the symbol alphabet. Thus, the complexity is very large when the number of channels or the symbol alphabet is large.

3.3 Bound on Probability of Error

An estimation error occurs if the received signal is such that the likelihood function is larger for a sequence other than the transmitted. Assuming that the vector sequence $\{c(k)\}$ is transmitted, the receiver then makes an error if there is a sequence $\{e(k)\}$ such that $\{c(k)\} + \{e(k)\}$ is a possible transmitted sequence and has greater likelihood than the transmitted sequence, i.e.

$$\Lambda(\{c(k)\} + \{e(k)\}) > \Lambda(\{c(k)\}) \quad (17)$$

To calculate the probability that this event occurs, the likelihood function is manipulated as follows.

The samples of the bank of matched filters, defined in (10), can be rewritten using (5)

$$\begin{aligned}\hat{r}_{k,n} &= \int_I r(t)\psi_n^*(t-kT)\} \, dt \\ &= \int_I y(t)\psi_n^*(t-kT)\} \, dt + \int_I n(t)\psi_n^*(t-kT)\} \, dt \end{aligned} \quad (18)$$

Now define variables for the noise part of the samples as

$$n_n(k) = \int_I n(t)\psi_n^*(t-kT)\} \, dt \quad (19)$$

Assuming the transmitted sequence is $\{c(k)\}$ and using the cross-correlation defined in (11), the samples of the filter bank can be written

$$\hat{r}_{k,n} = \sum_{k'}\sum_{n'} c_{k',n'} \rho_{n,n'}(k-k') + n_n(k) \quad (20)$$

Using vector notation this becomes

$$\hat{r}(k) = \sum_{k'} R(k-k')c(k') + n(k) \qquad (21)$$

where the vector $n(k)$ is defined as

$$n(k) = [n_0(k), n_1(k), \ldots, n_{N-1}(k)]^T \qquad (22)$$

The likelihood function in (15) can now be rewritten by replacing $\hat{r}(k)$ by the expression in (21). The likelihood function for the hypothesis that the transmitted sequence is $\{c'(k)\}$, when the actual sequence is $\{c(k)\}$, can be written

$$\Lambda(\{c'(k)\}) = 2Re\{\sum_k \sum_{k'} c'^H(k)R(k-k')c(k') + \sum_k c'^H(k)n(k)\}$$
$$- \sum_k \sum_{k'} c'^H(k')R(k'-k)c'(k) \qquad (23)$$

From (11), it can be shown that the cross-correlation matrix $R^H(k)$ satisfies

$$R^H(k) = R(-k) \qquad (24)$$

Using this symmetry, the likelihood function for $\{c(k)\}+\{e(k)\}$ can be written

$$\Lambda(\{c(k)\} + \{e(k)\}) = 2Re\{\sum_k (c^H(k) + e^H(k))n(k)\}$$
$$- \sum_k \sum_{k'} e^H(k')R(k'-k)e(k) + \sum_k \sum_{k'} c^H(k')R(k'-k)c(k) \qquad (25)$$

By rewriting the likelihood function for $\{c(k)\}$ correspondingly, the inequality in (17) can now be written

$$2Re\{\sum_k e^H(k)n(k)\} > \sum_k \sum_{k'} e^H(k')R(k'-k)e(k) \qquad (26)$$

The right hand side of this equation is the squared of a distance between the transmitted sequence and the sequence with error, defined as

$$d^2(\{e(k)\}) = \sum_k \sum_{k'} e^H(k')R(k'-k)e(k) \qquad (27)$$

The left hand side of (26) is a zero mean Gaussian random variable with variance

$$\sigma^2 = E\left[\sum_k \sum_{k'} \left(e^H(k)n(k) + n^H(k)e(k)\right)\left(e^H(k')n(k') + n^H(k')e(k')\right)\right] \qquad (28)$$

Using (11), the variance σ^2 can, after some algebraic manipulations, be expressed as

$$\sigma^2 = N_0 \sum_k \sum_{k'} e^H(k') R(k'-k) e(k) = N_0 d^2(\{e(k)\}) \qquad (29)$$

where $N_0/2$ is the power spectral density of $n(t)$. With this expression for the variance, the probability that the received signal is such that the inequality in (26) is satisfied is

$$Pr[\epsilon_2] = Pr[\Lambda(\{c(k)\} + \{e(k)\}) > \Lambda(\{c(k)\})] = Q\left(\frac{d(\{e(k)\})}{\sqrt{N_0}}\right) \qquad (30)$$

where the function $Q(\cdot)$ is the integral of the normalised Gaussian distribution. This is the probability that the likelihood function has a larger value for the sequence with error ($\{c(k)\} + \{e(k)\}$) than for the transmitted sequence $\{c(k)\}$. The probability that this error sequence is chosen does also depend on whether this is an allowable sequence and if there are other allowable sequences for which the likelihood function has an even larger value.

To simplify the notation, define the following events

ϵ: the event that $\{c(k)\} + \{e(k)\}$ is chosen as the estimated sequence

ϵ_1: the event that $\{c(k)\}$ is such that $\{c(k)\} + \{e(k)\}$ is an allowable sequence.

ϵ_2: the event that the received signal is such that the likelihood function is greater for $\{c(k)\} + \{e(k)\}$ than for $\{c(k)\}$ (i.e. (26) is satisfied).

Also, let E denote the set of error events ϵ which gives allowable sequences.

The event that the sequence $\{c(k)\} + \{e(k)\}$ has maximum-likelihood is included in the event ϵ_2 and the probability that ϵ occurs is therefore

$$Pr[\epsilon] \le Pr[\epsilon_1] Pr[\epsilon_2] \qquad (31)$$

The probability that any error event occurs is

$$Pr[E] = \sum_{\epsilon \in E} Pr[\epsilon] \le \sum_{\epsilon \in E} Pr[\epsilon_1] Pr[\epsilon_2] \qquad (32)$$

This bound becomes tight when N_0 is small. This formula can therefore be used to estimate the error rate in this case. When N_0 is small, the error is dominated by the event which corresponds to the smallest distance $d(\{e(k)\})$. In this case, the probability that any error event occurs is well approximated by

$$Pr[E] \simeq \sum_{\epsilon \in E_{min}} Pr[\epsilon] \qquad (33)$$

where E_{min} is the set of all error events with minimum distance.

4 Evaluation of Zero-Forcing Equaliser

The error probability derived for the maximum-likelihood receiver in the previous section can be used to evaluate the performance of other receiver structures. The maximum-likelihood receiver is optimal in the sense of minimum probability of error. By comparing other receivers to its performance it can be evaluated how close to optimum they perform. In this section the zero-forcing equaliser is evaluated by comparing its performance to that of MLSE for the same received signal $r(t)$.

Assume that the transmitter in Fig. 1 is transmitting 4-QAM symbols and that $h(t)$ is a unit energy rectangular pulse of length T. Also assume that the channel $c(t)$ has a transfer function that can be written

$$C(f) = 1 + \alpha e^{j\theta} e^{-j2\pi f \tau} \tag{34}$$

The noise $n(t)$ has power spectral density $N_0/2$.

Zero-Forcing Equaliser

This is the same case as was used in [7], where the bit error rate (BER) for a receiver with zero-forcing equaliser was calculated. Thus, from [7], the BER is

$$BER \approx \frac{1}{N} \sum_{n=0}^{N-1} Q\left(\sqrt{\frac{E_b}{\eta_n}}\right) \tag{35}$$

where E_b is the transmitted energy per bit. The noise variance η_n is defined as

$$\eta_n = \frac{N_0}{2} \int_{-\infty}^{\infty} \left|\frac{1}{C(f)} \cdot H(f - \frac{n}{T})\right|^2 df \tag{36}$$

where $H(f)$ is Fourier transform of $h(t)$.

Maximum-Likelihood Receiver

As a non-coded system is considered, vectors of arbitrary symbols may occur in any order. In this case, error events with minimum distance correspond to single symbol errors. These error events are vectors $e(0)$ with a single component not equal to zero. The position of the non zero component depends on the channel parameters τ and θ. The minimum distance error events correspond to errors in OFDM channels where the transfer function $C(f)$ has its minima. To get an expression for the BER that does not depend on where the minima are located, error events in all positions of $e(k)$ are included. For a single bit error in channel n, the squared distance from (27) is

$$d^2(\{e(0)\}) = \rho_{n,n}(0) \tag{37}$$

In each of the N position of $e(k)$, there are 4 possible error events; $1, -1, j$, and $-j$. The probability that any of these error event occurs is given by (32). As it was assumed that only one of $2N$ bits is wrong, the BER is

$$BER \approx \frac{1}{N} \sum_{n=0}^{N-1} Q\left(\sqrt{\frac{\rho_{n,n}(0)}{N_0}}\right) \qquad (38)$$

Evaluation

Fig. 4 shows the BER as a function of the level of the second path, α. The solid line is for a maximum-likelihood receiver and the dashed line for a receiver with zero-forcing equaliser. The number of channels is 16 and the transmitted energy per bit is such that $E_b/N_0 = 9$ dB. The channel has $\tau/T = 0.125$ and $\theta = \pi$.

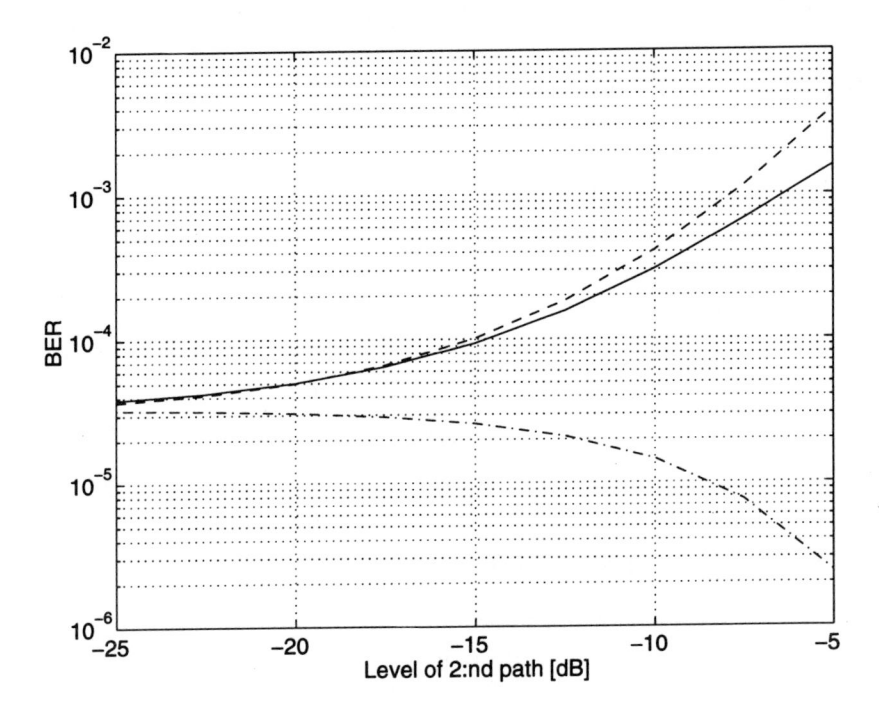

Fig. 4. BER as a function of the level of the second path of a two-path channel for OFDM system with MLSE (solid line), OFDM system with linear equaliser (dashed line) and single carrier system with MLSE (dash-dot line).

The performance is similar for both receivers. This shows that for the channel considered here, the zero-forcing equaliser performs close to optimum.

When designing the receiver, it is therefore not worth considering solutions that are more complex than the zero-forcing equaliser.

5 Comparison to Single Carrier Modulation

In this section, OFDM is compared to single carrier modulation for transmission over the two-path channel from the previous section. Both systems transmit the same power, use the same bandwidth and information rate. A maximum-likelihood receiver is used in both cases.

OFDM

The configuration is identical to the OFDM system with maximum-likelihood receiver used in the previous subsection.

Single Carrier Modulation

A block diagram of the transmitter and channel for the single carrier case is given in Fig. 5. The pulse shaping filter $g(t)$ was chosen to be a root raised

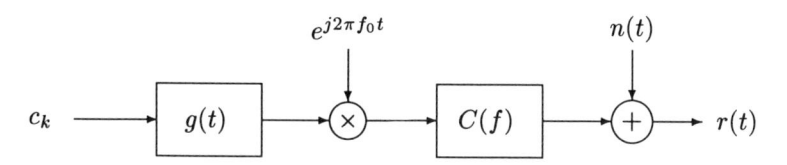

Fig. 5. Block diagram of single carrier system with two-path channel.

cosine filter with roll-off 0.5 [8]. The baseband signal is shifted to f_0 in order to occupy the same frequency band as the OFDM signal. The maximum-likelihood receiver and its BER for this case are found in e.g. [4, 5].

Using the same constellation as in the previous section, the BER is

$$BER \approx Q\left(\sqrt{\frac{s_0}{N_0}}\right) \qquad (39)$$

where

$$s_0 = \int_{-\infty}^{\infty} |G(f - f_0)C(f)|^2 \, df \qquad (40)$$

where $G(f)$ is the Fourier transform of $g(t)$.

Evaluation

The BER of the two systems are shown in Fig. 4. The solid line represents the OFDM system and the dash-dot line the single carrier system. Both systems have the same channel parameters as in the previous section.

The results show that for small levels of the second path, both systems have the same performance. That is as expected because the channel is close to an ideal white Gaussian channel.

When the level of the second path, α, is increased, the BER increases for the OFDM system but decreases for the single carrier system. This can be explained by considering the change in distances $d(\{e(k)\})$ when α is increased. In the OFDM system, some of the events which have minimum distance for $\alpha = 0$ have increasing distance and some have decreasing with increasing α. In the single carrier system all distances are increasing with increasing α. The increase is smaller than the largest increase in the OFDM system but as the minimum distance error events are dominating, the single carrier system performs better.

The single carrier system is able to utilise the additional energy that comes from the second path and thereby reduce the BER when α increases. In the OFDM system, some channels suffer from a large degradation in SNR and the BER is consequently increasing with an increasing α.

6 Conclusions

A maximum-likelihood sequence estimation (MLSE) algorithm for OFDM has been derived. An upper bound on probability of error, which is tight for high SNR, was also computed.

The error bound of the MLSE was used to assess the performance of an OFDM system with a two-path channel and a linear zero-forcing equaliser. The results show that the equaliser has close to optimal performance for this channel.

The OFDM system with MLSE was compared to a single carrier system which also used MLSE. The single carrier system gives a lower BER when a two-path channel is used. It should also be noted that MLSE is usually far more complex for OFDM than for single carrier modulation. Even if the MLSE for OFDM is prohibitively complex to implement, its performance is useful as a reference on what is possible to attain.

In the cases presented in this paper, the channel transfer function is not known in the transmitter. The performance of the OFDM system can be improved if the transmitted signal is properly adapted to the channel characteristics. It has been shown [9] that OFDM with this adaptation can give better performance than single carrier modulation.

The work presented in this paper shows attainable performance of non-coded OFDM systems. A further study should also investigate what can be achieved with error correction codes which use channel state information. This should be a suitable technique as the reliability of the symbols is different in different channels.

References

[1] J. A. C. Bingham, "Multicarrier modulation for data transmission: An idea whose time has come," *IEEE Communications Magazine*, vol. 28, pp. 5–14, May 1990.

[2] M. Alard and R. Lassalle, "Principles of modulation and channel coding for digital broadcasting for mobile receivers," *EBU Review*, pp. 47–69, Aug. 1987.

[3] J. S. Chow, J. C. Tu, and J. M. Cioffi, "A discrete multitone transceiver system for HDSL applications," *IEEE J. Select. Areas Commun.*, vol. 9, pp. 895–908, Aug. 1991.

[4] G. D. Forney, Jr., "Maximum-likelihood sequence estimation of digital sequences in presence of intersymbol interference," *IEEE Trans. Inform. Theory*, vol. IT-18, pp. 363–378, May 1972.

[5] G. Ungerboeck, "Adaptive maximum-likelihood receiver for carrier-modulated data-transmission systems," *IEEE Trans. Commun.*, vol. COM-22, pp. 624–636, May 1974.

[6] R. E. Blahut, *Digital transmission of information*. Reading, Massachusetts, USA: Addison Wesley, 1990.

[7] A. Vahlin and N. Holte, "Use of a guard interval in OFDM on multipath channels," *IEE Electron. Lett.*, vol. 30, pp. 2015–2016, 24th Nov 1994.

[8] E. A. Lee and D. G. Messerschmitt, *Digital Communication*. 3300 AH Dordrecht, The Netherlands: Klüwer Academic Publishers, 2 ed., 1990.

[9] I. Kalet, "The multitone channel," *IEEE Trans. Commun.*, vol. 37, pp. 119–124, Feb. 1989.

A New Analysis Method for the Performance of BandPass Sampling Digital Demodulators Considering IF Filter Effects

L. P. Sabel

Institute for Telecommunications Research,
SPRI Building, University of South Australia, The Levels, SA 5095, Australia

Abstract

This paper presents a method for the analysis of the performance of digital demodulators which use BandPass Sampling to perform both signal spectral shifting and the suppression of noise and spurious signals. The bandpass sampling technique can simplify the front–end processing of a digital receiver by eliminating RF downconversion stages. In addition, as the signal is sampled prior to downconversion to baseband, accurate conversion can be performed without the usual analog circuit problems of carrier I/Q phase offsets, amplitude imbalance and carrier feed–through. In this paper the effects of aliasing on the noise density at baseband and the effect of the BPF on the wanted signal characteristics are analysed.

1 Introduction

The use of BandPass Sampling (BPS) in communications receivers is receiving increasing attention due to its ability to simplify and reduce the analog portions of the receiver in the IF region, and the increasing availability of suitable hardware for implementation. For example [1] discusses performance implications while [2] demonstrates the use of the technique in a satellite communications receiver application, and [3] details the theory of the general use of bandpass sampling. A number of more descriptive articles have also been published, e.g. [4].

Figure 1 shows the block diagram of a bandpass sampling receiver. The IF signal at frequency f_I is sampled at a frequency of f_s which then generates an image at f_d. The required image at f_d is then downconverted to baseband through multiplication with a complex sinusoid of frequency f_d. The sampling rate must be selected so the images generated by the downconversion are not aliased into the wanted signal. We note that the use of the BPS technique eliminates the final analog downconversion process of an equivalent lowpass analog system and also allows the final baseband quadrature downconversion to be performed in the digital domain where exact quadrature and amplitude balances can be obtained. In addition only one ADC is required. The disadvantages are the need for a faster Sample and Hold (S/H) and digital processing circuitry although this last point is marginal due to the increasing speed and density of DSP hardware. Note that low–IF sampling techniques [5] which use a combination of the bandpass and lowpass sampling techniques offer trade–offs between these advantages and disadvantages.

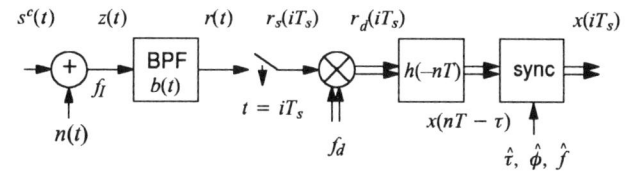

Fig. 1 Bandpass sampling receiver architecture.

This paper contributes to the body of knowledge in the field of BPS through the theoretical analysis of the process as applied to digital communications receivers. In particular, new results are derived for the description and analysis of aliased noise and the effects the IF BandPass Filter (BPF) has on BER performance considering both ISI and noise effects.

The structure of this paper is as follows: Section 2 establishes the time domain system model of the BPS receiver while Section 3 analyses the frequency domain relationships involved. Section 4 deals with the characterisation of aliased noise due to non–ideal IF BPF characteristics and Section 5 proposes a BER analysis method for BPS digital demodulators.

2 System Model

The system signal model is shown in Fig. 1. The complex transmitted signal, $y(t) = s^I(t) + js^Q(t)$, is the convolution of the coded data symbols, $\underline{a}$, and the transmit filter impulse response $h(t)$ where $a_i = a_i^I + ja_i^Q$ represents the ith symbol which is transmitted at $t = iT$, T is the symbol period, and $a_i^{I,Q} \in \{-A, A\}$. The superscripts I and Q indicate the Inphase and Quadrature channels respectively. If $a_i^Q = 0$ the transmitted signal is BPSK, whilst if the quadrature portion is included the signal is QPSK. Higher order modulation schemes, MPSK or QAM, can be generated by using multilevel complex data. Hence, for the I channel,

$$s^I(t) = \sum_{i=-\infty}^{\infty} a_i^I h(t-iT) \tag{1}$$

and similarly for the Q channel. The filtered data sequence is then modulated by a complex exponential with the real part of the result being the bandpass signal

$$s^c(t) = Re\{y(t)\sqrt{2}\,e^{j2\pi f_I t}\} \tag{2}$$

where f_I is the transmit oscillator frequency and $y(t) = s^I(t) + js^Q(t)$. For the purpose of this paper no effects at frequencies greater than IF, i.e. f_I, will be considered. The received signal, $z(t)$, is the sum of the transmitted signal, $s^c(t)$, and complex bandpass Additive White Gaussian Noise (AWGN) with double sided Power Spectral Density (PSD) $N_0/2$ such that $z(t) = s^c(t) + n(t)$. The signal is then filtered by a bandpass filter with impulse response $b(t)$ to form $r(t) = z(t) \otimes b(t)$ where $\otimes$ indicates convolution. $r(t)$ is then sampled at a rate of $f_s = 1/T_s$ where T_s is the sampling period, to obtain $r_s(iT_s) = r(t)|_{t=iT_s}$. The received bandpass sampled signal is then demodulated (frequency translated) by multiplying the signal by a complex exponential. The complex exponential generated by the receiver has frequency f_d and a phase relative to the transmit oscillator of ϕ. At this stage we assume that f_I, f_s and f_d have been appropriately chosen to ensure that a pair of complex conjugate images occur at baseband. The selection of appropriate frequencies is dealt with in [5]. The resultant signal is

$$r_d(iT_s) = r_s(iT_s)\sqrt{2}\,e^{-j(2\pi f_d iT_s - \phi)}. \tag{3}$$

A matched filter is then applied to remove unwanted image terms and provide optimal signal shaping. Assuming ideal frequency and phase synchronisation the I channel signal component can be described by

$$x^I(iT_s) = \sum_{j=-\infty}^{\infty} a_j^I g_b(iT_s - jT - \tau) + v^I(iT_s) \tag{4}$$

where $v^I(t)$ is receive filtered Gaussian noise in the I channel and τ is the sampler timing offset from the symbol centre. The impulse response $g(t)$ is the result of the convolution of the transmit and receive filter impulse responses such that $g(t) = h(t) \otimes h(-t)$. The raised cosine filter characteristic with roll-off factor α will be used for $g(t)$ throughout. In (4) $g_b(t)$ indicates that the normal receive signal impulse response has been distorted by $b_L(t)$ the lowpass equivalent of the bandpass filter $b(t)$ and is described as $g_b(t) = g(t) \otimes b_L(t)$.

As the sampled version of the coherent received signal will be referred to with respect to its relative symbol location we introduce the subscript n to denote the nth symbol period, as $x_n(\tau) = y_n(\tau) + v_n$ where $x_n(\tau) = x(nT - \tau)$ and

$$y_n(\tau) = \sum_{j=-\infty}^{\infty} a_{n-j} g_b(jT - \tau). \tag{5}$$

3 Frequency domain analysis

Observation of the nature of the signals in the frequency domain gives insight into the relationships between f_I, f_d and f_s. The effect of the noise component will be ignored to allow the operations on the signal to be clearly shown.

The spectrum of $r(t)$ can be described as $R(f) = Z(f)B(f)$ where $B(f) = \mathcal{F}\{b(t)\}$, and

$$Z(f) = \mathcal{F}\{z(t)\} = \frac{\sqrt{2}}{2}(H(f-f_I) + H^*(-f-f_I)) \tag{6}$$

[6] p406 where the * superscript indicates complex conjugate, $H(f) = \mathcal{F}\{h(t)\}$ is the lowpass spectrum of $s(t)$ and $\mathcal{F}\{.\}$ is the continuous time Fourier Transform. After the sampling process the spectrum of $r_s(iT_s)$ is

$$R_s(f) = \frac{\sqrt{2}}{2} \sum_{k=-\infty}^{\infty} (H_b(f-f_I + kf_s) + H_b^*(-f-f_I + kf_s)) \tag{7}$$

where $H_b(f) = H(f)B_L(f)$ and $B_L(f)$ is the lowpass equivalent spectrum of the BPF. Note that $H_b(f)$ is a lowpass equivalent spectrum and hence $H_b(0)$ corresponds to evaluating the response at the centre of the "passband", that is DC. Clearly (7) shows that the spectral images of each part of (7) are positioned at integer multiples of f_s. Careful selection of f_s will ensure that the images produced will not overlap. This is shown in Fig. 2.

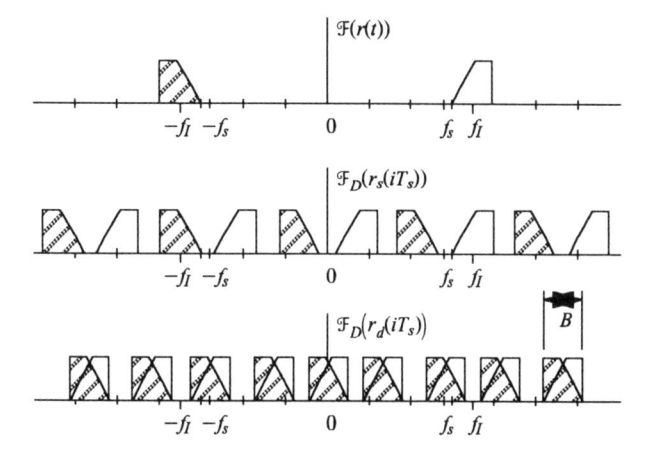

Fig. 2 Example spectra.

The spectrum of the downconverted sampled received signal, $r_d(iT_s)$, may be obtained by convolving the spectrum (7) with the frequency domain representation of the quadrature sinusoids used, that is

$$R_d(f) = R_s(f) \otimes \mathcal{F}_D\left\{\sqrt{2}\, e^{-j(2\pi f_d iT_s - \phi)}\right\}. \tag{8}$$

where $\mathcal{F}_D\{.\}$ is the Discrete Time Fourier Transform (DTFT). Assuming $\phi=0$, the real part of (8), that is the signal in the I channel, may be written as

$$R_d^I(f) = \frac{1}{2}\left\{\begin{array}{l} \sum_{k=-\infty}^{\infty} (H_b(f-f_d-f_I + kf_s) + H_b^*(-f-f_d-f_I + kf_s) \\ + H_b(f + f_d-f_I + kf_s) + H_b^*(-f + f_d-f_I + kf_s)). \end{array}\right. \tag{9}$$

A similar result may be obtained for the Q channel.

4 Filtering effects in BPS

In this paper we use two BPS systems to demonstrate the theory presented. The filters used in both systems are 5th order Butterworth BPFs with nominal 1dB bandwidths of 1.5MHz suitable for use with 1.024Msymbol/sec QPSK with $\alpha=0.4$ which has an ideal bandwidth of 1.434MHz. The frequencies used for f_I, f_s and f_d for each system are shown in Table 1 and have been calculated using the frequency selection method described in [5]. We note that in both cases $f_d=f_s/4$ to allow simplified downconversion techniques to be used. Figs. 3 and 4 show the frequency responses of the Butterworth BPFs for systems 1 and 2 respectively.

	f_I (MHz)	1dB Bandwidth (MHz)	f_s (MHz)	f_d (MHz)
system 1	71.680	1.5	8.192	2.048
system 2	72.704	1.5	4.096	1.024

Table 1: The frequencies used in systems 1 and 2.

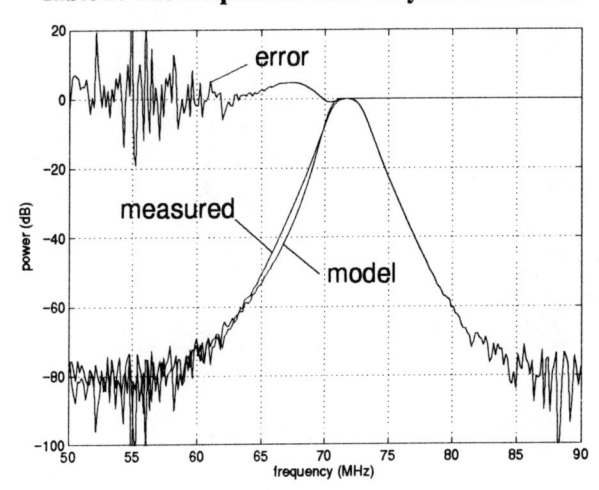

Fig. 3 Frequency response of the 5th order Butterworth filter centred at f_c=71.680MHz, the digital model and the resultant error response.

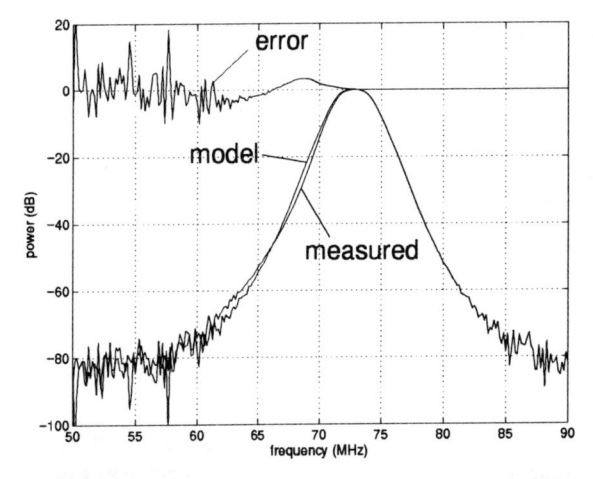

Fig. 4 Frequency response of the 5th order Butterworth filter centred at f_c=72.704MHz, the digital model and the resultant error response.

Figures 5 and 6 show overlay spectra for the system 1 and 2 BPFs respectively. The overlay spectra provide a visualisation of the way the BPF out-of-band regions impact on the passband and clearly demonstrate that the lower sampling rate of system 2 increases the out-of-band noise power which is aliased back into the passband. The resultant noise density can be calculated by summing the out-of-band BPF spectral values at frequencies which line up in the overlay spectra. Note that the near white out-of-band noise spectrum shown is due to the background noise of the network analyser used to measure the filter responses.

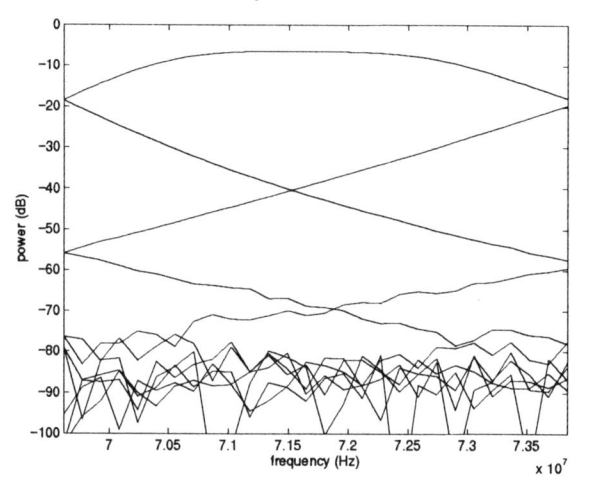

Fig. 5 Overlay spectrum of the f_c=71.680MHz filter with f_s=8r_s, system 1.

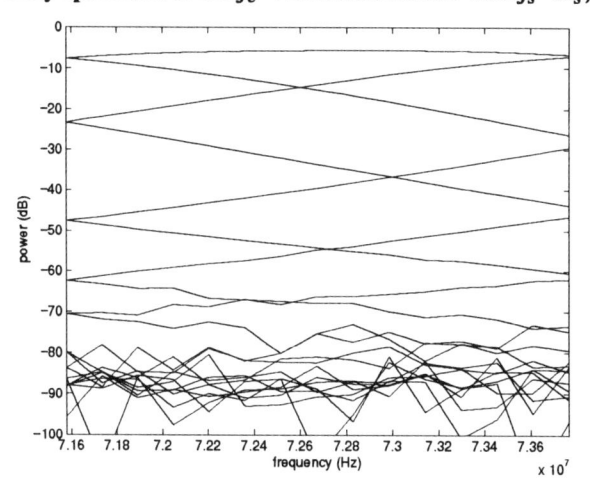

Fig. 6 Overlay spectrum of the f_c=72.704MHz filter with f_s=4r_s, system 2.

The aliased noise density can be calculated by modifying (7):

$$N_0^a(f) = \frac{N_0}{2} \sum_{k=-\infty}^{\infty} |B(f-f_I + kf_s)|^2 + |B^*(-f-f_I + kf_s)|^2 \qquad (10)$$

where $-\frac{f_s}{4} \leq f \leq \frac{f_s}{4}$ and $k \neq 0$. Figures 7 and 8 show the noise PSD $N_0^a(f)$ of systems 1 and 2 respectively where the image shown is centred at f_I. The noise spectra shown repeat along the frequency axis at intervals of $f_s/2$ with every second image being reflected about the image centre. The aliased noise PSD of system 2 is approximately 30dB worse than that of system 1. This is due to relatively low out–of–band attenuation near $f_c \pm f_s/4$ and is attributable to the low sampling frequency used. The total noise PSD $N_0^{aT}(f)$ is simply the sum of the passband filtered additive noise, $B(f)N_0$, and the aliased noise, and may be obtained by including the $k=0$ case in (10). This is also shown in Figs. 7 and 8.

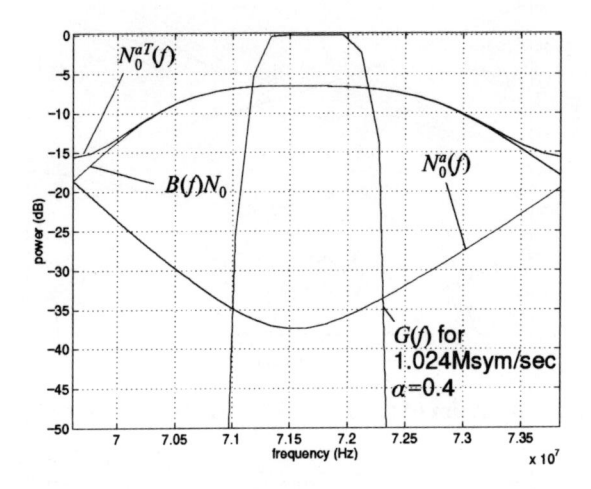

Fig. 7 Theoretical noise PSD for system 1.

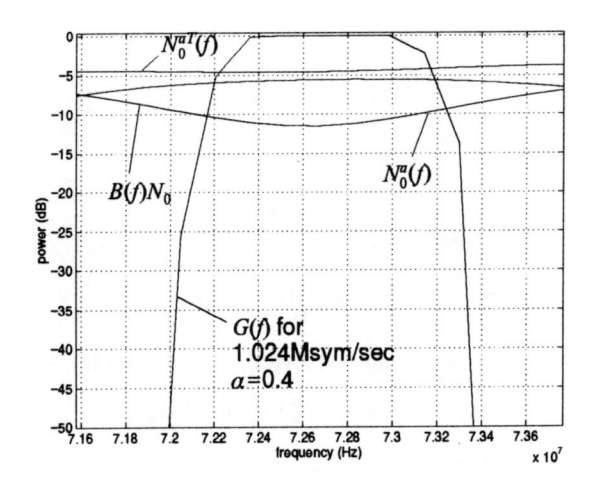

Fig. 8 Theoretical noise PSD for system 2.

The total noise PSD at baseband, including the aliased noise, after final downconversion from f_d, is

$$N_0^T(f) = \frac{N_0}{4}\left\{\begin{array}{l}\sum_{k=-\infty}^{\infty}\left(|B(f-f_d-f_I + kf_s)|^2 + |B^*(-f-f_d-f_I + kf_s)|^2\right. \\ \left. + |B(f + f_d-f_I + kf_s)|^2 + |B^*(-f + f_d-f_I + kf_s)|^2\right)\end{array}\right\} \quad (11)$$

where $-\frac{f_s}{4} \leq f \leq \frac{f_s}{4}$. Again the $k \neq 0$ case represents the aliased noise with the $k=0$ contribution being the in-band noise. If f_s, f_d and f_I are selected such that a complex image pair occurs at DC (with the other images being centred at integer multiples of $f_s/2$) then the resultant PSD is simply half the sum of two mirror imaged versions of the spectra shown in Figs. 7 and 8.

The theoretical results are supported by practical measurements of a bandpass signal sampled using an appropriate S/H and ADC combination with the frequency parameters of the implemented BPS systems shown in Table 1. The test signals consisting of a sinusoid at the filters' centre frequencies corrupted by bandpass filtered noise are shown in Figs. 9 and 10 respectively. The power spectra of the image centred at f_d obtained from the ADC are shown in Figs. 11 and 12 for systems 1 and 2 respectively. The power spectra were generated by averaging the modulus squared values of 4 8192 point FFTs incorporating Hanning windowing. To obtain estimates of the signal power the largest 16 bins where summed. The estimate of the noise PSD was obtained by averaging the values of the bins within a 1MHz bandwidth, offset from the sinusoid by approximately 100kHz. Similarly, the signal and noise PSD of the input signal where obtained by applying the same averaging procedure to the output of the spectrum obtained from a spectrum analyser. The resultant Signal to Noise density Ratios (SN$_0$R) are shown in Table 2.

	input to S/H (dB)	ADC output (dB)	practical difference	theoretical N_0 loss	practical – theoretical
system 1	74.25	73.09	1.16	0.004	1.156
system2	74.17	71.45	2.72	1.710	1.010

Table 2: SN$_0$R results.

The theoretical results indicate that the difference between the SN$_0$R of the input and output of the sampling system should be approximately 0.004dB and 1.71dB for systems 1 and 2. The results show that the difference between the measured and theoretical results is consistently approximately 1dB. This is considered to be within the accuracy of the measurements taken. It is clear that while the accuracy of the practical results shown do not completely verify the analysis method above, they do give strong support. In addition, comparison of both the shape of the spectra, and the peak to band edge values, in Figs. 11 and 12 with those of $N_0^{aT}(f)$ in Figs. 7 and 8 respectively shows good agreement. Note that the spectrum shape in Fig. 11 is reversed relative to that in Fig. 7, or indeed the general shape of the filter response shown in Fig. 9. This is due to a reversal of the spectrum due to the sampling process as predicted by (10). As digital communications signals are generally spectrally symmetrical this does not present a problem as demonstrated by the eye diagram of the received signal using the system 1 scenario in [5].

The results show that if the BPF bandwidth and f_s are chosen such that the BPF response is small at the passband edges, i.e. $f_c \pm f_s/4$, then simple filters can be implemented with very low resultant aliased noise and hence implementation loss.

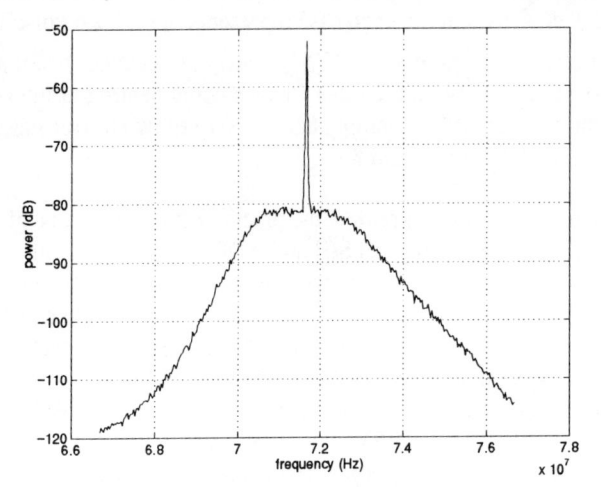

Fig. 9 Spectrum of the test signal for system 1.

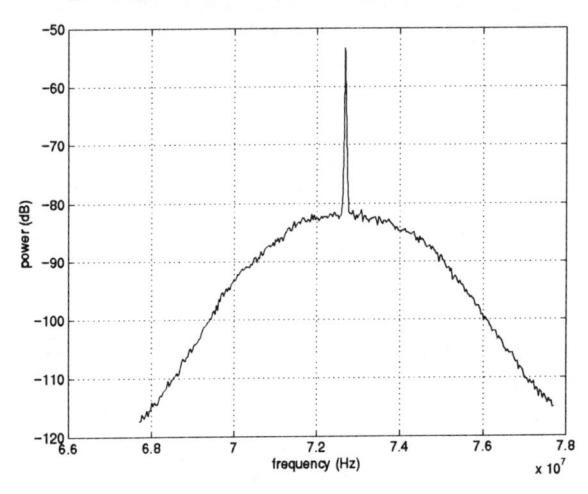

Fig. 10 Spectrum of the test signal for system 2.

5 BER analysis

As the noise density $N_0^T(f)$ varies with f, and an effective value is required for the evaluation of the probability of error, we determine the expected value of the total noise density using

$$\tilde{N}_0^T = E_f\{N_0^T(f)\} = \int_{-\frac{r_s}{2}(1+a)}^{\frac{r_s}{2}(1+a)} N_0^T(f)|G(f)|^2 df \tag{12}$$

where the raised cosine frequency response, $G(f)$, has been used to weight the significance of the relative frequencies. Note that the total area under the $|G(f)|^2$ equals unity.

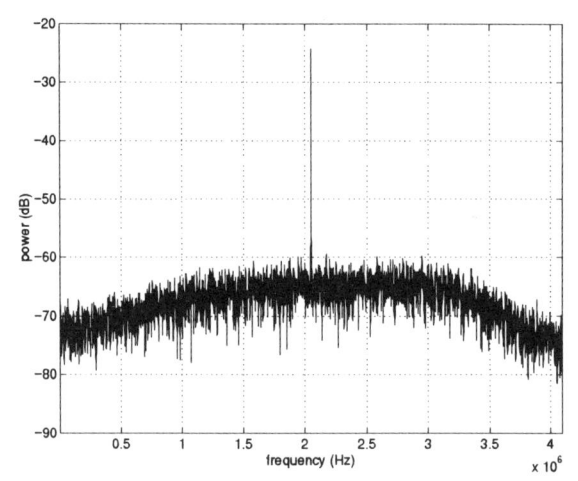

Fig. 11 Spectrum of the ADC output signal for system 1.

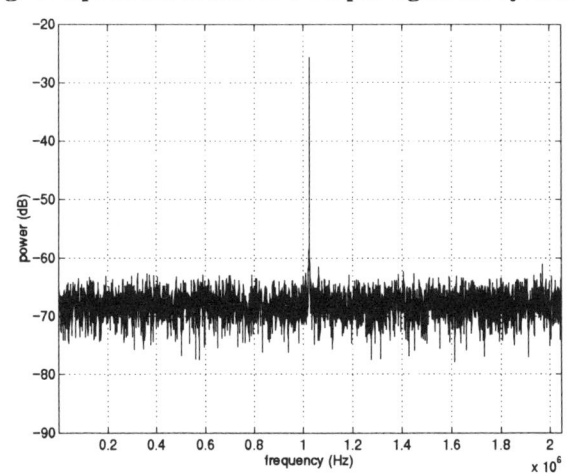

Fig. 12 Spectrum of the ADC output signal for system 2.

To determine the ISI introduced by the BPF we require $g_b(t)$ as in (4) and (5). In this paper $g_b(t)$ is obtained by applying the frequency sampling FIR design method [6] to a shifted copy of the bandpass magnitude spectrum located at $f_c=0$ where the negative frequency response is a reflected version of the positive frequency response. To test the result we take the FFT of the resultant $b_L(t)$, and shift the frequency response to a centre frequency of f_c. The resultant model spectra are shown in Figs. 3 and 4 along with the error resulting from subtracting the actual measured filter response from the model filter response. The error curve shows that perfect modelling is obtained for $f > f_c$, as expected, with some relatively minor errors, generally at large out-of-band frequencies for $f < f_c$.

Given $b_L(t)$ and hence $g_b(t)$, and considering the finite effective memory of the impulse response $g_b(t)$, (5) may be written as

$$y_n(\tau) = \sum_{i=-M}^{M} a_{n-i} g_b(iT - \tau). \tag{13}$$

Using the total average noise PSD, $\tilde{N}_0^T$, the probability of error for BPSK with no phase or frequency offset, that is, binary PAM, can now be determined using

$$P_e = \frac{1}{2^{2M}} \sum_{\forall \underline{a}} Q\left[\frac{y_n(0)}{A} \sqrt{\frac{2E_b}{\tilde{N}_0^T}}\right] \tag{14}$$

where $y_n(0)$ is the value of the nth symbol sampled at the centre of the symbol (and has a nominal value of A), that is, the ideal sampling location and $\underline{a}$ represents the set of all data sequences $\underline{a}$ of length $2M+1$ where the centre symbol value is A.

The probability of error results in the form of Implementation Loss (IL), that is the increase in E_b/N_0 relative to the ideal E_b/N_0 for a particular P_e, are shown in Fig. 13 for system 1. The lower curve is the case of no BPS, that is no aliased noise and no BPF ISI effects, and hence represents the losses incurred by the non-optimality of the transmit and receive filters only. These filters were both root raised cosine FIR filters with $\alpha = 0.4$ and 6 symbols in length. The upper curve shows IL for the BPS case. The loss due to aliased noise is immediately apparent as a vertical shift. The increasing loss as E_b/N_0 increases is due to ISI effects. The results show that although some additional loss is incurred it is insignificant even at $E_b/N_0 = 12$dB where IL=0.035dB. Note that the use of analog downconversion could quite easily incur more loss than that demonstrated by the BPS system here.

For the $f_s = 4.096$MHz case the IL due to aliased noise is 1.7dB. Clearly the losses in this case preclude its use. Note however, that this case has only been included to demonstrate the aliased noise effects discussed.

6 Conclusions

This paper has described a new analysis method for determining the amount of noise which is aliased into the passband during the bandpass sampling process. In addition measurements were taken using a practical system which demonstrate the effect of aliased noise and the validity of the analysis method. A method has been presented for the determination of the resultant BER for BPSK using a BPS receiver which takes into consideration the signal distortion produced by the BPF.

The results show that simple BPFs may be used to implement BPS schemes. As long as the correct combinations of f_i, f_s and f_d are selected system performance will probably be improved and analog complexity reduced while the increased complexity of the digital receiver will be minimal.

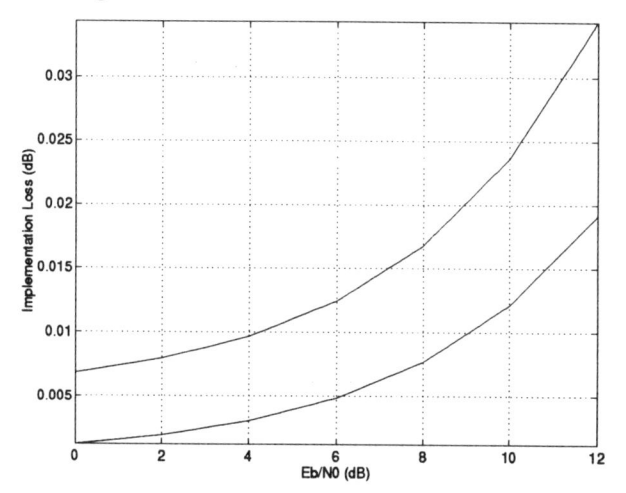

Fig. 13 Implementation Loss for system 1.

Acknowledgments

The author thanks Duong Tran, Gerald Bolding, Jeff Wojtiuk and Philip McIllree for there assistance and acknowledges the Australian Space Office colaborative grant.

References

[1] J. K. Cavers and S. P. Stapleton, "A DSP–Based Alternative to Direct Conversion Receivers for Digital Mobile Communications", *Proc. IEEE GLOBECOM*, San Diego, USA, pp. 2024–2029, December, 1990.

[2] L. P. Sabel, "A DSP Implementation of a Robust Flexible Receiver/Demultiplexer for Broadcast Data Satellite Communications", *Proc. Communications 90*, Melbourne, Australia, pp. 218–223, October 1990.

[3] R. G. Vaughan, N. L. Scott and D. R. White, "The Theory of Bandpass Sampling", *IEEE Trans. Signal Proc.*, Vol. 39, No.9, pp. 1973–1984, September 1991.

[4] R. Groshong and S. Rusack, "Undersampling Techniques Simplify Digital Radio", *Electronic Design*, May 23 1991.

[5] A. Guidi and L. P. Sabel, "Digital Demodulator Architectures for Bandpass Sampling Receivers", *Proc. 7th International Tyrrhenian Workshop on Digital Communications*, Viareggio, Italy, September, 1995.

[6] J. G. Proakis and D. G. Manolakis, "*Digital Signal Processing: Principles, Algorithms and Applications*", second edition, MacMillian, 1992.

Comparison of Demodulation Techniques for MSK

Uwe Lambrette, Ralf Mehlan and Heinrich Meyr

Lehrstuhl für Integrierte Systeme der Signalverarbeitung
RWTH Aachen

Abstract. For MSK, three demodulators are compared. The first demodulation algorithm, partially coherent demodulation, is based on a classical matched filter approach combined with feedforward phase synchronization, whereas the second algorithm, block demodulation, is based on minimizing a distance measure based on the symbol vector trial and the observed differential phase vector. The third algorithm is based on the same distance measure, however the minimization is carried out using the viterbi algorithm. We provide a derivation of the second algorithm. It is shown that the first approach is superior both in performance and computational complexity. The first algorithm also exhibits the best robustness properties in the case of signal impairments.

1 Introduction

MSK IS a modulation technique widely known to the research community. Although the difficulty of synchronizing MSK signals due to their stronger intersymbol interference (ISI) effects difficulties in finding optimum receivers, MSK *is* attractive in all cases where nonlinearities in the analogue signal path enforce using a modulation technique with minimal amplitude fluctuations.

This is the case especially when either the application domain has not yet been explored extensively for finding (almost) linear analogue receivers and transmitters or when receiver nonlinearities are accepted for reducing implementation costs. However, in all cases, one is still interested in finding optimum demodulators with an acceptably simple implementation.

In this contribution, we investigate three modulation methods for MSK, which are a partially coherent MSK demodulator (PC) and an incoherent block demodulation approach (BD) [1] and thirdly differential phase demodulation using the Viterbi algorithm (VA). In [2], a similar scheme is employed using a limiter discriminator, in [3] Viterbi sequence estimation is used based on the baseband signal. In here, we will focus on Viterbi detection considering the differential phase.

After briefly giving necessary definitions and describing the three demodulators mentioned already we compare the demodulators in terms of their computational efficiency and their performance. We derive the BD approach from coherent demodulation. The VA demodulator is motivated by observation of the MF outputs.

In any case, our analysis is limited to the effects of carrier phase distortion in a non (or only slowly) flat fading channel. Timing recovery is not discussed in here, but has been discussed previously in [4,5,6] and others.

2 Signal Models

2.1 MSK

MSK modulation can be described in two ways; firstly, we look at the signal phase only, in this case the *modulated signal* for $t \in [KT, (K+1)T]$ (symbol K is transmitted) and $\tau = t - KT$ is described as

$$s(t) = c \cdot \exp\left[j\left(\sum_{k=0}^{K-1} a_k \frac{\pi}{2} + a_K \frac{\pi}{2} \frac{\tau}{T}\right)\right] e^{j\Phi_0} \qquad (1)$$

where $a_k = \pm 1$ represent the data bits.

Quantities having an impact on the signal phase may be gathered in matrices: Let the signal phase vector $[\ldots \arg(s(KT)))\ldots]^T$ be denoted as $\boldsymbol{\Psi} + \boldsymbol{\Phi}_0$. The vector $\boldsymbol{\Psi}$ is expressed in terms of the symbol vector $\boldsymbol{a}$ as

$$\boldsymbol{\Psi} = \frac{\pi}{2} \underbrace{\begin{pmatrix} 1 & 1 & \cdots & 1 \\ 0 & 1 & \cdots & 1 \\ \vdots & \ddots & \ddots & \vdots \\ 0 & \cdots & 0 & 1 \\ 0 & \cdots & 0 & 0 \end{pmatrix}}_{\mathbf{W}} \boldsymbol{a} \bmod 2\pi \quad \begin{array}{l} \boldsymbol{a} \text{ is of size } L \times 1, \\ \mathbf{W} \text{ of size}(L+1) \times L. \end{array} \qquad (2)$$

Let us introduce the modulation pulse with energy E_B and duration $2T$

$$g(t) = \left\{ \begin{array}{ll} \frac{\sqrt{E_B}}{T} \cos \frac{\pi t}{2T} & t \in [-T, T] \\ 0 & \text{elsewhere} \end{array} \right\}. \qquad (3)$$

Using Eulers formula and performing some lengthy computation, we achieve a convenient baseband representation of the MSK signal for all t [7].

$$s(t) = c \cdot e^{j\Phi_0} \sum_{K=0}^{K=N-1} b_K j^K g(t - KT), \qquad (4)$$

where $b_K = \prod_{k=0}^{K-1} a_k$ for $K > 0$ and $b_K = \prod_{k=K-1}^{0} a_k$ for $K \leq 0$. For our analysis we assume a white additive gaussian noise channel described by a double sided noise spectral density N_0. The Phase offset Φ_0 is initially assumed constant, but during simulations, we investigated a dynamic phase, as well. For simplicity, a vector consisting of elements Φ_0, ie $[\ldots \Phi_0 \ldots]^T$ also is denoted with Φ_0.

Hence both I and Q channel are modulated alternatingly with the pulse $g(t)$ leading to a signal with constant amplitude. The received signal is thus expressed as

$$r(t) = s(t) + n'(t) \qquad (5)$$

In the following, let $c = 1$.

2.2 The Matched Filter Approach

The optimum approach to demodulate MSK is to employ a matched filter (MF). Filtering any pulse $g(t)$ with the matched filter $g(-t)$ will lead to the MF outputs

$$m'(t) = \int_{\max(0,t)}^{\min(2T, 2T+t)} g(\tau') \cdot g(t-\tau') d\tau' = \left(\left(1 - \frac{|t|}{2T}\right) \cos\frac{\pi t}{2T} + \frac{1}{\pi}\sin\frac{\pi|t|}{2T}\right) E_B \quad (6)$$

when $t \in [-2T, 2T]$ and $m'(t) = 0$ elsewhere. Hence $b_K m'(\pm T) = E_B b_K/\pi$, $b_K m'(0) = E_B b_K$ and $b_K m'(kT) = 0, |k| > 1$. The output of the matched filter applied to the *complex* noisy baseband signal (5) thus has the form

$$m(t) = e^{j\Phi_0} \sum_{K=0}^{K=N-1} b_K j^K m'(t - KT) + n(t) \quad (7)$$

using the time limitations of $m'(\cdot)$, we obtain at $t = KT$

$$m(KT) = E_B e^{j\Phi_0} b_{K-1} j^K \left(\frac{1}{\pi} + ja_{K-1} - \frac{1}{\pi} a_{K-1} a_K\right) + n(KT), \quad K > 0. \quad (8)$$

which is *not* ISI - free. Figure 1 depicts possible values of $m(KT)$. Detection consists of minimizing the norm of the vector

$$\mathbf{e}_{MF} = [m(kT) - E_B e^{j\hat{\Phi}_0} j^{k-1}(\frac{1}{\pi}\hat{b}_{k-1} + j\hat{b}_k - \frac{1}{\pi}\hat{b}_{k+1})], k = 0 \ldots N-1 \quad (9)$$

for a given MF output vector and requires a phase estimation.

For even K the actual symbol (A,B,C,J,K,L in figure 1) is transmitted on the Q channel. If the previous and the next symbol will be mapped both on the positive I axis, an ISI of $2E_B/\pi$ is observed at the correct sampling instant. Given that the adjacent symbols are mapped on positive and negative side of the In-Phase axis, no ISI is observed. However, as ISI and symbols of interest always are transmitted on separate axes (either I or Q), the ISI always can be cancelled by ignoring either real or imaginary part of the MF output. For even K, the real part of the MF output has to be removed.

If the signal phase is known, removal of the ISI will thus cause no problem. However, if the phase is not known, ISI will affect demodulation.

The noise $n(KT)$ in this case is colored but gaussian. Using (8), the (optimum) SNR is

$$\gamma_{MF} = E_B/N_0 \quad (10)$$

2.3 The Differential Phase Approach

Suboptimum receivers may be built based on detecting the sign of the L elements of the differential phase vector of the narrowband filtered ($h(kT)$, effective Bandwidth B_h) MSK signal, this technique is of an entirely different

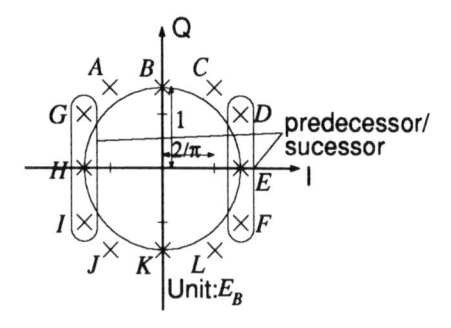

Figure 1 MF outputs for MSK signal for $\Phi_0 = 0$.

flavor and has its roots in the theory of suboptimum CPM receivers applicaple to any CPM signal [7]. In addition to the advantage of extendability to other modulation formats, a phase synchronization unit is avoided.

Focusing on the differential phase hence provides the receiver with observations

$$\Phi(KT) = \arg\left((s(KT) + n(KT)) * h(kT)\right) + \Phi_0 \tag{11}$$

$$\Phi =: \underbrace{\begin{pmatrix} 1 & -1 & 0 & \cdots & 0 \\ 0 & \ddots & \ddots & \ddots & \vdots \\ \vdots & \ddots & 1 & -1 & 0 \\ 0 & \cdots & 0 & 1 & -1 \end{pmatrix}}_{\mathbf{T}} \begin{bmatrix} \vdots \\ \Phi(lT) \\ \vdots \end{bmatrix} + \mathbf{T}n_\phi. \tag{12}$$

The (phase) noise n_ϕ perturbing the observations is neither gaussian nor white (due to the filtering with $h(KT)$ in the I/Q-domain). Focusing on the differential phase, the noise model has to be modified to the consideration of the *phase* perturbation, which is discussed eg in [8] [1] [9]. For our purposes, it is accurate enough to model the (mathematically complex) [8] phase perturbation $n_\phi(KT)$ as gaussian [5] with a variance of $1/2\gamma_h$, where $\gamma_h = E_B/(N_0 B_h T)$ is the SNR per bit. $\mathbf{T}$ is of size $L \times (L+1)$.

Employing the matched filter as $h(kT)$ will create strong ISI. When employing the matched filter only, differential phase values of 'transitions' as depicted in figure 1 are summarized in table 1. Hence, six different phase values

Trans.	Phase	Trans.	Phase
G-A	$\arg(-1 + j2/\pi)(-2/\pi - j)$	D-C	$\arg(-1 + j2/\pi)(-2/\pi - j)$
H-A	$-\arg(2/\pi + j)$	E-C	$\arg(2/\pi + j)$
G-B	$-\arg(2/\pi + j)$	D-B	$\arg(2/\pi + j)$
H-B	$-\pi/2$	E-B	$\pi/2$

Table 1 Differential phase observed at the output of the matched filter.

can be observed. Transitions to the terminal points A,B and C which are not mentioned cannot occur. Transitions E-C and D-B, for example, lead to the same phase difference values, since exchanging preceding and succeeding symbols of the actual symbol does not change its accumulated ISI. When transmitting a sequence of symbols, the sequence of differential phase values is best described using a trellis diagram, whose exploitation for data detection using the Viterbi algorithm is described in section 3.3.

Using a filter with a larger bandwidth, ISI in the differential phase will be smaller, however, the same splitting in six (or more) differential phase values can be observed eg as described for a limiter discriminator in [2].

In case a *mismatched* filter is employed, the filter impulse response $h(kT)$ is *not* optimized with respect to the modulation pulse but under consideration of the BER given the suboptimum demodulation method [10], [4], possibly including the effect of frequency offsets [11] [12], and the sign of $\phi(KT)$ is used for detection.

3 Demodulation

3.1 Partially Coherent Demodulation (PC)

The first approach to demodulation is to provide the receiver with a phase reference and to detect the data from the MF outputs. This is not a new approach, but is developed for providing a basis to understand the goal of the paper. Minimizing e_{MF} from eqn. (9) is achieved by computing a phase reference and performing serial detection:

Phase Reference. We firstly discuss computing the phase reference. *Feedforward* phase synchronization is best accomplished *before* the MF, as in this case ISI does not perturb the signal to the same extent. The price for pre-MF phase synchronization is to increase noise and to provide a second filter stage, on the other hand, phase dynamics may easier be compensated. To remove the modulation, we neglect any ISI caused by the prefilter, ie we assume

$$r(lT) \approx r(lT) * h(lT). \tag{13}$$

It has to be considered that at any time KT the MSK signal $r(KT)$ may assume two values which are located on the real axis at odd sampling instants and on the imaginary axis at even sampling instants. The twofold ambiguity is removed by squaring the samples[1], alteration between I and Q channel by multiplying any second sample with j. We obtain the phase estimate [14]

$$\hat{\Phi}_0 = \frac{1}{2} \arg \sum_{l=0}^{L-1} (-1)^l (r(lT) * h(lT))^2 \tag{14}$$

We have to compare the performance of this estimator to the CRLB [15]

$$Var\hat{\Phi}_0 \geq \frac{1}{2L\gamma_{MF}}. \tag{15}$$

---
[1] Other nonlinearities may be of interest [13]

Since the noise after having passed through the prefilter, only, has a higher noise variance, an 'optimum' estimator for this filter may only achieve a variance

$$Var\hat{\Phi}_0 \geq \frac{1}{2L\gamma_h} \qquad (16)$$

Simulation results are displayed in figure 2 for estimator lengths L of relevance. The phase estimate $\tilde{\Phi}_0$ is used to adjust the received signal's phase and ensures the first symbol to be real (but ambiguous due to the twofold phase ambiguity). As the performance bound (16) is asymptotically reached, the use of approximation (13) to obtain the estimator is justified.

Phase synchronization is performed using a gliding window.

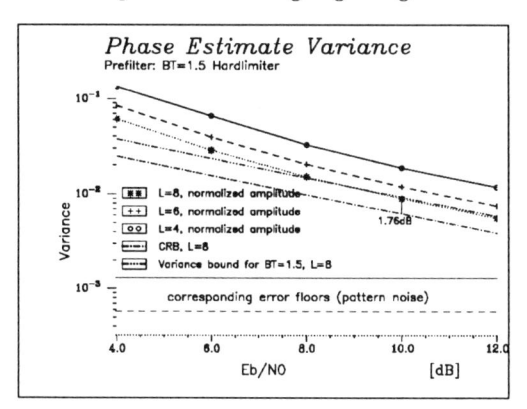

Figure 2 Performance of feedforward MSK phase synchronizer with hardlimiting, normalized bandwidth $BT = 1.5$.

Serial Detection. It is known [7] [16] that coherent detection of MSK is best performed in the bandpass domain choosing an intermediate frequency $f_D = 1/4T$. In brief, the idea of this detection approach is to introduce zeros in the MF outputs such that the ISI between I and Q channel is removed.

A suitable bandpass transform is

$$s_{BP}(KT) = \text{Re}\left(\exp\left(\frac{-j2\pi(K+1)}{4}\right) \cdot m(KT) \cdot e^{-j\tilde{\Phi}_0}\right) \qquad (17)$$

and assuming ideal phase recovery leads to

$$m_{BP}(KT) = E_B \cos\left(\frac{2\pi(K+1)}{4}\right) b_K + E_B \sin\left(\frac{2\pi(K+1)}{4}\right) b_K + \tilde{n}(KT) \qquad (18)$$

Each of the cos (sin) terms vanishes in even (odd) time instants and thus the sequence

$$m_{BP}(KT) = E_B b_K + \tilde{n}(KT) \qquad (19)$$

is obtained. Differential decoding produces data estimates $\hat{a}_K$. In fact, the real and imaginary part of the MF outputs are multiplied alternatingly with zeros and ± 1 which removes the real or imaginary part of the MF output and hence leads to ISI free reception. Of course, phase errors $\Phi_0 - \hat{\Phi}_0$ will introduce ISI again. It has been shown that serial detection is less sensitive to phase errors than detection in the lowpass domain [17].

Due to the differential decoding procedure, any data bit output is based on *two* decisions, the error performance for large L is given by [7,12]

$$p_{e,PC} = \text{erfc}\sqrt{\frac{E_B}{N_0}}. \tag{20}$$

3.2 Block Demodulation (BD)

We consider a vector of $L+1$ phase values $\phi(KT)$ from (11) and then approximate the rule $\min_{\hat{b}} \|\mathbf{e}_{MF}\|$ by observing the signal phase and neglecting ISI (which is commonly justified when not using the matched filter but a function h with a larger bandwidth.).

$$\mathbf{e}_\phi = [\boldsymbol{\Psi} + \Phi_0 - \hat{\boldsymbol{\Psi}} - \hat{\Phi}_0] + \mathbf{n}_\phi \tag{21}$$

This metric will for high SNR be minimal for the same $\hat{b}_k$ as the original metric $\mathbf{e}_{MF}$. As before, we assume $\mathbf{e}_\phi$ gaussian and thus have to minimize $\mathbf{e}_\phi^T (Var\mathbf{e}_\phi)^{-1} \mathbf{e}_\phi$ for approximating ML detection and maximizing the probability density of $\mathbf{e}_\phi$ conditioned on the symbol trials. Note that $\mathbf{e}_\phi$ is of size $(L+1) \times 1$.

How does the phase offset estimation error $\Phi_0 - \hat{\Phi}_0$ affect the statistics of $\mathbf{e}_\phi$? We assume an efficient estimate $\hat{\Phi}_0$ whose *effect* on the statistics is modeled by averaging over all noise samples belonging to the estimator window of length $L+1$. (Note that the CRB (15) may be interpreted in this way, as well.) The effect of introducing the phase estimate on the statistics of $\mathbf{e}_\phi$ is thus modeled by

$$\mathbf{e}_\phi = [\boldsymbol{\Psi} - \hat{\boldsymbol{\Psi}}] + \mathbf{n}_\phi + \underbrace{\frac{1}{(L+1)} \begin{pmatrix} 1 \\ \vdots \\ 1 \end{pmatrix} \sum_{i=0}^{L} n_{i,\phi}}_{\Phi_0 - \hat{\Phi}_0}. \tag{22}$$

After having removed the explicit occurrence of the phase offset, we can transform the metric into the differential phases using the transformation matrix $\mathbf{T}$ sized $L \times (L+1)$.

$$\mathbf{e}_{\Delta\phi} = \mathbf{T}\mathbf{e}_\phi \tag{23}$$

$\mathbf{e}_{\Delta\phi}$ is the vector whose probability conditioned on the symbol trials is to be maximized when detecting symbols using the differential phase.

Note that $\mathbf{T} \cdot \mathbf{W} = \mathbf{I}_{L \times L}$, the identity matrix (sized $L \times L$), and consequently

$$\mathbf{T}\hat{\boldsymbol{\Psi}} = \hat{a}\frac{\pi}{2}. \tag{24}$$

Assuming $\mathbf{n}_\phi$ to be white gaussian noise, the noise correlation $\mathbf{C}$ of $\mathbf{e}_{\Delta\phi}$ is given by

$$\mathbf{C} := \sigma^2 \mathbf{T} \left(I_{L+1 \times L+1} + \frac{1}{L+1} \begin{pmatrix} 1 & \cdots & 1 \\ \vdots & & \vdots \\ 1 & \cdots & 1 \end{pmatrix} \right) \mathbf{T}^T \quad (25)$$

$$= 2\sigma^2 \begin{pmatrix} 1 & -0.5 & 0 & \cdots & 0 \\ -0.5 & 1 & -0.5 & \ddots & \vdots \\ 0 & -0.5 & 1 & \ddots & 0 \\ \vdots & \ddots & \ddots & \ddots & -0.5 \\ 0 & \cdots & 0 & -0.5 & 1 \end{pmatrix} \quad (26)$$

which is the correlation matrix used in [1]. We have thus shown that neglecting ISI and considering an efficient phase estimate $\hat{\Phi}_0$ based on $L+1$ observations leads to the metric used in [1] when looking at the differential phase.

Providing a receiver with $L+1$ phase observations $\phi(KT)$ and an efficient phase estimate based on these observations is thus equivalent to providing L succeeding phase difference observations $\Delta\phi = \phi(KT) - \phi(KT - T)$.

Symbol detection is based on maximizing the conditional probability of the occurrence of a symbol vector, ie minimizing $\mathbf{e}_{\Delta\phi}^T \mathbf{C}^{-1} \mathbf{e}_{\Delta\phi}$ by varying $\hat{a}$.

Thus we have understood that BD and PC implicitly perform the same functions (phase estimation and ML symbol detection), but in different signal spaces. BD, however, is based on several approximations.

We assess the effect of making these approximations using simulations.

As BD requires an exhaustive search [1], we seek for a means to reduce computing effort. Errors in differential detection of MSK are likely to occur when the phase difference between two successive symbols is close to zero or $\pm\pi$, either caused by noise or intersymbol interferences. On the contrary phase differences close to $\pm\pi/2$ may be considered as more reliable. Taking the distance measure $\Delta\phi - \pi/2$ or $\Delta\phi + \pi/2$ as a reliability estimate the L_R most reliable phase differences may be used to detect the corresponding symbol by conventional DMSK threshold detection. Then, only the remaining $L - L_R$ symbols in Φ have to be detected with the block demodulation procedure. Note, that now the evaluation of the detection metric and the minimization search has to be performed only for a reduced set of possible data patterns. Choosing $L_R = 2$ seems possible without performance degradation.

However, a further reduction of computational complexity can be achieved using the Viterbi algorithm.

3.3 Viterbi detection (VA)

The representation of the symbol transitions in a trellis is exploited. Depending on the two previous symbols, there exist four states in the trellis. Any state may be expanded with any new symbol leading to the trellis depicted in figure 3. A suitable metric for the minimum–distance path is (for

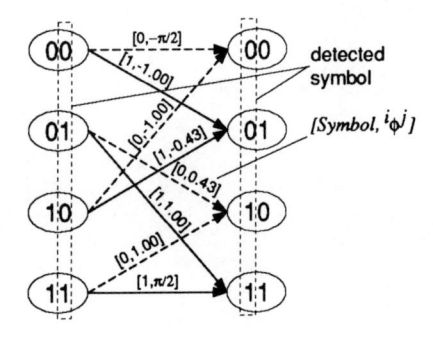

Figure 3 Trellis of observed phase differences

gaussian noise) the Euclidean distance. In [9], a different metric was used which, however, does at least for moderate to high SNR not differ too much from the Euclidean distance. Simulation results confirmed the similar performance using either metric. Let the path metrics be described by $\Gamma_k^{1...4}$ and the correct observations be denoted as $^{0,1}\phi_k^{1...4}$ leading to the metric increments $^i\lambda^k$. Now the minimum path metric is given by the well known add–compare–select operations of the VA.

$$^i\Gamma_{k+1}^j = \Gamma_k^i + {}^i\lambda_k^j \tag{27}$$

$$^i\lambda_k^j = (\Delta\phi_k - {}^i\phi^j)^2 \tag{28}$$

$$\Gamma_{k+1}^j = \min_i({}^i\Gamma_{k+1}^j) \tag{29}$$

We expect some degradations as the increments of the paths are not independent and noise is not gaussian.

4 Computational Complexity

We consider any operation following the IF filter and divide into multiplications and additions, where all operations are scaled with the window length and normalized to the processing of L symbols, respectively. We assume an oversampling rate of $4/T$ for computing the MF outputs. Both algorithms require a phase computation once per symbol.

From table 2 it is seen that PC demodulation requires less computations than BD when $L > 3$. Memory requirements for all algorithms are low. For PC and VA, the computing effort is independent of the observation interval. PC always requires less computations than VA.

Operation	PC Mult.	PC Add.	BD Mult.	BD Add.	VA Mult.	VA Add.
Diff. Phase				L		1
Phase Synch.	$9L$	$7L$				
Diff. Comp.				$2L-2$		
Metric Comp.			$4(L-1)L-2$	$2^{L-3}L(L+1)$	20	8
M. Filt.	$16L$	$16L$	$16L$	$16L$	16	16
BP Transf.	L					
Σ/Symb.	26	23	$4L+12-2/L$	$2^{L-3}(L+1)+$ $+19-2/L$	36	25
$L=3,5,7$			23,32,40	22,43,126		
Bit Oper.						
Diff. Dec.		$2L$				
Sign		$2L$				
RP-conv.		L		L		1

Table 2 Comparison of Computational Complexity. For BD, $L_R = 2$ was chosen. For VA, Operations / Symbol are given.

5 Signal Impairments

To compare the demodulation algorithms we look at the optimum synchronization scenario as well as at signal impairments, ie a residual frequency offset and random phase noise.

A frequency offset firstly shifts the signal spectra away from the zero frequency, hence the IF filter causes an asymmetric distortion of the signal spectrum leading to a degradation. For PC, the variance of the phase estimate is increased. For BD, the metric computation becomes biased. In general one may assume that the gliding window filter of PC will be able to cope better with frequency errors, as some sort of carrier tracking is established.

Finally, we investigated the influence of a hardlimiter in the signal path. The signal's amplitude was fixed after having passed the IF filter. Hence, in any case a further degradation was observed due to the fact that noise now becomes non-gaussian and secondly the IF filter introduced amplitude fluctuations.

6 Simulated Results

In figure 4, it can be seen that BD almost achieves the ideal performance for $L > 3$. PC achieves optimal performance for $L > 5$. VA also does not degrade for a static channel. The survivior depth proved as an uncritical parameter.

In case of a static frequency offset, BD is most sensitive, as seen in fig 4, right. Clearly this depends on the implicit assumption of a constant phase during the observation interval, whereas PC performs a tracking of the current phase and hence is more robust. On the other hand, VA is most sensitive against a random phase walk. Due to the slow change of the phase, a completely incorrect path may be selected in the trellis and cause many bit errors.

In figure 5, right, the degradations at 10dB E_B/N_0 are summarized (ie the ideal demodulator achieves the respective performance at ΔdB less E_B/N_0). It is seen that PC is the most robust demodulation method.

7 Conclusions

We have discussed three demodulation schemes for MSK. Partially coherent demodulation is based on a feedforward phase estimator and serial demodulation of the MSK matched filter outputs. Block demodulation decides upon symbols providing the minimum–distance differential phase trajectory. Viterbi demodulation exploits the ISI of the phase differences produced by the matched filter. We derived block demodulation from partially coherent demodulation. It was shown that partially coherent demodulation for MSK is less complex and exhibits a better performance and robustness than block demodulation and also than Viterbi demodulation. In cases, where low computational complexity is desired, but an MF receiver cannot be realized, Viterbi detection will be an interesting alternative.

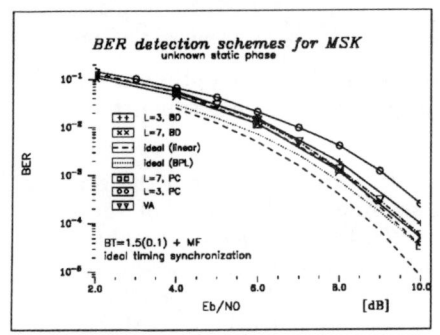

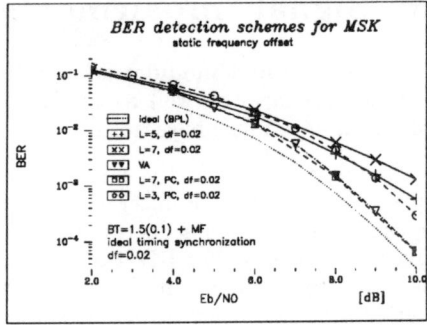

Figure 4 The detection schemes for a static channel and a frequency offset of $dfT = 0.02$.

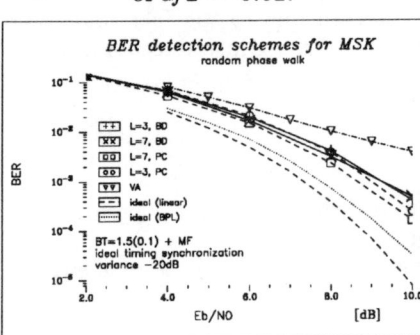

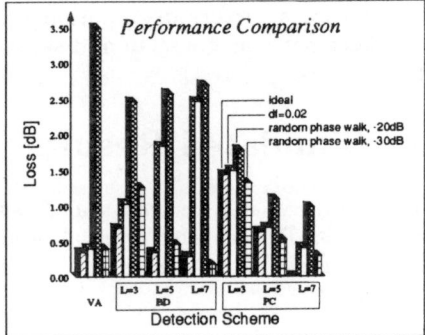

Figure 5 The detection schemes for a random phase walk with a variance of -20dB. Losses at 10dB.

References

[1] R. Mehlan and H. Meyr, "Differential Block Demodulation of Hardlimited MSK Signals," in *Proceedings of the IEEE Global Telecommunications Conference GLOBECOM*, (San Francisco, CA), pp. 918–922, IEEE, November 1994.

[2] Y. Iwanami, "Sequence Estimation Scheme for Narrowband Digital FM Signals with Limiter–Discriminator under Fading Environment," in *Proceedings Globecom*, pp. 1792–1797, 1993.

[3] G. Kaleh, "Differential Detection via the Viterbi Algorithm for Offset ¡odulation and MSK-Type Signals," *IEEE Transactions on Vehicular Technology*, pp. 401–406, 1992.

[4] R. Mehlan, Y. Chen, and H. Meyr, "A Fully Digital Feedforward MSK Demodulator with Joint Frequency Offset and Symbol Timing Estimation for Burst Mode Mobile Radio," *IEEE Transactions on Vehicular Technology*, vol. 42, pp. 434–443, November 1993.

[5] U. Lambrette, "A Novel Selection Diversity Technique for MSK using Hardlimiter Output," in *Proceedings of the IEEE Global Telecommunications Conference GLOBECOM*, 1993.

[6] A. N. D'Andrea, U. Mengali, and R. Reggiannini, "A Digital Approach to Clock Recovery in Generalized Minimum Shift Keying," *IEEE Transactions on Vehicular Technology*, vol. 39, pp. 227–234, August 1990.

[7] J. Anderson, T. Aulin, and C. Sundberg, *Digital Phase Modulation*. Plenum Press, 1986.

[8] J. Proakis, *Digital Communications*. McGraw Hill, 1989.

[9] S. Ariyavisitakul, "Equalization of a Hard–Limited Slowly–Fading Multipath Signal Using a Phase Equalizer with Time–Reversal Structure," *proceedings VTC 1990*, pp. 520–526, 1990.

[10] H. Suzuki, "Optimum Gaussian Filter for Differential Detection of MSK," *IEEE Transactions on Communications*, vol. 29, pp. 916–918, June 1981.

[11] U. Lambrette and H. Meyr, "A Digital Feedforward DMSK Receiver for Packet–Based Mobile Radio," in *Proceedings of the IEEE International Conference on Vehicular Technology*, 1994.

[12] T. Masamura, "Intersymbol interference reduction for differential msk by nonredundant error correction," *IEEE Transactions on Vehicular Technology*, vol. 39, February 1990.

[13] A.J. Viterbi, A.M. Viterbi, "Nonlinear Estimation of PSK–Modulated Carrier Phase with Application to Burst Digital Transmission," *IEEE Transactions on Information Theory*, vol. IT-29, pp. 543–551, August 1983.

[14] F. Gardner, "Demodulator Reference Recovery Techniques suited for Digital Implementation," tech. rep., Gardner Research Company, 1988.

[15] Viterbi. Andrew J., *Priciples of Coherent Communication*. McGraw–Hill Book Company, 1966.

[16] R. Ziemer, C. Ryan, and J. Stilwell, "Conversion and Matched Filter Approximations for Serial Minimum–Shift Keyed Modulation," *IEEE Transactions on Communications*, vol. COM-30, pp. 495–509, March 1982.

[17] A. Svensson and C. Sundberg, "Serial MSK–Type Detection of Partial Response Contious Phase Modulation," *IEEE Transactions on Communications*, vol. COM-33, pp. 44–52, January 1985.

[18] U. Lambrette: WWW page
http://www.informatik.rwth-aachen.de/ERT/Personen/lambrett.html.

Variable Symbol-Rate Modem Design for Cable and Satellite TV Broadcasting

G. Karam, K. Maalej, V. Paxal, and H. Sari

SAT, Telecommunications Division
11, rue Watt, BP 370, 75626 Paris Cédex 13, France

Abstract

We present the design and basic features of variable symbol-rate modems for digital TV broadcasting by cable and satellite. The modems are fully compliant with the DVB recommendations which were later published as ETSI standards. The satellite modem employs the QPSK signal constellation and can handle symbol rates up to 30 Mbaud. The cable modem is configurable to operate with all QAM signal formats from 16-QAM to 256-QAM, and handles symbol rates up to 7 Mbaud which is the maximum possible symbol rate in an 8 MHz channel. Both modems employ polyphase filtering for variable symbol rate operation and fully-digital carrier recovery loops with an extended frequency acquisition range. In addition, the cable modem includes a blind equalizer with smooth transition to a conventional adaptation algorithm after convergence.

1. INTRODUCTION

Over the past three years, there has been a strong activity in Europe to finalize standards for digital TV broadcasting by cable and satellite and then develop equipments based on these standards. The technical specifications are elaborated by the Digital Video Broadcasting (DVB) project conducted under the auspices of the European Broadcasting Union (EBU). They are next passed to the European Telecommunications Standards Institute (ETSI) for further procedures toward publication of the standard.

The specifications of channel coding and modulation for digital satellite TV broadcasting were finalized in December 1993 and the ETSI standard was published in January 1995. Similarly, the specifications of channel coding and modulation for digital cable TV broadcasting were finalized in January 1994, and the corresponding ETSI standard was published in early 1995.

The standard for satellite TV broadcast applications is based on Quaternary Phase Shift Keying (QPSK) with a concatenated error correction coding scheme. The inner code is a rate-1/2 convolutional code with constraint length $K = 7$ (the convolutional code which has become a de-facto industry standard), but the standard also requires higher code rates by puncturing this basic code. This allows to trade-off coding gain against useful data rate on a given satellite transponder.

In addition to modulation and channel coding, the standard also defines interleaving, scrambling, framing, and transmit and receive filtering. In summary, the standard defines all transmit and receive functions and characteristics, except the symbol rate at which the modems must operate. The reason for this is that there are a variety of satellites with different transponder bandwidths, and even for a given satellite transponder, the operators were reluctant to fix a symbol rate once for all. As a consequence, manufacturers had the options of either developing application-specific fixed symbol-rate modems or a variable symbol-rate modem which satisfies all applications intended.

In the cable standard, the modulation is a quadrature amplitude modulation (QAM) with M = 64, 32, or 16 states, and no convolutional code is employed. Although the cable channel width is 8 MHz virtually in all European countries, no symbol rate was fixed for cable TV either. The reason for this is the desire to be able to feed satellite signals directly to cable networks at cable head-ends. As a result, variable symbol-rate modem design was the best solution also for cable TV systems.

The purpose of this paper is to describe the design of variable symbol-rate modems for both satellite and cable TV systems and discuss their performance. First, in the next two sections, we briefly review the ETSI standards for satellite and cable TV broadcasting. Next, in Section 4, we describe our QPSK modem, highlight its basic features, and present its performance. The design of our QAM modem for cable TV systems is described and its performance results are reported in Section 5. Finally, we give our conclusions in Section 6.

2. SATELLITE TV STANDARD

The standard concerning modulation and channel coding for satellite TV/HDTV services to be used for primary and secondary distribution in fixed satellite services (FSS) and broadcast satellite services (BSS) in the 14 GHz/12 GHz band was published in January 1995 by the ETSI [1]. The system is intended to provide direct-to-home (DTH) services for consumer integrated receiver decoder (IRD) as well as for community reception (SMATV) and cable TV head-end stations.

The system uses a QPSK modulation and a concatenated forward error correction (FEC) coding scheme with a convolutional inner code and a shortened Reed-Solomon (RS) outer code. The transmission frame is based on the MPEG2 transport data stream [2], and prior to channel coding, a scrambler is used for energy dispersal purposes. The shortened RS outer code has a block length of 204 symbols, carries 188 information symbols per block, and can correct up to 8 symbol errors per block. The associated bandwidth expansion is approximately 8.5%. A convolutional interleaver with interleaving depth I = 12 is inserted between the inner and outer codes in order to uniformly distribute the errors which occur at the Viterbi decoder output at the receive side. The convolutional code rates are 1/2, 2/3, 3/4, 5/6, and 7/8.

Finally, the data at the inner encoder output are mapped onto QPSK symbols and pulse shaped by a square-root raised-cosine Nyquist filter before modulation. The roll-off factor is $\alpha = 0.35$. A simplified block diagram of the satellite QPSK modulator system is shown in Fig.1.

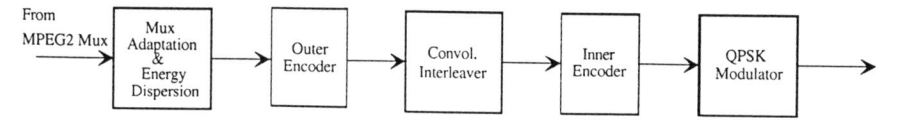

Figure 1: Block diagram of the satellite QPSK modulator system.

3. CABLE TV STANDARD

The basic modulation scheme for digital cable TV broadcasting is 64-QAM, but the standard also includes two lower-level modulation schemes, namely 16-QAM and 32-QAM, which increase system margin and robustness at the expense of reduced data throughput. Future extensions to higher-level modulation schemes such as 128-QAM and 256-QAM are also envisioned, although these are not a part of the current standard. Channel coding consists of the Reed-Solomon code used for satellite broadcasting, i.e., RS(204, 188, 8) which can correct up to 8 symbol erros per block of 204 symbols. The interleaver too is the same as that used for satellite TV. Unlike the satellite TV standard, however, the cable TV standard does not employ concatenated FEC coding (i.e., no convolutional code), which can be justified by the fact that typical signal-to-noise ratio (SNR) values encountered in cable networks are substantially higher than those corresponding to satellite links. Conversely, the cable channel suffers from echoes due to imperfect cable terminations and impedance mismatches, and this requires the use of an adaptive equalizer at the receiver. Further, since no training symbols are included in the transmitted data stream, the equalizer must be able to operate in a blind mode.

The roll-off factor of the Nyquist overall filtering is $\alpha = 0.15$, and is evenly split between transmitter and receiver. For an 8 MHz channel, the maximum symbol rate without spectral overlapping (and mutual interference) of adjacent channels is 6.96 Mbaud. With this maximum symbol rate, the useful bit rate is limited to 38.46 Mbit/s with 64-QAM, but this can be increased in the future to 44.87 Mbit/s with 128-QAM and to 51.28 Mbit/s with 256-QAM.

4. THE SATELLITE RECEIVER

4.1. Receiver Architecture

The receiver block diagram is shown in Fig.2. The signal at the output of the low-noise converter (LNC) has a center frequency in the range of 850 to 2050 MHz. This signal enters a tuner where channel selection and baseband conversion functions are performed. The in-phase (I) and quadrature (Q) signal components at the output of this stage are digitized using a free-running clock. The digitized I and Q signals are inputted to a fully-digital QPSK demodulator integrated in a single chip. This chip performs all baseband functions excluding those related to error correction. The latter functions are carried out in a second VLSI chip.

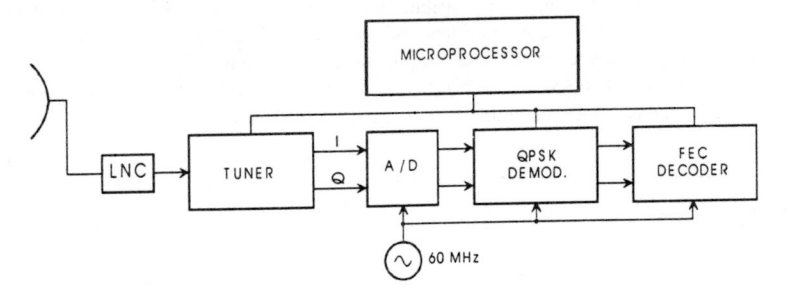

Figure 2: Block diagram of the QPSK satellite TV receiver.

The first function of the DEMOD chip (see Fig.3) is variable-rate clock synchronization using polyphase filtering. From the fixed-rate samples available at its input, the clock synchronization loop is capable of synthesizing the samples corresponding to any symbol rate from 2 to 30 Mbaud that may be used by the transmitter. The next function is to perform matched filtering using a finite impulse response (FIR) filter structure with programmable coefficients. This is followed by digital demodulation implemented as a complex multiplier operating at the symbol rate. The carrier recovery loop includes a phase/frequency detector which leads to an extended frequency acquisition range without degrading its steady-state properties.

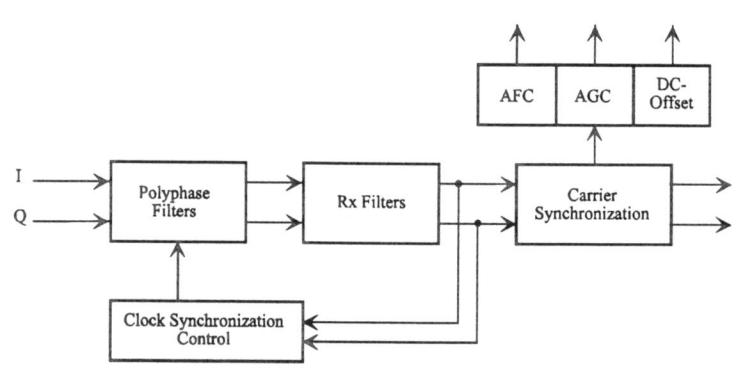

Figure 3: Block diagram of the DEMOD chip.

Frequency offsets on the order of 10% of the baud rate can thus be handled without resorting to automatic frequency control (AFC) in the tuner. The DEMOD chip also integrates other auxiliary functions including an AFC which can cope with frequency offsets as large as the baud rate, an automatic gain control (AGC), and a dc-offset compensation circuit. These control signals are all fed back to the analog front-end.

The block diagram of the FEC chip is shown in Fig.4. The inputs are 4-bit I and Q signals. The chip performs the functions of Viterbi decoding, convolutional deinterleaving, RS decoding, and descrambling. The main feature of the FEC chip is its ability to auto-detect the inner code rate, synchronize the Viterbi decoder, and perform byte, frame, and superframe synchronizations required for the operation of the deinterleaver, RS decoder, and descrambler. Both of the DEMOD and FEC chips were developed using 0.6μ CMOS technology.

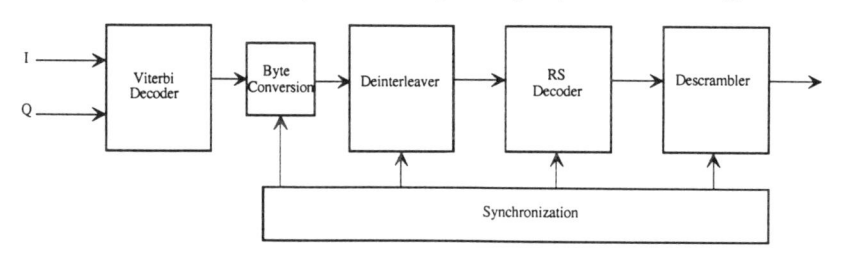

Figure 4: Block diagram of the FEC chip

4.2. Performance Analysis

Simulated BER performance of the QPSK receiver is depicted in Fig.5 which shows the BER at the Viterbi demodulator output as a function of the energy per bit to the noise spectral density ratio (E_b/N_0). These results take into account the finite-precision and finite-memory implementation of the Viterbi decoder, but assume ideal modulator/demodulator functions. They also neglect the degradation due to bandwidth expansion of the outer RS code, which is given by

$10 \cdot \log(204/188) = 0.36$ dB.

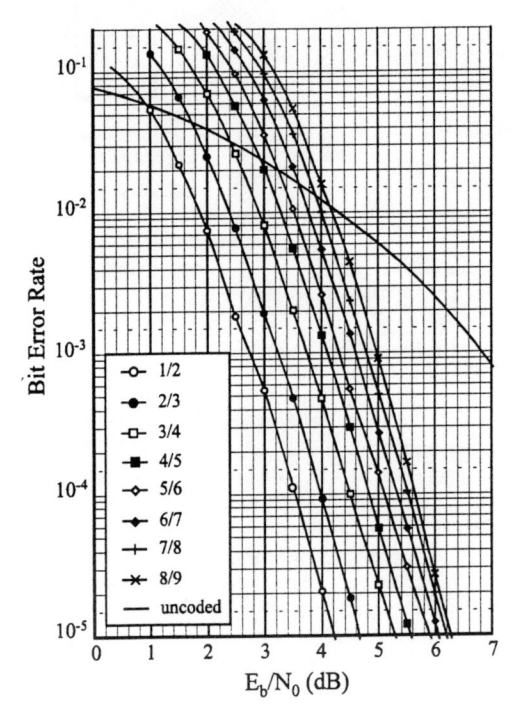

Figure 5: BER at the Viterbi decoder output vs. E_b/N_0.

Note that a BER of 10^{-4} at the Viterbi decoder output is translated to a BER less than 10^{-12} at the RS decoder output. From Fig.5, we can therefore conclude that the receiver virtually operates error-free with $E_b/N_0 = 4.6$ dB for a convolutional code rate of 3/4 and with $E_b/N_0 = 3.6$ dB for a code rate of 1/2.

Carrier frequency acquisition performance of the receiver was evaluated using a laboratory prototype with a baud rate of 25 Mbaud. It was found that the acquisition range is in excess of ±2.5 MHz, and that for an initial frequency offset of ±1 MHz, the acquisition time of the complete receiver is on the order of 50 ms. This figure accounts for clock, carrier, and frame synchronizations.

5. THE CABLE RECEIVER

5.1. Receiver Architecture

A simplified block diagram of the variable-rate QAM demodulator is depicted in Fig.6. The demodulator can handle all QAM signal formats from 16-QAM to 256-QAM, i.e., not only it offers all modulation formats included in the current standard, but also its future extensions to 128- and 256-QAM. The received signal, whose carrier frequency is in the range of 120 to 860 MHz, is fed to the tuner where channel selection is performed. Using a free-running local oscillator, the signal is then downconverted to an intermediate frequency (IF) of 7 MHz. The IF signal is next passed to an analog-to-digital (A/D) converter driven by a free-running clock whose nominal frequency is 28 MHz.

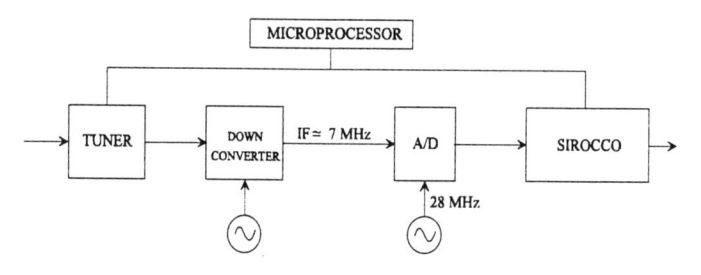

Figure 6: Block diagram of the QAM cable TV receiver.

Given the fact that the symbol rate does not exceed 7 Mbaud, the IF sampling operation gives at least 4 samples per symbol period, T. The digitized IF signal is passed to a VLSI chip which performs all signal processing functions required for QAM demodulation, variable symbol-rate timing recovery, channel equalization, and error correction. Using a serial bus, the tuner and the VLSI chip are interfaced with a microprocessor for configuration and monitoring of the receiver functions.

This receiver architecture has several attractive features: First, it uses a single A/D converter and overcomes the problem of amplitude and phase imbalances between the I and Q channels inherent to conventional quadrature demodulator implementations. Second, the demodulation operation is entirely digital, and no feedback loops are required to control the frequency and phase of the oscillators used for frequency downconversion.

We now describe the signal processing functions performed by the VLSI chip whose block diagram is depicted in Fig.7. The first function is to perform IF-to-baseband conversion and generate I and Q components. Next, as in the QPSK modem, polyphase filtering is employed to generate signal samples at an integer multiple of the symbol rate. The third function is to perform matched filtering using two FIR filters with programmable coefficients. This allows to synthesize other filter shapes than those specified in the DVB recommendations.

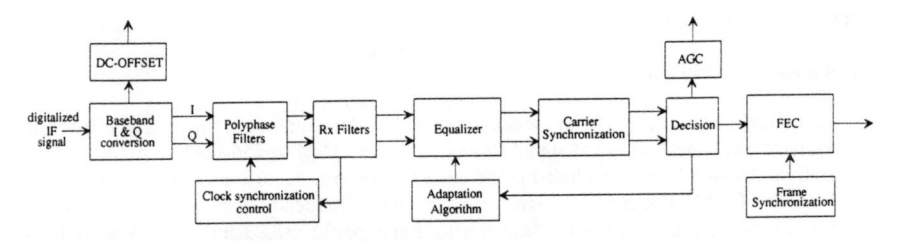

Figure 7: Block diagram of the QAM demodulator.

After matched filtering, the signal enters an adaptive equalizer whose function is to compensate for signal distortions caused by imperfect channel characteristics of cable TV networks. Note that the cable channel is characterized by the presence of echoes due to imperfect amplifiers, bridged taps, and cable terminations. Optimization of the equalizer structure in our design was performed using the worst-case channel characteristics measured in cable networks. Further, a blind algorithm (like the constant modulus algorithm [4] or the reduced-constellation algorithm [5]) was included since the equalizer has to converge in a blind mode. (As mentioned earlier, the data stream contains no training sequence.) Further, to ensure fine convergence and small steady-state error, the equalizer adaptation switches to a conventional LMS-type algorithm after initial convergence.

Carrier synchronization is performed at the equalizer output using a decision-feedback loop. This consists of a complex multiplier operating on symbol-rate equalizer output samples, a decision-directed phase/frequency detector, a digital loop filter, an accumulator, and a read-only memory (ROM) which provides the phase information to the complex multiplier. One of the major problems in carrier recovery loop design is the difficulty to achieve a large frequency acquisition range together with a small steady-state phase jitter. It is difficult to satisfy these two contradictory requirements, particularly in digital communication systems employing high-level QAM signal formats. The phase/frequency detector in our design behaves as a frequency detector during the initial acquisition period and as a phase detector in the steady state. The resulting loop leads to rapid frequency acquisition over an extended range while still guaranteeing a small steady-state phase jitter. This feature of the carrier recovery loop allows to use standard low-cost tuners and RF front-ends which may involve frequency offsets larger than ± 100 kHz between transmitter and receiver.

Finally, the VLSI chip performs FEC functions which include byte conversion, convolutional deinterleaving, Reed-Solomon decoding, and descrambling. This block is controlled by a synchronization circuit which performs byte, frame, and superframe synchronizations. The chip also integrates other auxiliary functions including automatic gain control (AGC) for fine amplitude adjustment of the digitized IF signal and dc-offset compensation. It also includes lock detectors which deliver alarm signals for the clock, carrier, and frame synchronization circuits, and BER estimation for quality monitoring.

5.2. Computer Simulation Results

In this section, we report simulated performance of the QAM receiver. First, BER performance at the Reed-Solomon decoder input vs. the transmitted energy per bit to noise spectral density ratio (E_b/N_0) is shown in Fig. 8 for 64-QAM. Curve 1 in this figure corresponds to an ideal 64-QAM system operating over an additive white gaussian noise

(AWGN) channel. Curve 2 is obtained by including the degradation due to the hardware implementation of the demodulator. Comparing these two curves, we observe that the degradation is approximately 0.3 dB at the BER of 10^{-4} at the Reed-Solomon decoder input, which corresponds to quasi-error-free transmission at the decoder output.

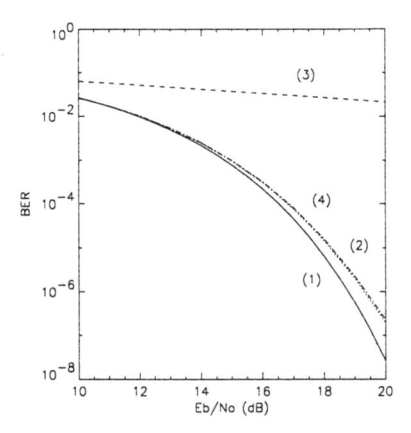

Figure 8: BER performance vs. E_b/N_0.

The same figure also shows the modem performance on a channel with two echoes (in addition to the main path), which typically corresponds to a severely distorted CATV channel The impulse response and transfer function of this channel are shown in Figs. 9 and 10, respectively. Deep notches can be observed in the channel whose transfer function also includes a significant group-delay distortion. A symbol rate of 6.96 Mbaud was assumed in the simulations using this channel.

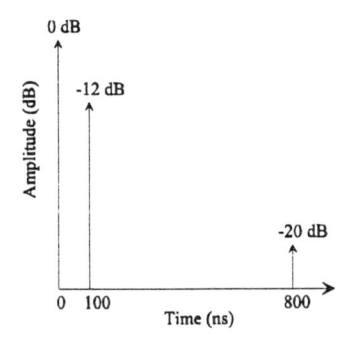

Figure 9: CATV channel impulse response.

Curve 3 shows the system performance without an equalizer. Clearly, there is a BER floor of 10^{-2}, which is unacceptable. The equalized 64-QAM system performance is illustrated by curve 4. Notice that the BER floor disappears, and the BER degradation due to residual distortion is less than 0.1 dB (at the BER of 10^{-4}).

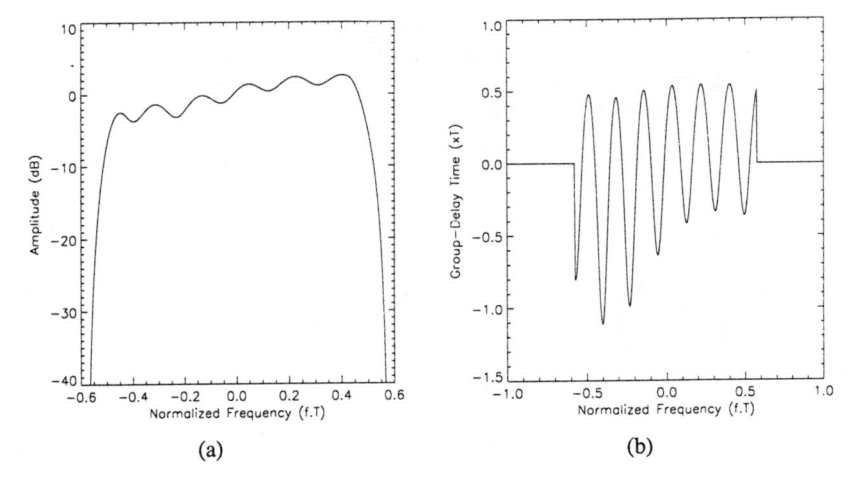

Figure 10: CATV channel transfer function.
(a) Magnitude response,
(b) Group-delay response.

To further highlight the equalizer performance, we show in Fig. 11 several constellations. The plot in Fig. 11.a corresponds to the distorted signal present at the equalizer input. Next, Fig. 11.b shows the signal constellation at the equalizer output. What we see here is essentially a set of 9 concentric circles, which indicates that the distortion has been compensated, but the signal is still not synchronized. Finally, Fig. 11.c shows the constellation at the demodulator output, where we can see a quasi-ideal 64-QAM signal constellation.

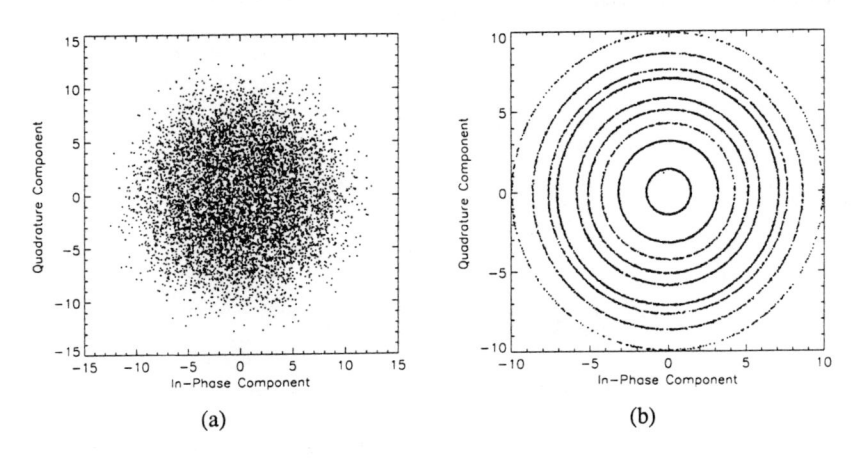

(c)

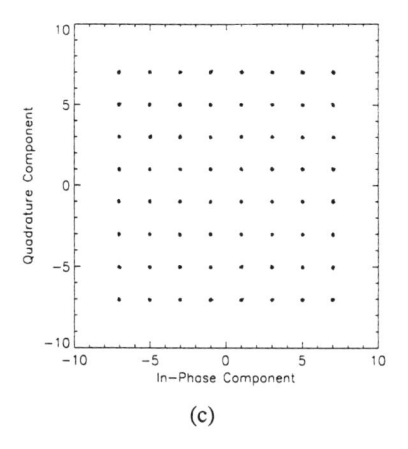

Figure 11: 64-QAM signal constellation.
(a) at the equalizer input,
(b) at the equalizer output,
(c) at the demodulator output.

We now investigate the performance of the carrier synchronization circuit. Assuming an AWGN channel and $E_b/N_0 = 16.5$ dB, Fig. 12 shows the acquisition time vs. frequency offset between transmit and receive oscillators normalized by the symbol rate ($\Delta f.T$). We can see that the normalized acquisition range for 64-QAM is approximately 0.07/T for a maximum acquisition time of $3 \times 10^6 T$. For a 6.96 Mbaud system, these figures read an acquisition range of ±487 kHz and an acquisition time of 430 ms.

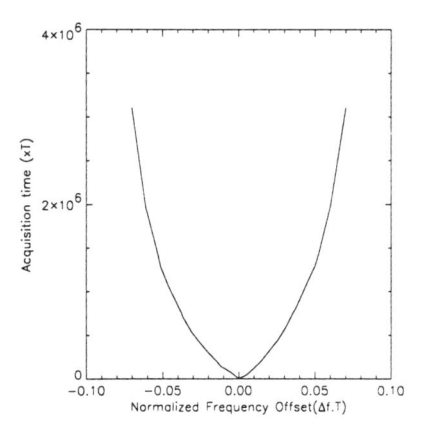

Figure 12: Acquisition time vs. normalized frequency offset.

Finally, to analyze the acquisition behavior of the overall system, we plot the squared instantaneous error at the decision circuit output. Fig. 13 shows the results corresponding to the channel with two echoes mentioned earlier. The frequency offset used in this simulation was $\Delta f = 50$ kHz and the SNR value was $E_b/N_0 = 30$ dB. The acquisition characteristic displayed in this figure accounts for both equalizer convergence and carrier synchronization. The equalizer starts its adaptation using a blind algorithm, and then switches to an LMS-type algorithm for fine convergence. The total acquisition time of the receiver, as read from this figure, is approximately 80000 T, i.e., 11 ms for a 6.96 Mbaud system.

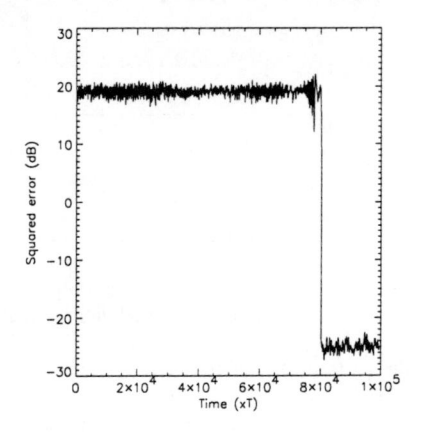

Figure 13: Squared error at the decision circuit output vs. time.

5.3. Field Test Results

Using a 6.96 Mbaud 64-QAM modem laboratory prototype, a series of measurements were carried out on the CATV network at the CCETT premises in Rennes, France, referred to as "site captif réseaux câblés 0G". This consists of a hybrid fiber/coax network which reproduces the different configurations encountered in France Telecom's CATV network in the region of Paris. The transport section of this network consists of optical transmission links, and the distribution section is a coax local network which includes 7 cascaded amplifiers.

The field test results on this network are given in [6]. Here, we only summarize the main results. In the field tests, we inserted the digital TV channel between two adjacent SECAM channels. Fig. 14 indicates their relative positions in the frequency plan. Defining the back-off (BO) as the ratio of the SECAM channel power to the average power of the digital channel, we first use a BO value of 8 dB. The analog channel quality remains essentially unchanged in these conditions, and the BER of the digital channel is on the order of 10^{-8} before Reed-Solomon decoding. Next, using an additive noise source, we measured a BER of 10^{-4} (also before Reed-Solomon decoding) for an E_b/N_0 of 20.2 dB. This represents a 3.4 dB SNR degradation from theory, but this degradation accounts for network noise, linear and nonlinear distortions, and adjacent channel interference, in addition to the 64-QAM prototype implementation margin. Further, when the BO value was increased by 3 dB, an additional SNR degradation of 1.4 dB was observed. Finally, for a BO value of 15 dB, the measured BER value was 10^{-4}, which corresponds to quasi-error-free transmission when Reed-Solomon encoding/decoding is included.

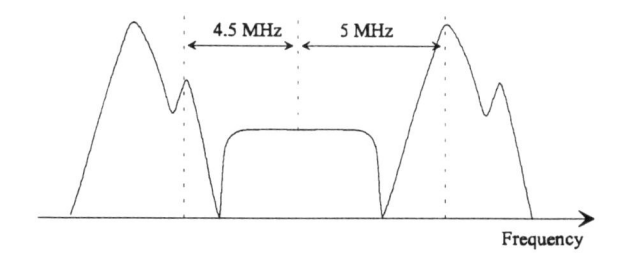

Figure 14: Position of the adjacent SECAM channels.

6. CONCLUSIONS

We have presented two variable symbol-rate receivers, one for digital satellite TV and one for digital cable TV broadcasting. Both modems are fully compliant with the DVB Recommendations which were later published as ETSI standards. The satellite demodulator was integrated in two chips using 0.6μ CMOS technology, and the cable demodulator was integrated in a single chip using 0.5μ technology.

The variable symbol-rate feature of these demodulators give network operators all required flexibility to initially fix the symbol rate and subsequently change it according to their needs. Another distinctive feature is the carrier synchronization technique used which makes it possible to use low-cost consumer-type tuners. Also, the cable demodulator incorporates an adaptive equalizer capable of converging in a blind mode and ensuring small output MSE in the steady state.

REFERENCES

[1] "Digital broadcasting systems for television, sound and data services: Framing structure, channel coding and modulation for 11/12 GHz satellite services," Draft prETS 300 421, August 1994.

[2] "Digital broadcasting systems for television, sound and data services: Framing structure, channel coding and modulation for cable systems," Draft prETS 300 429, August 1994.

[3] G. Karam, V. Paxal, and H. Sari, "A variable-rate QPSK demodulator for digital satellite TV reception," Proc. IBC '94, pp. 646-650, September 1994, Amsterdam.

[4] D. Godard, "Self-recovering equalization and carrier tracking in two-dimensional data communication systems," IEEE Trans. Commun., vol. COM-28, pp. 1867-1875, November 1980

[5] A. Benveniste and M. Goursat, "Blind equalizers," IEEE Trans. Commun., vol. 32, pp. 871-883, August 1984.

[6] M. Mathieu and J.M. Cardin, "Rapport de mesures effectuées au site captif 0G du CCETT avec un modem MAQ64 conforme DVB de la SAT," CCETT internal report, ref. SRL/DLR/T/67/94/MM, November 1994.

A New Class of Carrier Frequency and Phase Estimators Suited for MPSK Modulation

Ivar MORTENSEN, Marie-Laure BOUCHERET

Ecole Nationale Supérieure des Télécommunications (TELECOM Paris)
Site de Toulouse, BP4004, 31028 Toulouse Cedex, France
Tel: 33 - 62 17 29 87
e-mail: mortense@enst-tlse.fr, bouchere@enst-tlse.fr

Abstract

This paper deals with new feedforward synchronisation algorithms suited for digital implementation. We first describe a new gliding window phase estimator doing all the calculations in polar coordinates and its performances. Then we present the frequency estimator based on the same algorithm and its performances. It is shown that the hardware complexity is highly reduced by doing all the calculations in polar coordinates with only a small performance degradation.

1 Introduction

Numerous studies have dealt with the implementation of all digital feedforward demodulators. The main drawback of the previous methods resides in the implementation complexity because of the numerous cartesian-to-polar and polar-to-cartesian coordinates transforms required for frequency and phase estimation. A method [1] [2] [3] [4] has already been developed by TELECOM Paris and ALCATEL ESPACE which allows direct calculations in polar coordinates for carrier phase recovery. This method has recently been extended to gliding window carrier phase recovery and carrier frequency recovery.

2 Feedforward Carrier Phase Recovery

Principle of Phase Estimators: Timing has to be recovered prior to phase estimation. Phase estimation is performed in three steps :
 S1 : modulation removal
 S2 : averaging to improve the Signal-to-Noise Ratio (SNR).
 S3 : Phase estimation with M-fold ambiguity

2.1 The Viterbi & Viterbi Algorithm

The three different steps above are implemented in the V&V algorithm [5] as follows for the QPSK-case: Let (x_n, y_n) be the in-phase and in-quadrature coordinates of the samples at the output of the Nyquist filter. These samples are delivered at the symbol rate R_s.

S1 : The data are forwarded to a non-linearity which delivers a complex sample (x'$_n$,y'$_n$) such that :

$$x'_n + jy'_n = F(\rho_n)e^{j4\varphi_n} \tag{1}$$

where

$$\rho_n = \sqrt{x_n^2 + y_n^2} \text{ and } \varphi_n = \tan^{-1}(y_n/x_n) \tag{2}$$

Various possibilities for $F(\rho_n)$ have been investigated [5] and it has been shown that $F(\rho_n)=1$ leads to little degradation. This leads to the simplified equation:

$$x'_n + jy'_n = \cos 4\varphi_n + j\sin 4\varphi_n \tag{3}$$

S2 : Averaging is performed both on the in-phase (x'$_n$) and in-quadrature (y'$_n$) channels. Results are noted X and Y.

S3 : The phase estimate is :

$$\hat{\varphi} = \frac{1}{4}\tan^{-1}\left(\frac{Y}{X}\right) \tag{4}$$

2.2 The Barycenter Algorithm

The originality of this algorithm resides in the fact that all the calculations are performed in polar coordinates. The algorithm is thouroughly explained and the performances are analysed for the case of block-estimation in [2] and [3]. We will here give an outline of the algorithm and extend it to the case of gliding-window-estimation (for QPSK modulation):

S1: Modulation removal: x_n, y_n, φ_n, ρ_n are defined in the same way as for the V&V algorithm. Modulation removal is performed by reducing the definition domain of φ_n from $[-\pi,\pi[$ to $[-\pi/4,\pi/4[$. This can be seen as adding $\pi/2$ to φ_n until the final value is in the right domaine.

S2: Averaging: The core of the Barycenter algorithm is a function that we have named the BAR-function. This function calculates the average of two phase values in polar coordinates represented in two's complement. However averaging in polar coordinates is not as straightforward as with cartesian ones, where sinus and cosinus are continuous functions. A direct summation such as:

$$\hat{\varphi} = \frac{1}{N}\sum_{i=1}^{N}\varphi'_n \tag{5}$$

may lead to wrong results as can be seen from the following example (N=2):

$$\varphi'1 = \pi/4 - \pi/10, \quad \varphi'2 = -\pi/4 + \pi/10$$

$$\hat{\varphi} = 1/2(\varphi'1 + \varphi'2) = 0 \tag{6}$$

While the exact value is: $\hat{\varphi} = -\pi/4$.

This problem is overcome by using the properties of the representation in two's complement. The BAR-function allows the averaging of L (L being a power of two) phase values when implemented in a binary tree structure as shown in figure 1. This is the Barycenter algorithm for block-estimation and the averaged value in polar coordinates can be used directly as the final estimate when the phase-correction and symbol-decision are also made in polar coordinates.

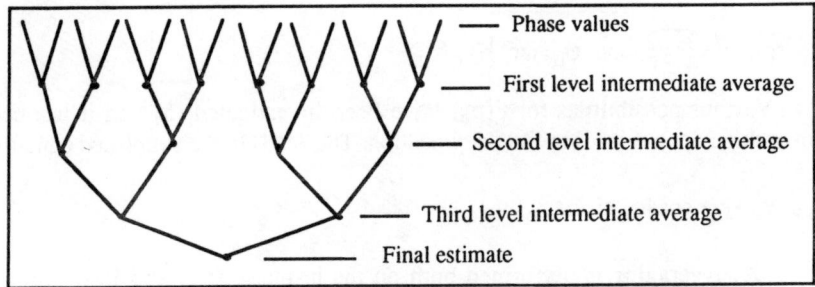

Figure 1: The tree structure in the Barycenter block estimator

2.3 The Enhanced Barycenter Algorithm

An improvement to the BAR function has been developed by TELECOM Paris and ALCATEL ESPACE. This improvement is split into two parts. The first part (The EBAR function, figure 2) is a function that not only calculates the average of two values but also calculates a significance bit assigned to the answer. This bit is zero if the two values that were averaged were further apart than a threshold "S", or if both the two values had their significance bit set to 0 from some previous passage through the EBAR function. Otherwise it is 1. The second part is the SEBAR function (figure 3). This function consists of the EBAR function plus a switch that performs after the following rules: If B_1 and B_2 are both "1" we operate as usual. If B_1="0" and B_2="1" we replace φ_1 by φ_2, if B_1="1" and B_2="0" we replace φ_2 by φ_1, and if both B_1 and B_2 are "0" the two values φ_1 and φ_2 change place. The tree structure that allows the calculation of the estimate remains the same. We have called the algorithm the Enhanced Barycenter (EBAR) algorithm.

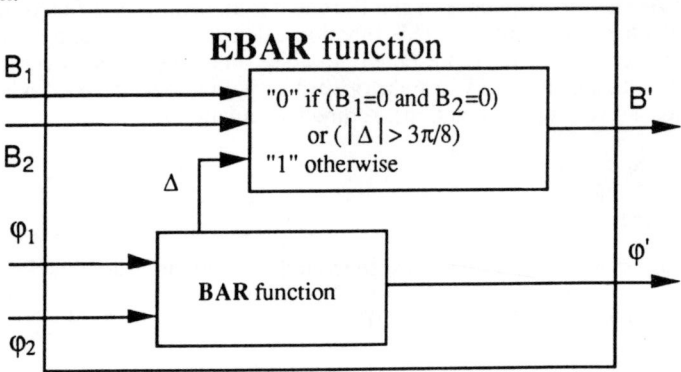

Figure 2: The EBAR function

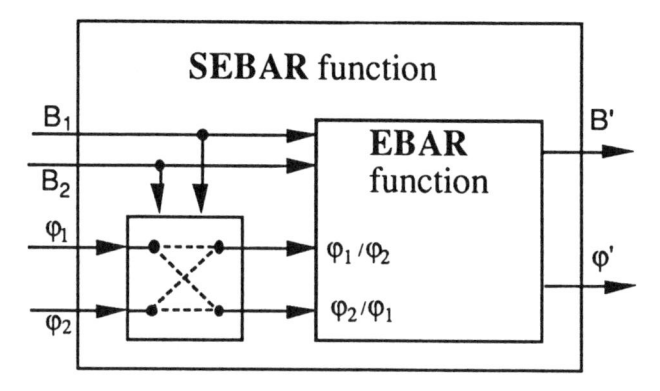

Figure 3: The SEBAR function

2.4 Gliding Window Estimation

The EBAR phase estimator as it has been presented earlier is a block-estimator. A more common form of estimation is the so called Gliding Window estimation (GlW estimation). The GlW Viterbi &Viterbi estimator is relatively simple to implement. In the averaging, step S2, the two integrate and dump functions are replaced by two gliding window functions, which can be implemented recursively.

This is not possible with the Barycenter algorithm because the BAR-function is not a linear operation. At first glance a gliding window needs the calculation of an estimate for every symbol. Nevertheless a closer inspection of the tree structure allows major simplifications to be made. A barycenter block estimation on L symbols requires L-1 calculations of the SEBAR function. So for one symbol, the number of SEBAR functions to be calculated is:

$$\frac{L-1}{L} \approx 1 \text{ SEBAR/SYMBOL} \qquad (7)$$

A direct recalculation to perform a GlW would need L-1 SEBAR/symbol. A GlW algorithm has been developed by TELECOM Paris and ALCATEL ESPACE requiring only:

$$\frac{Log_2(L)}{2} \text{ SEBAR/SYMBOL} \qquad (8)$$

2.5 The Gliding Window Tree structure

For the gliding window structure, we use the fact that an even L will necessarily introduce a bias in the estimation. So instead of performing one estimate for each symbol, we will perform one estimate for two adjacent symbols. The resulting tree structure is described hereafter, where "i" stands for the i'th phase value of the processed block. The structure is best explained with a detailed example as follows:

We use the values "i" to "i+L-1" to calculate the estimate for the two values "i+L/2-1" and "i+L/2". This structure for the example L=16 is shown in figure 4.

We have called the intermediate phase values Intgliss(k,j) where "k" denotes the level in the structure and "j" the number of the sample. For example we can see that:

- Intgliss(1,8) = SEBAR[i=15,i=16]
- Intgliss(2,3) = SEBAR[Intgliss(1,3),Intgliss(1,7)]
- etc. (9)

The value of Intgliss(4,1) is hence used to correct the phase of the two symbols related to "i=8" and "i=9". Now to calculate the estimate that we need to correct the two following symbols, "i=10" and "i=11" (Intgliss(4,2)), we must replace the two values "i=1" and "i=2" with "i=17" and "i=18". It can be seen from the figure 4 that the following values need to be calculated:

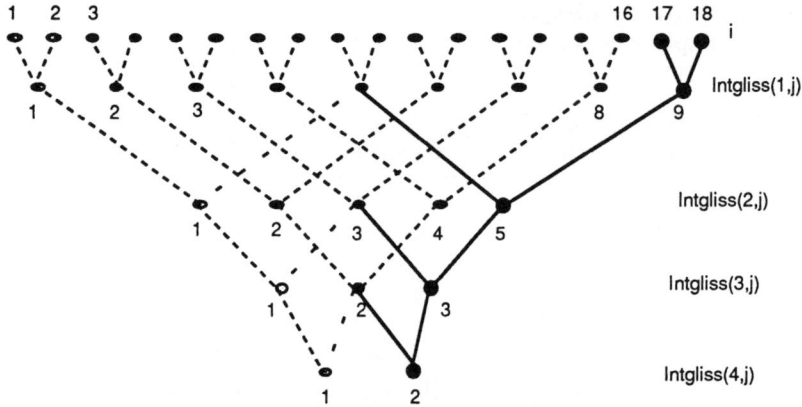

- Intgliss(1,9) = SEBAR[i=17,i=18]
- Intgliss(2,5) = SEBAR[Intgliss(1,5),Intgliss(1,9)]
- Intgliss(3,3) = SEBAR[Intgliss(2,3),Intgliss(2,5)]
- Intgliss(4,2) = SEBAR[Intgliss(3,2),Intgliss(3,3)] (10)

Figure 4: The tree structure for the Barycenter gliding window estimator

We deduce that we need to perform the SEBAR-function "$Log_2(L)$" times for every two symbols ($Log_2(L)$ is the number of stages in the tree structure), and that we need to memorise the L/2-1 intermediate phase values in the structure.

2.6 Implementation

We will here give a short description of two possible implementations of the Gliding Window Enhanced Barycenter algorithm for QPSK modulation (extension to MPSK is straightforward).

S1: Modulation removal in the case of QPSK is performed by removing the two most significant bits of the phase values in polar coordinates (Two's complement).

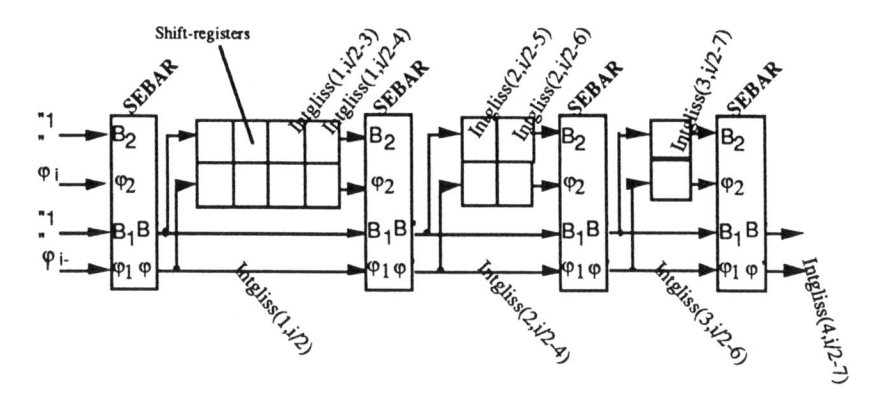

Figure 5: The gliding window Enhanced Barycenter algorithm.

S2: The BAR-function consists of two 6 bit adders, one shift-right, one shift-left and one sign-extension. The SEBAR-function consists of, in addition to the BAR-function, five simple one-bit gates to calculate the significance bits and two simple switches with two inputs and only one output, these switches are controled by one bit each.

The two possible implementations of the gliding window tree structure (with L=16) are shown in the figures 5 and 6. In the first version the calculations are performed in parallell so that we need to implement $\log_2(L)$ SEBAR-functions. The second version is a looped implementation where we need only one SEBAR-function. The significance bits are not shown on the last version.

The looped structure

Figure 6: The gliding window Enhanced Barycenter algorithm, looped structure.

2.7 Performances

The performance of the Enhanced Barycenter algorithm in the block estimation version has been thouroughly analysed in [3]. The losses with respect to the Cramer-Rao bound are given in the figures 7 and 8 for the three algorithms: Viterbi & Viterbi, Barycenter and Enhanced Barycenter.

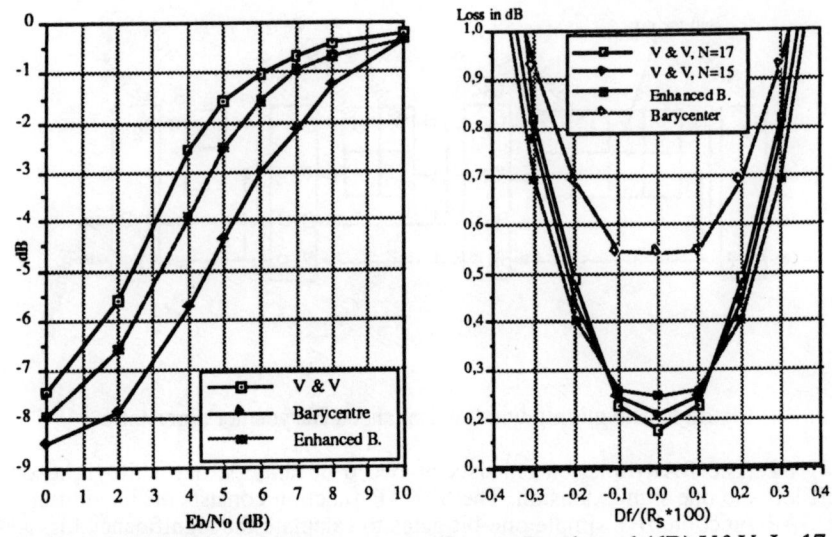

Figures 7 and 8: Loss with respect to the Cramer Rao bound (dB) V&V: L=17, Bary: L=16, and performance with respect to frequency deviation. [Df/(R_S*100)] (E_b/N_0=7dB)

The innovation presented in this paper with respect to the phase estimation is linked to the gliding window structure. This allows the Enhanced Barycenter estimator to cope with a higher frequency error. A comparison between the V&V-estimator and the EBAR-estimator is shown in the figure 9. The curve shows the bit-error-rate for a QPSK receiver with perfect timing at E_b/N_0=7dB. The parameter on the abscissa is the frequency error as % of R_S, where R_S is the symbol rate. We note very similar performances.

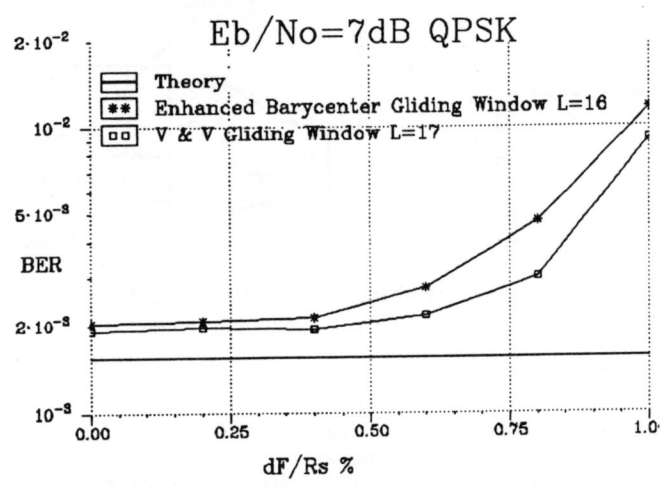

Figure 9: BER simulations for the two estimators V&V and EBAR in the presence of frequency deviation.

3 Feedforward Carrier Frequency Recovery

Principle of frequency estimation with only one sample/symbol: Timing has to be recovered prior to carrier-frequency estimation. The frequency estimation is performed in four steps:
- *S1* - Modulation removal
- *S2* - Numerical differentiation
- *S3* - Filtering and averaging
- *S4* - Frequency estimation

3.1 The Rotational Frequency Estimator

The algorithm commonly known as the NDAFF Rotational Frequency Estimator is described and analysed in [6]. The frequency estimate for MPSK is given by:

$$\hat{\Omega}_n T = \frac{1}{dM} \arg\left\{ \sum_{m=-(L-1)/2}^{(L-1)/2} b_m \, F(|z_{n+m+d}|) \, F(|z_{n+m}|) \, e^{jM(\arg\{z_{n+m+d}\} - \arg\{z_{n+m}\})} \right\} + i\frac{2\pi}{dM} \quad (11)$$

This <u>frequency estimator</u> in its proposed realisation differs from the Viterbi & Viterbi <u>phase estimator</u> in only three points:
- The difference between two phase values is used and not the phase values themselves.
- A simple averaging is not optimal and therefore a filter with the coefficients b_m is recommended.
- The frequency range is limited to $|\Delta f T| \leq \frac{1}{2dM}$ \quad (12)

As one can see from equation 11 modulation removal and numerical differentiation are best performed in polar coordinates. In [6] it is shown that there is only negligible loss when simplifying equation 11 down to equation 13 by selecting $F(\rho_n)=1$ (as in the Viterbi & Viterbi phase estimator).

$$\hat{\Omega}_n T = \frac{1}{dM} \arg\left\{ \sum_{m=-(L-1)/2}^{(L-1)/2} b_m \, e^{jM(\arg\{z_{n+m+d}\} - \arg\{z_{n+m}\})} \right\} + i\frac{2\pi}{dM} \quad (13)$$

The differentiated phase-values are translated back into cartesian coordinates where filtering is performed. The optimal filter coefficients b_n (In the Maximum Likelihood sense) have been derived in [6]. The implementation of this filter function requires 2L extra multiplications and brings only a moderate gain. Therefore the filter is often replaced with an integrate and dump function ($b_m=1$).

3.2 The Rotational Frequency Estimator in Polar Coordinates.

The proposed frequency estimation algorithm in polar coordinates is based on the Enhanced Barycenter algorithm. The main difference from the algorithm described in chapter 3.1 is that averaging is performed in polar coordinates. This means only one translation from cartesian to polar coordinates is necessary. The algorithm is shown in figure 10.

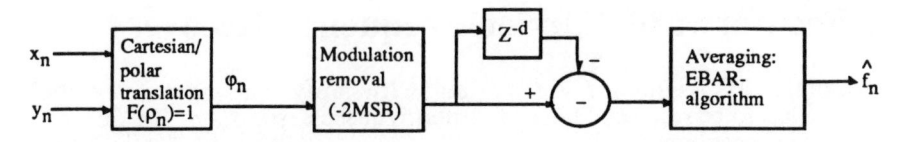

Figure 10: The rotational frequency estimator in polar coordinates (PRFE)

Estimation is performed in four steps:

S1: Modulation removal is done by removing the two most significant bits.

S2: The numerical differentiation is a simple subtraction which output is restricted to $[-\pi/M, \pi/M[$. "d" is the same as in equation 13

S3: Filtering is not straightforward because multiplication in cartesian coordinates is not easily transposed to polar coordinates. Nonetheless several schemes approaching the effect of the optimal filter have been tried with success. One of these will be presented later. The averaging is done with the Enhanced Barycenter algorithm in its block version.

S4: The estimate is directly obtained from the averaging in polar coordinates.

The effect of the parameter "d": The parameter "d" is the distance in symbols between the two phase values to be subtracted. Increasing d leads to a lower variance at the expense of a reduction of the frequency estimation interval.

3.3 Differentiation and Averaging, an Alternative Approach

The performance results of the two frequency estimators that have been presented will show that it is not a simple task to obtain a satisfactory frequency estimate for a SNR below 10 dB using a reasonable length (L≤128) averaging. As many systems today are specified for a lower SNR, this seems to be a relevant problem. Some different schemes to improve the performances of our frequency estimator have been investigated and the most promising one is presented hereafter:

A "d" greater than 1 is used, reducing the frequency estimation interval. (For instance, for d=4 and QPSK modulation, the maximum frequency deviation is +/- 3,125% of the symbol rate.) The idea is simple, we use a gliding window averaging of length "d" before doing the numerical differentiation. The method is shown in figure 11 and will be referred to as "presmoothed" in the following.

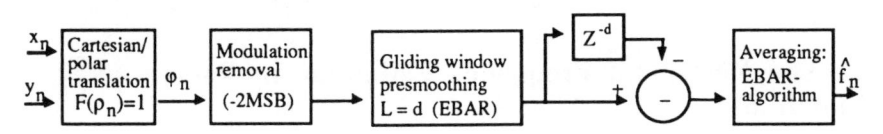

Figure 11: The presmoothed EBAR (SPRFE)

3.4 Performances

Three frequency estimators have been evaluated with COSSAP, a powerful tool to simulate communication chains; The Rotational Frequency estimator (RFE) in cartesian coordinates (based on the V.&V. algorithm), the Rotational Frequency estimator in polar coordinates (PRFE, based on the Enhanced Barycenter algorithm) and the latter in its "presmoothed" version (SPRFE). We first compare the estimator variances to the Cramer Rao bound as a function of the SNR with different averaging lengths "L". (Figure 12).

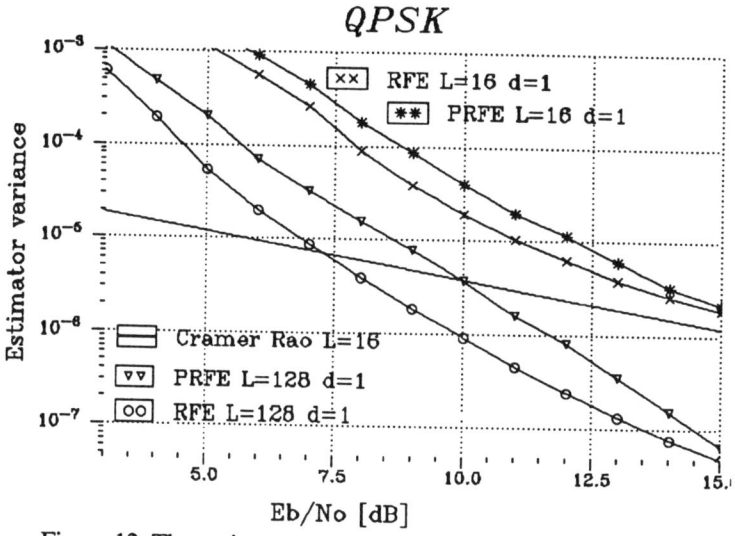

Figure 12: The variances compared to the Cramer-Rao bound

The following remarks can be made:
- Both algorithms have performances far from the Cramer-Rao bound.
- RFE performs better than PRFE, increasingly with large L.

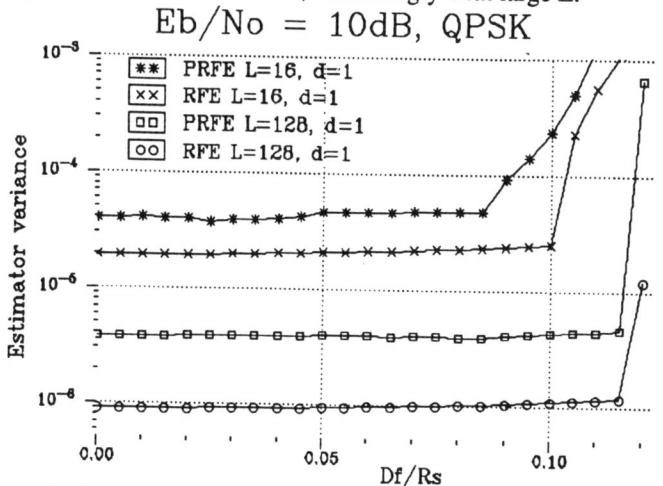

Figure 13: Estimator variances as functions of the frequency deviation

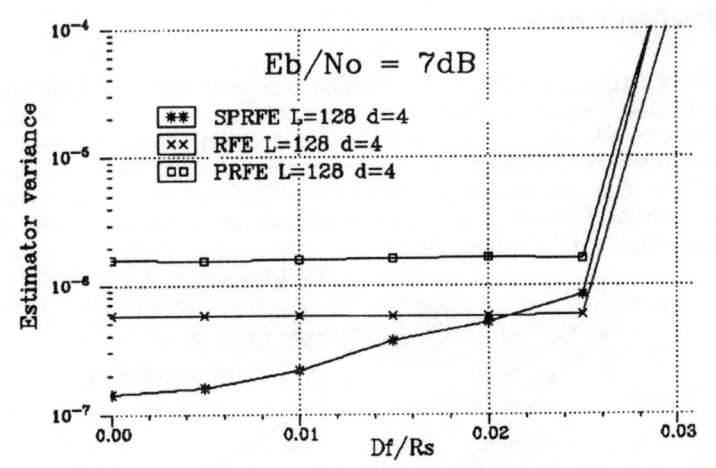

Figure 14: Estimator variances as functions of the frequency deviation, d=4

We now investigate the influence of the following parameters:
- Frequency error, relative to the symbol rate.
- Distance between the two phase values in the differentiation "d"

We note that the SPRFE estimator is dependent on the frequency deviation. This limits the frequency range, but the estimator is still very interesting.

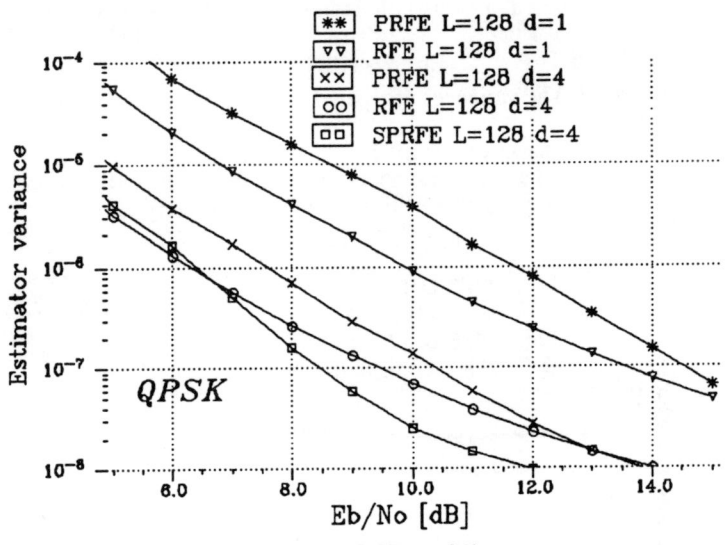

Figure 15: Df/Rs = 2%

The results shown in figure 15 have been obtained with a frequency deviation of 2% of Rs (QPSK).: We remark that the SPRFE estimator and the RFE estimator have near identical performances.

4 Conclusion

In this paper, frequency and phase estimators performing calculations in polar coordinates have been described and their performances have been compared to those of the V&V estimator (for phase) and RFE (for frequency).

Phase estimation in polar coordinates leads to a reduced hardware complexity with performances close to those of the V&V estimator.

For frequency estimation, the results are different because of the very low frequency error variance required at the input of the phase estimator. For medium to high SNR, both algorithms achieve acceptable variance, PRFE leading to a lower complexity. For low SNR, a value for d greater than 1 is necessary to meet the required variance. SPRFE and RFE have similar performances, SPRFE being less complex to implement.

Acknowledgment:

The authors gratefully acknowledge the support from the European Union, Human Capital & Mobility Program, Contract: CHRX -CT93-0405

References:

[1] ALCATEL ESPACE, ESTEC contract nr: 8744/90/NL/RE, MF/TDMA multicarrier demodulator, phase 1 report, october 1992

[2] ML BOUCHERET et al., A new algorithm for nonlinear estimation for PSK-modulated carrier phase, proceedings of ECSC3 (Third European Conference on Satellite Communications), Manchester, november 1993

[3] MORTENSEN I. et al., An improvement to a digital algorithm for nonlinear estimation of PSK-modulated carrier phase, ESA workshop on DSP for space applications. London 26-28 sept. 1994

[4] Patent pending: FRANCE TELECOM, "Procédé d'évaluation numérique en coordonées polaires de la phase des signaux utilisant des informations de fiabilité et adapté à des moyennes glissantes. Dispositif de mise en oeuvre de tel procédé."

[5] VITERBI, A.J, VITERBI, A.M, Non-linear estimation of PSK-modulated carrier phase with application to burst digital transmission, 1983,IEEE Trans. on Information Theory, vol IT-29, 543-551

[6] CLASSEN F.,MEYR H.,SEHIER P., "Maximum Likelihood Open Loop Carrier Synchronizer for Digital Radio", ICC93

Part 4

Signal Processing in Satellite Networks

Part 4

Signal Processing in Satellite Networks

Payload Digital Processor Hardware Demonstration for Future Mobile and Personal Communication Systems

A D Craig[1] and F A Petz[2]
[1] Matra Marconi Space, Stevenage, UK
[2] ESA/ESTEC, Noordwijk, The Netherlands

Abstract

Future advanced mobile and personal satellite communications systems, such as the ICO orbit Inmarsat-P system and GEO regional systems, require digital signal processing to provide the necessary flexibility in the routing of channels to/from the multiple beam mobile link coverage. The paper describes a hardware demonstrator, developed under ESA contract, which is capable of demonstrating both switch-based wideband beamforming and digital narrowband beamforming architectures. The central elements of the demonstrator are 4 ASICs, developed in flight representative radiation hard technology, which may be flexibly configured to demonstrate various architectural options.

1. Introduction

The paper summarises work performed under an ongoing ESA contract [Ref 1] aimed at developing digital processor hardware relevant to advance mobile and personal satellite communications systems.

There are a number of ongoing developments of advanced satellite based mobile and personal communications involving both global coverage, with a constellation of low or intermediate orbit satellites (eg Inmarsat-P), and regional coverage with geostationary satellites. Such systems are characterised by the use of very low G/T and EIRP terminals with handsets being the extreme example. This in turn necessitates the use of multiple high gain spot beam coverage on the mobile link in order to achieve the necessary link performance; typically 50 to 300 beams may be required. A changing traffic distribution between beams results from statistical, diurnal and longer terms effects in geosynchronous systems and, in the case of global systems, utilising lower orbits there are traffic changes resulting from the natural scanning of beams along the ground track. Thus there is the need for the payload to flexibly route traffic between beams. Bandwidth limitations will generally preclude frequency addressing as a means of routing such that an active onboard mechanism is required. In this respect the term granularity block denotes the smallest unit of capacity that may be routed between beams and is characterised by a bandwidth which may contain a single or multiple carriers and a given carrier may typically carry a small number of time multiplexed user channels. The large number of granularity blocks and beams typical of such systems implies the need to utilise digital signal processing technology in order to achieve the required flexibility in channel to beam routing. The digital technology, in the form of high integration radiation hard ASICs, capable of meeting these requirements has only recently become available.

ESA has supported a series of R/D contracts since 1987 aimed at developing digital technology for transparent payloads for such advanced mobile and personal communications systems. Early study contracts demonstrated the feasibility of such systems [Refs 2 and 3]. A major ongoing contract has developed key hardware elements in representative ASIC technology and has assembled them into a configuration which has demonstrated processor feasibility in hardware terms. It is this hardware demonstrator development that forms the main topic of this paper.

Section 2 describes candidate payload and processor architectures to meet the requirements of typical planned systems. Section 3 describes the ASIC developments, demonstrator configurations and the tests that have been performed. Section 4 provides some conclusions.

2. Payload And Processor Architectures

2.1 Requirements

The requirements that characterise a typical system of interest may be summarised as follows:

• A satellite system (single or multiple satellites) is required to provide duplex user channels linking a small number of feeder terminals to a large number of mobile terminals. Some systems may, in addition, be required to provide direct mobile to mobile and feeder to feeder channels .
• Single feeder link beam.
• Multiple spot beam mobile link; the number will depend on gain requirements and size of coverage but may typically be 50 to 300 for a personal communications system.
• Feeder bandwidth is divided into a number of granularity blocks where each block contains a number of user channels (on a single or multiple carriers). Total bandwidth may typically be 100 to 200 MHz with granularity block bandwidth typically being 30 to 200 kHz. Block bandwidth should ideally be a small as possible to minimise bandwidth inefficiency associated with partially filled blocks allocated to beams.
• Mobile link bandwidth is divided into a number of frequency slots which may potentially be occupied by feeder link granularity blocks. Total mobile bandwidth may be typically between 10 and 34 MHz.
• Mobile link uses extensive frequency reuse between spatially separate beams.

The fundamental flexibility requirement is that the payload should be capable of routing a given feeder link block to/from any available frequency slot within any mobile link beam. In some cases this requirement may be relaxed but it will be assumed within the subsequent discussion.

2.2 Payload Architectures

Two different payload and processor generic architectures have emerged during the studies of the last few years and represent candidates both within the ongoing hardware developments and within planned satellite systems. The architectures are characterised according to the bandwidth of the beamforming on the mobile link:

• Narrowband beamforming systems in which beamforming is on the basis of the individual granularity blocks using digital beamforming techniques with routing being on the basis of beam steering.
• Wideband beamforming systems in which each mobile link beam is formed across the full mobile link bandwidth using either analogue or digital processing techniques and routing is performed by spatial switching.

Narrowband Architectures. The narrowband architecture was first described in an ESA sponsored contract performed by British Aerospace Space Systems (now part of MMS) in 1987 [Ref 2]. Fig 1 shows a representative forward link architecture, ie from the feeder terminals to the mobile terminals. In Fig 1(a) the architecture is drawn to emphasise the key processor functions whilst Fig 1(b) shows a practical implementation architecture. The architecture, as shown, assumes the use of a phased array antenna on the mobile link but is equally applicable to a multimatrix reflector antenna system.

The signal flow through the system is described as follows initially with reference to Fig 1(a):

• Feeder link signal, following the receiver and initial downconversion, is filtered by an anti-alias filter (AAF), implemented at a suitable IF, to isolate the required band and is further downconverted to a low IF (lower end of band close to zero frequency).
• Feeder band is sampled by an A/D convertor. Sampling may be in quadrature or real form; real sampling may require subsequent digital processing for real to complex conversion. The complex sample rate must at least equal the bandwidth of the feeder signal (together with the AAF transition width with signals in the transitions subsequently rejected in the digital processing). In practice the feeder bandwidth may be such that it has to be divided into a number of subbands with multiple AAFs prior to sampling.
• Digital demultiplexing of the feeder link band with the output being in the form of K separate signals corresponding to the individual feeder link blocks again in digitally sampled form but decimated to a rate consistent with the individual block bandwidth.
• Individual blocks may have independent level control (LC) applied as a simple scaling of the digital samples.
• Each feeder link block signal has a separate digital beamforming network (DBFN) function the purpose of which is to determine mobile link beam properties for that specific block. The block signal is split N ways, where N is

the number of elements of the phased array, and a complex weight multiplies samples for each of the element paths, ie to in effect provide amplitude and phase control. Thus to create a spot beam in a given direction for a given block the weights would be such as to create the required linear phase gradient across the phased array. Thus the beamforming is implemented at the decimated sample rate characteristic of the narrow bandwidth of the individual block.

• A switch function is required to provide flexible mapping to mobile link frequency slots.

• The same phased array is used to provide all the beams and therefore signals must be combined on an element basis. For blocks which are mapped to the same frequency slot on the mobile link samples are combined by simple addition in a frequency reuse concentrator (FRC). Samples for all mobile frequency slots must then be frequency multiplexed in a digital multiplexer. For a given element multiplexer the number of the inputs will be the number slots on the mobile link (not all necessarily occupied) and will be less than K because of the frequency reuse; the output will be a sampled form of the full element signal at a rate consistent with the full mobile bandwidth.

• Each element signal is D/A converted (in either real or quadrature form), anti-image filtered and is upconverted to RF and amplified prior to going to the phased array element.

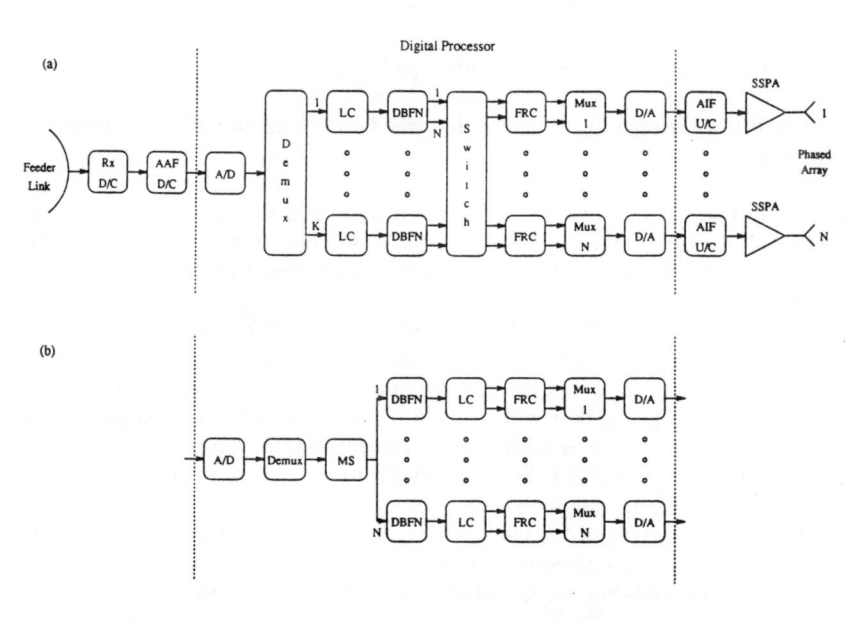

Fig. 1 Forward Link Narrowband Beamforming Architecture

Channel to beam routing is therefore, in effect, implemented by beam steering through control of the beamforming weights. Each block may potentially be individually optimised in terms of beam location to offer maximum gain to the users concerned; alternatively valid beam locations may be limited to a finite regular grid in which case peak gain may not be achieved.

The signal flow is next described with respect to Fig 1(b) in order to emphasise practical implementation issues following from the need to partition the processing between ASICs and noting the key advantage of a digital implementation that samples may be time multiplexed to limit interconnections:

- Demux output is time multiplexed (TDM) onto a single bus with the output in the form of frames of length K where a given frame contains a single complex sample for each feeder block; in practice multiple TDMs with shorter frame length may be required in order to limit bus sample rate.
- Flexible frequency mapping is implemented in a memory switch (MS) function. The order of samples in the input frame from the demux follows the feeder link block frequencies. These are input into RAMs within the MS function and read out in a different order to reflect the required mapping to the mobile link; this may involve grouping together frequency reuse blocks in the output time frames from the MS function.
- Beamforming processing is partitioned such that weight multiplication is performed on an element basis; thus the time multiplexed output from the MS is input into a series of element processors which, for a given time frame, sequentially apply the required complex weight multiplication to the block samples. Amplitude scaling of individual blocks can be included within the weights.
- Frequency reuse concentration is implemented by addition of samples following the beamforming weighting on an element path basis the output being in the form of reduced dimension time frame with samples ordered according to downlink frequency. Element based digital multiplexers form a sampled representation of the full element signals across the full mobile bandwidth.

The return link essentially involves the inverse functions. Within the return digital beamforming function there is the requirement to add samples for a given block following complex weighting. If direct mobile to mobile channels are required a linkage is provided from the return link after receive beamforming into the forward link path prior to transmit beamforming.

The discussion above assumes the use of digital processing to implement the demux and mux functions. An interesting alternative in terms of technology is to use chirp Fourier transform (CFT) processing for these functions. CFT technology is being developed within the ongoing ESA contract but will not be discussed in any detail here.

Wideband Architectures. Fig 2 shows the forward link wideband architecture for a case in which the beamforming is implemented using analogue techniques. The approach offers the potential to significantly decrease the complexity of the digital processing relative to the narrowband approach if it is acceptable to limit the occupied contiguous bandwidth of a given beam to be a relative small fraction of the overall mobile bandwidth; such a limitation may be acceptable within certain practical systems where the upper limit on beam traffic can be defined.

The signal flow is described as follows:

- Same as for the narrowband case up to the digital demultiplexing and level control of the feeder link blocks.
- Beam routing in this case is performed by explicit spatial switching of blocks to beams. Flexible frequency mapping is included within the switch function using RAMs to map between input and output time frames in the same way as for the narrowband architecture.
- Digital multiplexing is on a beam port basis in this case. The dimension of the beam digital mux is determined by the maximum contiguous bandwidth of the beam and may be significantly lower than the dimension of the element based digital mux of the narrowband architecture which must cover the full mobile bandwidth.
- Beam port signals are D/A converted and upconverted to RF; upconversions will in general be different between beams to place the beam band at the required location within the overall mobile band.
- Beamforming is performed using analogue technology; for example a Butler matrix can efficiently generate the necessary phase gradients to generate a uniform grid of beams. Note that in an alternative approach the beamforming may be implemented digitally again on a wideband basis using efficient FFT processing techniques.

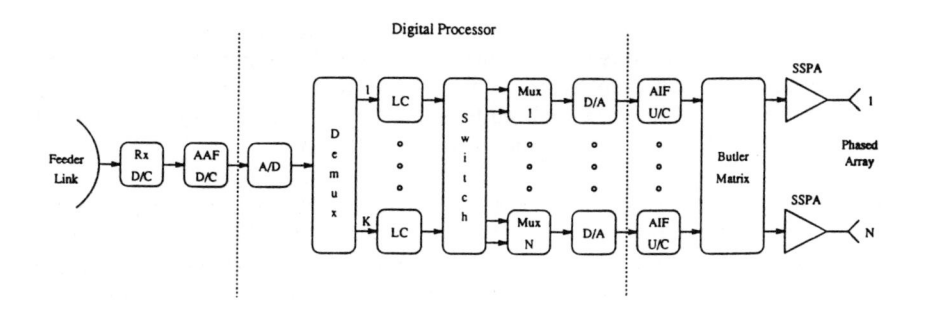

Fig. 2 Forward Link Wideband Beamforming Architecture

The overall digital processing rate is largely determined by the mobile link demux/mux functions and scales with the number of beams and beam bandwidth; this compares with the narrowband case where the scaling is with the number of elements and the total mobile bandwidth. Given that the number of beams is typically comparable with the number of array elements it is clear that a reduced beam bandwidth leads to reduced digital processing when compared with the narrowband approach. The detailed tradeoff must be made on an individual mission basis.

2.3 Implementation Feasibility

The feasibility of realising such architectures in practice depends on the ability to achieve an implementation with sufficiently low mass and power; this in turn depends on:

• Use of efficient DSP algorithms which seek to minimise processing rates and word lengths whilst achieving acceptable performance in terms of signal degradation. Considerable work has been performed on optimising algorithms and overall architecture and has been complemented by extensive emulation to assess the tradeoff between processing and signal degradation within the processor. Mechanisms for degradation include quantisation noise within the A/D convertor and the various processing stages and adjacent channel interference resulting from aliasing of incompletely rejected adjacent channels due to decimation within the demux processing, etc. Degradation can be expressed in terms of an effective carrier to noise ratio where noise includes all noiselike inband signals due to the processing; the allowable value of this ratio will depend on specific mission requirements and processor architecture but 20dB is typical for the TDM/FDMA systems under development.
• Availability of appropriate technology, in particular highly integrated, low power, radiation hard ASIC technology and mass efficient multichip module (MCM) packaging. A recent review of key relevant technology issues is given in Ref 4.

Significant study activities have been performed on estimating implementation parameters, in particular power and mass, based on the use of radiation hard ASIC technology and an MCM packaging approach. Estimates are specific to particular missions and choice of architecture and technology but typical estimates are for processors of several tens of kilograms and several hundreds of watts.

3. Demonstrator Configuration and Test Results

The objective of the demonstrator work has been to develop a functional demonstration of both wideband and narrowband beamforming digital processor architectures utilising flight representative, radiation hard ASICs.

The following ASICs have been developed within the contract:

• Digital demux/mux ASIC utilising an efficient weighted overlap and add (WOLA)-FFT algorithm. As well as providing the flexibility to operate in either demux or mux modes the ASIC also provides the flexibility to channelise 112, 56, 28 or 14 channels within a processed bandwidth of up to 8 MHz. The ASIC has been designed by MMS and manufactured in Thomson 0.8 micron radiation hard silicon on insulator (SOI) technology and utilises approximately 70 kgates.
• Tight feeder link filter ASIC which, in combination with the demux/mux ASIC, performs the full feeder link demux and mux function but is not required on the mobile link. It utilises a patented imaged halfband filter approach which

performs tight filtering of all the channels as a single processing function. This ASIC has also been designed by MMS in the same technology as the demux/mux ASIC and again utilises approximately 70 kgates.

• Memory switch ASIC which performs the required frequency mapping of the narrowband architecture of up to 224 channels and is also representative of the routing switch requirement of the wideband architecture. The ASIC is also applicable to both forward and return links and various granularity options. The ASIC has been designed by Alcatel Espacio and manufactured in Temic 1 micron radiation tolerant technology and utilises about 25 kgates.

• Digital beamforming ASIC which performs forward and return beamforming processing and frequency reuse concentration/distribution for up to 224 mobile link channel slots for 5 element paths. In its return link mode of operation the ASIC includes the sample addition function of the beamforming. The ASIC has been designed by Alcatel Espace and is being manufactured in Temic 0.6 micron radiation tolerant technology with in excess of 100 kgates; delivery is imminent.

The ASIC design process has relied heavily on the use of VHDL modelling at an overall system and individual ASIC level to ensure correct interfacing of ASICs within the demonstrator and within a practical processor.

The overall configuration of the hardware demonstrator is shown in Fig 3 for both forward and return link tests and comprises the following major elements:

• The unit under test (UUT) comprising groups of ASICs, mounted on a set of PCBs, which may be flexibly configured to demonstrate various processor functions and architectures, including both wideband and narrowband architectures.

• Specialised test equipment to provide representative input signals to the UUT and to capture outputs from the UUT for analysis. Input and output interfaces are currently digital.

• Control PC and special purpose control PCB. Special purpose software has been written to control the test equipment and to configure the UUT and ASIC parameters including digital filter coefficients, MS frequency mapping RAM and DBFN weights.

The specialised test equipment has been developed within the contract by ERA and comprises the following units:

• Digital signal synthesiser (DSS). Required input signal sequences are generated offline and downloaded into large memories within the DSS. Data is then cyclically read out of memory in real time to provide the necessary input signals to exercise the processor.

• Digital antenna simulator (DAS). This unit is only relevant to return link tests and generates up to 7 digital signals representative of signals received at elements of a phased array on the basis of a number of incoming signals from different directions.

• Data capture unit (DCU) which captures output data from the processor demonstrator into large memories for offline analysis.

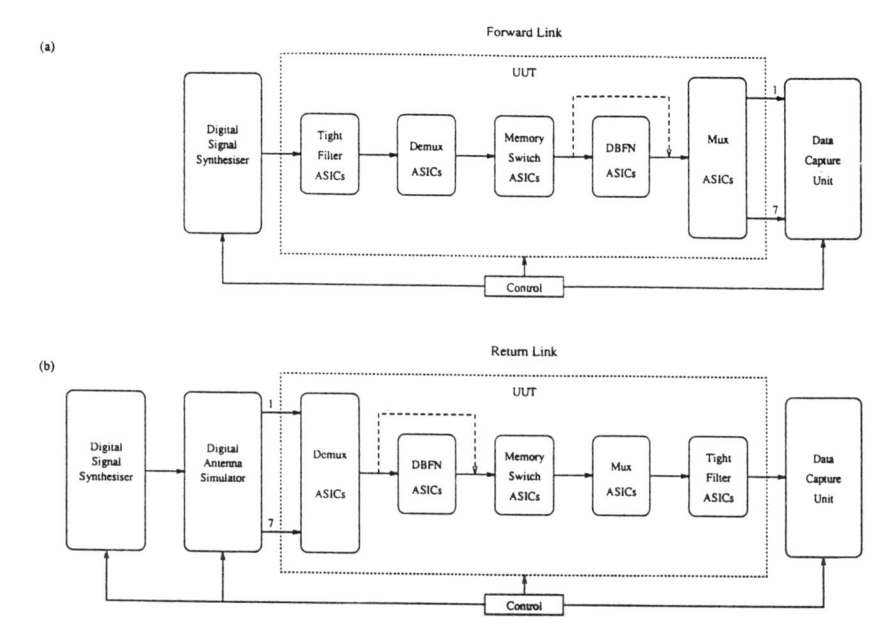

Fig. 3 Hardware Demonstrator Configuration

A typical forward link test corresponding to a wideband architecture is described as follows and is illustrated by Fig 4.

• A sequence of finite wordlength samples is computed offline representing a specific time period of a feeder uplink signal containing a frequency multiplex of 112 QPSK modulated carriers with a spacing of 36 kHz. Such samples are therefore representative of the output from the feeder A/D convertor. The samples are downloaded into the memory within the DSS; typically a sequence of several thousand samples is used. The samples are cyclically read out from the DSS to form a real time sampled input into the UUT. Fig 4(a) shows the spectrum of the sampled signal at this point; there are 112 peaks representing the carriers.
• Feeder link demultiplexing is performed by a combination of the tight filter and demux ASICs. The output may be captured by the DCU and analysed offline. The output at this point is in the form of time frames of length 112 each containing 1 sample per input channel. Fig 4(b) shows spectra of a selection of 4 of the individual carriers formed by Fourier transforming corresponding samples from the time frames.
• Output from the memory switch may be similarly captured and analysed and confirms the specified reordering of channel sample locations within the output frames according to the required frequency mapping.
• Frequency multiplexing is implemented by an ASIC of the same design as the demux but in the different mode of operation. Fig 4(c) shows the spectrum of the mux output; in this case the full number of input channels have been multiplexed (corresponding to the case of all channels being switched to a given beam in the context of the wideband case) but with the channels reordered by the MS ASIC;

again the individual QPSK carriers are clear.

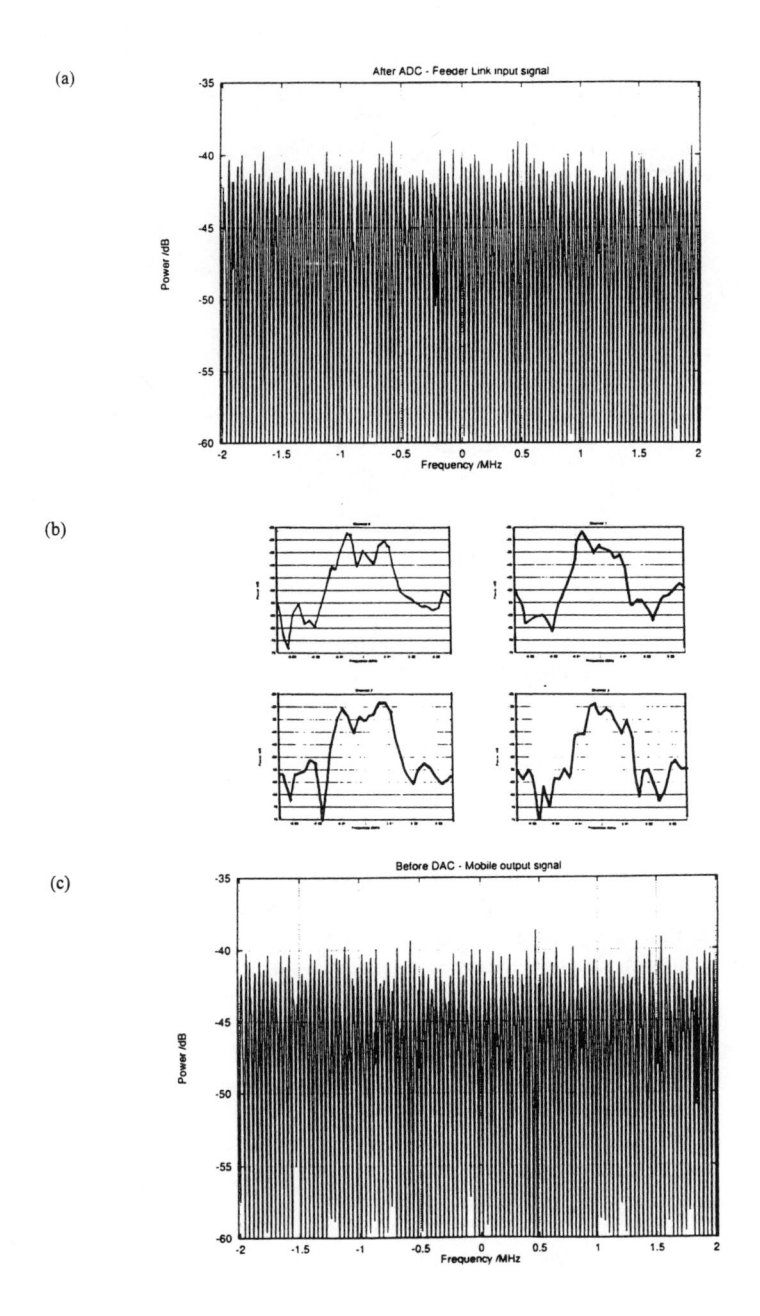

Fig. 4 Spectra from Forward Link Wideband Architecture Test

Tests of this kind have been carried out on individual ASICs, sub-chains of ASICs and complete processing chains, as in the above example. Whilst showing signal spectra is useful for demonstration purposes, the essential element of the test has been to confirm, at the bit level, that the ASICs and processing chains perform as predicted against simulated results from separate emulation and VHDL models. The demonstrator has successful demonstrated agreement in all modes for all ASICs and combinations of ASICs in both forward and return directions for this wideband architecture.

Power measurements have been made on the demux/mux and tight filter ASICs with approximately 600 mW in each case for 5V operaration; operation has also been demonstrated at 3V with a corresponding power reduction.

A live demonstration has also been implemented in which a continuous PSK modulated audio channel, placed within a multiplex of 112 channels, has been demultiplexed and routed to an audio output section by means of the tight filter, demux and MS ASICs.

The following enhancements to the demonstrator are being implemented or are planned:

- Inclusion of the digital beamforming ASIC when it becomes available in order to demonstrate the narrowband beamforming architecture.
- Inclusion of a representative MCM utilising 4 demux/mux ASICs and 4 feeder tight filter ASICs in bare die form. This is currently being manufactured in thin film technology by Dynamics Microcircuits. This MCM will be functionally demonstrated within the context of the overall processor demonstrator.
- Inclusion of A/D and D/A convertors and associated IF processing chains (comprising frequency conversion and AAF/AIF filtering)
- Development of a processor engineering model.

4. Conclusions

The hardware demonstrations described represent a significant step in establishing the practical feasibility in the use of digital signal processing to provide the necessary channel to beam routing within future multibeam mobile and personal communications systems. Such systems represent examples within a general trend to the use of digital signal processing within satellite payloads which may be expected to progress from the relatively narrow bandwidth mobile systems to the larger bandwidths of fixed and broadcast services of the future.

Acknowledgements

The work described has been performed by a large number of colleagues at the prime contractor, Matra Marconi Space, and at the various subcontractors, Alcatel Espacio, Alcatel Espace, Frobe Radio, AME Space and ERA and has benefited from the advice and support of colleagues at ESTEC.

References

1. ESA ITT AO/1-2671, Digital beamforming techniques, 1992.
2. Final report on ESA contract 8087/88/NL/JG(SC), Study of digital beamforming networks, British Aerospace, 1990.
3. Final report of ESA contract 8972/90/NL/RE, Study of applicability of different onboard routing and processing techniques to a mobile satellite system, British Aerospace, 1992.
4. M. Hollreiser 'VLSI for DSP in Space' Proc Fourth Int. Workshop on Digital Signal Processing Techniques Applied to Space Communications, London, September 1994.

Frequency Domain Switching: Algorithms, Performances, Implementation Aspects

G. Chiassarini , G. Gallinaro

Space Engineering S.p.A., Via dei Berio, 91, I-00155 ROMA ITALY
tel: +39 6 225951, fax: +39 6 2280739, e-mail: chiassarini@space.it, gallinaro@space.it

Abstract
This paper presents a new approach for implementing an SS-FDMA payload based on the use of Digital Signal Processing. Main feature of the proposed technique is the fact that channels switching is performed in the frequency domain instead of the more conventional time domain thus justifying the name of Frequency Domain Switching for the proposed method. The technique lends itself to a very efficient hardware implementation, whilst allowing very large flexibility in the channel routing and the rearrangement of the up and down link frequency, due to the utilization of the Fast Fourier Transforms (FFT) for conversion of the signals in the frequency domain. The paper addresses the principles of the Frequency Domain Switching by introducing the relevant algorithms and a possible payload architecture. Then an analysis of the degradation produced by the algorithms and their actual hardware implementation follows. Finally, the results obtained analytically, are compared with those obtained via simulation.

1. Introduction

This paper concerns the utilization of the Frequency Domain Switching (FDS) technique for implementing an SS-FDMA (Satellite Switched FDMA) repeater. The objective is the identification of an efficient architecture able to route the individual channels from spot to spot. In a SS-FDMA system implemented with traditional technologies, each signal coming from a Rx spot feeds a bank of filters where the individual channels (to be routed to different down-link spots) are separated. Then each channel is routed to the appropriate Tx spot where it is combined with other channels also directed to the same Tx spot but originated from different Rx spots. Large complexity, due to the need of a large number of synthesized up and down-converters as well as of filters, and a reduced flexibility, due to the impossibility to change the bandwidth of the individual filters, generally results. With the proposed approach instead, no synthesized up and down converter is needed, due to the possibility to perform frequency shifting in a trivial way directly in the frequency domain. Moreover, full flexibility is obtained thanks to the possibility to change the channel bandwidth, with very small granularity and on a very wide range, on a channel by channel basis.

2. FDS Payload architecture

A reference architecture of a SS-FDMA payload based on the proposed Frequency Domain Switching concept is shown in fig. 1 where the case of a 16 beam up-link and 8-beam down-link coverage is shown. The rationale behind the overall repeater architecture can be found in ref. [1] and [2]. Here, we will concentrate on the Digital Processor portion of the repeater, whose purpose is:
- to demultiplex the channels from each up-link beam;
- to route (switch) each channel group according to its destination;
- to multiplex the different channel groups directed to the same downlink beam.

The Digital Processor is actually composed by a number of FFT and IFFT Processors interconnected by an S-stage. In particular, the IF analog input signal coming from an Rx spot, after anti-aliasing filtering via a SAW filter, is firstly converted into a stream of digital samples via an Analog to Digital Converter.

Then transformation to the frequency domain via an FFT Processor takes place on partially overlapping frames of the input samples. Individual channels are easily separated in the frequency domain and can thus be routed to the appropriate output processor by a conventional "time-division Space-switching stage" (*S-stage*). Finally an inverse FFT (IFFT) is performed each output by an IFFT Processor for channel recombination. The architecture of the FFT and IFFT processors may be identical (thus further simplifying the development effort) although this is not mandatory. The only real constraint is that the FFT and IFFT processors share the same frame length, since the frequency resolution of the processors (which is the inverse of the frame length) must be identical. The FFT Processors also include a phase rotation function, to compensate for the channel phase rotation in the period between two successive FFT computations.

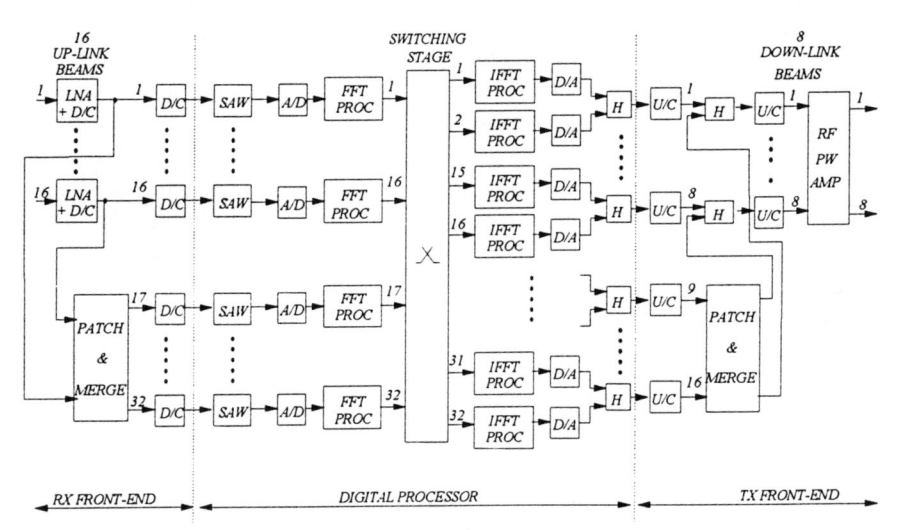

Figure 1: Repeater Architecture using FDS

The FFT Processors also include a phase rotation, to compensate for the channel phase rotation between two successive frames. By appropriate selection of the FFT processing parameters such a phase rotation may be made trivial.
Finally, the IFFT output is fed to a DA converter for analog output reconstruction. The switching stage can be implemented by a conventional S-stage, because the FFT (and IFFT) processors can include also the functions of a T-stage.

3. Algorithms Overview

In order to formalize the FDS algorithm we start from the conventional approach requiring a demultiplexer, to separate the input channels, and a multiplexer to recombine the output channels. We start from the demultiplexer algorithms, then we generalize to include the complete FDS.

Demultiplexer. A demultiplexer for separation of channels with different bandwidths is basically a bank of passband filters; each filter can be implemented using a fast convolution algorithm. Now fast convolution is based on the three basic steps: translation of the signal in the frequency domain via the Discrete Fourier Transform (DFT); filtering in the frequency domain via multiplication with the filter transfer function; translation of the result again in the time domain. It is worthy noticing that the initial translation in the frequency domain can be identical for all the filters to be implemented and can thus be shared by all filters in the filter bank.

We recall that the convolution via DFT is a circular convolution between the filter impulse response and the block of input samples over which the DFT is performed. This circular convolution differs from the aperiodic convolution we are interested in, by the fact that *time aliasing* is present. A way to solve this problem is the utilization of the overlap and save technique. According to such a technique, the input signal stream is divided in frames partially overlapping; then circular convolution is computed and the edge portions of each frame, where *time aliasing* is concentrated, are discarded.

To formalize the algorithm let us indicate with $x(.)$ the demultiplexer input sample stream (assumed complex) and with $y(.)$ the signal obtained by translating to baseband the output of the filter, centered on f_0, whose impulse response is $f(.)$. It is:

$$y(m_0 T_y) = W_1^{m_0 T_y f_0} * \sum_{n=-\infty}^{\infty} f(m_0 T_y - n_0 T_x) * x(n_0 T_x) \tag{1}$$

where:

$$W_a = e^{-j\frac{2\pi}{a}};$$

m_o and n_o are integer indexes;
T_x and T_y are the input and output sampling periods (they are typically different);

Since in the fast convolution approach the processing is performed by dividing the signals in frames, let us express the sample index as follows:

$$n_o T_x = nT_x + dDT_x \qquad d = ..,-1,0,+1,.. \qquad n = 0,..,N-1$$
$$m_o T_y = mT_y + dET_y \qquad d = ..,-1,0,+1,.. \qquad m = P,..,M-P-1$$

where the meaning of the symbols is as follows (see fig. 2):
- N and M are the number of samples over which FFT is computed, respectively direct and inverse (they are generally different because decimation is typically performed before inverse FFT), i.e., they represent the frame length expressed in input and output samples;
- D and E are the number of useful samples, respectively in the input FFT and output IFFT;
- P represents the samples in the output IFFT that are discarded at each edge of the frame.

D,E,P fulfill the following conditions:
- $MT_y = NT_x = T$; where T is the FFT processing frame length;
- $DT_x = ET_y = V$; where V is the portion of the frame length containing valid samples;
- $2P = M-E$; is equal to the number of samples to be discarded.

Within this context the frame of input samples is the following:

$$x_d(nT_x) = x(nT_x + dDT_x) \qquad n = 0,1,..,N-1$$

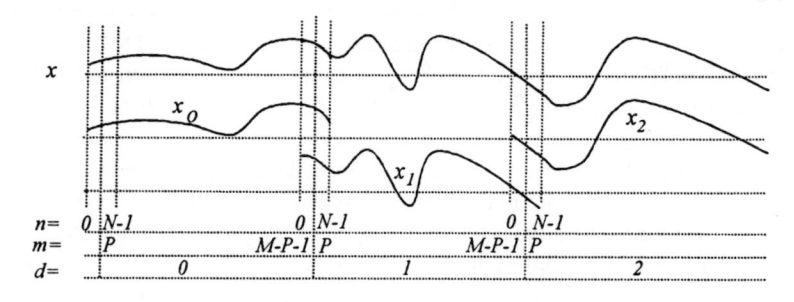

Figure 2: Frame overlapping timing

Equation (1), with the new notations, becomes:

$$y(mT_y + dDT_x) = W_1^{(mT_y + dDT_x)fo} * \sum_{n=0}^{N-1} f(mT_y - nT_x) * x_d(nT_x) \qquad (2)$$

Notice that the sum can be limited to the interval $0,..,N-1$ provided that the filter impulse response is relevant only in the interval $[T_y(M-P-1)-T_x(N-1),..PT_y]$ or about $[-PT_y,..,PT_y]$.

Now, we assume $X_d(i)$ as the DFT of $x_d(nT_x)$, i.e.: $x_d(nT_x) = \dfrac{1}{N}\sum_{i=0}^{N-1} W_N^{-ni} * X_d(i)$, and the

identity: $f_p(mT_y - nT_x) = \dfrac{1}{M}\sum_{k=0}^{M-1} W_M^{-mk} * \sum_{l=0}^{M-1} W_M^{lk} f_p(lT_y - nT_x)$ where $f_p(.)$ is $f(.)$

periodic of period M.
With these assumptions we can write:

$$y(.) = W_1^{(mT_y + dDT_x)fo} * \dfrac{1}{MN}\sum_{k=0}^{M-1} W_M^{-mk} \sum_{i=0}^{N-1}\sum_{n=0}^{N-1} W_N^{-n(i-k)} * X_d(i) * W_N^{-nk} \sum_{l=0}^{M-1} W_M^{lk} f_p(lT_y - nT_x)$$

Hence, observing that, $F(k) = W_N^{-nk} \sum_{l=0}^{M-1} W_M^{lk} f_p(lT_y - nT_x)$ is not dependent on n, and the factor $\frac{1}{N} \sum_{n=0}^{N-1} W_N^{-n(i-k)}$ is I if $i=k$ and zero otherwise, the $y(.)$ signal becomes :

$$y(mT_y + dDT_x) = \frac{W_1^{(mT_y+dDT_x)f_o}}{M} \sum_{k=0}^{M-1} W_M^{-mk} * X_d(k) * F(k) \quad (3)$$

Complete FDS. After having demultiplexed several x signals, the recombination of channels extracted from different x signals follows. This recombination can be very effective if performed at DFT level, where the single channels are still in the frequency domain.

The recombination is a sum of a set of $y(.)$ signals suitably translated in frequency in order to not overlap and labeled by the index c. From (3), including frequency translation f_c, we have:

$$z(mT_y + dDT_x) = \sum_{c=1}^{C} y_c(mT_y + dDT_x) = \sum_{c=1}^{C} \frac{W_1^{(mT_y+dDT_x)f_c}}{M} \sum_{k=0}^{M-1} W_M^{-mk} * X_{dc}(k) * F_c(k)$$

We assume that:
$H_c = M\, T_y f_c = N\, T_x f_c = T f_c$ and $k = h + Hc$;
$F_c(i) = 0$ for $H_c > 0$ and $i = -H_o, ..., -1$;
$F_c(i) = 0$ for $H_c < 0$ and $i = M, ..., M-H_c-1$;
The Hc terms are selected in such a way that the channels do not overlap.

The final expression, which gives the FDS output for each spot is the following :

$$z(mT_y + dDT_x) = \frac{1}{M} \sum_{h=0}^{M-1} W_M^{-mh} * \sum_{c=1}^{C} W_1^{dH_c D/N} * X_{dc}(h+H_c) * F_c(h+H_c) \quad (4)$$

I.e. the computation of the downlink spot signal $z(.)$ turns out to be in:
- direct N-point FFTs applied to the input slots $x_{dc}(.)$ to obtain the DFT $X_{dc}(.)$;
- spectral translations of $X_{dc}(.)$ by H_c frequency bins;
- spectral weighting via the $F_c(.)$ functions to select the wanted carriers;
- samples rotations via the factor $W_1^{dHcD/N}$;
- a combination of the C spot channels and an inverse M-point FFT.
In the formula: m is the sample index inside the frame, d is the frame index, T_x and T_y are input and output sampling periods; DT_x is the interval from frame to frame.

$F_c(.)$ functions do not perform normally any modification of the spectrum, then their samples are either zeros or ones. The samples rotation changes from frame to frame and, in each frame from channel to channel. Assuming D/N=3/4, the above rotations are trivial and only 4 different rotation sequences are possible.

4. System Parameters

Several parameters affect the system performance. The most important ones are:

- the FFT size;
- the frame overlapping factor;
- the time window;

The effect of these parameters on the system performance will be addresses below.

FFT size. The FFT size is surely the most important system parameter. As already mentioned, the sizes of the direct and inverse FFTs can be different, the only constraint being the bandwidth corresponding to a single frequency bin, which has to be fixed, since it corresponds to identical time slots.

As it will appear from the following discussion, large FFT sizes are desirable to maximize the performances, since they offer high frequency resolution. Hence greater flexibility is achieved because the minimum channel bandwidths which can be individually routed may be smaller. Moreover a greater efficiency of the frequency plan is also achieved because of the smaller transition region obtained.

In fact, the FDS system basically selects a frequency slot (useful channel) from an input signal and combines this to other channels to form an output signal. The main FDS function is then a filtering action, which has of course two transition regions, where the signals can be significantly distorted.

In our approach, since for computational reasons we assumed a very simple spectral weighting (only zero and one samples allowed), the transition band is one frequency bin. However, in absence of windowing, the region near to the transition band is affected by the Gibbs phenomenon: it is useless and has to be considered as part of the transition band.

In this way, instead of shaping the filter we have just introduced a forbidden zone for the input signal. Clearly a time window may considerably alleviate the problem.

Although large FFT sizes are better for performance optimization, it is also evident that small FFTs are preferable from the implementation point of view. Increasing the FFT size produces not only an increase of the number of operations per input sample, but also an increase of the required wordlengths for internal arithmetic and memories, given the same implementation losses. To this end the possibility of integrating the memories and the processor internally to a single ASIC is very important, in that relevant power and size savings can be obtained with respect to a multi-chip architecture. The presence of external memories requires in fact the utilization of standard RAMs, which are designed normally for size, speed and driving capability normally higher then those required in this application. The power consumption of these standard memories can be much higher than the power for a memory tailored to the application and integrated within the ASIC.

In conclusion we can say that the FFT size should be the smallest compatible with the specifications in terms of band resolution and efficiency. We verified that a FFT with about 500 points is adequate to process about 10 MHz band with a 20 KHz resolution and a band efficiency ranging between 62.5 % (for individual channel bandwidth of about 320 KHz) to 94 % (for 2 MHz channel bandwidth).

The selection of the exact FFT size depends essentially on the implementation complexity. After an analysis of the best candidates, see ref. [1-2], we selected a 512- point FFT implemented with a pipeline of 3 stages of 8-point FFTs.

Frame Overlapping. The frame overlap factor is defined as the ratio $O_V = 1 - D/N = 1 - E/M$ and is related to the performances of the FDS system. One can roughly say that, the larger is the overlapping, the more power hungry is the processor, thus the O_V parameter should be small. We recall also that the presence of frame overlapping requires a phase rotation of $2\pi D/N$ after the direct FFT and an appropriate selection of O_V can greatly simplify this step. The most appropriate O_V values are 1/2 and 1/4, since they only require 2-complement and I-Q-exchange operations. Unfortunately O_V also affects the FDS system losses because, given the same FFT size, a smaller O_V implies a larger loss.

An approximate estimation of this degradation is derived below, where for simplicity the assumption that $T_x = T_y$ has been retained. We recall that the FDS system is based on the fast convolution algorithm and the fast convolution is an exact algorithm for the implementation of a Finite Impulse Response (FIR) filter, provided that no sample in the frequency domain is discarded. It is also known that, for a FIR filter, the number of required taps Tap is approximately given by:

$$Tap \approx \frac{2}{3} * \frac{F_s}{F_t} * \log_{10}\left[\frac{1}{10 R_i R_o}\right] ; \quad \text{where:}$$

- $F_s = 1/T_x$ is the sampling rate;
- F_t is the filter transition band;
- R_i is the filter in-band ripple;
- R_o is the filter out-of-band ripple.

Now assuming a filter with finite pulse response $f(n)$ and $R_i \approx R_o \approx R$, then the in-band distortions L_1 from filter ripple would be approximately $L_1 \approx R^2/2$.

To avoid time aliasing the FIR filter impulse response shall, at most, have a number of taps $Tap = 2P+1 \approx N * O_v$. Hence, it follows:

$$F_s = N/T \quad \text{and} \quad TF_t \approx \frac{2}{3 * O_v} * \log_{10}\left[\frac{1}{20 L_1}\right]; \quad \text{then:} \quad L_1 \approx 10^{-(O_v TF_t 3/2)/20}$$

where TF_t is the transition band expressed in frequency bins.

Actually, the filter we are interested here has a transfer function whose frequency samples are ones or zeros. The corresponding impulse response is not actually finite. Hence time aliasing will be present. The above relation however has been found as a first guess of the introduced degradation.

Time Window. We will now consider the fact that the input samples are shaped by a simple rectangular window and we will evaluate its effect. This window generates a ripple which turns out in interferences in the useful band after the channels recombination; it is approximately :

$$L_2 \approx TF_d^{-2} / 2\pi^2$$

where TF_d is the frequency distance (expressed in frequency bins) of the useful channel with respect to the nearest interfering frequency bin.

Analytic evaluation of Losses. Now we can use the above discussions to evaluate the overall interferences due to the FDS algorithm, considering these interferences as additive thermal noise. Assuming the TF_t and TF_d values, the E_b/N_o then in dB and the attenuation Att in dB with respect to nominal value, we can evaluate the corresponding expected loss due to the FDS algorithm;

$$Loss_{FDS} \approx E_b/N_o + 10 * log_{10}(10^{-E_b/N_o/10} + L_1 + 10^{Att/10} *L_2)$$

Although the above interference model is quite approximate, nevertheless it gave estimations close to the simulation results on a wide range of operating conditions. In the simulations of figure 5, as shown later on, we assumed $TF_t=5$, $TF_d=7.22$, $O_V=1/4$, $E_b/N_o=10$, $Att=10$ and the resulting $Loss_{FDS}$ is 0.42 dB.

5. Processor Arithmetic

As mentioned before we have selected a 512-Point FFTs for direct and inverse domain translation. The Processor is implemented by 3 stages of 8-Point FFTs in cascade, as shown in figure 3.

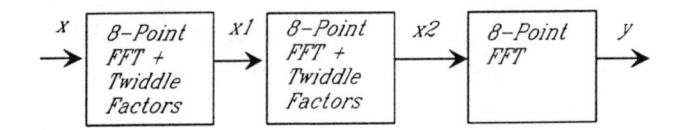

x → [8-Point FFT + Twiddle Factors] → $x1$ → [8-Point FFT + Twiddle Factors] → $x2$ → [8-Point FFT] → y

Figure 3: FFT/IFFT Processor

The structure is based on the following (Cooley-Tukey) algorithm:

$$X(k_1+8k_2+64k_3) = \frac{1}{\sqrt{8}}\sum_{i_1=0}^{7}W_8^{k_1 i_1}*W_{512}^{(k_1+8k_2)i_1}\frac{1}{\sqrt{8}}\sum_{i_2=0}^{7}W_8^{k_2 i_2}*W_{64}^{k_1 i_2}\frac{1}{\sqrt{8}}\sum_{i_3=0}^{7}W_8^{k_1 i_3}*x(i_1+8i_2+64i_3)$$

Each 8-point FFT is computed in parallel using approximated coefficients and requires in turn five sets of 16 adders-accumulators in cascade. The multiplications by twiddle factors are (but the last) implemented via CORDIC processors.

The following *parameters* characterize the finite arithmetic resulting from this structure, with the numerical values used in the simulation example of paragraph 6:
- Margin form the nominal input to the maximum dynamic range (DRM = 4);
- Wordlength of the ADC and DAC (ADW = 8);
- Wordlength used to store the intermediate results from stage to stage (IRW = 8);
- Wordlength for the arithmetic internal to the 8-point FFT stages (FAW = 13);
- Wordlength for the arithmetic of the CORDIC Algorithm (CAW = 13);
- Number of Stages of the CORDIC Algorithm (CAS = 9).

Besides that, the following *arithmetic operation rules* are observed:
- rounding is performed whenever a division by 2^n is required;
- overflow is accepted inside the 8-point FFTs;
- saturation is performed on the intermediate results from stage to stage;
- 4 LS bits and 1 MS bit are added at the input of each stage;
- 5 LS bits are discarded at the output of each stage.

Finite Arithmetic Effects. The finite arithmetic effects have been evaluated assuming the linear noise-like model of the quantization. Following this model, every AD operation and rounding operation corresponds to the introduction of a noise source. This source affects several output samples via coefficient gains which depend on the path from the source to the output sample considered. It is well-known that for an AD of given quantization step Q, the corresponding noise power is $Q^2/12$. For simplicity we assumed the same model for the rounding process: this assumption is the more accurate the higher is the number of bits discarded. We analyzed all the possible noise sources throughout the FDS arithmetic, in order to evaluate the related power and gain at the output; finally we combined these powers assuming them statistically independent.

First of all we have the AD quantization, which produces a quantization power of $Q^2/12$ for each component of the complex input sample. It would be easy to verify that the resulting quantization noise superimposed on each complex sample component at the output of the inverse FFT is still $Q^2/12$.

Regarding the quantization noise sources introduced by the finite arithmetic of each stage, it appears that the gain of each is exactly compensated by the number of sources, so that the contribution of the output noise source of each stage, to the final quantization noise, is identical to the power of the individual source.

With the assumed arithmetic, the resulting overall quantization noise power, which affects the each FDS output component, was found to be approximately:

$$QNP \approx 25{,}4 \, Q^2/12$$

Having 8 bits for quantization, considering a signal dynamic range 4 times smaller than the allowed maximum (say 1 for normalization), finally considering a sign bit, we have $Q = 1/2^{(8-1-2)} = 0{,}03125$, then: QNP= 0,00207.

Finally, considering that the useful signal may be attenuated with respect to the nominal value of say 10 dB, the arithmetic loss, to be applied to each signal component, is : ARITloss = $10 + 10*\log_{10}(0{,}1207) = 0.82$ dB.

Although approximated this quantization model matches with the simulation results.

6. FDS System Simulator

A simulation program in MATLAB environment has been developed to check the system performances. Block diagram and frequency plan are shown in figure 4.

The end-to-end performance is evaluated in terms of Bit Error Rate degradation with respect to the theoretical performances. The BER computation can be done either with the semi-analytic method or with the error counting technique. The former technique is faster but, in principle, less accurate (especially in presence of strong up-link noise) due to the non-linearities in the on-board processor.

The key features of the *simulated architecture* were the following:
- Three distinct up link carrier groups are simulated (FFT1, FFT2, FFT3);
- 512-point Direct FFT and Inverse FFT in floating point or finite arithmetic;
- 25% FFT frame overlap;

The *frequency plan* is basically characterized by:
- 16-bin channel under test in FFT2, including a 5-bins gap from channel to channel:
- 1.48-bin useful QPSK carrier at the edge of the useful channel for worst case;
- spectral power density of useful carrier 10 dB lower than the other signals.

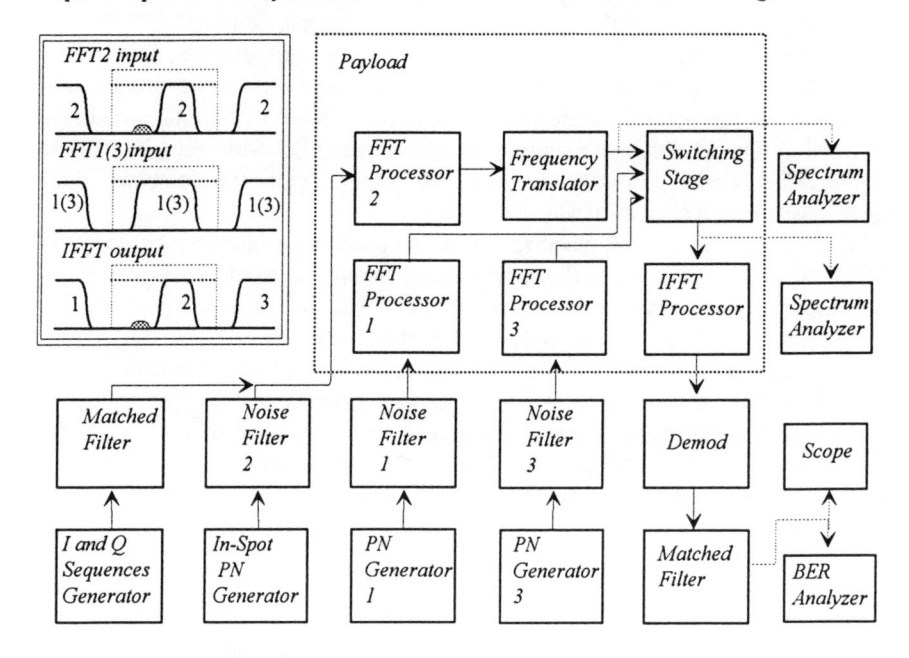

Figure 4: Simulator Block Diagram and signals spectra

Floating point simulations have shown :
- negligible distortions result for a carrier situated in the center of the channel;
- largest interferences for useful carrier at the channel edge;
- losses of 0.4 dB for Eb/No < 10 dB with carrier 10 dB below adjacent signals.
Finite arithmetic simulations, 8 bits for ADW and IRW, 13 bits for FAW and CAW, and 12 dB of DRM, guarantee an overall degradation of about 1.2 dB, for Eb/No = 10 dB and 10 dB fading of the wanted carrier (figure 5).

7. Conclusions.

Frequency Domain Switching appears to be an attractive technique for implementing, with fully digital technology, a flexible SS-FDMA system. A detailed hardware complexity evaluation (see ref. R1) has shown the feasibility of the concept for implementing, with the presently available technology, a multibeam (16 spots up-link and up to 16 spots also for the down link) SS-FDMA payload with a total bandwidth of about 300 MHz with large flexibility of bandwidth reallocation according to the traffic distribution.

5.a: UPLINK FFT OUTPUT *5.b: DOWNLINK FFT INPUT*

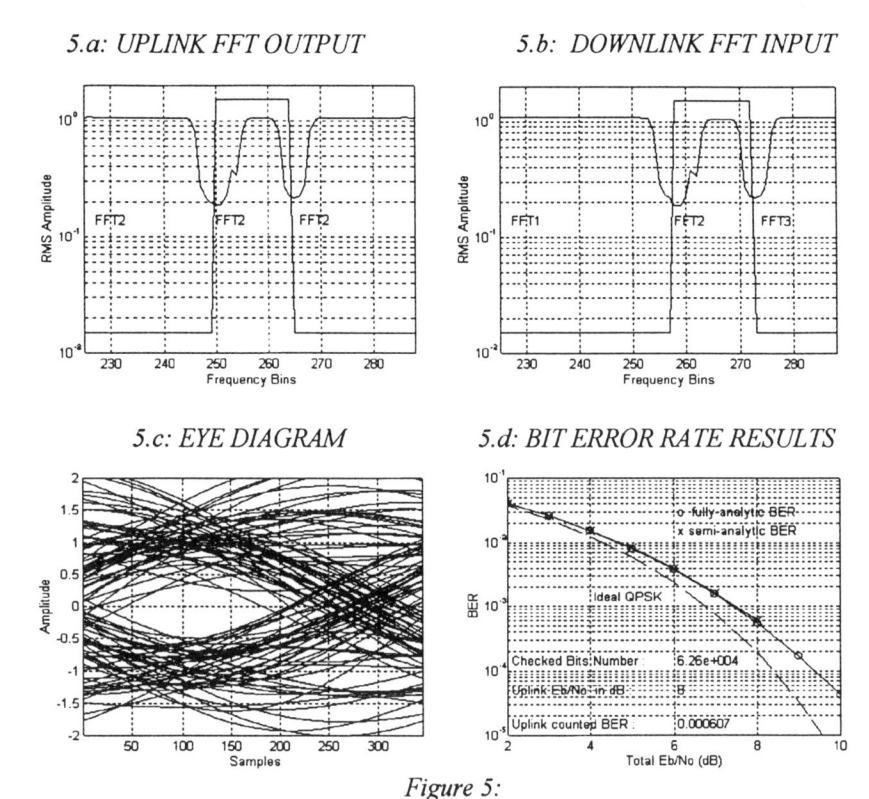

5.c: EYE DIAGRAM *5.d: BIT ERROR RATE RESULTS*

Figure 5:
Simulation results with the useful carrier at the left edge of the channel; adjacent signals 10 dB higher than useful carrier in uplink and downlink; arithmetic with 13 bits for FFT operations, 8 bits for memories and ADC

In particular such a payload may be based on a number of Processing Units (PU) interconnected by an S-stage. Each PU manages a bandwidth of up to 10.8 MHz and includes complete FFT and IFFT processors. A complete PU is implemented by 2 identical ASICs, each of them has 117,000 gates and is housed in a 68-pin package. The corresponding power consumption is 0,73 W (the 0.6 μm MG1 ASIC family from MHS has been considered for the estimation). Hence a 300 MHz payload would require 64 ASICs with an overall power consumption of 46,7 W. The overall power consumption may be reduced to 17.9 W in case of operation at 3V.

References
1. ESTEC Contract 2705 "Study for the definition of a satellite communication system for support to VSAT networks at 20/30 GHz", Final Report.
2. G. Chiassarini, G. Gallinaro, C. Soprano, A. Vernucci, "An on-board Digital Processor for Ka-band Transparent Payload", submitted to Globecom '95.

Design Study for a CDMA Based LEO Satellite Network: Downlink System Level Parameters

S. G. Glisic, J. Talvitie, T. Kumpumäki, M. Latva-aho and J. Iinatti
University of Oulu, Telecommunication Laboratory, P.O.Box 444, FIN- 90571, Oulu, Finland

Abstract

Performance analysis of a new concept of CDMA system for LEO satellite network is presented and discussed.

Due to extremely high Doppler, which is characteristic of LEO satellites, code acquisition is significantly simplified by using a CW pilot carrier for Doppler estimation and compensation. The elements for the analysis presented in this paper are: pilot carrier frequency estimation for Doppler compensation, and multipath and multisatellite diversity combining. The main results of the analysis can be summarised as follows:

- Doppler estimation based on using a pilot carrier is possible with accuracy ±100 Hz and probability better than 0.999 for all elevation angles with the pilot carrier to all user signals power ratio (worst case) −0.4 dB for chip rate R_c = 1.8 MHz and −5.8 dB for chip rate 9.8 MHz. If the minimum elevation angle is 20°—30° this ratio is negligible.
- The use of multipath RAKE combining would be beneficial only in urban environment, at low elevation angles (≤ 30°) and at chip rates of the order of 10 MHz. In these conditions the combining gain is in the range 3…8 dB. In rural environment a multipath RAKE would not improve the system performance significantly.
- Using multisatellite diversity combining, the line of sight signal component availability can be increased in the range from 1 to 7 times depending on the satellite constellation.

1 Introduction

The modern trend in the analysis of mobile satellite communications is very much oriented towards the implementation of CDMA concepts. Most of the work is considering LEO (Low Earth Orbit) constellations due to limited link power budget. Unfortunately this type of satellite generates another problem due to severe Doppler in carrier and chip frequency. In addition to this, due to low minimum elevation angles, multipath and shadowing are also a problem that has to be carefully addressed.

The starting point of this work was to analyse the feasibility of implementing multisatellite and multipath diversity reception in CDMA network for LEO satellites. The results will be used to specify the design parameters for a system experimental testbed.

The existence of carrier and chip Doppler results in a prolonged acquisition process or increased hardware complexity due to the need for two dimensional (delay and frequency) search of code synchronisation positions. A number of papers have been published addressing this issue [1—6]. In our system we suggest Doppler compensation prior to code synchronisation and analyse conditions under which this solution is feasible. The approach is based on a CW pilot carrier located at the spectral null of the wideband signal. The analysis, based on frequency estimation using FFT, provides values for the minimum power of the pilot carrier needed to provide reliable Doppler estimation. This is equivalent to the information about how much system capacity is to be sacrificied for the pilot carrier.

The multipath and shadowing problem can be solved by using multipath and multisatellite diversity reception. Based on the channel delay profile we analyse how much improvements we can expect from using a multipath RAKE receiver. To combat the shadowing problem we consider feasibility of using multisatellite diversity reception. The overall analysis demonstrates that CDMA for LEO satellites is a feasible approach, and detailed system parameters are presented along with discussion of the possible system improvements.

In this initial analysis we have not considered all aspects of the receiver performance. Code acquisition, code tracking, and carrier frequency tracking and phase estimation are issues that we have not addressed in this paper. Also, the overall BER performance of the receiver with multisatellite combining is not discussed. However, all these issues are currently being analysed in detail, and the results will be presented later, along with the final analysis and simulation results following the work presented in this paper.

Connected to the current work, a channel modelling study has also been carried out. The channel models applied in this work are taken from the preliminary results of the channel model study, available at the time of work. The model parameters are based on measurement results obtained for internal use in the study. The final results of the channel modelling study will be presented separately later.

Within the project a number of schemes for fast acquisition of the long code have also been suggested and analysed. These and all other synchronization aspects are presented within a separate paper.

The paper is organised as follows: A general system model is presented in Section 2. Pilot carrier frequency estimation is discussed in Section 3. Multipath diversity (RAKE) and multisatellite diversity reception are discussed in Section 4. The results are presented and discussed in Section 5.

2 Model of Multisatellite CDMA Diversity Reception

In this paper we analyse the down link and a mobile receiver. So, we assume a synchronous channel. The form of the received signal is depicted in Fig. 2.1, assuming three visible satellites.

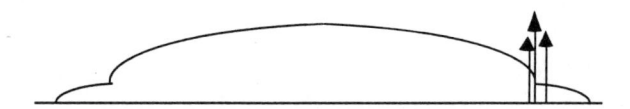

Fig. 2.1. Received signal spectrum.

Analytically the transmitted downlink signal for the mth satellite can be expressed as

$$\begin{aligned} s_m(t) &= s'_{tm} + s_c + s_{sm} \\ &= \sum_{k=1}^{K} \text{Re}\{C(k,m)\exp[j(\omega_0 t + \varphi_m)]\} \\ &\quad + \text{Re}\{\exp[j(\omega_0 + \omega_c)t + \varphi_m]\} + P_m(t) \end{aligned} \qquad (1)$$

where the first term represents K traffic channels, the second term the pilot carrier for Doppler estimation and the third term a pilot channel signal for synchronization purposes. Parameter $C(k,m)$ is a complex signal that contains a long code used to separate signals coming from different satellites (index m), and a short code used to separate different users (index k) within the same satellite.

A CW pilot is placed in the spectral null (see Fig. 2.1) so that zero or low level of interference between the traffic channels and pilot tone should be expected. In order to be able to operate with a low level of the pilot, the CDMA signal can be pre-notched so that the estimation of the pilot signal frequency will not be effected by the variation of the number of users in the network.

Instead of using a pilot tone, information about the Doppler can be distributed in the network from the central node by using an FSK signal with a level higher than the level of CDMA signal. This approach requires a complex coordination within the network. Under these conditions the receiver structure for three visible satellites is shown in Fig. 2.2. The operation of the receiver can be described as follows.

The first step is to detect the Doppler shift for each satellite. Independent of whether this information is sent by the network in the form of an FSK or similar signal, or a pilot carrier is used, the detector may be a simple FFT circuit. This block is designated in Fig. 2.2 as "frequency compensation data extraction" block. Three separate frequency downconversions are performed by using local carriers $f_0 + \hat{f}_{Dm}$ where $\hat{f}_{Dm}$ is the estimated Doppler for the mth satellite ($m = 1, 2, 3$).

Prior to frequency downconversion, frequency compensation data (FSK signal or CW pilot) can be suppressed by using an adaptive narrowband interference canceller based on LMS algorithm. If the Doppler range is narrow compared with the signal bandwidth, a simple passive notch filter can be used for these purposes. The third option is not to suppress this signal but to use only the inherent processing gain of the system. This will slightly reduce the system capacity.

Each frequency downconversion will produce a sum of three signals, one of which will be with essentially no Doppler, and another two will contain a residual Doppler. If the residual Doppler is larger than the bit rate, these two signals will be suppressed in the correlation process so that in the further processing only one signal from a corresponding satellite will be dominant and signals from the other satellites will be suppressed. The additional separation between two signals from different satellites is based on different delays so that the long code will be able to separate such signals. In general further processing is based on a RAKE receiver. Once a sample of decision variable is formed at the output of a RAKE unit, this sample is combined in the multisatellite combiner with the outputs of the other similar units to get the final decision about the bit being received.

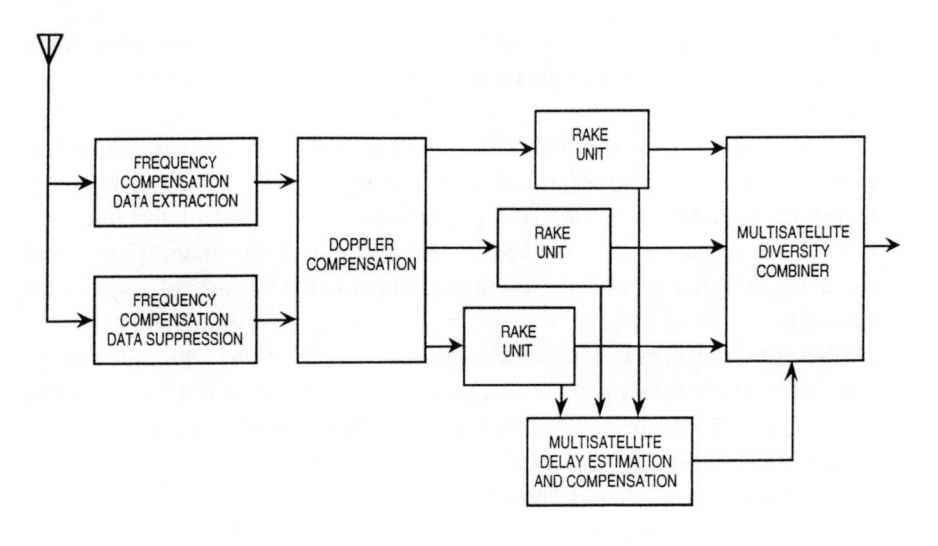

Fig. 2.2. General block diagram of CDMA satellite receiver.

Prior to the combining, the differential delays between the satellites have to be estimated so that the samples of the same bit are combined for the final decision. This function is formally represented as a separate block. In practice this means proper synchronization of the long code in each RAKE unit.

3 Pilot carrier frequency estimation

The proposed system initialisation is based on pilot carriers which are placed at a spectral null frequency of the transmitted wideband satellite signal. Different satellites are seen at different carrier frequency offsets at the mobile station. Hence the Doppler shifts from different satellites can be estimated from the pilot carriers. Modern spectral analysis provides several advanced techniques for spectral line estimation, which are shown to be powerful especially for short data records. In practice the classical spectral estimation methods are satisfactory in most applications. This leads to FFT algorithms which are quite practical and commercially available.

Regardless of the implementation of IF stages and sampling in the receiver front end, the pilot carriers have to be filtered before spectral analysis. The filter bandwidth has to be at least 100 kHz, which is approximately twice the maximum Doppler shift range. The filter should also include some kind of windowing in order to avoid spectral leaking due to finite length data records.

From the code acquisition point of view the frequency should be estimated at the accuracy of few hundred Hz.

A 1024 point FFT with a 200 kHz sampling frequency will be used in the analysis, which gives the frequency estimate with ±100 Hz accuracy at a certain probability. New frequency estimates are produced once in 5 ms when processing the samples on a block-by-block basis. The maximum frequency change during the processing due to satellite motion-induced Doppler rate is 1 Hz. Since the frequency change is very small compared with the accuracy requirements, some extra post-processing can be used to further decrease the probability of error in the frequency detection. The post-processor has also to select three strongest frequencies for receiving signals from different satellites.

Let's start with probability of pilot carrier detection from one satellite. Each out of $M = 1024$ estimated frequency components contains either pure noise or signal plus noise. If we are interested only in pilot carrier detection from one satellite, we will start with an assumption that there will be only one spectral component including signal and all the other components are pure noise. Later on we will modify the analysis to include the fact that, due to terrestrial Doppler spreading and spectral leakage, even when one tone is transmitted the received signal will consists of a number of spectral lines. So, the probability of correct detection is

$$p_{c,i} = \Pr\{|s + n_i| > |n_j|, \forall j \neq i\}, \tag{2}$$

where i,j are indeces of spectral lines. The probability of detecting one correct spectral line out of M lines is

$$\begin{aligned} p_c &= \{1 - \Pr[|s + n_i| < |n_j|]\}^{M-1} \\ &= \{1 - p_f\}^{M-1} \end{aligned} \tag{3}$$

where p_f is the probability of false detection.

For this analysis, a Rician distribution has been assumed for the received pilot, with the Rice factor (giving the line-of-sight-to-multipath power ratios) depending on the elevation angle of the satellite. The overall estimation error probability has to be evaluated by averaging p_f over the Rician distribution

$$p_\varepsilon = 1 - \left\{1 - \int_0^\infty \left[\frac{1}{2}\exp\left(\frac{-x^2}{2\sigma^2}\right) \cdot \frac{x}{A_{mp}^2} \cdot \exp\left(\frac{-(x^2 + A_{los}^2)}{A_{mp}^2}\right) \cdot I_0\left(\frac{2xA_{los}}{A_{mp}^2}\right)\right]dx\right\}^{M-1} \tag{4}$$

Equation (4) was used in numerical analysis to find out the performance of the FFT based frequency estimator in the LEO satellite channel for different elevations angles. In (4), the first term under the integral is probability p_f

which is then averaged over the Rician distribution of the signal envelope. The total received signal power is given by $A_{mp}^2 + A_{los}^2$, and A_{los}^2 / A_{mp}^2 is the Rice factor.

Let's now derive the equation for pilot carrier to user signals ratio. The FFT input signal to noise ratio is

$$SNR = \frac{P}{I' \cdot (A_1 + A_2 + A_3) KS + N_0 W_d}, \tag{5}$$

where P is the power of pilot carrier, A_n is the signal power ratio of the nth satellite compared to the satellite whose frequency is being estimated, S is the power of one user signal, N_0 is the thermal noise density, W_d is the filter bandwidth preceding the FFT (twice the maximum Doppler shift), and I' is the relative power of wide band user signal near the spectral null. For a wideband signal bandwidth $R_c = 1.8$ MHz $I' = 6.087 \cdot 10^{-4}$ and for $R_c = 9.8$ MHz $I' = 5.408 \cdot 10^{-7}$.

The pilot carrier to user signals power ratio is

$$\frac{P}{KS} = SNR \cdot \left(I + \frac{G}{K} \frac{W_d}{R_c} \frac{N_0}{E_b}\right), \tag{6}$$

where $I = I' \cdot (A_1 + A_2 + A_3)$ and G is the processing gain. At $E_b/N_0 = 0$ dB (minimum value specified for the current work) equation (6) becomes

$$\frac{P}{KS} = SNR \cdot \left(I + \frac{G}{K} \frac{W_d}{R_c}\right). \tag{7}$$

4 Multipath and multisatellite combining

The benefit of using RAKE combining of terrestial multipath signals is analysed. The analysis is based on the calculation of the average bit error probabilities with and without RAKE. A channel model with a line-of-sight (LOS) signal and $M_p - 1$ multipaths is used. The delay profiles were defined, based on preliminary results of the channel modelling study, in such a way that the strengths of the $M_p - 1$ multipaths were decreasing with increasing elevation angle. Furthermore, a double state behaviour similar to [7], but without the log-normal shadowing effect, has been assumed as follows. The LOS signal has been assumed to be fixed during the good channel state ($1-A$ part of the time) and Rayleigh distributed during the bad channel state (A part of the time). All other paths are assumed to be Rayleigh fading. Maximum ratio combining is assumed in the RAKE receiver.

The amplitudes of M_p propagaton paths are denoted as α_i, $i = 0, 1, ..., M_p -1$. The first path ($i = 0$) is the direct path (LOS), the one the receiver tries to lock on in the no RAKE case.

In the no RAKE case the instantaneous SNR is given by

$$SNR_{nr} = \frac{Y\alpha_0^2}{1 + \frac{Y}{G}\sum_{i=1}^{M_p-1} \alpha_i^2} \qquad (8)$$

where $Y = E_b/N_0$, $N_0 = N_{th} + N_{mu}$ (thermal noise power density + multiuser interference power density), and G = processing gain. The channel thermal noise density is assumed to be such that equal noise power is received in each RAKE arm.

With maximum ratio combining the signal amplitude in each RAKE arm is multiplied by a weighting factor w_i equal to the corresponding path strength α_i. Hence the instantaneous signal-to-noise ratio (SNR) with a RAKE becomes

$$SNR_r = \frac{Y\left(\sum_{j=0}^{M_p-1} w_j \alpha_j\right)^2}{\sum_{j=0}^{M_p-1} w_j^2 + \frac{Y}{G}\sum_{j=0}^{M_p-1}\sum_{\substack{i=0 \\ i \neq j}}^{M_p-1} (w_j \alpha_i)^2} \qquad (9)$$

Equations (8) and (9) were used to define the instantaneous bit error probability for a given SNR. For coherent PSK the error probability is given by $P_e = 0.5 \cdot \text{erfc}\sqrt{0.5 \cdot SNR}$.

When the path strengths are not fixed but have a certain distribution, the average bit error probability can be calculated by averaging the instantaneous bit error probability over the probability density functions of the path strength coefficients. Since we assumed that the LOS signal during the bad channel state, and all other paths are Rayleigh fading, the probability density function of α can be defined as $p(\alpha) = (\alpha / \sigma_\alpha^2)\exp(-\alpha^2 / 2\sigma_\alpha^2)$, where $2\sigma_\alpha^2$ corresponds to the average power of each path.

For simplificity only two or three strongest paths ($M_p = 2$ or 3) were taken into account in our analysis. To get the final average bit error probability, the bad and good channel states were also taken into account. Thus the final form used in the calculations for three paths becomes

$$P_a = (1-A) \cdot \int_0^\infty \int_0^\infty P_e(\alpha_1, \alpha_2) \cdot p(\alpha_1) \cdot p(\alpha_2) d\alpha_1 d\alpha_2$$
$$+ A \cdot \int_0^\infty \int_0^\infty \int_0^\infty P_e(\alpha_0, \alpha_1, \alpha_2) \cdot p(\alpha_0) \cdot p(\alpha_1) \cdot p(\alpha_2) d\alpha_0 d\alpha_1 d\alpha_2 \quad (10)$$

and for the two path case in a similar fashion.

A multipath diversity combining gain is then defined as

$$G_{mp} = \frac{SNR_0}{SNR_{mp}} \quad (11)$$

where SNR_0 and SNR_{mp} are the signal-to-noise ratios required to achieve the same BER without and with multipath combining, respectively.

An initial estimate of the gain obtained by utilizing multisatellite diversity can be evaluated as follows. Let us define the probability of a single link to be unavailable as the parameter A (time-share of bad state) [7]. A is a function of the elevation angle θ, $A = A(\theta)$ and thus also a function of time t, $A = A(t)$. Then the probability that all links are not available is given by

$$A_{M_s} = \prod_{m=1}^{M_s} A_m(\theta_m) \quad (12)$$

where $A_m(\theta_m)$ is the probability of link m at elevation θ_m to be unavailable, and M_s is the number of satellites whose signals are combined. For a given satellite constellation the mutual relation between the elevation angles θ_m is well defined, and if one of these parameters is known at each time instant, then all of them are known. Therefore the parameter A_{M_s} should be understood as a function of time, although not shown explicitly by the notation used.

By defining the availability of the overall link to be equal to the probability of having the LOS component available in at least one of the satellite links, the result is a system availability given as

$$av_{M_s} = 1 - A_{M_s} \quad (13)$$

The multisatellite availability gain is then defined as

$$G_{ms} = \frac{av_{M_s}}{av_1}. \quad (14)$$

5 Results

5.1 Pilot carrier frequency estimation

From the preliminary results of the channel model study, the Rice factor 5 dB corresponds to 14° and 10 dB corresponds to 27° elevation angle. The pilot carrier to all user signals ratios from one satellite, required to get a certain FEER for different chip rates, are presented in Tables 5.1 (14°) and 5.2 (27°). The relative powers were assumed to be: $A_1 = 0$ dB, $A_2 = 10$ dB, $A_3 = 10$ dB. The processing gains $G = 64$ and $G = 512$ were used for $R_c = 1.8$ MHz; $G = 512$ and $G = 8192$ for $R_c = 9.8$ MHz.

Table 5.1. The required pilot carrier to all user signal ratios to achieve a certain FEER for elevation angle 14°, $M=1024$.

FEER	SNR [dB] (before FFT)	P/KS [dB] $R_c=1.8$MHz $K=100, G=64$	P/KS [dB] $R_c=1.8$MHz $K=100, G=512$	P/KS [dB] $R_c=9.8$MHz $K=1000, G=512$	P/KS [dB] $R_c=9.8$MHz $K=1000, G=8192$
10^{-3}	22	7.5	16.5	−0.8	11.2
10^{-2}	12	−2.5	6.5	−10.8	1.2
10^{-1}	2	−12.5	−3.5	−20.8	−8.8

Table 5.2. The required pilot carrier to all user signal ratios to achieve a certain FEER for elevation angle 27°, $M=1024$.

FEER	SNR [dB] (before FFT)	P/KS [dB] $R_c=1.8$MHz $K=100, G=64$	P/KS [dB] $R_c=1.8$MHz $K=100, G=512$	P/KS [dB] $R_c=9.8$MHz $K=1000, G=512$	P/KS [dB] $R_c=9.8$MHz $K=1000, G=8192$
10^{-3}	−2	−16.5	−7.6	−24.8	−12.8
10^{-2}	−8	−22.5	−13.6	−30.8	−18.8
10^{-1}	−13	−27.5	−18.6	−35.8	−23.8

The final decision about the frequency value can be obtained by taking D successive primary decisions and using majority logic for the final decision. Let's suppose that during the time interval needed for D primary decisions the value of the frequency is in the same frequency slot. So, the best decision would be if we take the slot with the largest number of primary positive decisions.

If the probability of error of primary decision is p_f, the final decision will be correct with the probability of

$$P_c = \sum_{k=\frac{D}{2}+1}^{D} \left(\frac{D!}{k!(D-k)!} \right) (1-p_f)^k p_f^{D-k} \tag{15}$$

Hence the probability of incorrect final decision can be expressed as

$$P_e = 1 - P_c \tag{16}$$

In Table 5.3 below probabilities of incorrect final decisions P_e are presented as a function of parameter D and probability of error for primary decision (p_f). It can be seen that even for a rather high probability of incorrect primary decision we can have reliable final estimation of the frequency.

We should be aware that due to Doppler within D observation intervals, the frequency can move from one slot to the next (this is equivalent to having a few spectral components in the signal). For that reason we will be looking not only into one slot but into a number of adjacent slots (cluster) at the same time.

Table 5.3. Probabilities P_e as a fuction of D and p_f

D	p_f	P_e
10	0.1	$1.6 \cdot 10^{-3}$
	0.2	$3.2 \cdot 10^{-2}$
	0.3	$1.5 \cdot 10^{-1}$
20	0.1	$7.1 \cdot 10^{-6}$
	0.2	$2.6 \cdot 10^{-3}$
	0.3	$4.8 \cdot 10^{-2}$
30	0.1	$3.6 \cdot 10^{-8}$
	0.2	$2.3 \cdot 10^{-4}$
	0.3	$1.7 \cdot 10^{-2}$
40	0.1	$1.9 \cdot 10^{-10}$
	0.2	$2.2 \cdot 10^{-5}$
	0.3	$6.3 \cdot 10^{-3}$

5.2 Multipath and multisatellite combining

In the analysis four delay profiles for urban environment, corresponding to elevation angles 10°, 25°, 35° and 45°, were chosen based on the preliminary results of the channel model study connected to this work. The value of A for each elevation angle was selected accordingly. The resulting SNR gains G_{mp} as a function of elevation angle are presented in Table 5.4.

Table 5.4. SNR differences to achieve average BER of 10^{-2} (no RAKE vs. RAKE)

Environment	Elevation angle (degrees)	G_{mp} for average BER of 10^{-2}	
		Three branch RAKE	Two branch RAKE
Urban	10	8.41 dB	6.89 dB
Urban	25	4.32 dB	3.27 dB
Urban	35	1.50 dB	1.12 dB
Urban	45	0.59 dB	0.33 dB

In the rural case the improvement obtained by using RAKE is insignificant even at low elevation angles. Thus it would be beneficial to use maximum ratio RAKE receiver for multipath combining only in urban environment, at low elevation angles.

It should be noted that the calculations made in this section are based on channel delay profiles in which the delay differences between taps are 0.1 µs. Basically this means that if we want to resolve multipath components, the chip rate should be at least 10 MHz. On the other hand, the maximum excess delay defined in the channel model equals 0.5 µs. This means that it does not make sense to use RAKE if the chip rate is of the order of 2 MHz or below.

The evaluation of the multisatellite availability gain G_{ms} is complicated by the large amount of different situations (satellite geometries) offered by any satellite system. It should be remembered that, for a given constellation, the elevations of the visible satellites as functions of time are governed by the location (latitude and longitude) of the mobile terminal on the globe. The satellite data used in this study is applicable to one point on the globe and thus is only an example.

Given a constellation and a point on earth, the value obtained for the multisatellite combining gain G_{ms} is thus a function of two parameters: time (giving the elevations of the visible satellites), and the choice of the single satellite with availability av_1 (see Equation (14)), against which the multisatellite availability is compared. The gain could also be considered as a function of the elevation of the single satellite.

An example of the multisatellite availability gains in a realistic LEO constellation is shown below in Figure 5.1.

In Fig. 5.1, the plot represents the availability gains (G_{ms}) vs. time, the solid line corresponding to the maximum gain and the dashed line to the minimum gain. The maximum gain is given by choosing for the single satellite availability av_1 at each time instant the lowest visible satellite. The minimum gain is given by choosing for the single satellite availability av_1 at each time instant the highest visible satellite. At the time instants where the gain range goes to unity (0 dB), the maximum and minimum gains coincide, and there is only one satellite visible to the mobile terminal.

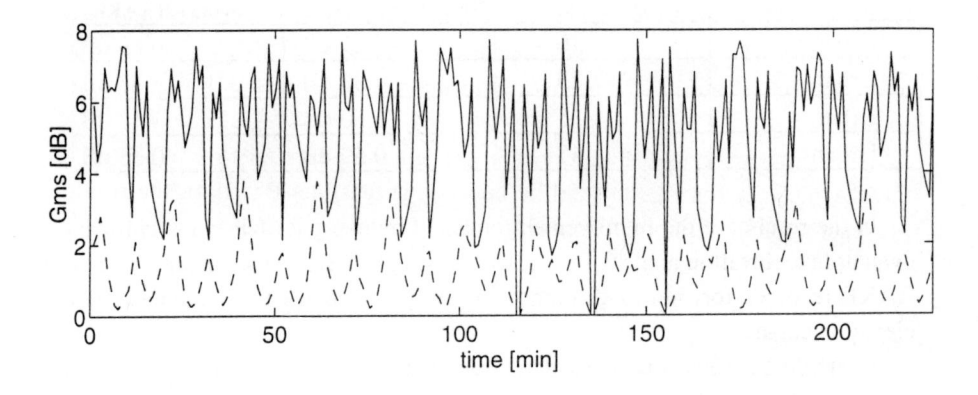

Fig. 5.1. Example of multisatellite availability gains in a realistic LEO constellation.

6 Conclusion

Within this paper a study of the design parameters for a CDMA-based LEO satellite network have been presented and discussed. In order to simplify the acquisition process, Doppler compensation based on a CW pilot carrier is proposed. The minimum pilot carrier power needed for a certain quality of Doppler estimation is studied.

It was shown that Doppler estimation based on a CW pilot carrier located at the first spectral null of the CDMA signal is possible with accuracy ±100 Hz with probability better that 0.999 for all elevation angles. This was achieved with a pilot carrier to all user signals power ratio (worst case) of −0.4 dB for a chip rate of 1.8 MHz and −5.8 dB for a chip rate of 9.8 MHz. For elevation angles above 20° this ratio becomes negligible.

The analysis also investigates possible gains when using multipath and multisatellite diversity reception. Using preliminary results of the channel model study connected to this work, it was found that a multipath RAKE would be beneficial only in an urban environment, at low elevation angles, and at chip rates of the order of 10 MHz or above. The expected combining gain is in the range 3...8 dB for elevation angles 30°...10°. For rural environments the gains are insignificant.

The evaluation of possible gains obtained by combining multisatellite signals is based on an equivalent availability of LOS signal component av_{Ms}, which is by definition the probability that at least one out of M_s signals coming from the visible satellites has the LOS component. In the example LEO case available for this study, the link availability is increased 1 to 7 times by using multisatellite reception, corresponding to multisatellite availability gains of 0 to 8 dB.

As a final conclusion, the system concept proposed in this paper for a CDMA receiver for a LEO satellite network is feasible. Based on the results presented, it is possible to specify initial system parameter values for a receiver testbed design

Acknowledgements

This work was done in co-operation with Elektrobit Ltd., Oulu, Finland. The comments and advice given by Mr. Torsti Poutanen and Dr. Hannu Hakalahti of Elektrobit Ltd. are also gratefully acknowledged.

References

1. A. Fuxjaegov and R. Iltis - Acquisition of Timing and Doppler-shift in a Direct-Sequence Systems - IEEE Transactions on Communications, Vol. 42, No. 10. October 1994, pp. 2870-80.
2. W. Hurd et al - High Dynamic GPS Receiver Using Maximum Likelihood Estimations and Frequency Tracking - IEEE Transactions Aerospace, Vol. 25, September 1987.
3. M. Cheng et al - Spread Spectrum Code Acquisition in the Precence of Doppler Shifts and Data Modulation - IEEE Transactions on Communications, Vol. 38, No. 2, February 1990, pp. 241-250.
4. A. Attelak et al - Design and Implementation of Spread Spectrum Demodulator for Data Delay-Systems - IAF-92-0416-1992.
5. M. Thompson et al - Non-Coherent PH Code Acquisition in Direct Sequence Spread Spectrum Systems Using a Neural Network - Milcom'93 Conference Record, Vol. 1, pp. 30-39.
6. W. K. M. Ahmed, P. J. Mclane, A Simple Method for Coarse Frequency Acquisition Through FFT, VTC'94, pp. 297 - 301.
7. E. Lutz & al. The Land Mobile Satellite Communication Channel, recordings, Statistics, and Channel Model, IEEE Transactions on Vehicular Technology, Vol. 40 (1991), No. 2, p. 375—386.

A Multi-user Approach to Combating Co-channel Interference in Narrowband Mobile Communications *

J. Ventura-Traveset,[1] G. Caire,[2] E. Biglieri,[2] and G. Taricco[2]

[1] European Space Agency (ESA/ESTEC) • RF Systems Division
PO Box 299, 2200 AG Noordwijk, The Netherlands.
e-mail: jventura@estec.bitnet

[2] Politecnico di Torino • Dipartimento di Elettronica
c.so Duca degli Abruzzi 24, I-10129, Torino, Italy
e-mail: <name>@polito.it

Abstract. *We study the impact of diversity on coded digital communication systems operating over channels affected by co-channel interference and by independent flat Rayleigh fading. For the conventional single-user coherent receiver diversity improves dramatically the performance of coded modulation schemes. In particular, diversity eliminates the error floor due to co-channel interference asymptotically as the diversity order increases. Based on the similarity of a system with diversity, perfect channel state information and co-channel interference to a CDMA system with Gaussian spreading sequences, we introduce a multi-user approach to the detection of coded modulation. We show that this new detector can achieve large gains over the conventional single-user coherent receiver and it is able to eliminate the error floor also for finite diversity order.*

---

* This work was supported in part by the Human Capital and Mobility Program of the Commission of the European Union.

1 Introduction

The ability of a transmission scheme to cope with co-channel interference (CCI) has a direct impact on its spectrum efficiency. The performance of uncoded modulation with CCI has been studied in [2], and the effect of diversity (with selection combining) has been recently considered in [3]. In [4], CCI is studied for systems using pilot symbols assisted modulation. Recently, the performance of coherent and differential PSK with coded modulation, diversity and CCI for the mobile fading channel was addressed by the authors in [7]. Here, we examine the effect of CCI on coded modulation for two different receivers, namely, the conventional coherent diversity receiver and a new receiver based on the multi-user detection approach. For simplicity, we restrict our analysis to the case of independent diversity, Rayleigh fading, ideal interleaving and perfect channel state information (CSI). In order to separate the intrinsic effect of diversity from the obvious increase in the signal-to-noise ratio (SNR) due to the combining of M diversity signals, we consider normalized diversity [5], where the average SNR per diversity branch is $1/M$ of the total SNR. This can be realized in practice by decreasing the transmitted power by a factor M.

The first part of this paper is devoted to the performance analysis of the conventional coherent receiver. This receiver treats the co-channel interfering signal as additional AWGN. For this reason it suffers from an error floor. In Section 4 we prove that, as $M \to \infty$, the coherent receiver, operating with maximum-ratio diversity combining, turns the channel into an AWGN CCI-free channel with the same total SNR. For finite diversity order, we show that the error floor caused by CCI decreases exponentially with the product of the code diversity L (i.e., the minimum Hamming distance of the coded-modulation scheme) and of the space diversity order M.

A narrowband system with diversity order M, CCI and fading is similar to a CDMA system where each user is given a Gaussian spreading sequence of length M, which becomes asymptotically orthogonal to the other users' sequences as $M \to \infty$. This observation directs us to the second main contribution of this paper: by assuming perfect CSI of the fading affecting the wanted signal and that affecting the interferer, we derive a new receiver for narrowband communications based on the multi-user detection approach [10]. Similarly to CDMA multi-user architectures, our receiver exploits the statistics of the CCI signal. This way, it is able to eliminate the error floor also for finite diversity order (non-perfect orthogonality). Since the error floor due to CCI is a main limitation to capacity, our receiver emerges as a promising candidate to increase the capacity of a mobile communication system.

2 System model

The basic transmission system under analysis is described in [5]. We assume that the TCM modulator produces the encoded sequence $\mathbf{x} = (\ldots, x_{k-1}, x_k, \ldots)$. Since we examine q-ary PSK modulation, $x_k \in \mathcal{X}_q = \{e^{j2i\pi/q} : i = 0, 1, \ldots, q-1\}$.

The sequence **x** is first interleaved and then modulated. The resulting sequence is then shaped by the signal $\sqrt{E} \cdot p(t)$, where $p(t)$ is a unit-energy pulse chosen in such a way that, when passed through the matched filter $p^*(-t)$, the output is intersymbol-interference free. The interleaver is assumed to be ideal. The transmitted signal

$$x(t) = \sqrt{E} \sum_k x_k p(t - kT) \tag{1}$$

is sent through M different fading channels (diversity branches). Each diversity branch is affected by AWGN, flat slow Rayleigh fading, and CCI. We assume CCI to be of the same type as the wanted signal and that there is no correlation among the diversity branches. Thus, our model reflects the case of ideal interleaving, Rayleigh channel and uncorrelated branches, which is selected for a better understanding of the effect of diversity under the CCI transmission.

By combining the analysis carried out in this paper with that in [5], our results could be extended to the more general case of correlated branches, non-ideal interleaving and Rician channel.

After demodulation, matched filtering, and sampling (with ideal timing), the output sample of the i-th diversity branch can be written as

$$y_k^i = \sqrt{\gamma_0} g_k^i x_k + \sqrt{\gamma_1} h_k^i I_k + n_k^i, \tag{2}$$

where i) n_k^i, g_k^i, and h_k^i are complex independent Gaussian random variables with zero mean and variance $1/2$, ii) $I_k = \sum_\ell w_\ell a_{k-\ell}$ is the CCI sample (w_ℓ are samples of $w(t) = p(t) * p^*(-t)e^{j\theta}$ sampled with an arbitrary delay with respect to the optimal sampling time and rotated by the phase difference θ between the signal and the interference carriers, and $a_k \in \mathcal{X}_q$ are q-ary PSK symbols from the CCI source, and iii) γ_0 and γ_1 are the average SNR and the interference-to-noise ratio per branch, respectively. We define the signal-to-interference ratio (SIR) as $\beta = \gamma_0/\gamma_1$ and the total average SNR as $\Gamma = M\gamma_0$.

We define diversity vectors as follows: $(g_k^i)_{i=1}^M \rightarrow \mathbf{g}_k = (g_k^1, \ldots, g_k^M)^T$. The signal samples y_k^i are combined to form the diversity-channel output $r_k = \mathbf{g}_k^\dagger \mathbf{y}_k$. The sequence of symbols r_k is deinterleaved, then fed to the metric computer to provide the branch metrics for the Viterbi decoder. In this paper we consider two different metrics, for coherent and multi-user receivers.

3 Error-probability analysis

Let us denote by Δ the difference between the metrics of two sequences **x** and $\hat{\mathbf{x}}$. Then, the pairwise error probability (PEP) is given by

$$P(\mathbf{x} \to \hat{\mathbf{x}}) = P(\Delta \leq 0). \tag{3}$$

The right-hand side of (3) can be computed exactly [1] by first evaluating the Laplace transform

$$\Phi_\Delta(s) = \mathrm{E}[e^{-s\Delta}] \tag{4}$$

of the probability density function of Δ, then by using the inversion formula [1]

$$P(\mathbf{x} \to \hat{\mathbf{x}}) = \frac{1}{2\pi j} \int_{c-j\infty}^{c+j\infty} \frac{1}{s} \Phi_\Delta(s) ds \qquad (5)$$

where c is chosen so that (5) converges. Exact evaluation of (5) can be easily performed by numerical integration following the procedure described by the authors in [6, appendix A].

An alternative widely used simple method for the evaluation of the PEP is the Chernoff bound, which is given by

$$P(\mathbf{x} \to \hat{\mathbf{x}}) \leq \min_{\lambda \geq 0} \Phi_\Delta(\lambda) = C(\mathbf{x}, \hat{\mathbf{x}}). \qquad (6)$$

The PEP is used to obtain bounds to bit error probabilities [9].

4 Coherent detection

Following the approach of [7] with ISI-free CCI ($|I_k|^2 = 1$), we obtain

$$\Phi_\Delta(s) = \prod_{k \in \mathcal{K}_L} E\left[(1 + \sqrt{\gamma_0}|d_k|^2 s - (\gamma_1 + 1)|d_k|^2 s^2)^{-M}\right] \qquad (7)$$

where $\mathcal{K}_L = \{j_1, \ldots, j_L\} = \{j : x_j \neq \hat{x}_j\}$ and $d_k = x_k - \hat{x}_k$.

Similarly, we can compute the Chernoff bound on the PEP as

$$C(\mathbf{x}, \hat{\mathbf{x}}) = \prod_{k \in \mathcal{K}_L} \left(1 + \frac{1}{4} \frac{\gamma_0}{\gamma_1 + 1} |d_k|^2\right)^{-M} \qquad (8)$$

Error floor analysis. If we let $\Gamma \to \infty$ in bound (8), we obtain a bound on the irreducible error floor of the PEP due to fading and CCI for the coherent detection case. We get

$$\lim_{\Gamma \to \infty} P(\mathbf{c} \to \hat{\mathbf{c}}) \leq \prod_{k=1}^{L} \left(1 + \frac{1}{4}\beta|d_k|^2\right)^{-M} \leq \left(1 + \frac{1}{4}\beta|d_{\min}|^2\right)^{-LM},$$

We note that the error floor decreases exponentially with the product of the code diversity L and the space diversity M. In this way, space diversity emerges as a very effective countermeasure in CCI-limited channels [3]. This effect is mostly remarkable when weak codes (up to uncoded modulation) are employed.

Asymptotic analysis. A deeper insight in the behavior of the diversity channel with CCI can be obtained by taking a closer look at the combined channel output r_k. By dividing the combined channel output by $\sqrt{M}$, we obtain

$$r_k = \sqrt{\Gamma} u_k x_k + \sqrt{\frac{\Gamma}{\beta}} v_k I_k + z_k \qquad (9)$$

with $u_k = \frac{1}{M}|\mathbf{g}_k|^2$, $v_k = \frac{1}{M}\mathbf{g}_k^\dagger \mathbf{h}_k$, and $z_k = \frac{1}{\sqrt{M}}\mathbf{g}_k^\dagger \mathbf{n}_k$. By the Khinchine strong law of large numbers, we have that as $M \to \infty$, $u_k \to 1$ and $v_k \to 0$ with probability 1. By the central limit theorem, we have that z_k is asymptotically distributed as a zero-mean complex Gaussian random variable with variance 1/2. Then, asymptotically, the statistics of r_k converge in distribution to the output statistics of a CCI-free AWGN channel with the same SNR. We conclude that, for the case of coherent detection with perfect CSI, diversity is able to remove both Rayleigh fading and CCI.

It is interesting to note the similarity between (9) and the output sample of a CDMA system after de-spreading. Under the independent Rayleigh channel assumption, we can assume that each user is assigned a complex Gaussian signature spreading sequence of length M. As the spreading factor becomes large (i.e., for large diversity order), the "spreading sequences" associated to different users become mutually orthogonal. From this analogy, we may think to apply the multi-user detection approach to the detection of narrowband modulation.

5 Multiuser receiver

In this section we introduce the concept of multi-user narrowband communication for CCI cancellation. We assume that the receiver has perfect knowledge of both $\mathbf{g}_k$ and $\mathbf{h}_k$. From the practical point of view several schemes can be envisaged to approach this assumption, including non-overlapping pilot tones, multiplexed pilot symbols or multibeam receivers in the return link of a satellite mobile communication system.

Metric derivation. The ML decoder branch metric relative to the signal x_k is the log-likelihood function

$$m(\mathbf{y}_k, x_k) = \alpha_1 \log P(\mathbf{y}_k | x_k, \mathbf{g}_k, \mathbf{h}_k) + \alpha_2 \qquad (10)$$

where α_1 and α_2 are suitable constants and where $\mathbf{y}_k$, $\mathbf{g}_k$ and $\mathbf{h}_k$ are M-component vectors (M is the diversity order) representing the received vector, the CSI vector associated to the wanted signal and that associated to the interfering signal, respectively. Given the Gaussian distribution of $P(\mathbf{y}_k | x_k, \mathbf{g}_k, \mathbf{h}_k)$, by suitable selection of α_1 and α_2, (10) can be written as

$$m(\mathbf{y}_k, x_k) = \ln \left(E_{I_k} \left[\exp \left\{ 2\Re[\mathbf{r}_k^\dagger \mathbf{A} \mathbf{b}_k] - \mathbf{b}_k^\dagger \mathbf{H}_k \mathbf{b}_k \right\} \right] \right) \qquad (11)$$

where $\mathbf{r}_k = (\mathbf{y}_k^\dagger \mathbf{g}_k, \mathbf{y}_k^\dagger \mathbf{h}_k)^T$, $\mathbf{b}_k = (x_k, I_k)^T$, $\mathbf{A} = \text{diag}(\sqrt{\gamma_0}, \sqrt{\gamma_1})$ and $\mathbf{H}_k = \mathbf{A} \cdot \mathbf{R} \cdot \mathbf{A}$, where $\mathbf{R}$ is defined by

$$\mathbf{R} = \begin{bmatrix} |\mathbf{g}_k|^2 & \mathbf{g}_k^\dagger \mathbf{h}_k \\ \mathbf{h}_k^\dagger \mathbf{g}_k & |\mathbf{h}_k|^2 \end{bmatrix} \quad (12)$$

We further assume that i) the interfering signal is not affected by ISI, and therefore wanted and interfering channels are essentially symbol-synchronous, and ii) the two channels are not frame-synchronous, and therefore the (coded) interfering sequence of I_k is received as a sequence of independent, uniformly distributed, symbols from $\mathcal{X}_q$. Then we obtain

$$m(\mathbf{y}_k, x_k) \propto \ln \left[\sum_{I_k \in \mathcal{X}_q} \exp \left\{ 2\Re[\mathbf{r}_k^\dagger \mathbf{A} \mathbf{b}_k] - \mathbf{b}_k^\dagger \mathbf{H}_k \mathbf{b}_k \right\} \right] \quad (13)$$

which can be approximated via the log-sum approximation $\ln \left[\sum_{z_i} \exp(z_i) \right] \simeq \max_i z_i$ by the sub-optimum metric

$$m'(\mathbf{y}_k, x_k) = \max_{I_k \in \mathcal{X}_q} \{\Omega(\mathbf{b}_k)\} \quad (14)$$

where we defined

$$\Omega(\mathbf{b}_k) = 2\Re[\mathbf{r}_k^\dagger \mathbf{A} \mathbf{b}_k] - \mathbf{b}_k^\dagger \mathbf{H}_k \mathbf{b}_k \quad (15)$$

Note that (15) is formally identical to the multi-user metric derived in [10] for K-user CDMA in AWGN (with $K = 2$) and Gaussian spreading sequences $\mathbf{g}_k$ and $\mathbf{h}_k$. Thus, implementing (15) in our receiver is equivalent to follow the multi-user decision. The general multi-user metric (15), will be used in the next section to derive an upper bound to the PEP of this receiver.

Performance analysis. Evaluation of $P(\mathbf{x} \to \hat{\mathbf{x}})$ through the computation of the inversion formula (5) is difficult in this case beacuse of the metric definition (14) and the max($\cdot$) operation involved. A possible way to solve this problem is to resort to Monte Carlo statistical integration. However, we used a different approach consisting of bounding the PEP and then use the inversion formula (5) to evaluate it.

Consider an error event that spans L *consecutive* symbols (note that now we consider the whole error event and not only the set $\mathcal{K}_L$). Let $\mathbf{x} = (x_1, \ldots, x_L)$ and $\mathbf{I}^j = (I_1^j, \ldots I_L^j)^T$ denote the transmitted sequence and a generic interfering sequence, respectively, of L consecutive symbols. There are q^L possible sequences $\mathbf{I}^j$, each with the same probability q^{-L}. Applying the law of total probability to $P(\mathbf{x} \to \hat{\mathbf{x}})$, we obtain

$$P(\mathbf{x} \to \hat{\mathbf{x}}) = \frac{1}{q^L} \sum_{j=0}^{q^L - 1} P(\mathbf{x} \to \hat{\mathbf{x}} | \mathbf{I}^j) \quad (16)$$

Let us now consider $P(\mathbf{x} \to \hat{\mathbf{x}}|\mathbf{I}^j)$. Recalling (14) we can write

$$P(\mathbf{x} \to \hat{\mathbf{x}}|\mathbf{I}^j) = P\left\{\sum_{k=1}^{L} \max_{\mathbf{b}_k} \Omega(\mathbf{b}_k) - \sum_{k=1}^{L} \max_{\hat{\mathbf{b}}_k} \Omega(\hat{\mathbf{b}}_k) \leq 0 | \mathbf{I}^j\right\}$$

$$= P\left\{\bigcup_{l=0}^{q^L-1} \left(\sum_{k=1}^{L} \Omega(\hat{\mathbf{b}}_k^l) \geq \sum_{k=1}^{L} \max_{\mathbf{b}_k} \Omega(\mathbf{b}_k)\right) | \mathbf{I}^j\right\} \quad (17)$$

where we introduced $\hat{\mathbf{b}}_k = (\widehat{x}_k, \widehat{I}_k)^T$ and $\hat{\mathbf{b}}_k^l = (\widehat{x}_k, I_k^l)^T$. Equation (17) can be simply upper bounded as follows

$$P(\mathbf{x} \to \hat{\mathbf{x}}|\mathbf{I}^j) \leq \sum_{l=0}^{q^L-1} P\left\{\sum_{k=1}^{L} \Omega(\hat{\mathbf{b}}_k^l) \geq \sum_{k=1}^{L} \max_{\mathbf{b}_k} \Omega(\mathbf{b}_k)|\mathbf{I}^j\right\}$$

$$\leq \sum_{l=0}^{q^L-1} P\left\{\sum_{k=1}^{L} \left(\Omega(\mathbf{b}_k^j) - \Omega(\hat{\mathbf{b}}_k^l)\right) \leq 0\right\} \quad (18)$$

where we defined $\mathbf{b}_k^j = (x_k, I_k^j)^T$ and the first upper bound is a consequence of the application of the union bound and the second of relaxing the inequality condition inside the probability function.

The metric difference $\Omega(\mathbf{b}_k^j) - \Omega(\hat{\mathbf{b}}_k^l)$ can be written as

$$\Omega(\mathbf{b}_k^j) - \Omega(\hat{\mathbf{b}}_k^l) = 2\Re[(\mathbf{d}_k^{j,l})^\dagger \mathbf{z}_k] + (\mathbf{d}_k^{j,l})^\dagger \mathbf{H}_k \mathbf{d}_k^{j,l} \quad (19)$$

where we defined $\mathbf{d}_k^{j,l} = \mathbf{b}_k^j - \hat{\mathbf{b}}_k^l$ and $\mathbf{z}_k = \mathbf{r}_k^\dagger \mathbf{A} - \mathbf{H}_k \mathbf{b}_k^j$. Note that, given the random vectors $\mathbf{g}_k$ and $\mathbf{h}_k$, $\mathbf{z}_k$ is a zero-mean complex Gaussian random vector with covariance matrix $\frac{1}{2}\mathbf{H}_k$, and therefore, given $\mathbf{g}_k$ and $\mathbf{h}_k$, the RV $(\mathbf{d}_k^{j,l})^\dagger \mathbf{z}_k$ is a zero-mean complex Gaussian RV with variance $\frac{1}{2}(\mathbf{d}_k^{j,l})^\dagger \mathbf{H}_k \mathbf{d}_k^{j,l}$. Hence, introducing (19) in (18), and applying iterated expectations, we finally get

$$P(\mathbf{x} \to \hat{\mathbf{x}}|\mathbf{I}^j) \leq \sum_{l=0}^{q^L-1} E_{\mathbf{g},\mathbf{h}} \left[Q\left(\sqrt{\sum_{k=1}^{L}(\mathbf{d}_k^{j,l})^\dagger \mathbf{H}_k \mathbf{d}_k^{j,l}}\right)\right] \quad (20)$$

where $Q(x) = \frac{1}{\sqrt{2\pi}} \int_x^\infty \exp(-t^2/2)\, dt$. For a given interfering vector sequence $\mathbf{I}^j$, (20) depends only on the q^L possible distances $\mathbf{d}_k^{j,l}$. Thus $P(\mathbf{x} \to \hat{\mathbf{x}}|\mathbf{I}^j)$ does not depend on j and we can write

$$P(\mathbf{x} \to \hat{\mathbf{x}}) \leq P(\mathbf{x} \to \hat{\mathbf{x}}|\mathbf{I}^0) \leq \sum_{l=0}^{q^L-1} E_{\mathbf{g},\mathbf{h}} \left[Q\left(\sqrt{\sum_{k=1}^{L}(\mathbf{d}_k^{0,l})^\dagger \mathbf{H}_k \mathbf{d}_k^{0,l}}\right)\right] \quad (21)$$

where $\mathbf{I}^0 = \mathbf{1}$, the all-one sequence. The above expectations can be exactly evaluated by noting that the square of argument of $Q(\cdot)$ is a quadratic form of

Gaussian random variables and by applying (5). We get

$$E_{\mathbf{g},\mathbf{h}}\left[Q\left(\sqrt{\sum_{k=1}^{L}(\mathbf{d}_k^{0,\ell})^\dagger \mathbf{H}_k \mathbf{d}_k^{0,\ell}}\right)\right] =$$
$$\frac{1}{2\pi j}\int_{c-j\infty}^{c+j\infty}\frac{1}{s}\prod_{k=1}^{L}\left[1+(|d_k|^2+\frac{1}{\beta}|d_k'|^2)(s-\frac{s^2}{\gamma_0})\right]^{-M} ds \quad (22)$$

where we have defined $d_k' = 1 - I_k^\ell$. By applying the Chernoff bound to the RHS of (18), we get the bound

$$P(\mathbf{x} \to \hat{\mathbf{x}}) \leq \sum_{l=0}^{q^L-1}\prod_{k=1}^{L}\left[1+\frac{1}{4}\gamma_0(|d_k|^2+\frac{1}{\beta}|d_k'|^2)\right]^{-M} \quad (23)$$

In spite of the looseness of this bound, (23) reveals that the multi-user receiver is able to suppress the error floor, irrespectively of the value of the SIR β and for any finite diversity order M. Actually, for a very powerful interfering signal, i.e., for a very low value of SIR, (23) indicates that the multi-user detection approaches the performances of the CCI-free system. The reason for this behavior is that if the interfering user is very strong, then the primary source of errors in the multi-user receiver is the background noise, rather than the randomness of the information carried by the interferer, which can be well estimated. Similarly, when the interfering signal is weak, i.e., for a large value of β, the performance approaches the CCI-free case, since the effect of CCI becomes negligible.

However, because of the union bound in (18), and since for large β the intersection of the events

$$\left\{\sum_{k=1}^{L}\Omega(\mathbf{b}_k^\ell) \geq \max_{\mathbf{b}_k}\Omega(\mathbf{b}_k)\right\}$$

becomes large, the bound (21) is not tight for large β unless the SNR becomes large too. From the above discussion, a tighter general bound than (21) can be obtained as follows

$$P(\mathbf{x} \to \hat{\mathbf{x}}) \leq \min\left\{\text{bound (21), PEP of the coherent receiver with CCI}\right\} \quad (24)$$

which will proof to be tight when compared to the simulation results in the next section.

6 Results

We consider in this section some examples illustrating the performance of the systems studied and the main conclusions. Results were obtained for the following TCM schemes. **U4,U8:** Ungerboeck's rate-2/3 coded 8-PSK with 4 and 8 states, respectively [9]. **Q64:** TCM obtained by mapping the "standard" rate-1/2 64-state binary convolutional code with generators 171, 133 (octal notation) onto a Gray-encoded 4-PSK. Table 1 shows the error events considered to bound the BER for U4.

Span	Error phases					Path coefficient
1	$\Delta\phi_4$	—	—	—	—	1
3	$\Delta\phi_2$	$\Delta\phi_3$	$\Delta\phi_2$	—	—	16
3	$\Delta\phi_2$	$\Delta\phi_1$	$\Delta\phi_2$	—	—	12
4	$\Delta\phi_2$	$\Delta\phi_1$	$\Delta\phi_3$	$\Delta\phi_2$	—	24
4	$\Delta\phi_2$	$\Delta\phi_3$	$\Delta\phi_1$	$\Delta\phi_2$	—	24
4	$\Delta\phi_2$	$\Delta\phi_1$	$\Delta\phi_1$	$\Delta\phi_2$	—	24
4	$\Delta\phi_2$	$\Delta\phi_3$	$\Delta\phi_3$	$\Delta\phi_2$	—	24
5	$\Delta\phi_2$	$\Delta\phi_3$	$\Delta\phi_0$	$\Delta\phi_3$	$\Delta\phi_2$	20
5	$\Delta\phi_2$	$\Delta\phi_1$	$\Delta\phi_0$	$\Delta\phi_3$	$\Delta\phi_2$	16
5	$\Delta\phi_2$	$\Delta\phi_3$	$\Delta\phi_0$	$\Delta\phi_1$	$\Delta\phi_2$	16
5	$\Delta\phi_2$	$\Delta\phi_1$	$\Delta\phi_0$	$\Delta\phi_1$	$\Delta\phi_2$	12
5	$\Delta\phi_2$	$\Delta\phi_1$	$\Delta\phi_3$	$\Delta\phi_3$	$\Delta\phi_2$	28
5	$\Delta\phi_2$	$\Delta\phi_3$	$\Delta\phi_3$	$\Delta\phi_1$	$\Delta\phi_2$	28
5	$\Delta\phi_2$	$\Delta\phi_3$	$\Delta\phi_3$	$\Delta\phi_3$	$\Delta\phi_2$	28
5	$\Delta\phi_2$	$\Delta\phi_1$	$\Delta\phi_3$	$\Delta\phi_1$	$\Delta\phi_2$	28
5	$\Delta\phi_2$	$\Delta\phi_3$	$\Delta\phi_1$	$\Delta\phi_1$	$\Delta\phi_2$	24
5	$\Delta\phi_2$	$\Delta\phi_3$	$\Delta\phi_1$	$\Delta\phi_3$	$\Delta\phi_2$	24
5	$\Delta\phi_2$	$\Delta\phi_3$	$\Delta\phi_4$	$\Delta\phi_3$	$\Delta\phi_2$	24
5	$\Delta\phi_2$	$\Delta\phi_1$	$\Delta\phi_1$	$\Delta\phi_1$	$\Delta\phi_2$	24
5	$\Delta\phi_2$	$\Delta\phi_1$	$\Delta\phi_1$	$\Delta\phi_3$	$\Delta\phi_2$	24
5	$\Delta\phi_2$	$\Delta\phi_3$	$\Delta\phi_4$	$\Delta\phi_1$	$\Delta\phi_2$	20
5	$\Delta\phi_2$	$\Delta\phi_1$	$\Delta\phi_4$	$\Delta\phi_3$	$\Delta\phi_2$	20
5	$\Delta\phi_2$	$\Delta\phi_1$	$\Delta\phi_4$	$\Delta\phi_1$	$\Delta\phi_2$	16

Table 1. Pairwise error events considered for U4. $\Delta\phi_i$ stands for an absolute phase difference of $i\pi/4$ radians. Path coefficient is the average number of information bit errors divided by the number of information bits in a trellis step.

Example 1: Impact of diversity on the error floor. Here, we consider a Rayleigh channel limited by co-channel interference with perfect CSI coherent detection. As we showed in Section 4, this transmission system is affected by an error floor, i.e., an irreducible BER for $E_b/N_0 \to \infty$ that is reduced by increasing the diversity order M. Fig. 1 shows the error floor vs. SIR performance for U4 with $M = 1, 2, 4, 8, 16$. Similar results were obtained for U8 and Q64. The curves show that the error floor roughly squares for each doubling of the diversity order, as expected from our asymptotic error analysis. Fixing an acceptable error floor level to 10^{-5} we show in Table 2 the minimum SIRs required.

Example 2: Impact of multi-user detection. Finally, we consider the effect of multi-user detection with perfect CSI of both wanted and interfering signal. The results are obtained by simulation and by using the bounds of Section 5. Fig. 2 shows the performance for SIR= 0 dB and $M = 1, 2, 4$. The bounds are looser than in the case of coherent detection but they are still in good agreement with

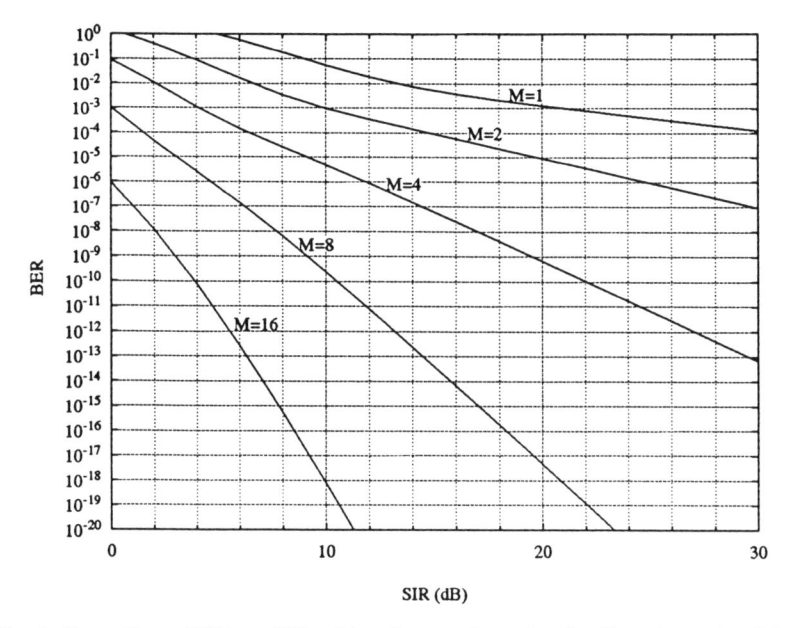

Fig. 1. Error floor of U4 vs. SIR with coherent detection for diversity order $M = 1, 2, 4, 8, 16$.

Diversity order	Code		
	U4	U8	Q64
1	> 30 dB	23 dB	8 dB
2	20 dB	12 dB	2 dB
4	9 dB	6 dB	< 0 dB
8	3 dB	2 dB	< 0 dB
16	< 0 dB	< 0 dB	< 0 dB

Table 2. Minimum SIR required to achieve an error floor equal to 10^{-5} with coherent detection over the Rayleigh channel.

simulation results for BER lower than 10^{-3}. The error floor has been eliminated and a lower diversity order is necessary to achieve the same results as with coherent detection. Below 10^{-2}, the BER is almost independent from the SIR and we found that the worst performance is obtained with SIR=5 dB. This is also reported by Figs. 3 and 4, which show the minimum E_b/N_0 required to achieve BER=10^{-3}. From these results we can outline the main advantages of the multi-user receiver.

1. The system can cope with all values of SIR while the error floor prevents coherent detection to operate below a threshold SIR depending on the diver-

sity order. This has a major impact on the capacity of a narrowband mobile communication system.
2. A smaller diversity order is sufficient to achieve the same performance that sometimes was unreachable due to the error floor with coherent detection.
3. The multi-user receiver is most effective when the SIR is well below 20 dB (a value commonly used for narrowband systems), thus ruling out one of the most important reasons to prefer CDMA systems which are able to work at SIR as low as 6 dB [11].

Even though ideal CSI for both wanted and interfering signal is a strong assumption, these results indicate a possible way to operate narrowband systems at SIR levels which standard coherent detection is unable to cope with.

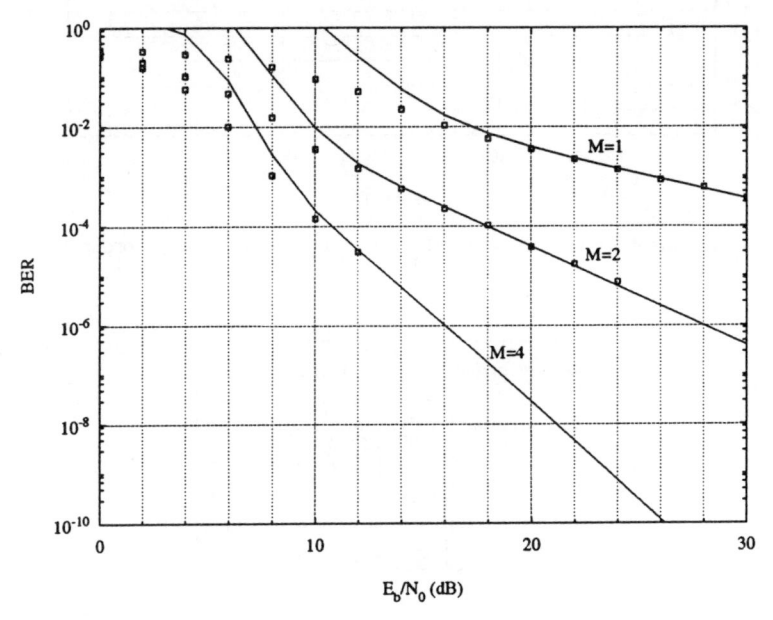

Fig. 2. BER vs. E_b/N_0 of U4 for SIR=0 dB and $M = 1, 2, 4$ with multi-user detection. Solid curves: minimum upper bounds, dots: simulation results.

7 Conclusions

In this paper we have studied the impact of diversity on PSK coded-modulation systems affected by CCI and independent flat Rayleigh fading. We focused our analysis on two different receiver strategies, namely, conventional single-user coherent detection and a new original multi-user receiver.

By deriving bounds on the error probability we show that diversity provides substantial performance improvements with both receiver strategies.

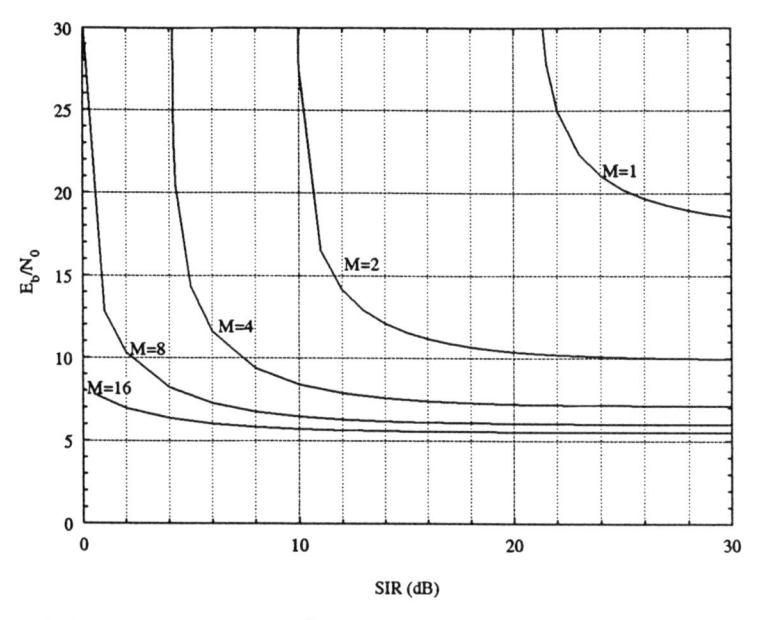

Fig. 3. E_b/N_0 vs. SIR at $P_b = 10^{-3}$ of U4 with coherent detection for different values of M (diversity order).

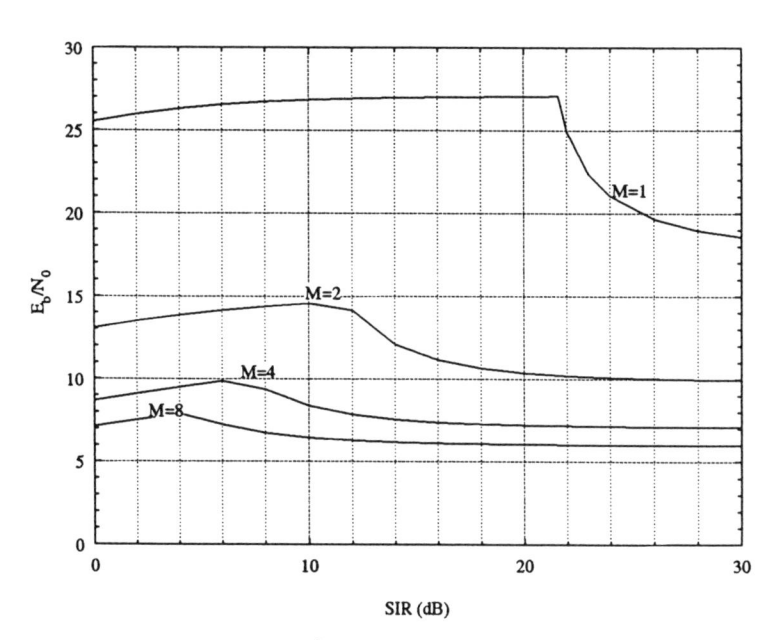

Fig. 4. E_b/N_0 vs. SIR at $P_b = 10^{-3}$ of U4 with multi-user detection for different values of M (diversity order).

For coherent detection with perfect CSI, as the number of diversity branches increases, the channel converges asymptotically to a CCI-free AWGN channel with the same SNR. However, with finite diversity order, conventional single user receivers do always suffer from error floor when the channel is affected by CCI.

On the other hand, our new receiver does consider the interfering signals as modulated signals affected by fading. Then, assuming that the CSI associated to the signal can be estimated, a simple decoder branch metric can be derived. This is related to a multi-user receiver developed for CDMA signals in AWGN. This new receiver exhibits a dramatic improvement over the the conventional coherent approach, including the suppression of the error floor.

In conclusion, use of diversity and multi-user approach emerge as promising candidates to increase the capacity of a narrowband mobile communication system.

References

1. J. Proakis, *Digital Communications*, New York: McGraw-Hill, 1983.
2. H. Suzuki, "Canonic receiver analysis for M-ary angle modulation in Rayleigh fading environment," *IEEE Trans. Vehic. Technol.*, Vol. VT-31, pp. 7-14, Feb. 1982.
3. F. Adachi and M. Sawahashi, "Error analysis of MDPSK/CPSK with diversity reception under very slow Rayleigh fading and co-channel interference," *IEEE Trans. Vehic. Technol.*, Vol. 43, pp. 252-263, May 1994.
4. J. Cavers and J. Varaldi, "Cochannel Interference and Pilot Symbol Assisted Modulation," *IEEE Trans. Vehic. Technol.*, Vol. 42, pp. 407-413, Nov. 1993.
5. J. Ventura-Traveset, G. Caire, E. Biglieri and G. Taricco, "Impact of diversity reception on fading channels with coded modulation. Part I: Coherent detection," submitted to *IEEE Trans. on Commun.*, July 1995.
6. ——, "Impact of diversity reception on fading channels with coded modulation. Part II: Differential block detection," submitted to *IEEE Trans. on Commun.*, July 1995.
7. ——, "Impact of diversity on mobile radio systems with coded modulation, fading and co-channel interference," *IEEE Globecom'95*, Nov. 13-17, 1995, Singapore.
8. J. K. Cavers and P. Ho, "Analysis of the error performance of trellis-coded modulations in Rayleigh fading channels," *IEEE Trans. Commun.*, Vol. 40, pp.74-83, Jan. 1992.
9. E. Biglieri, D. Divsalar, P. J. McLane, and M. K. Simon, *Introduction to Trellis-Coded Modulation with Applications*, New York: MacMillan, 1991.
10. H. V. Poor and S. Verdu, "High-speed Digital Signal Processing for Satellite Communications," *Final Report European Space Agency P.O. 134422*, Feb. 1995.
11. A. J. Viterbi, *Principles of Spread Spectrum Communication*, Addison Wesley, 1995.

Digital Radio for the World: High-Grade Service Quality Through On-Board Processing Techniques

G. Losquadro
Alenia Spazio S.p.A.
Via Saccomuro 24, Rome , Italy

Abstract

A Digital Audio Broadcasting (DAB) Satellite System is here presented providing optimized service and guaranteeing an high-grade quality service for some hundreds of channels.
This with minimum cost of the radio set and of the broadcasting station, greatly improving the radio broadcasting service throughout the world, especially in areas where deployment of conventional relay infrastructures is difficult and not yet cost-effective.
Potential broadcasters located in the visible hemisphere can directly up-link their programming to the DAB satellites.
Broadcasters can also copy each up-link channel and steer flexibly to any beam or combination of satellite beams.
The paper illustrates the advantages and the feasibility of a DAB system based on geostationary satellites with on board processing generating signals of MCPC (Multiple Channel Per Carrier) type. This multiplexing technique allows large system margins. This solution allows the satellite
L band TWTA amplifiers to be operated in saturation and eliminates the intermodulation noise associated with the transmission of FDMA channels. A powerful coding scheme has been selected to create a "robust" down-link.
In this way several high power DAB channels of radio programming, with selectable data rate, are directly delivered to the users.

1 Introduction

The inadequacy of shortwave and medium-wave radio broadcasting for continental wide coverage led to search for a satellite-based solution to distributing audio programming to roughly more than two-thirds of the world's nations.
A new global satellite digital audio system has been defined to serve, with high quality channels, a very large population. A constellation of identical satellites is foreseen, any one providing three wide spot beams nearly providing a continental coverage.
For market success low transmission costs are essential specially for limited resource broadcasters. For this reason the programmability of the data rate on each

braodcasted channel allows broadcasters to pay for the minimum they actually need, increasing the attraction of the satellite service. The envisaged data rates and corresponding applications for each broadcasted channel are:

- 16 Kbps: Mono Audio Voice, basic system channel block
- 32 Kbps: Mono Audio Music, FM quality
- 64 Kbps: Stereo Audio Music
- 128 Kbps: CD Stereo Audio Music Equivalent

with following compression algorithm:

- MPEG 2, Layer 3
- support for ancillary data

The here presented solution provides large link margins (for stationary environment: up to 12 dB, for mobile environment: $\cong$ 8 dB) compared to other, more conventional, transmission schemes without increasing the payload complexity and costs of the broadcaster up-link stations.
The optimization issues and trade-off results presented in the paper are:

- access techniques and bandwidth occupation a novel access solution has been defined, (patented by the author) providing Multiple Channel Per Carrier per spot beam coverage, exploiting on-board regeneration and multiplexing operation ("superpacket" structures)
- coding and interleaving techniques (a novel concatenated coding scheme involving mainly the broadcaster station and transparently routed on the satellite repeater is considered to cope with down link propagation impairments i.e. building penetration, tree blockage and multipath),
- service area coverage solutions for three continental areas
- low-cost (50 - 100 $) radio set (stationary, portable table top and mobile)
- low cost broadcast stations (plug-in and forget concept)

2 The DAB Satellite System

The Digital Audio Broadcasting Satellites provide complete coverage of the service area with consistent reception quality.
Each satellite supports the broadcasters with the capability to receive up to 288 channels on a "global" coverage up link to be routed on as many down link channels. The up feeder link stations should be within the part of the earth visible from the satellite. Broadcasters can also copy each uplink channel and steer to any beam or combination of beams.
The broadcasting service consists of 288 broadcasted audio channels divided on three spots with 96 channels per spot and independent access from the individual broadcaster site. The channels may be grouped at reception side to provide the required rate for high quality reception.
The "initial" system configuration, shown in Fig. 1, has been conceived based on three identical geostationary satellites any one of which can cover either:

a) Africa and Middle East
b) Mexico, Caribbean and Latin America
c) India, Southeast Asia, Korea and China.

The satellite concept is presented in Fig. 2

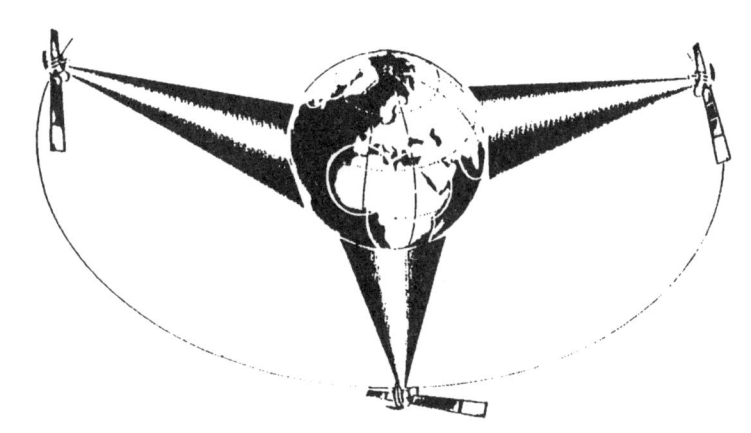

Fig. 1 DAB Satellite System

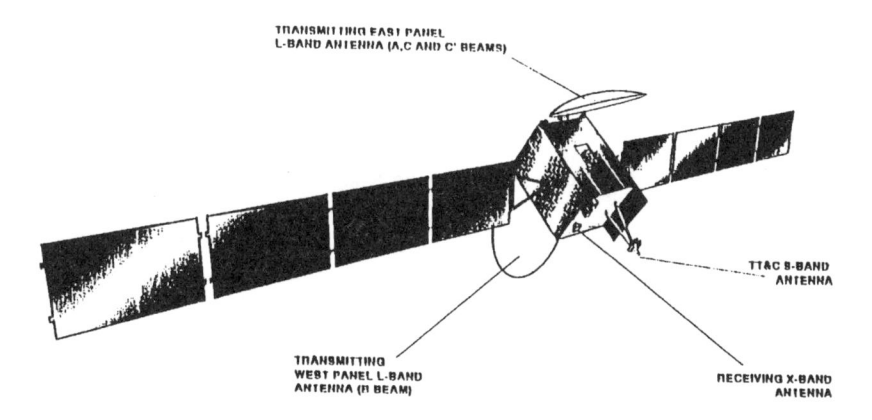

Fig. 2 DAB Satellite

2.1 DAB System concept

The primary pursued goal has been to provide a performance optimized service providing an *high-grade quality service, a large number of channels* with minimum cost of the receiver set. This in combination with an inherent satellite high reliability, long life and low schedule risk.

The communication system concept is presented in Fig.3: the proposed solutions creates from the up link SCPC-FDMA channels three Multi Channel Per Carrier (MCPC) time multiplexed single carrier.
The benefits of this solution are:

• programmable creation of MCPC signal to be steered to a given spot beam

• satellite L band TWTA amplifiers operation in saturation: more than 3 dB link margin, as compared to the FDMA approach, maximizing the power utilization

• elimination of the intermodulation noise issues and relevant losses (about 1 dB) associated with transmission of FDMA channels

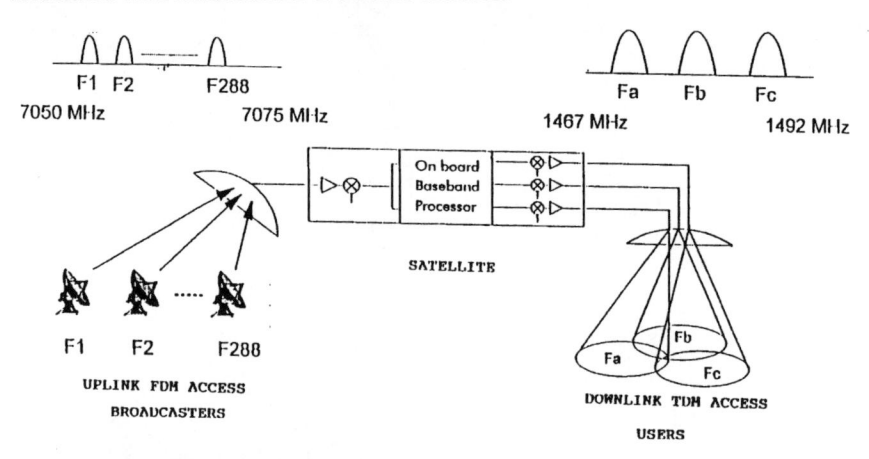

Fig. 3 DAB Communications System architecture

• reduced broadcaster up link station EIRP due to the improved sensitivity of the satellite repeater

• ability of the service provider, to control directly on ground the access to the L band down link channels; each broadcaster access can be either authorized or denied by the control center via telecommand

• elimination of the uplink power coordination: up link broadcaster's signal power unbalance, received at the spacecraft, will not affect the down link L band channels

• transparent regeneration, which means that modifications/evolutions of the structure of the multichannel signals are allowed on ground to cope with service demand and future objectives; this without affecting the overall complexity, availability reliability and schedule of the satellite payload

2.2 DAB system design

The satellite design combines a payload providing up to 288 high power 16 Kbps channels with three axis stabilized satellite bus for an operational lifetime of 12 years.

The down link coverages provided by the DAB system are shown in fig.s 4, 5, 6. The net link margin contours have been evaluated according to link budget results shown in Table I. The data formatting operations shown in Fig. 7 will be here following described.

Fig. 4 AsiaSpace Coverage
Link Margin Contours: A = 12 dB B= 10.5 dB C = 8.5 dB

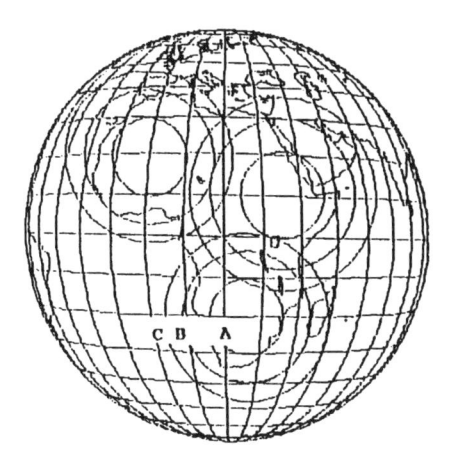

Fig. 5 AfriSpace Coverage
Link Margin Contours: A = 12 dB B= 10.5 dB C = 8.5 dB

Tab. I Link Dimensioning

World Space DAB Link		Feeder Link	User Link
Availability Feeder Link	%		99.9
Availability User Link	%		99.9
Bit Error Rate	dB		1E-05
Coding Gain (RS + Conv Enc.: r=1/2)	dB		7.0
Required Uncoded Eb/No	dB		9.6
Demod losses	dB		1.8
Net Eb/No required	Kbps		2.6
Information Bit Rate			64
Frequency (Lower)	GHz	7.050	1.467
Frequency (Upper)/used below	GHz	7.075	1.492
TWTA Output Power (e.o.l.)	W	30	300
No of channels		1	1
TWTA O/P Back-off & multicarrier deg.			0.0
Equivalent Power/Channel	dBW	14.8	24.8
Waveguide etc. Losses	dB	2.0	1.0
Antenna Diameter	mt	2.5	2.8
Antenna Efficiency	%	0.65	-
Antenna Gain (EOC)	dB	43.5	26.0
Total maximum EIRP	dBW		49.8
Equivalent EIRP per carrier (EOC)	dBW	56.3	49.8
Space Distance	km	41000.0	41000.0
Free Space Loss	dB	201.7	234.2
Atmos. Att./Propagation Loss	dB	4.0	0.1
Polarization loss	dB		
Antenna Diameter	cm	20.0	-
Antenna Gain (EOC)	dBi	15.3	11.5
Rx chain loss	dB	1.5	1.1
Temp. seen by Antenna	K	290.0	96
Receiver Noise Figure	dB	1.8	1.3
System Noise Figure (at Ant.)	dB	4.966	3.157
System Temperature (at Ant.)	dBK	27.9	24.9
G/T towards Tx	dB/K	-12.6	-13.4
C/No	dBHz	66.6	76.7
LNA Input Power (Signal+N)	dBm	-96.0	-96.8
PFD (per carrier)	dBW/m2	-111	-113.6
Net. C/No	dBHz		76.7
Implementation loss	dB		1.80
Eb/No achieved			14.8
Eb/No required	dB		2.6
SYSTEM MARGIN (audio channel)	dB		10.4

Fig. 6 CaribSpace coverage Link Margin Contours: A = 12.5 dB B = 10.5 dB
C = 8.5 dB

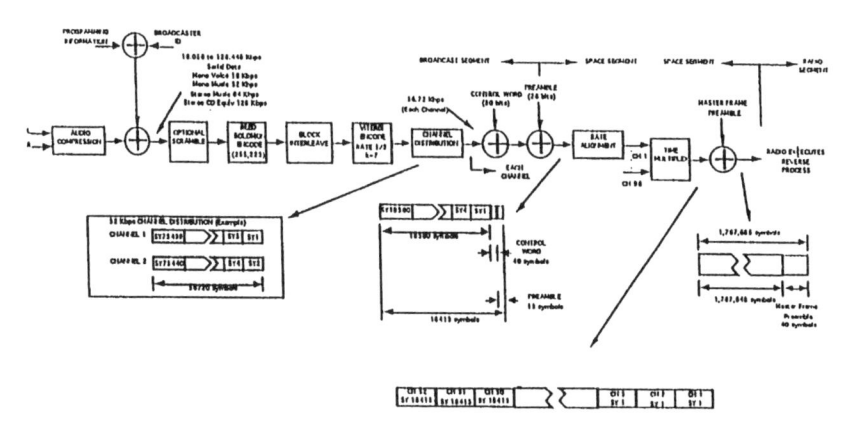

Fig. 7 Digital Formatting Operation

3 System Alternatives: Conventional Solutions and MCPC

SCPC transmission on the down link combined with convolutional encoding was initially considered. This solution presents disadvantages in terms of reduced link margin, intermodulation noise and bandwidth efficiency.
The selected solution, which is based on FDMA/TDM conversion can be implemented with
two different approaches:

- regenerative repeater solution, shown in Fig. 3 and later on illustrated

- the double hop solution shown in Fig. 8.

The double hop solution implements on ground at the TT&C communication monitoring master station,the control and signal processing functions foreseen inside the base band processor (BBP) equipment relevant to the on board processing solution. The possibility to enter in the DAB system is completely under control of the TT&C and Communication Control& Monitoring Station (TCMS).

The uplink broadcaster station, installed at the radio studios, transmits on one of the 288 frequency channels with SCPC (FDM) access technique at X band frequency.

The satellite contribution repeater routes the contribution signals toward the TCMS station. The foreseen frequency for the down link is 17.4 GHz in the band allocated for Broadcasting Service.

The TCMS station demodulates the signals extracting the information data; hence the baseband section of the station creates three digital signals, to be modulated on three carriers, containing 96 signals at 36.72 Kdigit/s rate, to be steered to one of the three envisaged L Band DAB Satellite down link beams.

A digital switch selects the channels to be routed to each beam and creates a time multiplexed bit stream containing 96 individual channels.

The near baseband analog signal is then up converted to IF and sent to the X band up converter for translation to the 7.050 - 7.075 GHz spectrum.

The three carriers are power amplified and transmitted on the up feeder link to three active DAB satellite repeaters.

Each DAB repeater on board selects (through an IF processor module) and high power amplifies one MultiChannel signal (96 Audio Channels). Also in this case the L Band repeater transmits one MultiChannel 300 W signal per beam.

A top level description of the regenerative solution is here given.

Each up link broadcaster station transmits the signals with FDM technique to the DAB satellite repeater. Each signal is not routed to the TCS station, as in the previous case, but is directly demodulated on board and successively combined inside the BBP to obtain the Multiple Channel signal to be diffused at L band.

It is underlined that the processing and control functions foreseen for the Multi Channel Per Carrier generation are performed on board, *while in the transparent case, the <u>same</u> function is performed at the ground station.*

This approach on one side eliminates the hop to the TCS but, on the other side, requires the development of the BBP assembly, with expected increased program risks and costs.

The payload converts the received X band (7.050 - 7.075 GHz) up-link to a common IF centered at about 150 MHz. The IF/synthesizer then converts the signal to near baseband where the spectrum will be converted from the analog domain into the digital domain by an analog to digital converter.

Using digital signal processing techniques based in ASIC, the digital spectrum is demultiplexed and demodulated in blocks of 24 channles.

A digital switch then selects the channels to be routed to each beam and creates a time multiplexed bit stream containing up to 32 individual Audio channels.

Also in this case the digital bit stream is coded using a concatenated approach: Reed Solomon (255,223), interleaving and then Viterbi rate 1/2 K=7. The bit stream will then be modulated and converted back to analog domain with a digital to analog converter.

The near baseband analog signal is then up converted to the common IF and passed to the L band up converter for translation to the 1467 to 1492 MHz spectrum.

The signal is then steered to the appropriate down link antenna beam.

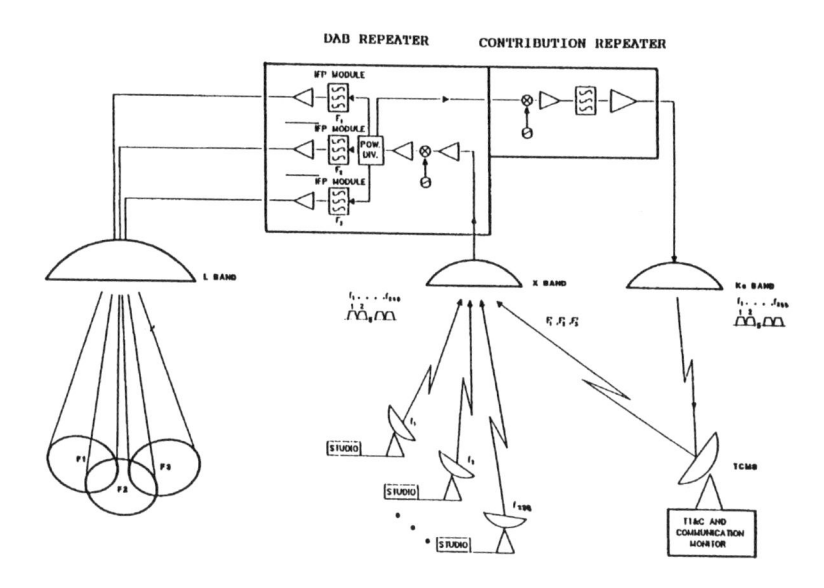

Fig. 8 On ground Processing Solution

4 Link Dimensioning and Elements Characteristics

4.1 Link Dimensioning

The links have been computed according to the calculated L band satellite antenna gain figures at the Edge Of Coverage (- 3dB) including the satellite pointing errors and assuming End Of Life TWTA power. The actual margin on the major part of the spot would be better.

The assumption of the TWTA saturated power of 300 Watts is consistent with the data of flight proven existing tubes.

The broadcasting of a 64 Kbps channel is here considered as reference dimensioning case.

The link dimensioning results, presented in Table 2.1 refer to the broadcaster station and radio set parameter values indicated in Tables II and 4.2 respectively.

Table II - Broadcaster transmitting station parameters

Frequency range	7050 to 7075 MHz
Antenna diameter	2.5 m
Antenna Gain	42.7 dB
Beamwidth	1.2 x 1.2 degr
Polarization	Circular
Power per 64 Kbps channel	30 Watt effective
TWTA power	75 Watt (OBO: 4dB)

Table III - Characteristics of the portable radio set

Frequency range	1467 to 1492 MHz
Polarization	circular
Antenna type	printed circuit
Pre selector loss	1.1 dB
Receiver noise figure	1.3 dB
Demodulation loss	1.8 dB
Antenna gain	high / moderate
	12 dBi / 9 dBi
Antenna Pointing Loss	0.48 dB / 0.14 dB
G/T (minimum)	-13.4 dB/K / - 16.dB/K

4.2 Payload

The reference payload is composed of:

- X band receiving antenna
- X band receiving section
- Base band processor
- L band transmitting section
- L band transmitting antenna
- Common equipment

The payload architecture is shown in Fig. 9. It has been directly derived from existing and proven designs minimizing design risk and avoiding critical or new items definition. Apart the BaseBand Processor all adopted hardware make use of already space qualified and flight proven technologies and is derived from flight units with minimum changes to take into account frequency, bandwidth and special performances. The redundancy concept adopted provides a reliability figure that assures the achievement of mission objective. This, together with the safe design approach allows the required 12 years of full flight operations.

4.3 Radio Set

The key requirements are:

- For the low-end radio product:
foreseen cost = in the range 50 to 100 $
- For the high-end radio product family:
a) RS 232 / RJ-11 interface
b) Line output
c) Supplementary digital service,
i.e. low rate video

Antenna:

- Integral part of the transportable radio
Option: remote location of the antenna

- Gain:
12 dBi - maximum in building penetration
   3 to 6 dBi - to support mobile service

4.4 Broadcaster Station

The key requirements of the broadcaster station are:

- single rack plus antenna
- low operation and maintenance
- modular and scaleable design
supporting 1 up to 50 uplink channels
(16 Kbps, information rate)
- antenna size fixed and TWTA power selectable
- support various studious interfaces
i.e. digital, analog
- adequate link margin to support link availability of 99.9% availability
- $10^o \div 90^o$ elevation

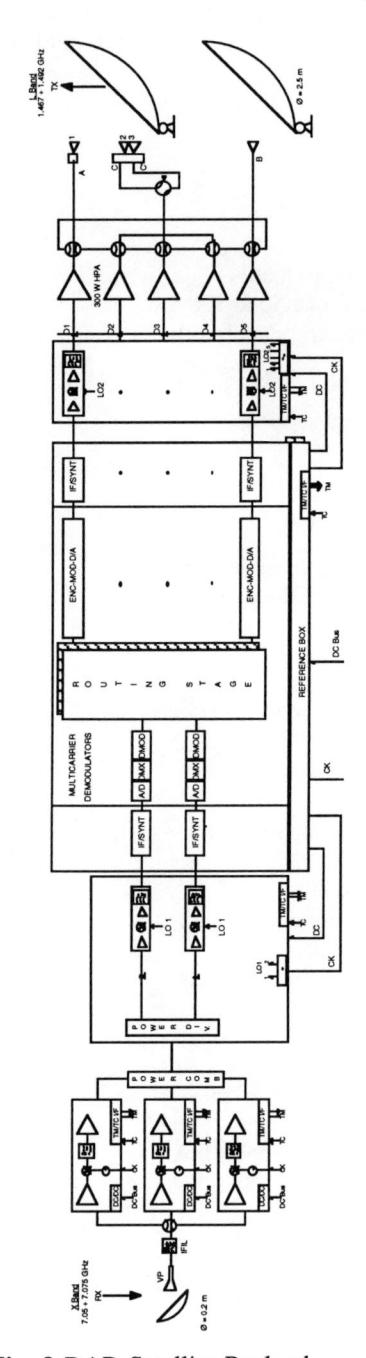

Fig. 9 DAB Satellite Payload

5 Concluding Remarks

The presented DAB system provides large link margin, mandatory to penetrate buildings, to cope with tree blockage, multipath and bursty nature of the channel, specially for mobile service.

Moreover the on board processing solution allows a controlled access to the resource via TT&C channel.

The presented system is feasible using to day existing technology increasing in this way the confidence to be first to market.

References

1. N. A. Samara, "The AfriSpace digital audio broadcasting system" - 15th AIAA International Communication Satellite System Conference

Part 5

Advanced Signal Processing Techniques

Digital Equalization Using Modular Neural Networks: an Overview

Jesús Cid-Sueiro[1] Aníbal R. Figueiras-Vidal[2] *

[1] ETSI Telecomunicaci'on-UV, Valladolid, 47011 Spain
jesus@tel.uva.es
[2] DSSR, ETSI Telecomunicación-UPM, Madrid, 28040 Spain
anibal@gtts.ssr.upm.es
Tel. 341-336-7226
Fax 341-336-7350

Abstract. Neural networks have a high potential to improve the performance of linear equalizers, but they are limited by a series of practical difficulties: the design of the adequate architecture, the application of algorithms for a fast training, and the selection of an appropriate objective function are the most important.

The recently proposed modular structure of Jacobs and Jordan offers a good alternative to solve the two first difficulties, because it combines a variable number of one layer nets, thus allowing easy ways to establish powerful global architectures as well as a fast training (because each element is an one layer net).

This paper reviews how to proceed with the third question: discussing the effects of different objective functions and showing the results of using them in practical examples. Both the discussion and the examples serve to conclude that these schemes plus an appropriate objective funcion, such as that proposed by Amari and others, provide a practical option for improved digital equalization.

1 Modular Classifiers for Equalization

The application of non-linear structures to equalize digital communication channels has been proposed by several authors ([2] and [8] are first examples). The current research has shown that neural networks outperform other conventional structures, such as the Linear Equalizer (LE) and the Decision Feedback Equalizer (DFE), but complexity and learning problems remain and are difficult to solve: for instance, a Multi-Layer Perceptron (MLP) can equalize linear non-minimum-phase channels without delaying decisions [2]; however, the computational load an the size of the structure required to do that is usually too high for most practical applications. Furthermore, as the length of the channel response increases, complexity and learning problems become intractable for a real time application.

The previous difficulties have motivated the search of non-linear equalizers that, keeping a moderate size and an acceptable convergence time, reduce the error rates of conventional approaches. A promising strategy consists of using structures that include the LE as a particular case: for instance, nonlinear neural networks in parallel with a linear filter or networks growing from the LE. When doing this, the performance of the LE can be guaranteed.

Gelfand [6] and Jordan [10, 12] have applied the previous ideas in a modular approach to classification: different linear networks (modules) are combined in such a way that every module is specialized in different kinds of input patterns: in other words, the input space is divided in subregions where each linear filter is specialized; as every module is a linear filter, conventional learning algorithms can be used or easily generalized. The schemes have been succesfully applied to equalization [7, 4].

---

* This work has been partially supported by CICYT grant TIC 92 #0800-C05-01

In this paper we go deeper in discussing the application of modular structures to the equalization of digital communication channels. First, we note that Gelfand and Jordan approaches do not differ in the structure they propose (which is essentially the same), but in the nature of the decisions: while Gelfand uses a hard-decision scheme (only one filter is selected at each time), the Jordan approach is based on soft decisions (the filters make soft decisions, the final decision being computed as a combination of them). This motivates our comparative work, with several goals: exploring the efficiency of the modular structure, searching adequate cost functions for stochastic gradient learning, comparing hard and soft decision strategies and, finally, getting a further insight on the network behaviour.

Although we restrict ourselves to binary transmissions, delaying the multi-level case to a future work, some of the final conclusions must be valid for other cases.

2 Modular Structures

A simple modular architecture based on (biased) linear filters is the Modular Neural Network (MNN), proposed by Jordan [12]: it is shown in Fig. 1. The final output is a weighted sum of the outputs proposed by different adaptive experts: at time k, the i-th expert output is

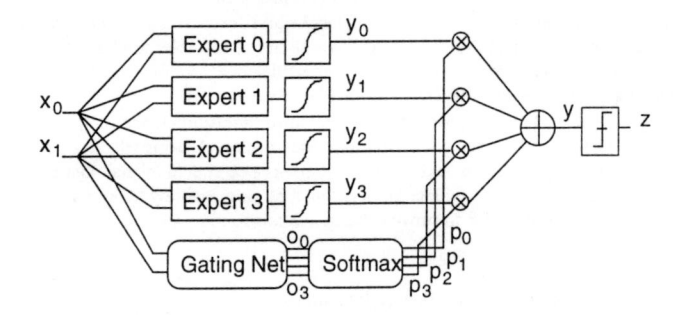

Fig. 1. Adaptive experts supervised by a gating network

$$y_i(k) = \text{sigm}\left(\mathbf{w}_i^T \mathbf{x}(k)\right) \quad (1)$$

where $\mathbf{w}_i$ is the weight vector of the i-th expert and $\mathbf{x}$ is the input vector with the received samples and a constant term biasing the filter. The input to the slicer is

$$y(k) = \sum_{i=1}^{N} p_i(k) y_i(k) = \mathbf{p}(k)^T \mathbf{y}(k) \quad (2)$$

To simplify the notation, the time dependance will be assumed implicitly in the following.

Weights p_i are computed by a gating network, with the goal of inhibiting the effect of the experts that are less appropriated for the current pattern. They are usually constrained to lie between 0 and 1, summing 1 alltogether, by using the normalizing "softmax" function

$$p_i = \frac{\exp o_i}{\sum_{i=0}^{N-1} (\exp o_j)} \quad (3)$$

where o_i are also outputs of linear filters

$$o_i = \mathbf{v}_i^T \mathbf{x} \quad (4)$$

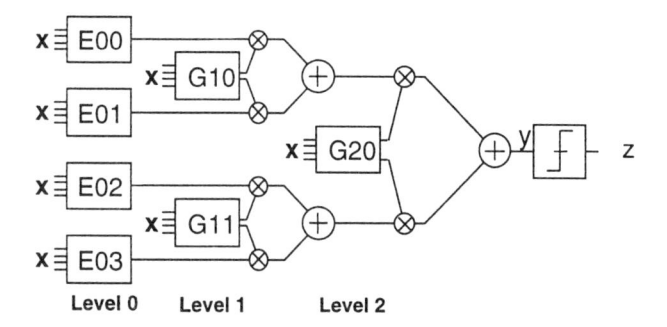

Fig. 2. Hierarchical structure of adaptive experts

The soft decision nature of the MNN is clear. A generalized arrangement of the filters is the Hierarchical MNN (HM, in the following) of Fig. 2. The gating net is now divided in an collection of linear filters partitioning sucesively the input space.

The soft nature of both MNN and HM networks makes easy the application of stochastic gradient learning rules. However, a hard working mode is also possible: if we replace the sigmoidal nonlinearities by hard slicers, and the "softmax" (which is nothing but a multidimensional sigmoid) by a "winner-take-all" scheme, only one expert is selected at each time by the gating nets. In such a case, the HM becomes the tree-structured Piecewise-Linear Equalizer (PLE) proposed by Gelfand in [7]. Thus, comparing the HM and the PLE we can appreciate the differences between hard and soft decision strategies.

The main advantage of the PLE is that only the selected filters are active at each time, reducing the computational load of the scheme during operation. However, the hard slicers difficult the application of stochastic gradient learning rules. As we will see, Gelfand has proposed an effective learning algorithm for this situation, that we have used in the simulations of this paper. This makes difficult our comparative work because, the learning algorithm of hard and soft schemes being different, it is not evident to determine if different performances are caused by decision strategies or by learning algorithms. To overcome this problem, we proceed in two steps: first, we try to optimize the learning algorithm for each network independently, and, after this, we compare the networks for the best case. Here, optimizae, optimization means the selection of the adequate cost funcion and the adaptation step for stochastic learning.

3 Cost Functions for Soft-Modular Networks

The search of a stochastic gradient learning rule to adapt the weights of a modular network to a particular channel has been addressed in [4]. We resume here the results: we will assume that there are only two classes, a_0 and a_1, and the slicer output is the index of the class, 0 or 1.

Consider the MNN network. We are interested in stochastic gradient learning rules for different cost functions $C(y)$; it is easy to see that

$$\nabla_{\mathbf{v}_i} C = p_i(y_i - y)\frac{\delta C}{\delta y}\mathbf{x} \tag{5}$$

$$\nabla_{\mathbf{w}_i} C = p_i y_i (1 - y_i)\frac{\delta C}{\delta y}\mathbf{x} \tag{6}$$

The final updating rules are determined by the particular cost selection. A similar procedure follows for the HM.

Four different cost functions have been selected, which seem to be representative enough of a large familiy of possible functions: the L_1 norm (Telfer and Szu [17]),

$$C_1 = |d - y| = (1-y)d + y(1-d) \tag{7}$$

$$\frac{\delta C_1}{\delta y} = 1 - 2d \tag{8}$$

where d is the desired output; the conventional quadratic error function

$$C_2 = (d-y)^2 \tag{9}$$

$$\frac{\delta C_2}{\delta y} = 2(y-d) \tag{10}$$

the logarithmic cost (El-Jaroudi and Makhoul [11], Hopfield [9] and Amari [1])

$$C_3 = -d\log(y) - (1-d)\log(1-y) \tag{11}$$

$$\frac{\delta C_3}{\delta y} = -\frac{d}{y} + \frac{1-d}{1-y} = \frac{1}{1-d-y} \tag{12}$$

and, finally, a heuristic cost proposed and studied in [14]

$$C_4 = 0.5((1+dz) + \gamma(1-dz))(d-y)^2 \tag{13}$$

$$\frac{\delta C_4}{\delta y} = ((1+dz) + \gamma(1-dz))(y-d) \tag{14}$$

Replacing the derivatives in Eqs. (5) and (6), the corresponding rules result.

4 Application to Channel Equalization

The previous learning rules were applied to the equalization of the non-minimum phase channel $H(z) = 0.5 + z^{-1}$ with additive, white Gaussian noise, which has been widely used in the literature for simulation purposes.

Fig. 3 is the average of 20 simulations comparing the performance of the 5-experts MNN for the objective functions C_1 to C_4. In the simulations, the adaptation step μ of the learning rules was decreased with time according to the formula

$$\mu_{k+1} = (1-a)\mu_k + a\frac{\mu_k}{1+\mu_k} \tag{15}$$

wich guarantees that the reduction of μ_k is not faster than $\frac{1}{k}$. The results, as well as those of other similar experiments with different channels, show better convergence properties for objective C_3: it is fast and local minima problems do not appear; quadratic cost C_2 showed isolated cases with bad convergence; costs C_1 and C_3 did not converge to a reliable solution in more than 30 % simulations. This number can be reduced using a smaller a (the decreasing rate of the adaptation step); but, in this case, the convergence is slower.

It is also noticeable that MNN networks show better convergence properties than those found when using other general-purpose networks. In our simulations, we have found that when a MLP is used to solve the same problem, a bigger network is needed in order to get the same results, the convergence problems being very difficult to solve. In this sense, our results agree with those shown in [2], for example.

Other simulations have also shown that learning is very sensitive to the evolution of both the initial value and the decreasing rate of the adaptation step. In order to simplify the analysis, we proceed to evaluate performances using a constant step (which is, on the other hand, the usual working mode in practical equalizers, because the channel may be subject to slow variations). Using a constant step, we found two interesting results:

- the optimal adaptation step is almost independent on the SNR of the channel
- the adaptation step must be proportional to the total number of linear filters in the network

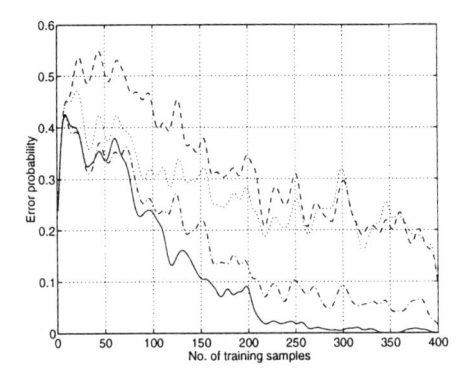

Fig. 3. Convergence of the symbol error probability using 5 experts and different objectives. Linear channel $H(z) = 0.5 + z^{-1}$ with noise variance 0.01. Dotted line: C_1; dashdot: C_2; solid: C_3; dashed: C_4.

5 Optimizing a Piecewise Linear Tree

Gelfand has proposed an iterative algorithm to adapt the PLE: let **x** be the input vector to the filter; the nodes of the tree are numbered consecutively from up to down and from left to right, as Fig 4 shows. The linear filter at node i is determined by weights $\mathbf{w}_i$ and offsets d_i, in such a way that output y_i is related with the input vector according to

$$y_i = \mathbf{w}_i^T \mathbf{x} + d_i \tag{16}$$

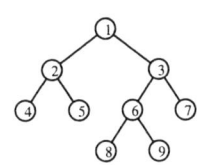

Fig. 4. Tree node numeration for the Gelfand algorithm

Unless those of the terminal nodes, every filter is related with an adaptive threshold $\theta_i(k)$: if $y_i \leq \theta_i$ the equalizer propagates the input vector through the left node with branch out from node i; otherwise, the right node is selected.

During training, the frequency of an input vector passing through node i is iteratively estimated using the rule given by

$$\Delta p_i = \mu(I_i - p_i) \tag{17}$$

where I_i is an indicator that is equal to one if **x** passes through node i and it is equal to 0 otherwise. As usual, μ is the adaptation step.

The filter coefficients and the threshold are updated by means of a stochastic gradient learning rule minimizing the cost functions

$$\epsilon_i = E_i\left\{(d - y_i)^2\right\} \tag{18}$$

$$\beta_i = E_i\left\{(d - \theta_i)^2\right\} \tag{19}$$

respectively. The following rules result

$$\mu_i = \frac{\mu}{p_i} \tag{20}$$

$$\Delta \mathbf{w}_i = \mu_i I_i (d - y_i) \mathbf{x} \tag{21}$$

$$\Delta d_i = \mu_i I_i (d - y_i) \tag{22}$$

$$\Delta \theta_i = \mu_i I_i (d - \theta_i) \tag{23}$$

In the Gelfand algorithm, only the filters selected by the tree at each time are updated, by means of a conventional LMS rule, to minimize the Mean Square Error (MSE) between the output of the filter and the desired output. For training purposes, we have tested the alternative of using a sigmoid nonlinearity in the selected filters, updating them with the goal of minimizing a quadratic or logarithmic error between the output of the sigmoid and the desired output. However, we could not improve the convergence speed of the former algorithm. As the parameters being optimized have a different nature, we presume that using a specific adaptation step for each one of them could improve the learning task; but, the network optimization using multiple steps is not easy and, thus, we decided to compare hard and sooft decision schemes training the PLE with the Gelfand algorithm.

6 Comparing the Networks

We have used the channel test $H(z) = 0.5 + z^{-1}$ with AWGN noise; the adaptation step was optimized independently for each network. Structures with 7 linear filters were applied: HM and a PLE networks with 3 levels and a MNN with 4 experts. In the last case, 4 experts require 4 gating filters and, thus, the MNN has 8 filters, not 7; but, because of the "softmax" normalization function, one of the gating filters is redundant, and there are just 7 effective filters.

Fig. 5 compares the error rate vs SNR for these structures. We have found that this simulation is representative of a general situation: the soft-decision based schemes perform much better that the PLE scheme, being the MNN and HM performances very similar when the same number of filters are used. To a certain extent, this is not a surprising result: the PLE is a limit case of the HM network when the moment of the sigmoid non-linearities tend to infinity. Thus, the potential capabilities of the PLE are inferior to that of a HM scheme with the same size; besides of this, simulations show that training a hard decision structure is a more difficult task.

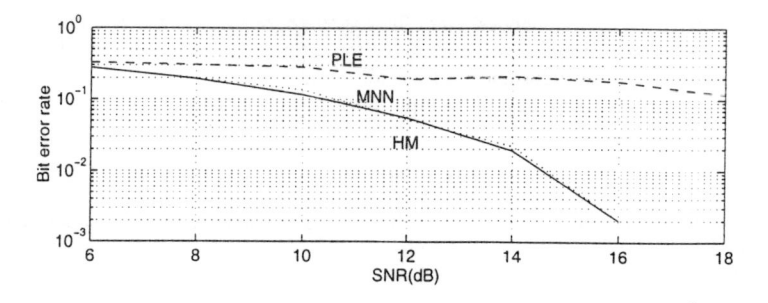

Fig. 5. Error performance of hard and soft structures vs. SNR. Linear channel $H(z) = 0.5 + z^{-1}$ with AWGN noise

7 A Probabilistic Interpretation

The behaviour of the soft-decision based scheme can be well understood by means of a probabilistic interpretation. Consider, for simplicity, the MNN network, and assume that the input data are generated according to the following process: in a first step, a distribution D_i is selected randomly between a collection of N distributions; second, a random sample $\mathbf{x}$ is generated according to distribution $D_i \mid y$, where y is the current class (in our application, the transmitted symbol). Note that this data model is not artificious of unrealistic: for instance, it is not difficult to see that the samples received through a digital FIR channel follow it.

The classifier has to determine d given $\mathbf{x}$. The optimal decision depends on the a posteriori probabilities of the symbols

$$p(d \mid \mathbf{x}) = \sum_{i=1}^{N} p(D_i \mid \mathbf{x}) p(d \mid \mathbf{x}, D_i) \qquad (24)$$

Note the similarity between this and Eq. (2). If the gating net outputs estimate conditional probabilities $p(D_i \mid \mathbf{x})$ and the experts estimate $p(d = 1 \mid \mathbf{x}, D_i)$, for every i, then MNN output y estimates $p(d = 1 \mid \mathbf{x})$; as the final decision is 1 only if $y > 0.5$, the network approximates the optimal *maximum a posteriori* (MAP) detector.

In practice, this does not necessarily occur, because gating and expert networks have linear constraints. In any case, a question aries: does learning lead to estimates of probabilities?

Some theoretical results have shown that minimizing a quadratic or a logarithmic cost, output y is in fact an estimate of conditional probability distributions $p(d \mid \mathbf{x})$. Proofs are not difficult, but we refer the interested reader to [1, 16].

Besides of this, when the logarithmic cost is used, some interesting and alternative expressions result from the probabilistic interpretation. If the outputs are estimates of the conditional probabilities stated above

$$\frac{y_i}{z} = \frac{\hat{p}(d = 1 \mid \mathbf{x}, D_i)}{\hat{p}(d = 1 \mid \mathbf{x})} = \frac{1}{2}\hat{p}(D_i \mid d = 1, \mathbf{x}) \qquad (25)$$

and also

$$\frac{1 - y_i}{1 - z} = \frac{\hat{p}(d = 1 \mid \mathbf{x}, D_i)}{\hat{p}(d = 1 \mid \mathbf{x})} = \frac{1}{2}\hat{p}(D_i \mid d = 0, \mathbf{x}) \qquad (26)$$

where symbol $\hat{p}$ means "estimate of probability". Using this expression, it is easy to show that the learning rules can be written as

$$\Delta \mathbf{w}_i = -\mu \hat{p}(D_i \mid d, \mathbf{x})(d - y_i)\mathbf{x} \qquad (27)$$

$$\Delta \mathbf{v}_i = -\mu \left(\hat{p}(D_i \mid d, \mathbf{x}) - p_i\right)\mathbf{x} \qquad (28)$$

Note that the adaptation rule for weights $\mathbf{w}_i$ is a generalized LMS rule, where every expert is updated as a function of the estimated a posteriori probability of distribution D_i generating input sample $\mathbf{x}$. The rule for weights $\mathbf{v}_i$ admits a similar interpretation. If we could know the distribution generating each sample, we would know what of the filters should be updated using a conventional LMS rule; the previous formulas show that this lack of knowledge is overcame by updating every filter depending on its expectations about the current distribution. This result is very similar to that found in the blind equalizers proposed in [3, 15], where the uncertainty is not about a distribution, but about the transmitted symbol.

In summary, the updating rules agree with the probabilistic interpretation which has been also pointed by Jordan (see [13], for example). In spite of this, although the logarithmic cost ensures that output y estimates a probability, there is no theoretical result ensuring that the expert or gating outputs are conditional probability estimates.

8 Conclusions and Further Work

In this paper we show that the modular approach is an efficient way of improving the error rates of conventional equalizers with a moderate increase in the network complexity, showing a better performance than that found in other neural equalizers. After noting that HM and PLE networks are essentially the same structure, we show that the soft decision based schemes are more efficient during both detection and learning. Also, we find that MNN and HM networks with the same number of filters have a similar performance, showing that the main aspect for learning purposes is not the network configuration, but the number of filters.

On the other hand, we have shown that, when the logarithmic cost (which has shown to be the more effective) is used, the network behaviour during learning and detection has an intuitive probabilistic interpretation which, unfortunately, has not a complete theoretical support.

There are many open questions arising from the work done up to now: for instance, we know that, if the adequate cost funtion is used, the network outputs are estimates of probabilities; but we do not know nothing about its quality; our preliminary work suggests that a large trainig set and network size are required to get a good estimate, but further work needs to be done to extract definite conclusions. Knowing the quality of these estimates is important for several applications: the estimation of the symbol error probability in the receiver [5] or learning in blind mode.

On the other hand, the soft-decision approach has shown to be more efficient, but it is computationaly expensive on a serial implementation, because all the filters work during both detection and learning. We are investigating ways of constraining the adaptive algorithm to get a piecewise linear or quasi-piecewise linear solution, in order to switch to a hard-decision mode after learning.

More complex channel responses and other classification problems must be considered. The multiclass problem is specially important, because there are different ways of generalizing the logarithmic and quadratic costs.

Finally, the probabilistic interpretation suggests the application of other learning algorithms. As we have seen, the updating formulas for the filters are generalized LMS rules; in a similar way, generalized RLS rules can be obtained; their justification and performance evaluation is another topic for further work.

References

1. S. Amari, Backpropagation and Stochastic Gradient Descent Method; *Neurocomputing*, No. 5, pp. 185-196, June 1993.
2. S. Chen, G.J. Gibson, C.F.N. Cowan: Adaptive Equalization of Finite Non-linear Channels Using Multilayer Perceptrons; *Signal Processing*, Vol. 20, No. 2, pp. 107-119, 1990.
3. J. Cid-Sueiro, A.R. Figueiras-Vidal: Recurrent Radial Basis Function Networks for Optimal Symbol-by-Symbol Equalization; *Signal Processing*, Vol. 40, No. 2, pp.53-63, Oct. 1994.
4. J. Cid-Sueiro, A.R. Figueiras-Vidal: The Role of Objective Functions in Modular Classification (with an Equalization Application); *Proc. 1st Int. Conf. on Neural, Parallel and Scientific Computations*, pp. 110-115; Atlanta, GA, May 1995.
5. G. Paton-Garcia, J. Cid-Sueiro: Estimacion de la Probabilidad de Error en Receptores Digitales; accepted for presentation at X Symposium Nacional de la URSI; Valladolid, Spain, Sept. 95.
6. S.B. Gelfand, C.S. Ravishankar, E.J. Delp, An Iterative Growing and Pruning Algorithm for Classification Tree Design; *IEEE Transactions on Pattern Analysis and Machine Intelligence*, Vol. 13, No. 2, pp. 163-174, Feb. 1991.
7. S.B. Gelfand, C.S. Ravishankar, E.J. Delp, Tree-Structured Piecewise Linear Adaptive Equalization; *IEEE Transactions on Communications*, Vol. 41, pp. 70-82, Jan. 1993.
8. G.J. Gibson, S. Siu, C.F.N. Cowan: The Application of Nonlinear Structures to the Reconstruction of Binary Signals; *IEEE Transactions on Signal Processing*, Vol. 39. No. 8, pp. 1887-1894, Aug. 1991.

9. J.J. Hopfield: Learning Algorithms and Probability Distributions in Feed-Forward and Feed-back Networks; *Proc. of the Nac. Academy Sci. USA*, Vol. 84, pp. 8429-8433; 1987.
10. R.A. Jacobs, M.I. Jordan: A Competitive Modular Connectionist Architecture; in D.S. Touretzky (Ed.), *Advances in Neural Information Processing Systems 2*, pp. 767-780, Morgan Kaufmann, San Mateo, CA, 1990.
11. A. El-Jaroudi, J. Makhoul: A New Error Criterion for Posterior Probability Estimation with Neural Nets; *Proc. Intl. Joint Conf. on Neural Networks*, Vol. I, pp. 185-192; San Diego, CA, 1990.
12. M.I. Jordan, R.A. Jacobs: Hierarchies of Adaptive Experts; in J. Moody, S. Hanson, R. Lipmann (Eds.), *Advances in Neural Information Processing Systems 4*, pp. 985-992, Morgan Kaufmann, San Mateo, CA, 1992.
13. M.I. Jordan, R.A. Jacobs: Hierarchical Mixtures of Experts and the EM Algorithm; *Tech. Report 9203, MIT Brain and Computat. Sci.*; MIT, Cambridge, MA, 1993.
14. S. Makram-Ebeid, J.A. Sirat, J.R. Viala: A Rationalized Back-Propagation Learning Algorithm; *Proc. Intl. Joint Conf. on NN*, Vol. 2, pp. 373-380; Washinghton, DC, 1989.
15. S.J. Nowlan, G.E. Hinton: A Soft Decision-Directed LMS Algorithm for Blind Equalization; *IEEE Transactions on Communications*, Vol. 41, No. 2, pp. 275-279, Feb. 1993.
16. D.W. Ruck, S.K. Rogers, M. Kabrisky, M.E. Oxley, B.W. Suter: The Multilayer Perceptron as an Approximation to a Bayes Optimal Discriminant Function; *IEEE Trans. on Neural Networks*, Vol. 1, No. 4, pp. 296-298, Dec. 1990.
17. B.A. Telfer, H.H. Szu: Energy Functions for Minimizing Missclassification Error with Minimum Complexity Networks; *Neural Networks*, Vol. 7, No. 5, pp. 809-818; 1994.

Vector Quantization using Artificial Neural Network Models

Aristides S. Galanopoulos
James E. Fowler, Jr.
Stanley C. Ahalt
Dept. of Electrical Engineering
Ohio State University
Columbus, OH 43210

ABSTRACT This paper describes our ongoing research into the construction of an Adaptive Vector Quantization (AVQ) image encode. We provide the motivations behind an AVQ encoder and report on our progress-to-date in realizing a Vector Quantization (VQ) encoder. We describe the hardware that has been built to compress video in real time using full-search vector quantization. This hardware implements a differential-vector-quantization (DVQ) algorithm which employs entropy-biased codebooks designed using an Artificial Neural Network (ANN). The theoretical properties of this codebook design method are discussed. We conclude with a description of the framework we are using to study various AVQ techniques with the goal of modifying our existing VQ hardware in order to realize an AVQ encoder.

1 Introduction

Ideally we would use *lossless* data compression for all compression tasks. Indeed, lossless compression is *required* for compression of many kinds of textual data, e.g., computer programs, documents and numerical data. Unfortunately, when applied to images lossless compression techniques achieve only modest compression rates (2:1 to 8:1). Furthermore, digital data has become more prevalent and the demand for real-time image-coding hardware, particularly for use with images, has increased dramatically.

Fortunately, for many image applications *lossy* data compression can be used as perfect reproduction is not necessary. Lossy compression techniques can achieve compression ratios can be as high as 100:1.

Vector quantization (VQ) has long been recognized as a useful technique for lossy data compression, attracting attention for its efficient compression of digitized image and speech data. However, the design of real-time vector quantizers for image coding has been challenging due to the inherent computational complexity of VQ encoders and the extremely fast speeds demanded by real-time video applications. Furthermore, the computational complexity of traditional codebook-design methods has also hindered real-time use of VQ. It has recently been shown that artificial neural networks (ANN's) can be used to suggest real-time VLSI architectures as well as provide algorithms for the design of VQ codebooks which circumvent limitations of traditional algorithms [2].

This paper describes our ongoing research into the construction of a Adaptive Vector Quantization (AVQ) encoder for images. We describe motivations behind an AVQ encoder and our progress to-date in realizing one vector-quantization encoder. In particular we describe hardware that has been built to compress video in real time using full-search vector quantization. This architecture implements a differential-vector-quantization (DVQ) algorithm which features codebooks designed using frequency-sensitive competitive learning (FSCL) [3], an ANN algorithm that attempts to maximize codebook entropy while minimizing distortion. We then describe some of the theoretical properties of this codebook design method. We conclude with a description of a framework we are currently using to study various AVQ techniques in order to modify our existing VQ hardware to realize an AVQ encoder.

2 Vector Quantization

A VQ system consists of an encoder, a decoder, and a transmission channel. The encoder and the decoder each have access to a fixed codebook, **Y**. The codebook **Y** is a set of Y codewords (or codevectors), **y**, where each **y** is dimension k and has a unique index, j, $0 \leq j \leq Y - 1$. We describe our codebook design method in Section 4.3.

The image is broken into blocks of pixels called tiles. Each image tile of $n \times m$ pixels can be considered a vector, **u**, of dimension $k = mn$. For each image tile, the encoder selects the codeword **y** that yields the lowest distortion by some distortion measure $d(\mathbf{u}, \mathbf{y})$. The index, j, of that codeword is sent through the transmission channel. If the channel is errorless, the decoder retrieves the codeword **y** associated with index j and outputs **y** as the reconstructed image tile, **û**. A block diagram of a VQ system is shown in Fig. 1.

An extensive discussion of vector quantization techniques and applications has been given by Gersho and Gray[9], and the basic theory has been summarized in the context of image applications by others [6].

2.1 Why use VQ rather than MPEG?

The MPEG standard has emerged as an effective standard for image compression. However, we believe there are four reasons why VQ should still

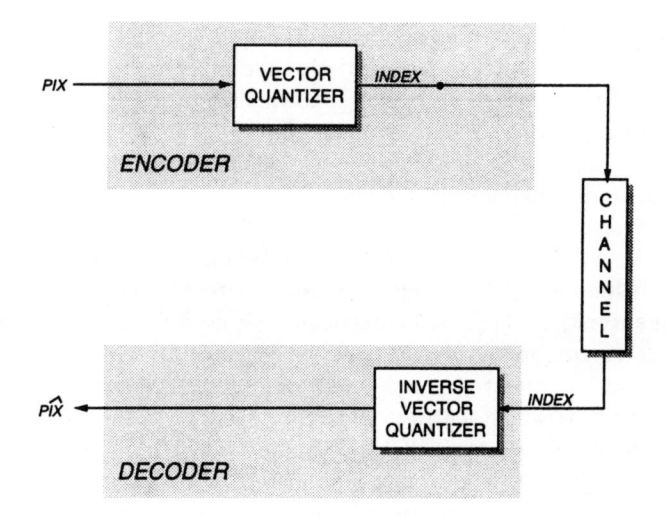

FIGURE 1. Block diagram of VQ system.

be considered as a viable candidate for wide-spread use as an image coding method:

- **Errors**: because of run-length encoding and transmission ordering, nominal-MPEG is quite "error sensitive." VQ, in contrast, is relatively error-insensitive, and can be made more error-insensitive via codebook ordering techniques [4, 11].

- **Regularity**: VQ design has been *inexpensively* implemented in real-time hardware.

- **Decoder cost**: MPEG is a symmetric algorithm, and thus both the encoder and decoder require equivalent computation. In contrast, VQ decoders are very simple.

- **Adaptive Coders**: MPEG does not provide an explicit adaptation mechanism, but Adaptive VQ (AVQ) has been under investigation for some time and adaptation mechanisms are easily incorporated into VQ.

These factors led us to build a prototype VQ encoder, which we describe below.

3 Differential Vector Quantization

Differential Vector Quantization (DVQ) is a combination of VQ and DPCM which replaces scalar quantization in DPCM framework with vector quantization. Consequently DVQ has many of the compression advantages of

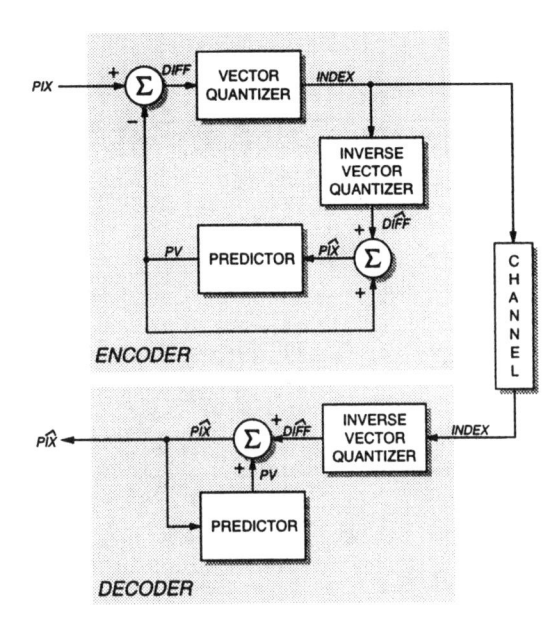

FIGURE 2. Block diagram of DVQ algorithm

both VQ and DPCM. DVQ has been presented previously in [14], where it was called vector DPCM, and in [9], where it was called predictive VQ (PVQ). Our DVQ algorithm has been reported in detail previously [6], so only a brief overview is given here.

Fig. 2 shows the general block diagram of our DVQ algorithm. In the encoding process, the predictor uses previously reconstructed tiles to predict the pixel values of the current tile. This predicted tile, PV, is subtracted pixel by pixel from the actual tile, PIX. The resulting difference tile, $DIFF$, is vector-quantized and the index, $INDEX$, is broadcast via the transmission channel to the decoder. The encoder inverse vector-quantizes $INDEX$, producing a reconstructed tile, $\widehat{PIX}$, to be used in later predictions. Note that, since the vector quantizer processes difference tiles, the VQ codebook must be appropriately derived from "difference images."

DVQ has several advantages over both scalar DPCM and VQ. Primarily, the quantization of vectors yields better compression performance than that of scalars. Additionally, since the VQ is performed on difference values rather than on the image itself, the resulting image is less "blocky" [14]. Finally, the codebooks for DVQ tend to be more robust and more representative of many images than codebooks designed for VQ because the difference tiles in a DVQ codebook are more generic than the image tiles in a VQ codebook [14].

The Vector-quantizing Associative Memory Processor Implementing Real-time Encoding (VAMPIRE) is a special-purpose, digital associative memory designed for video-rate vector quantization. The details of the design

TABLE 1.1. Summary of the VAMPIRE Chip

Die size	4.6 × 6.8mm
Technology	$2\mu m$ CMOS n-well
Vector Rate	3.57×10^6 vectors/sec
Encoding delay*	approx. $1\mu s$
Codebook size	32 codewords on one chip; expandable to 256 with 8 chips
Vector dimension	Four 8-bit components
Power supply	5V

*Encoding delay is for the VAMPIRE chip operating in the DVQ architecture

and operation of this chip are found elsewhere [1], so only a brief overview is given here. Table 1.1 presents a summary of the VAMPIRE chip's characteristics.

The VAMPIRE chip is designed to quantize vectors at video rates. The input to the chip is 32 bits representing a 4-dimensional vector with each vector component having 8 bits of resolution. Since these vectors are composed of four video samples, the designed throughput is that of the NTSC colorburst (3.579545MHz, or one vector every 280ns). Each VAMPIRE chip holds 32 codewords. The chips can be operated alone (for codebooks of 32 or less codewords) or can be linked together to accommodate codebooks of greater than 32 codewords.

4 Codebook Design and Optimal Quantizers

Obviously, since the codebook holds the vectors used to replicate all images (or at least all of the images for some period of time), the design of the codebook for a VQ encoder is quite critical to the overall effectiveness of any VQ-based coding scheme.

It has been known for some time that the necessary conditions for minimizing the average distortion, assuming a convex distortion measure, are:

- Nearest neighbor selection rule

$$\mathbf{y}(\mathbf{x}) = \{\mathbf{y}_i \mid \|\mathbf{y}_i - \mathbf{x}\| <= \|\mathbf{y}_j - \mathbf{x}\| \text{ for all } j\}, and$$

- Centroid condition

$$\mathbf{y}_i = \frac{\int_{C_i} \mathbf{x} p(\mathbf{x}) d\mathbf{x}}{\int_{C_i} p(\mathbf{x}) d\mathbf{x}}.$$

Thus, analytical methods for VQ codebook design require knowledge of the data pdf, which are not known in realistic cases. For unknown data pdf, a set of M **training** data vectors must be used for the design of the N quantizer codevectors, where $M \gg N$.

The classic and most commonly used batch codebook design algorithm is the LBG (Linde-Buzo-Gray) algorithm[10], a generalization of the Lloyd-Max algorithm. This is a descent algorithm, thus it converges to a local minimum depending on the initial state.

Unfortunately the computational complexity and memory requirements of such traditional VQ codebook design methods has restricted their use in real-time applications [10]. It has been shown that ANN's can be used to design VQ codebooks and circumvent the limitations of traditional algorithms [2], for the following reasons. First, ANN's consist of a large number of simple, interconnected computational units that can be operated in parallel. Second, ANN-codebook-design algorithms do not need access to the entire training data set at once during the training process. It should also be noted that these features make ANN algorithms ideally suited for the design of adaptive vector quantizers [6], which are discussed in Section 5.

4.1 Competitive Learning

The simplest ANN algorithm for VQ codebook design is Competitive Learning (CL). At time t (discrete) a data vector $\mathbf{x}(t)$ is presented and the following steps are executed:

- find the *winner*, the codevector $\mathbf{y}_w$ which is closest to the input vector:

$$\|\mathbf{y}_w(t) - \mathbf{x}(t)\| <= \|\mathbf{y}_i(t) - \mathbf{x}(t)\| \text{ for all } i.$$

- update the winner as:

$$\mathbf{y}_w(t+1) = \mathbf{y}_w(t) + a(t) * (\mathbf{x}(t) - \mathbf{y}_w(t))$$

where $a(t)$ is the *learning rate* at time t.

It can be shown that, for CL, the necessary conditions for convergence to a local equilibrium,

$$\lim_{t \to \infty} a(t) = 0, and \sum_{t=1}^{\infty} a(t) = \infty,$$

are sufficient for convergence to a local equilibrium if the initial state is inside the domain of attraction of some equilibrium (from stochastic approximation theory).

Unfortunately, this simple algorithm is easily trapped in local minima. In many realistic practical applications, these minima are the result of *codeword underutilization*. Codeword underutilization occurs when a codeword is never designated as the closest codeword to an input vector, and is never updated. Two mechanisms have been proposed to overcome this problem: Kohonen's Self-Organizing Feature Maps (KSFM), and Conscience techniques including Frequency Sensitive Competitive Learning (FSCL). Both of these techniques are described below.

4.2 Kohonen Self-Organizing Feature Maps

In a KSFM, a topology $\mathcal{B}$, (usually two-dimensional) is associated with the codevectors. Once again a data vector, $\mathbf{x}(t)$, is presented and the following steps are executed:

- the winner $\mathbf{y}_w(t)$ is selected as the codevector closest to the input vector $\mathbf{x}(t)$
- *All* codevectors inside a neighborhood of the winner, defined on $\mathcal{B}$, are updated towards the input vector as:

$$\mathbf{y}_n(t+1) = \mathbf{y}_n(t) + a(t) * h(n,w) * (\mathbf{x}(t) - \mathbf{y}_n(t)),$$

where $h(n,w)$ is usually a decreasing function of the distance $\|\mathbf{y}_n - \mathbf{y}_w\|_\mathcal{B}$, and the size of the winner's neighborhood also decreases with time.

The KSFM technique has been shown to utilize all codevectors, and thus solves the underutilization problem. Further, it can be shown that necessary and sufficient conditions for convergence, as in the case of CL, are: [5, 12]

$$\lim_{t \to \infty} a(t) = 0, and \sum_{t=1}^{\infty} a(t) = \infty.$$

Finally, equilibrium results for one dimensional data spaces have been derived [13], and the equilibrium codevector density is proportional to a power of the data pdf which depends on the size of the winner's neighborhood.

Unfortunately, while KSFMs solve the underutilization problem and have been shown to have attractive theoretical properties, they exhibit certain problems. First, because of the use of neighborhoods, $\mathcal{B}$, they are computationally more expensive. Second, the selection of the neighborhood topology can be difficult for many dimensional data spaces. The FSCL algorithm circumvents these problems.

4.3 Frequency Sensitive Competitive Learning

The Frequency Sensitive Competitive Learning (FSCL) method is an ANN method which features a modified distortion measure that ensures all codewords in the codebook are updated equally frequently during iterations of the training process.

This is accomplished by assigning a counter $c_n(t)$ to each codevector $\mathbf{y}_n(t)$ that is incremented every time $\mathbf{y}_n(t)$ wins the competition. The winner, $\mathbf{y}_w(t)$, is selected as the codevector that minimizes the product of the distortion measure and a **fairness function** F, i.e.,

$$F(c_n(t))\|\mathbf{y}_n(t) - \mathbf{x}(t)\|.$$

The usual form of the fairness function can be expressed as:

$$F(c) = c^{\beta(t)}, \text{ with } \lim_{t \to \infty} \beta(t) = 0.$$

It has been shown that codebooks designed with FSCL yield mean squared errors and signal-to-noise ratios comparable to those of the locally optimal LBG algorithm [2]. Also, FSCL yields codebooks with sufficient entropy so that Huffman coding of the VQ indices would not provide significant additional compression [6]. Finally, FSCL overcomes underutilization problems, is relatively efficient, and is well-suited for adaptive applications.

Recently we have been able to establish the convergence and equilibrium properties of the FSCL algorithm [7, 8]. We outline those results below.

FSCL convergence

First, in order to establish FSCL convergence, we can model the FSCL network as a Markov process with state

$$\sigma(t) = \begin{bmatrix} \mathbf{y}_1(t) & \cdots & \mathbf{y}_N(t) \\ f_1(t) & \cdots & f_N(t) \end{bmatrix}$$

where $f_i(t) = c_i(t)/t$ is the update frequency, $c_i(t)$ is the update count, and $\mathbf{y}_i(t)$ is the position of codevector i at time t. The winning codevector minimizes

$$F(f_i(t))\|\mathbf{y}_i(t) - \mathbf{x}(t)\|$$

where we consider fairness functions of the form

$$F(f_i) = f_i^{\beta}$$

with β a positive parameter. The counter $c_w(t)$ of the winner is incremented and its position $\mathbf{y}_w(t)$ updated as

$$\mathbf{y}_w(t+1) = \mathbf{y}_w(t) + a(t)(\mathbf{x}(t) - \mathbf{y}_w(t))$$

where $\mathbf{x}(t)$ is the input data vector at time t, and $a(t)$ is the learning rate.

We then formulate the evolution of an ensemble of systems in terms of the probability $Q(\sigma, \sigma')$ of transition from state σ to state σ' and describe the evolution of the process through a linear time-dependent Fokker-Plank equation.

With suitable approximations it can be shown that the mean and the covariance matrix of the codevectors' deviation from the equilibrium must vanish as $t \to \infty$ with necessary and sufficient conditions:

$$\lim_{t \to \infty} a(t) = 0, \quad \text{and} \quad \int_{t_0}^{\infty} a(t)dt = \infty.$$

Once again we note that these conditions are, satisfyingly, the same as those needed for CL and KSFM.

FSCL Equilibrium

In order to establish the equilibrium properties of the FSCL codebook design technique, we simplify the analysis to one-dimensional input data and a large number of codevectors. While we omit the details here, we can then show that, for fairness functions of the form $F(f_i) = f_i^\beta$, the equilibrium codeword density is proportional to a *power* of the input data pdf, i.e.,

$$(\text{codevector density}) \propto (\text{data density})^{\frac{3\beta + 1}{3\beta + 3}}.$$

and FSCL minimizes the L_q-norm distortion measure with

$$q = \frac{2}{3\beta + 1} \in (0, 2].$$

These results show that the FSCL algorithm can be used to optimize a larger class of distortion measures compared with previous VQ clustering algorithms, at least for a one-dimensional input space. It should be noted that, as far as we know, only one-dimensional equilibrium results have been calculated for any ANN clustering algorithm.

5 Adaptive Vector Quantization

Our ultimate objective, which we are now investigating, is to modify our VQ encoder with a suitable adaptation mechanism such that an Adaptive Vector Quantizer (AVQ) is realized. More precisely, we wish to realize an encoder that codes a non-stationary source with unknown statistics at an average of R bits per symbol, such that the average distortion per symbol is minimal. Among the open questions we hope to answer are 1) what is the optimal distortion that can be realized (by any coding), and 2) how close can AVQ come?

To structure the discussion, assume a Large (possibly infinite) *universal* codebook, $\mathcal{C}^* \subseteq \Re^k$.[1] We'll assume that the AVQ encoding uses a *local* codebook, $\mathcal{C}_t \subset \mathcal{C}^*$ and that the AVQ encoding is time-variant:

---

[1]Please note that we have changed notation here. The codebook, $\mathcal{C}^*$, plays the same role as that denoted **Y** previously. We adopt the use of $\mathcal{C}^*$ to emphasize that this is an *adaptive* codebook.

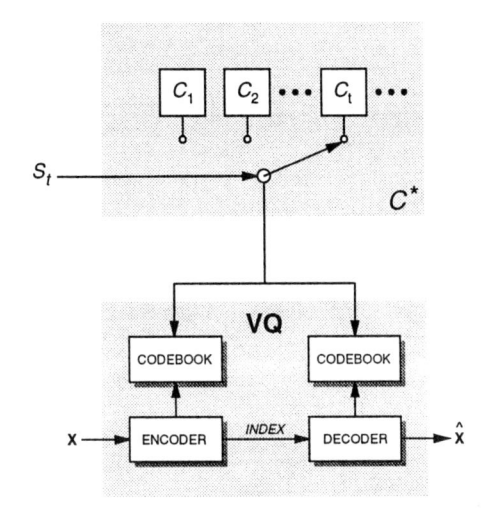

FIGURE 3. Block diagram of AVQ algorithms

$Q_t : \Re^k \to C_t$

$\hat{\mathbf{X}}_t = Q_t(\mathbf{X}_t) \in C_t$

This requires some local codebook selection mechanism:

$s_t : C^* \to \{0, 1\}$

such that $C_t = \{\mathbf{y} \mid s_t(\mathbf{y}) = 1, \mathbf{y} \in C^*\}$.

Now, assuming that C^* is finite, let $N = |C^*|$ and $S_t = s_t(\mathbf{y}_1) \cdots s_t(\mathbf{y}_N)$, where $\mathbf{y}_i \in C^*$. Then S_t, the codebook-selection process, is an integer-valued random variable $0 < S_t \leq 2^N - 1$.

A pictorial view of this model is shown below in Figure 3.

We have found it useful to classify all of extant AVQ encoders into two classes of universal codebook encoders. We refer to these as:

- *A priori* codebooks in which:

  - Both encoder and decoder have complete knowledge of C^*, and
  - we assumes we know enough about the source to set C^* before coding commences, or

- *Inductive* codebooks in which:

  - Encoder and decoder only know initial local codebook C_1
  - The rest of C^* is *induced* from observing the source.

With this fundamental differentiation, we have established a taxonomy of current AVQ encoders, as shown in Fig. 4. We are currently reviewing

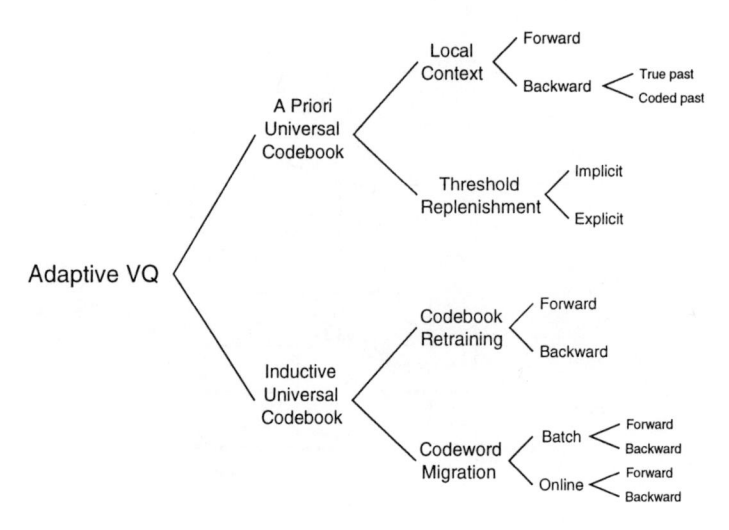

FIGURE 4. Taxonomy of AVQ algorithms

the literature to determine if this taxonomy is sufficient for all previously proposed AVQ techniques. We anticipate that this taxonomy will prove useful for our subsequent efforts at implementing an AVQ system consistent with our current DVQ algorithm.

6 Conclusions

In this paper, we have described our DVQ algorithm and presented a hardware architecture implementing the algorithm. We have shown that the codebook design technique we employ, FSCL, has attractive theoretical properties, and appears to be suitable for use with non-stationary sources. Finally, we have shown that previous AVQ investigations can be described in a convenient mathematical framework, and classified according to a simple taxonomy.

Our ongoing work is concentrated on determining what AVQ mechanisms are needed to approach optimal AVQ, and determining if these mechanisms are computationally feasible in real-time, inexpensive hardware.

Acknowledgments: S. Ahalt gratefully acknowledges partial support from the Air Force Maui Optical Station.

7 References

[1] Kenneth C. Adkins. *The VAMPIRE Chip: A Vector-quantizer Associative Memory Processor Implementing Real-time Encoding.* PhD thesis, The Ohio

State University, 1993.

[2] Stanley C. Ahalt, Prakoon Chen, and Ashok K. Krishnamurthy. Performance Analysis of Two Image Vector Quantization Techniques. In *Proceedings of the International Joint Conference on Neural Networks*, volume I, pages 169–175, Washington, D.C., June 18–22, 1989.

[3] Stanley C. Ahalt, Ashok K. Krishnamurthy, Prakoon Chen, and Douglas E. Melton. Competitive Learning Algorithms for Vector Quantization. *Neural Networks*, 3:277–290, 1990.

[4] Da-Ming Chiang and Lee C. Potter. Minimax Non-Redundant Channel Coding for Vector Quantization. In *Proceedings of the International Conference on Acoustics, Speech, and Signal Processing*, pages V–617–V–620, Minneapolis, MN, April 1993.

[5] M. Cottrell and J. C. Fort. A stochastic model of retinotopy: a self organizing process. *Biological Cybernetics*, 53:405–411, 1986.

[6] James E. Fowler, Matthew R. Carbonara, and Stanley C. Ahalt. Image Coding Using Differential Vector Quantization. *IEEE Transactions on Circuits and Systems for Video Technology*, 3(5):350–367, October 1993.

[7] Aristides S. Galanopoulos and Stanley C. Ahalt. Codeword Distribution for Frequency Sensitive Competitive Learning with One Dimensional Input Data. *tnn*, 1995.

[8] Aristides S. Galanopoulos, Randolph M. Moses, and Stanley C. Ahalt. Convergence Conditions for Frequency Sensitive Competitive Learning. *tnn*, 1995.

[9] Allen Gersho and Robert M. Gray. *Vector Quantization and Signal Compression*. Kluwer international series in engineering and computer science. Kluwer Academic Publishers, Norwell, MA, 1992.

[10] Y. Linde, A. Buzo, and R. M. Gray. An Algorithm for Vector Quantizer Design. *IEEE Transactions on Communications*, COM-28(1):84–95, January 1980.

[11] Lee C. Potter and Da-Ming Chiang. Minimax Nonredundant Channel Coding. *IEEE Transactions on Communications*, 43(2/3/4):804–811, February/March/April 1995.

[12] J. Ramanujam and P. Sadayappan. Parameter Identification for Constrained Optimization using Neural Networks. In *Proceedings of the 1988 Connnectionist Models*, pages 154–161, Carnegie Mellon Univ., June 1988.

[13] Helge Ritter. Asymptotic level density for a class of vector quantization processes. *IEEE Transactions on Neural Networks*, 2(1), January 1991.

[14] Charles W. Rutledge. Vector DPCM: Vector Predictive Coding of Color Images. In *Proceedings of the IEEE Global Telecommunications Conference*, pages 1158–1164, September 1986.

Neural Networks for Communications and Signal Processing: Overview and New Results*

Mohamed IBNKAHLA and Francis CASTANIE
National Polytechnics Institute of Toulouse
ENSEEIHT, 2, Rue Camichel, 31071 TOULOUSE, France
e-mail: ibnkahla@len7.enseeiht.fr

Abstract

Neural networks are able to give solutions to complex problems in signal processing and communications due to their non linear processing, parallel distributed architecture, and capacity of learning and generalization. This paper gives an overview of the applications of neural networks to digital communications and signal processing. It presents new results concerning the identification of digital satellite channels equipped with non linear memoryless devices (travelling wave tubes (TWT)).

1 Introduction

Artificial neural networks (ANN) are inspired from the computation in the nervous system. They can be viewed as an assembly of simple processing units (neurons) acting in parallel. These units are connected to each other by synaptic weights. Several neural network architectures can be constructed from these processors allowing a diversity of computational models such as self organizing maps [21], multi-layer NN [23], recurrent NN [13], stochastic NN [13], operator-valued NN [25], etc.

The last few years have known a great growth in the applications of neural networks to digital communications and signal processing. The non linear computation of neural networks, their ability to learn, their capability to self organize, and their efficient hardware implementation, make them well adapted to offer solutions to numerous problems in communications and signal processing. This paper deals only with multi-layer neural networks which are the most widely used in these domains [13, 23, 31].

The paper is organized as follows. In section 2 we present the multi-layer neural network structure, the real-valued back propagation algorithm

---

*This work has been supported in part by the Centre National d'Etudes Spatiales (CNES), Toulouse, France, under contract 962/94/CNES/1232/00.

(BP), and the complex-valued BP algorithm. Section 3 gives an overview of the applications of NN to linear and non linear adaptive signal processing. Section 4 is devoted to new results concerning NN-based modelling of digital satellite communication channels equipped with memoryless TWT amplifiers.

2 Multi-layer feed forward neural networks

2.1 Structure

The multi-layer neural network (MLNN) is one of the most popular neural network architectures used in signal processing. Its basic unit, the neuron (figure 1), is composed of a linear combiner and an activation function. The neuron receives inputs from others processors to which it is connected. The linear combiner output is the weighted sum of the inputs plus a fixed weight (bias):

$$net = \sum_{j=1}^{N} W_j x_j + b.$$

where x_j is the jth input value of the neuron, W_j the corresponding synaptic weight, and b the bias term. The activation function gives then the neuron output:

$$y = f(net) = f\left(\sum_{j=1}^{N} W_j x_j + b\right)$$

The activation function may be a linear or non linear function (e.g. the identity function, the hyperbolic tangent function, the sign function, a radial basis function (RBF), etc.). The choice of activation function depends on the application at hand [13].

A multi-layer neural net is composed of neurons connected with each other (figure 1). The input information is processed from the input layer to the output layer (feed forward). The network inputs are the inputs of the first layer. The outputs of the neurons in one layer form the inputs to the next layer. The network outputs are the outputs of the output layer.

Figure 1 denotes the layer index by i. x_{ik} is the output of neuron k of layer i. W_{ijk} is the weight that links the output x_{i-1j} to neuron k of layer i. $N(i)$ is the number of neurons in layer i. With these notations, the output x_{ik} of neuron (i, k) is given by:

$$x_{ik} = f(net_{ik}) = f\left(\sum_{j=1}^{N(i-1)} W_{ijk} x_{i-1j} + b_{ik}\right).$$

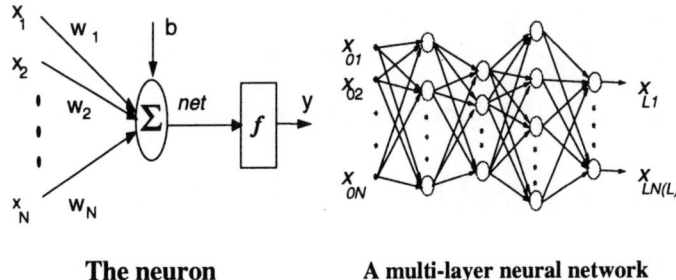

Figure 1: Multi-layer neural network structure.

It has been demonstrated that a two-layer feed forward perceptron with sigmoidal activation function and a scalar output can approximate arbitrarily well continuous functions, provided that an arbitrarily large number of neurons is available (see e.g. [15]).

2.2 The back propagation (BP) algorithm

Most neural network applications do not have a priori knowledge of the correct weights for the network to perform a desired mapping. Thus, a learning (training) procedure is necessary to obtain these weights. For that purpose, the BP algorithm [23] is used. A set of input-output pairs $\{x(n), d(n)\}$ train the network to implement the desired mapping. The BP algorithm adjusts the MLNN weights so as to minimize the squared error energy function (the error power between the network output and the desired output), $E(n) = \|d(n) - x_L(n)\|^2$, where $x_L(n)$ is the MLNN output vector at time n and $d(n)$ is the desired output. The BP algorithm performs a gradient descent on the energy function to arrive at a minimum:

$$W_{ijk}(n+1) = W_{ijk}(n) - \mu \frac{\partial E(n)}{\partial W_{ijk}(n)} \quad (1)$$
$$b_{ik}(n+1) = b_{ik}(n) - \mu \frac{\partial E(n)}{\partial b_{ik}(n)}.$$

These equations can be expressed explicitly as:

$$W_{ijk}(n+1) = W_{ijk}(n) + \mu \delta_{ik} x_{i-1j} \quad (2)$$
$$b_{ik}(n+1) = b_{ik}(n) + \mu \delta_{ik}.$$

The error term δ_{Lk} of the output layer is given by:

$$\delta_{Lk} = 2f'(net_{Lk})(d_k - x_{Lk}).$$

The calculation of the error term of the hidden unit (i,k), δ_{ik}, can be expressed easily as a function of the next layer error terms:

$$\delta_{ik} = f'(net_{ik}) \sum_{j=1}^{N(i+1)} W_{i+1kj}\delta_{i+1j}.$$

Thus, the weight update is performed by propagating the error terms from the output layer to the input layer.

2.3 The Complex-valued BP algorithm

In digital communications and signal processing, signals are often complex-valued and need to be processed in complex vector spaces. Complex-valued neural networks (CNN) have been proposed in order to represent and process complex-valued signals.

Some authors have generalized the MLNN and the BP algorithm to the complex plane [10, 17, 22, 28]. For example, Leung and Haykin [22] have proposed a CNN which is similar to the real-valued MLNN. Its weights are complex-valued and the activation function is an extension of logistic function to the complex plane: $f(z) = \frac{1}{1+\exp(-z)}$.

This activation function is not bounded everywhere. In order to avoid this problem of singularity, Leung and Haykin proposed to scale the input data to some region in the complex plane.

The complex BP (CBP) algorithm updates the complex weights by

$$W_{ijk}(n+1) = W_{ijk}(n) + \mu \delta_{ik} x^*_{i-1j}. \tag{3}$$

where $*$ denotes the complex conjugate, δ_{ik} is the error term which is computed by error back propagation. For the output layer, the output error term is given by

$$\delta_{Lk} = 2f'(net^*_{Lk})(d_k - x_{Lk}).$$

For the hidden layers, δ_{ik} is computed by

$$\delta_{ik} = f'(net^*_{ik}) \sum_{j=1}^{N(i+1)} W^*_{i+1kj}\delta_{i+1j}.$$

In [17], we have proposed a CBP algorithm derived from vector neural networks. This algorithm does not present singularities and preserves the universal approximation property.

3 Applications to adaptive signal processing

3.1 Linear adaptive filtering

Multi-layer neural nets with linear activation functions have been used as linear adaptive filters [16, 30, 31]. The network is composed of p input values and a scalar output. The input consists of p (the filter length) values of a sampled signal $x(n)$

$$\underline{x}(n) = [x(n)\; x(n-1)\; ...\; x(n-p+1)]^t.$$

Since all the activation functions are linear, the MLNN output is a linear function of the input signal (FIR filtering).

In [11], a linear neural net has been used as a line tracking estimator. It was shown from simulation results, that it gives better performance than LMS for both frequency resolution and line tracking. In [16], we have studied, under some simplifying hypothesis, the influence of the network parameters (the number of layers, learning rate, the initial weight values, etc.) on the BP algorithm behavior and performance. Let's take for example the problem of filter identification. The input signal $\underline{x}(n)$ is a white gaussian noise. The desired output is $\underline{x}(n)$ filtered with an unknown linear filter plus noise:

$$d(n) = \underline{h}^t \underline{x}(n) + N(n).$$

The purpose of the linear neural net is to give an approximation to the unknown filter $\underline{h}$. We have superimposed, in figure 2, the learning curves (mean squared error versus time) obtained by several neural networks (with different number of layers). The respective neural network parameters were chosen such that the steady state performance is the same for each network. In this simulation, the greater the number of layers, the faster the convergence speed. Note that the LMS algorithm corresponds to a single layer neural net.

Similarly, Beaufays and Widrow [3] have used a two-layer linear neural network for fast adaptive filtering. They showed that the two-layer adaptive filter achieves better speed performance than pure LMS while retaining its low computational cost and its extreme robustness. A tutorial overview on adaptive filtering using neural networks can be found in [30] and [31].

3.2 Non linear signal processing

Multi-layer neural networks with non linear activation functions have been used to solve non linear problems in signal processing. Recently, Widrow and Billelo [32] applied two neural network structures to inverse control. In particular, they showed that Newton's method for solving non linear equations, combined to standard BP algorithm, has better performance than recurrent BP.

Haykin and Li [12] applied a neural network-based method to detect signals in chaos. Chaos is a non linear deterministic dynamic system. The authors

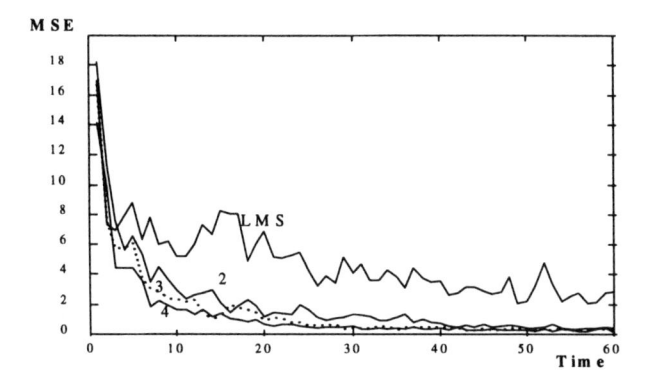

Figure 2: Learning curves of NNs with different number of layers.

point out that neural networks can be used to build a model which will predict the behavior of the chaotic back ground. This model can be applied to a received signal that contains a signal of interest plus a chaotic component. The prediction error produced by a such model is then used to detect the signal of interest. This method was applied to radar detection of a small target in sea clutter.

Recently, numerous real-valued neural network-based adaptive equalizers have been introduced to overcome nonlinear time varying distortion of non linear digital communication channels [7, 8, 9, 20, 24]. These equalizers employing various neural network structures (MLNN, radial basis function networks, recurrent NN) have been shown to equalize non linear channels successfully and outperform classical linear equalizers. Kirkland and Taylor [20] presented a comparative study on the use of neural network architectures for channel equalization. They showed their advantages over conventional equalization schemes.

Complex-valued neural networks have been used for channel equalization with two dimensional signaling, for instance QAM (quadrature amplitude modulation) and PSK (phase shift keying) modulation channels. Lo and Hafez [24] have studied the performance of conventional and complex perceptron-based equalizers of linear channels in the presence of intersymbol interference, additive noise, and co-channel interference. They concluded that the performance of the neural network equalizer match that of the conventional equalizer under all noise and interference conditions.

A complex-valued version of radial basis function networks (RBF) has been introduced by Chen et al. [8, 9].They have shown that this structure is able to generate complicated non linear decision surfaces and to approximate an arbitrary non linear function in complex multi-dimensional space. They ap-

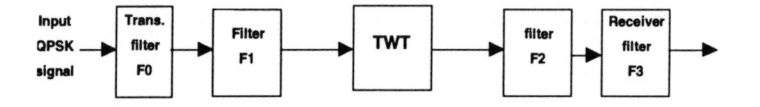

Figure 3: A simplified satellite channel.

plied the complex RBF network to adaptive realization of a bayesian solution for 4-QAM digital communication channel equalization. It was shown that the optimal bayesian equalizer is structurally equivalent to the complex RBF network. This intimate connection was exploited to develop fast training algorithms for implementing a bayesian equalizer based on RBF networks.

These adaptive equalizers view channel equalization as a classification problem and are based on the traditional mean squared error (MSE) performance criterion. Adali et al. [1, 2] have introduced a new approach in which conditional probability distribution function (PDF) of the transmitted signal given the received signal is parametrized by a general neural network structure. The PDF parameters are estimated by minimization of the accumulated relative entropy cost function. It was shown by simulations that this approach can track abrupt changes in the non linear channel response whereas the mean square error-based multi-layer perceptron equalizer can not.

4 Neural networks for modelling satellite channels

In this section, we present new results concerning modelling digital satellite channels with neural networks. Satellite communication channels [4] are equipped with non linear devices such as TWT amplifiers. This section deals only with **memoryless** TWT amplifiers. A satellite channel consists of two earth stations connected by a repeater (satellite) through two radio links (uplink and downlink). As an example, consider the simplified scheme of figure 3 modelled in the complex base band. The transmission filter F0, the IMUX (input multiplexing) filter F1, and the OMUX (output multi-plexing) filter F2 are linear. The TWT represents the satellite's on board amplifier. It acts as a memoryless non linearity with a complex transfer function which depends only on the input complex envelope. In this section we describe briefly some analytical models of TWT amplifiers and present our neural network-based model. We will then extend this approach to model the TWT cascaded with F1 and F2.

4.1 Analytical models of TWT amplifiers

TWT amplifiers exhibit two kinds of non linearities, amplitude distortion (AM/AM conversion) and phase distortion (AM/PM conversion). Two equivalent frequency-independent representations have been proposed for these non linearities [4, 26]: The amplitude-phase (A-P) representation and the in-phase and quadrature (I-Q) representation.

Let the input wave $x(t)$ be expressed as $x(t) = r(t)\cos(\omega_0 t + \theta(t))$, where ω_0 is the carrier frequency, and $r(t)$ and $\theta(t)$ are the modulated envelope and phase, respectively.

In the A-P representation, the TWT output is given by

$$y(t) = A(r(t))\cos(\omega_0 t + \theta(t) + \phi(r(t)))$$

where $A(r)$ and $\phi(r)$ are the amplitude and phase non linear distortions, respectively. For several purposes (e.g. analysis of the communication channel performance, simulations, etc.), these functions were approximated by analytical expressions. For example, Saleh [26] proposed a two-parameter formula for each conversion:

$$A(r) = \frac{\alpha_a r}{1 + \beta_a r^2} \;,\; \phi(r) = \frac{\alpha_p r^2}{1 + \beta_p r^2} \tag{4}$$

where the parameters $\alpha_a, \beta_a, \alpha_p$, and β_p are fitted to experimental data using a minimum mean square error procedure.

Several other analytical models have been proposed in the literature (e.g. Berman-Mahle phase formula [5], Thomas et al. amplitude formula [29]).

In the I-Q representation, the TWT output is represented by the in-phase and quadrature components,

$$\begin{aligned} p(t) &= P(r(t))\cos(\omega_0 t + \theta(t)) \\ q(t) &= -Q(r(t))\sin(\omega_0 t + \theta(t)), \end{aligned}$$

$P(r)$ and $Q(r)$ are the in-phase and quadrature non linearities, respectively.

Several analytical expressions have been proposed to approximate these non linearities (e.g. Saleh formulas with rational functions [26], Hetrakul-Taylor model which uses Bessel functions [14], and Shimbo-Pontano model which uses a complex Bessel function series approximation [27]).

The analytical models proposed for both A-P and I-Q representations are parametrized with relatively few parameters. The evaluation of the parameter values requires an optimization process to fit the data. These models have some limitations. For example, Saleh's model (equation 4) requires that, for large input amplitude r, $A(r)$ is proportional to $\frac{1}{r}$ and $\phi(r)$ approaches a constant. Furthermore, these models are not adaptive. A change in the TWT characteristics cannot be tracked by these models unless a new optimization is performed. In the section below, we propose a new approach based on neural networks.

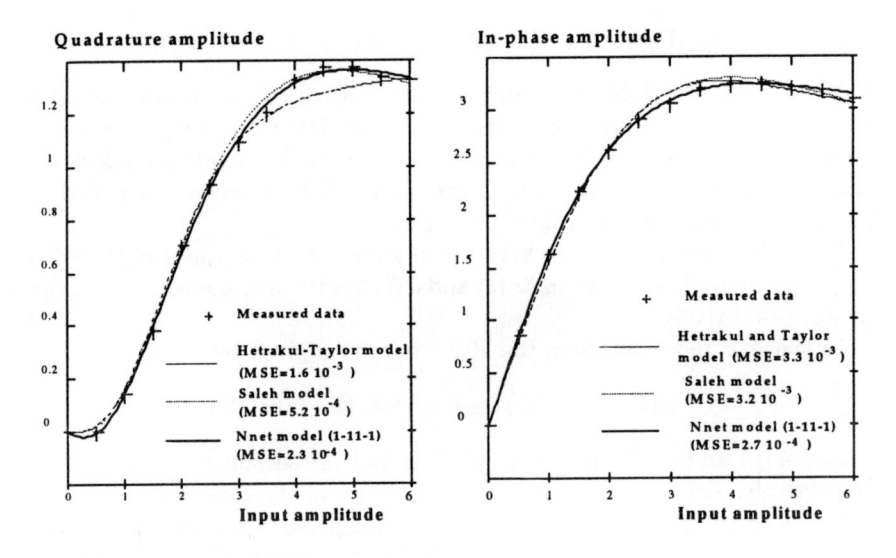

Figure 4: Quadrature representation: Comparison between analytic and NN models.

4.2 Neural network approach

To model the TWT characteristics, we use a two-layer structure with a scalar input, 11 neurons in the first layer, and a scalar output $(1, 11, 1)$. We present to the neural net a pair of measured data (input-output, for example $(r_0, A(r_0))$). The BP algorithm adjusts the neural weights so as to reduce the error between the network output and the desired output. This procedure is repeated (by selecting randomly a pair of measured data) until the error reaches a minimum and the weights no longer change.

Figure 4 compares simulation results for modelling the quadrature representation of Hughes 261-H Intelsat IV TWT amplifier. The measured data (13 points) is that given by Hetrakul and Taylor [14]. At each step of the algorithm, one of the 13 pairs of data points is selected randomly and presented to the network. The neural net approximation performance (MSE) is better than that of classical analytic models (Saleh's model [26] and Hetrakul-Taylor model [14]). Figure 5 shows the learning curve obtained for the in-phase non linearity. It can be learned that the MSE decreases rapidly at the beginning of the learning process, then it reaches an asymptote. For more details about the convergence and the learning behavior of the BP algorithm see [6], which studies a simple, but interesting example.

Figure 6 shows simulation results obtained for modelling the A-P representation of an Intelsat IV TWT amplifier. The measured data (13 points) is that given by Berman and Mahle [5]. The neural net approximation outper-

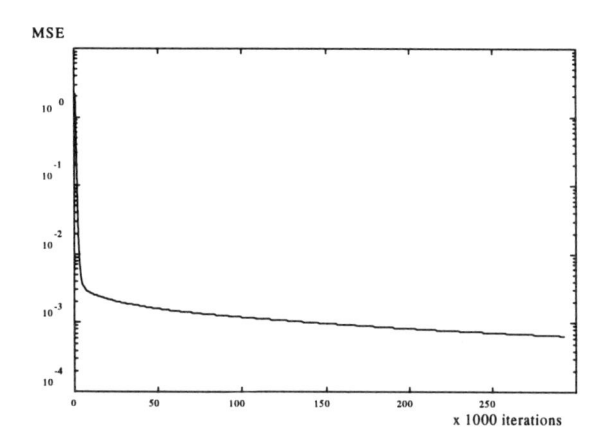

Figure 5: Learning curve of the BP algorithm (modelling the in-phase non linearity).

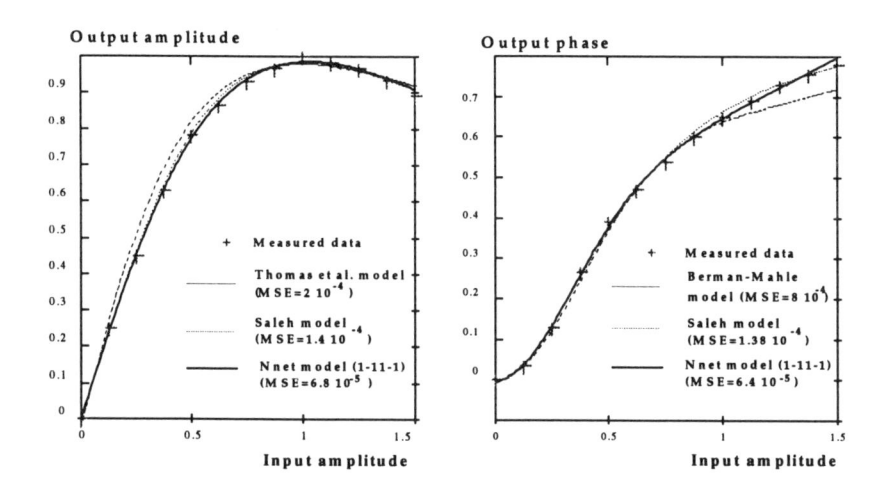

Figure 6: A-P representation: Comparison between Analytic and NN models.

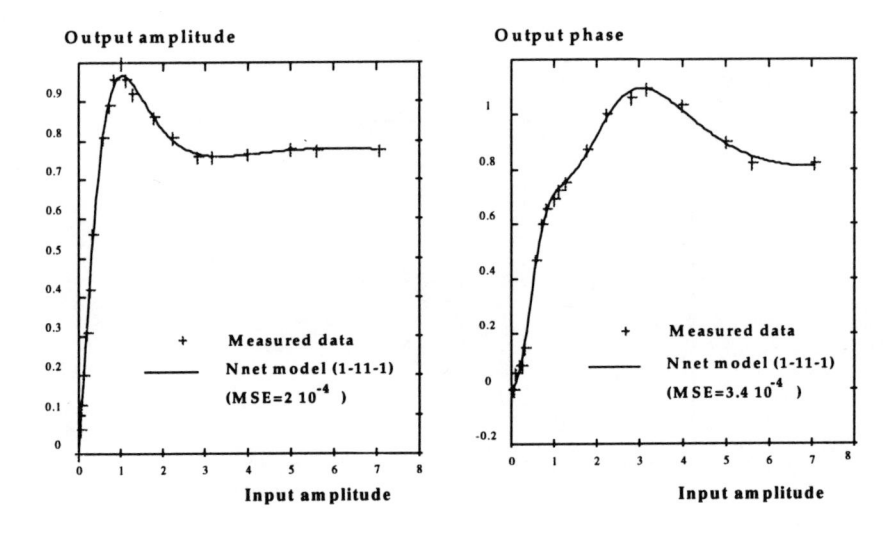

Figure 7: NN model of a TWT with complicated non linearities.

forms classical analytic models (Saleh's model [26], Thomas et al. amplitude formula [29], and Berman-Mahle phase formula [5]).

The neural net gives also good approximation for TWTs which were not modeled in the literature because of their complicated non linearities. Figure 7 shows an example of complicated non linearities. Another approach based upon the odd and even BP algorithms was also used to model TWT characteristics [18].

4.3 Identification of the TWT cascaded with linear filters

In this section, the satellite channel (figure 3) has the following characteristics: The signals are QPSK modulated, the transmission filter F0 is a four-pole Chebychev with $3dB$ bandwidth $\frac{1.66}{T}$, F1 is a four-pole Chebychev with $3dB$ bandwidth $\frac{2}{T}$, and F2 is a four-pole Chebychev with $3dB$ bandwidth $\frac{3.3}{T}$. The TWT is modeled by Saleh's formula (equation 4, with $\alpha_a = 2, \beta_a = 1, \alpha_p = 4.0033$, and $\beta_p = 9.104$).

4.3.1 Modelling the block F1-TWT

In this part, we model the block composed of the linear filter F1 and the TWT (working at saturation). The neural net is composed of a single-layer linear network with memory (linear adaptive combiner) followed by a memoryless non linear neural net, with the architecture 60-2-15-2 (figure 8). Thus, we

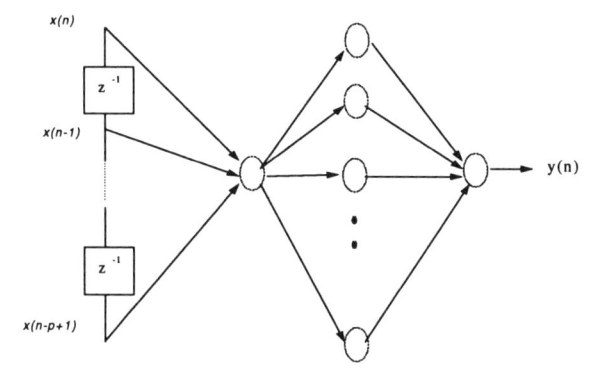

Figure 8: NN architecture used to model the block F1-TWT.

copy the structure to be modeled (a linear system with memory followed by a memoryless non linear system). In figure 9, we have superimposed the real part of the block F1-TWT output and the real part of the NN output (generalization). The generalization MSE is 1.210^{-3} (i.e. $RMS = 0.0346, SNR = 29.2dB$). We define the MSE performance in dB as $SNR = 10\log_{10}\left(\frac{P_s}{MSE}\right)$, where P_s is the signal power at the TWT input. Note that the input signal used for both learning and generalization is a white gaussian noise. Note also that, if we use for generalization a binary signal (filtered with F0), then the MSE reduces to 7.410^{-4}.

4.3.2 Modelling the block F1-TWT-F2

In this part, we model the block composed of the linear filter F1, the TWT (working at saturation) and the filter F2. The neural net is composed of a single-layer linear network with memory followed by a memoryless non linear neural net, and a single layer linear network, with the architecture 40-2-10-2-40-2 (figure10). In figure 11 we have superimposed the real part of the block F1-TWT-F2 output and the real part of the NN output (generalization). The generalization MSE is 210^{-3} (i.e. $RMS = 0.0444, SNR = 27dB$, the input signal is a white gaussian noise). Note that, if we use for generalization a binary signal (filtered with F0), then the MSE reduces to 10^{-3}.

5 Conclusion

This paper gave a overview on the recent applications of neural networks to digital communications and signal processing. It presented new results concerning the identification of digital satellite channels equipped with memory-

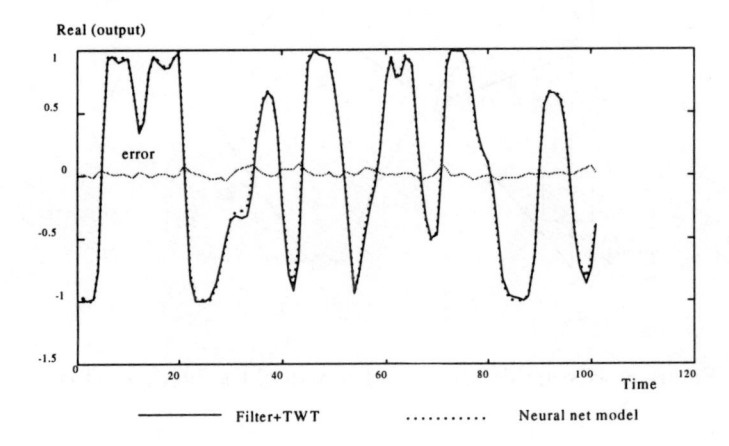

Figure 9: Generalization performance of the NN used to model the block F1-TWT.

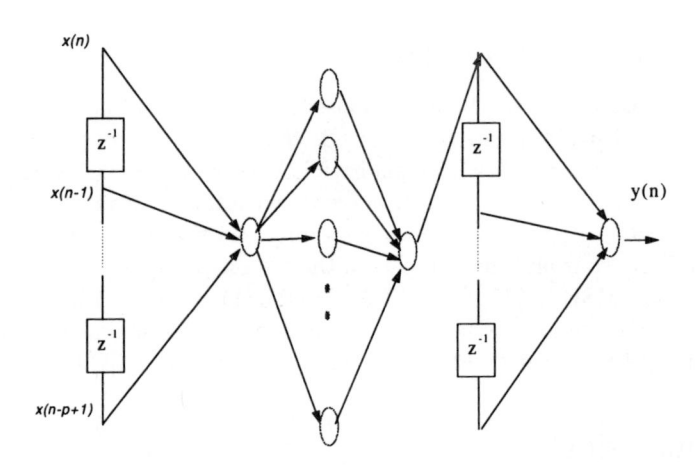

Figure 10: NN architecture used to model F1-TWT-F2.

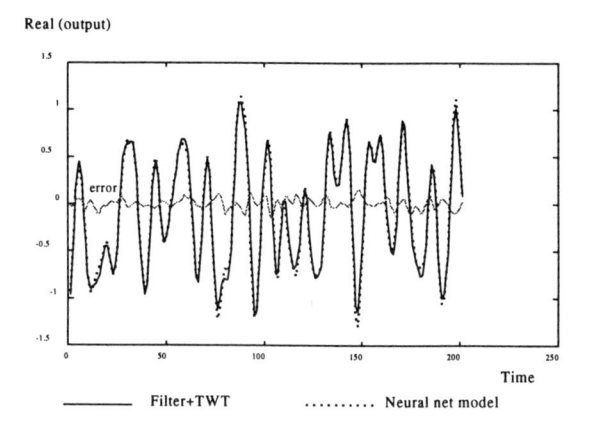

Figure 11: Generalization performance of the NN used to model the block F1-TWT-F2

less TWT amplifiers. It was shown in particular that neural network models of TWTs match better the experimental data than classical TWT analytical models. The NN approach allows to model several kinds of TWT characteristics with a unique and simple architecture (only the weights change from a model to another). Moreover, NN are adaptive, therefore any change in the TWT characteristics can be tracked. The NN identification of the TWT cascaded with linear filters gives good results too.

References

[1] T. Adali, X. Liu, and M. K. Sonmez, "Channel equalization by conditional distribution learning: a partial likehood framework", submitted to IEEE Trans. Signal Processing, 1995.

[2] T. Adali, M. K. Sonmez, and K. Patel, "On the dynamics of the LRE algorithm: A distribution learning approach to adaptive equalization", Proceedings of IEEE ICASSP'95, Detroit, MI, pp. 929-932, April 1995.

[3] F. Beaufays and B. Widrow, "Two-layer structures for fast adaptive filtering", Proceedings of WCNN'94, pp. III-87-93, San Diego, CA, June 1994.

[4] S. Benedetto, E. Biglieri, and V. Castellani, *Digital Transmission Theory*, Printice Hall International, Englewood Cliffs, New Jersey, 1987.

[5] A. Berman and C. Mahle, "Non linear phase shift in traveling-wave tubes as applied to multiple access communications satellites", IEEE Trans. Communications, Vol. 18, No. 1, February 1970.

[6] N. J. Bershad, M. Ibnkahla, and F. Castanié, "Statistical analysis of a two-layer perceptron back propagation algorithm used for modelling non linear memoryless channels: The single neuron case", submitted to IEEE Trans. Signal Processing, 1995.

[7] S. Chen, G. Gibson, C. Cowan, and P. Grant, "Adaptive equalization of finite non-linear channels using multilayer perceptron", Signal Processing, 20, pp. 107-119, 1990.

[8] S. Chen, S. McLaughlin, and B. Mulgrew, "Complex-valued radial basis function networks, Part I: Network architecture and learning algorithms", Signal Processing, Vol. 35, No. 1, pp. 19-31, January 1994.

[9] S. Chen, S. McLaughlin, and B. Mulgrew, "Complex-valued radial basis function networks, Part II: Application to digital communication channel equalization", Signal Processing, Vol. 36, No. 2, pp. 175-188, March 1994.

[10] T. Clarke, "Generalization of neural networks to the complex plane", Int. Joint Conf. on Neural Networks, IJCNN'90, Vol. 2, pp. 435-440, 1990.

[11] Z. Faraj and F. Castanié, "Line tracking using multi-layer neural estimator", Proc. IEEE ICASSP'91, pp. 3173-3176, Canada, 1991.

[12] S. Haykin and X. Li, "Detection of signals in chaos", IEEE Proceedings, Vol. 83, No. 1, pp. 95-122, January 1995.

[13] S. Haykin, *Neural Networks: a Comprehensive Foundation*, IEEE Press, 1994.

[14] P. Hetrakul and D. Taylor, "The effects of transponder nonlinearity on binary CPSK signal transmission", IEEE Trans. Communications, Vol. 29, No. 11, pp. 546-553, May 1976.

[15] K. Hornik, M. Stinchcombe, and H. White, "Multilayer feedforward networks are universal approximators", Neural Networks, Vol. 2, pp. 359-366, 1989.

[16] M. Ibnkahla, Z. Faraj, F. Castanié, and J. C. Hoffmann, "Multi-layer adaptive filters trained with back propagation : a statistical approach", Signal Processing, Vol. 40, pp. 65-85, 1994.

[17] M. Ibnkahla and F. Castanié, "Vector neural networks for digital satellite communications", Proc. of IEEE International Conference on Communications, ICC'95, Seattle, USA, June 1995.

[18] M. Ibnkahla, F. Castanié, and N. J. Bershad, "Neural networks for modelling nonlinear memoryless communication channels", submitted to IEEE Trans. Communications, 1995.

[19] A. Kaye, D. George, and M. Eric, "Analysis and compensation of bandpass nonlinearities for communications", IEEE Trans. Communications, Vol. 20, pp. 965-972, October 1972.

[20] W. R. Kirkland and D. P. Taylor, "Neural network channel equalization", in B. Yuhas and N. Ansari Eds. *Neural Networks in Telecommunications*, pp. 143-172, Kluwer Academic Publishers, 1994.

[21] T. Kohonen, *Self-Organization and Associative Memory*, 3rd ed., Springer-Verlag, Berlin, Heidelberg, New York, 1989.

[22] H. Leung and S. Haykin, "The complex back propagation algorithm", IEEE Trans. Signal Processing, Vol. 39, No. 9, pp. 2101-2104, September 1991.

[23] R. P. Lippmann, "An introduction to computing with neural nets", IEEE ASSP Magazine, pp. 4-22, April 1987.

[24] N. Lo and H. Hafez, "Neural network channel equalization", proc. of Int. Joint Conf. on Neural Networks, IJCNN'92, Vol. 2, pp. 981-986, 1992.

[25] S. Puechmorel and M. Ibnkahla, "Operator-valued neural networks", Proceedings of WCNN'95, Washington D.C., July 1995.

[26] A. Saleh, "Frequency-independent and frequency-dependent nonlinear models of TWT amplifiers", IEEE Trans. Communications, Vol. Com-29, No. 11, November 1981.

[27] O. Shimbo and B. Pontano, "A general theory for intelligible crosstalk between frequency-division multiplexed angle-modulated carriers", IEEE Trans. Communications, Vol. 24, No. 9, pp. 999-1008, September 1976.

[28] M. Soo Kim and C. Guest, "Modification of backpropagation networks for complex-valued signal processing in frequency domain", Int. Joint Conf. on Neural Networks, IJCNN'90, Vol. 3, pp. 27-31, 1990.

[29] M. Thomas, M. Weidner, and S. Durrani, "Digital amplitude-phase keying with M-ary alphabets", IEEE Trans. Communications, Vol. 22, No. 2, pp. 168-180, February 1974.

[30] B. Widrow and R. Winter, "Neural nets for adaptive filtering and adaptive pattern recognition", IEEE Computer Magazine, pp. 25-39, March 1988.

[31] B. Widrow and M. Lehr, "30 years of adaptive neural networks: perceptron, madaline, and back propagation", Proc. IEEE, Vol. 78, No. 9, pp. 1415-1442, September 1990.

[32] B. Widrow and Bilello, "Non linear adaptive signal processing for inverse control", Proceedings of WCNN'94, pp. III-3-13, San Diego, CA, June 1994.

Transform-Domain Signal Processing in Digital Communications

H. Sari (*) and P.Y. Cochet (**)

(*) SAT, Telecommunications Division, 11 rue Watt, B.P. 370, 75626 Paris cedex 13, France.

(**) ENSTB France Telecom Bretagne, Signal & Communications, BP 832, 29285 Brest cedex, France.

1. Introduction

The Discrete Fourier Transform (DFT) is undoubtedly the most popular tool in digital signal processing, and its popularity can be attributed to the strong intuition engineers and scientists have developed about the frequency-domain representation of signals. In digital communications, the DFT forms the basis of multicarrier transmission commonly known under the names of Discrete MultiTone (DMT) and Orthogonal Frequency-Division Multiplexing (OFDM). It is also the basic tool for frequency-domain channel equalization in single-carrier systems.

Multicarrier transmission has proven to be very efficient for high-speed data transmission over the twisted-pair telephone network [1] as well as for terrestrial audio/video broadcasting [2]-[4]. The first application consists of transmitting a high data rate (typically 6 Mbit/s) from the Central Office to the subscribers, and a low data rate on the return channel. These transmission systems are referred to as Asymmetric Digital Subscriber Loops (ADSL).

Another area where multicarrier transmission has strongly developed is the terrestrial broadcasting of digital audio and TV services. Coded-OFDM (COFDM) is now a standard for Digital Audio Broadcasting (DAB) in Europe, and is on the way to becoming a standard for terrestrial digital TV broadcasting [4].

One of the basic virtues of multicarrier transmission is that it can efficiently handle long impulse response channels. In [5] and [6], it was shown that the same type of channels can also be handled by single-carrier systems with a frequency-domain equalizer while alleviating the carrier synchronization problems and reducing the system sensitivity to nonlinear distortion, which is primarily due to the transmit power amplifier.

The DFT represents a major part of the receiver complexity both in multicarrier systems and in single-carrier systems with a frequency-domain equalizer. To reduce complexity, it is appealing to replace the DFT by a real and possibly binary transform. The work reported in this paper was motivated by these considerations. In a first step, we investigated binary orthogonal transforms, and particularly, the Walsh-Hadamard Transform (WHT). It turned out that the convolution properties of these transforms do not make them suitable for the problem at hand, and therefore, we switched our investigation to number theoretic transforms which translate time-domain convolution into term-by-term product.

The paper is organized as follows: In the next section, we briefly review multicarrier transmission and its application to static as well as to nonstationary channels. Next, in Section 3, we recall frequency-domain equalization in single-carrier systems. Section 4 discusses the application of binary orthogonal transforms, and particularly that of the WHT, to both transform-domain channel equalization and transform-domain signal transmission. Application of number theoretic transforms to this problem is discussed in Section 5. Finally, we summarize our results and give our conclusions in Section 6.

2. Multicarrier Transmission

2.1. Basic principle

Multicarrier transmission is a parallel transmission technique in which the parallel data streams are transmitted at different carrier frequencies. More specifically, in a multicarrier system with an overall symbol rate of 1/T and N carriers, the carrier spacing is 1/NT Hz. The respective spectra of adjacent carriers overlap, but orthogonality is preserved in the sense that there is no interference between modulated adjacent carriers at the sampling instants.

The classical representation of multicarrier transmission consists of N parallel transmitters and receivers, but in practice, the number of carriers is large and implementation of the transmitter and receiver employs the DFT as shown in Fig. 1. The transmitted signal is of the form

$$s(t) = \text{Re}\left\{ \sum_{n=-\infty}^{+\infty} b_n f(t-nT) e^{j(\omega_0 t + \varphi)} \right\} \qquad (1)$$

where Re{.} designates the real part of complex variables, $f(t)$ is the transmit filter impulse response, T is the symbol period, ω_0 is the carrier radian frequency, and φ is the carrier phase. This expression is identical to that of single-carrier systems. The difference is that the $\{b_n\}$ sequence in single-carrier systems represents the data sequence to be transmitted, whereas it is obtained using the inverse DFT (IDFT) operator in multicarrier transmission systems.

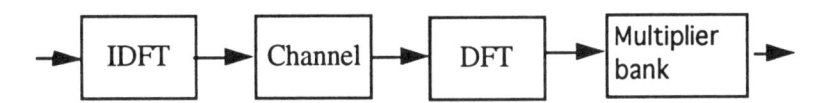

→ | IDFT | → | Channel | → | DFT | → | Multiplier bank | →

Fig. 1. Multicarrier transmission simplified block diagram

In order to distinguish successive N-point DFT blocks, it is convenient to write the index n in (1) as $n = m.N + k$ with $k = 0, 1, ..., N-1$, and m integer. The $\{b_n\}$ sequence in (1) is then given by

$$b_k(m) = \frac{1}{N} \sum_{l=0}^{N-1} a_l(m) . e^{j 2\pi kl / N}, \quad k = 0, 1, ..., N-1. \qquad (2)$$

In this 2-index representation, $a_l(m)$ represents the lth input symbol of the mth IDFT block, and $b_k(m)$ is the kth output sample of the same block. After this transformation, the N parallel output samples are converted into a serial form, lowpass filtered, and passed to a quadrature modulator which shifts the signal to center it on the center frequency $f_0 = \omega_0 / 2\pi$.

Note that the DFT is a transformation between two infinite-length periodic sequences. Since only one period of the output signal is transmitted, the spectra of adjacent carriers take a Sinc(x) = sin(x)/x form. Indeed, with 1/T denoting the overall symbol frequency and N denoting the number of points in the transform, transmitting one period of the IDFT output is equivalent to multiplying the infinite length periodic output by a rectangular window of NT seconds, which we denote Rect(t/NT). The input signal is a comb of Dirac delta pulses in the frequency domain with a separation of

1/NT Hz. Windowing the IDFT output replaces each Dirac delta by a $Sinc(\pi NfT)$ pulse centered on the same frequency. Since these pulses have regular zero crossings spaced by $1/NT\ Hz$, they do not have any interference at the center frequencies of individual pulses.

The second observation to make at this point is that only one period from the periodic input spectrum needs to be transmitted over the channel. In order to pass one period and cancel all other periods of the spectrum, the transmit filter needs to have an ideal brickwall lowpass characteristic which is physically not realizable. This difficulty is circumvented in practice by assigning a number of zero-valued carriers on both edges of the center period. This allows to use a spectral roll-off on the band edges and make the transmit filter physically realizable. The zero-valued carriers placed on band edges for spectral shaping are usually referred to as "virtual carriers".

On the receiver side, the received signal is coherently demodulated, sampled at the (overall) symbol rate 1/T, and passed to an N-point DFT operator which converts the signal back to the frequency domain. The receiver comprises a lowpass filter which limits noise and interference from adjacent channels without distorting the received signal.

The block diagram of Fig. 1 is a very simplified representation of multicarrier systems: It does not show serial-to-parallel and parallel-to-serial conversions inherent to this technique, frequency upconversion at the transmitter, and frequency downconversion at the receiver. Also not shown in this figure is the insertion of a "guard interval" between transmitted successive IDFT output blocks. This usually consists of a circular (cyclic) extension of the blocks at the IDFT operator output. The inserted circular prefix is dropped before the DFT at the receiver. Provided that the circular prefix length is larger than the channel impulse response, the linear convolution of the channel becomes identical to circular convolution (as far as the useful data is concerned) which is inherent to the discrete Fourier domain.

2.2 ADSL Applications

Multicarrier transmission turns out to be particularly attractive for static channels as in ADSL applications. The problem here is to transmit a high data rate from the central office to the subscribers using the existing twisted-pair copper lines. Those lines are band-limited and are subject to amplitude and group-delay distortions as well as to interference, and other types of impairments. For ADSL applications, a single-carrier system needs a powerful adaptive equalizer which may take the form of a nonrecursive linear equalizer or a decision-feedback equalizer. As it is discussed in the following paragraphs, multicarrier transmission offers a more elegant solution to the problem.

To begin with, suppose that the multicarrier transmission system has N carriers and that for $k = 1, 2, ..., N$, H_k designates the channel transfer function at the center frequency of the kth carrier. Further, assume that $\gamma_a(\omega)$ is the power spectrum density of the transmitted multicarrier signal, and N_0 is the spectrum density of the additive white Gaussian noise (AWGN). In the sequel, $\gamma_a(k)$ will denote the value of $\gamma_a(\omega)$ at the center frequency of the kth carrier.

In the discrete frequency-domain, the received signal spectrum density is of the form

$$\gamma_r(k) = |H_k|^2 \cdot \gamma_a(k) + N_0, \qquad (3)$$

and the signal-to-noise ratio (SNR) corresponding to the kth carrier is

$$SNR(k) = \frac{|H_k|^2 \cdot \gamma_a(k)}{N_0}. \qquad (4)$$

If the spectrum density $\gamma_a(\omega)$ is a constant (independent of the index k), *SNR(k)* will be proportional to $|H_k|^2$, and the system performance will asymptotically follow that of the most attenuated carrier.

An inspection of (4) reveals that the overall system performance will be improved if power allocation to different carriers is made such that $|H_k|^2 \cdot \gamma_a(k)$ is a constant, i.e., if

$$\gamma_a(k) = \frac{\lambda}{|H_k|} \tag{5}$$

where λ is a positive-valued real number. With respect to evenly splitting the transmitted signal power between the N carriers, this type of power allocation results in an asymptotic SNR improvement given by

$$\Gamma = \frac{N}{\sum_{k=1}^{N} H_{min}^2 / |H_k|^2} \tag{6}$$

where H_{min} designates the minimum value of $|H_k|$ over the index k.

This optimum power allocation strategy requires the knowledge of the channel transfer function which can be measured during an initial training period over twisted-pair telephone lines or all other static transmission mediums. The idea here resembles the preemphasis/deemphasis technique used in analog communications to maximize performance over channels with a nonwhite additive noise. The unequal power allocation technique is very appealing, but its implementation involves an excessive complexity, and in practice, a suboptimum technique is preferred which employs different signal constellations for different carriers.

In this technique, the power assigned to different carriers is in principle the same, but carriers subjected to a high channel attenuation employ small signal constellations, and carriers which suffer a smaller attenuation employ larger signal constellations. That is, the bit rate per carrier is a function of the channel attenuation affecting that carrier. This "water pouring" concept is known to maximize the overall data rate that can be transmitted on a given channel.

If we focus on bandwidth-efficient quadrature amplitude modulation (QAM) schemes, a given 2^M-point signal constellation has a power penalty of approximately 3 dB with respect to the 2^{M-1}-point constellation which transmits one bit less data per symbol. In other words, a 3 dB power penalty is the price paid to transmit one additional bit per symbol in QAM signal constellations. Assigning a signal constellation that is a function of the channel attenuation is equivalent to quantizing the channel SNR by steps of 3 dB, and then assigning a transmit power. This technique is used in the ANSI standard for ADSL in North America.

2.3 Application to Nonstationary Channels

The unequal power or bit allocation techniques are not applicable when the channel is not known beforehand and it varies with time. Now, we discuss this general case assuming that equal power and the same signal constellation are allocated to all carriers. To proceed further, we let h(t) designate the channel impulse response, $H(\omega)$ its transfer function, and for $k = 1, 2, .., N$, H_k the value of $H(\omega)$ at the center frequency of the *k*th carrier. The influence of multipath propagation on a given carrier is simply an attenuation and a phase rotation. But the fact that each carrier has a different attenuation and phase rotation implies that the channel needs to be equalized.

Channel equalization consists of multiplying the kth carrier by a complex coefficient

$$C_k = 1 / H_k \qquad (7)$$

so that the resulting overall transfer function has constant amplitude. This optimization is analogous to the zero-forcing (ZF) criterion [7] in time-domain equalizers which cancels ISI regardless of the noise level. To minimize the combined effect of ISI and additive noise, the equalizer coefficients can be optimized under the minimum mean square error (MMSE) criterion. This optimization yields

$$C_k = \frac{H_k{}^*}{|H_k|^2 + \sigma_n^2 / \sigma_a^2} \qquad (8)$$

where σ_n^2 is the variance of additive noise, and σ_a^2 is the variance of the transmitted data symbols. Note that the MMSE solution reduces to the ZF solution for $\sigma_n^2 = 0$.

Channel equalization in OFDM systems thus takes the form of a complex multiplier bank at the DFT output in the receiver. If the modulation used is a phase-shift keying (PSK) signal format, the channel does not need amplitude equalization, because the information is entirely carried by the signal phase. In addition, phase equalization can be made differentially, provided that differential encoding is used at the transmitter. Differential channel equalization for OFDM systems employing PSK signal sets was discussed in [5] and [6].

In [6], it was also highlighted that for fading channels, coding is compulsory in OFDM systems, and that only when it employs channel coding, frequency-domain interleaving, and weighted maximum-likelihood decoding, OFDM is more efficient than single-carrier transmission. Such a system was standardized for DAB applications in Europe, and is on the way to becoming a standard for future terrestrial digital TV broadcasting systems.

3. Frequency-Domain Equalization

The most popular approach to combating ISI is to employ an adaptive equalizer at the receiver. Channel equalization in digital communication systems is usually implemented in the time domain using a linear transversal filter. One of the problems of time-domain equalizers is that they can not easily handle transmission channels with a long impulse response.

Strictly speaking, the optimum receiver for ISI channels is a maximum-likelihood sequence estimator which has an *a priori* knowledge of the channel. The number of states of this receiver is an exponential function of the symbol alphabet size and of the channel impulse response. For long impulse response channels, the optimum receiver is, therefore, not realizable and in practical applications, a suboptimum solution is preferred. This takes the form of a nonlinear decision-feedback equalizer (DFE) or a linear transversal equalizer. In both cases, there is a limit to the channel memory size that can be reasonably handled. Indeed, equalization of long impulse response channels requires a large number of taps in the equalizer, but long equalizers are prone to convergence and stability problems, because the adaptation algorithm self-noise is proportional to the number of taps. These considerations have so far precluded the use of adaptive time-domain equalizers from some applications, such as audio and video broadcasting.

As discussed in [5] and [6], frequency-domain channel equalization [8] alleviates these problems and turns out to be a good alternative to COFDM on nonstationary multipath fading channels with a long impulse response. The virtue of frequency-domain equalization is that the number of taps (number of points in DFT) can be chosen large without compromising convergence and stability. The equalizer taps can converge separately without interacting with each other. A simplified block diagram of a frequency-domain equalizer is depicted in Fig. 2. As shown in this figure, the equalizer is comprised of a forward DFT, a complex multiplier bank, and an inverse DFT. The adaptation loop is not included for simplicity.

→ Channel → DFT → Multiplier bank → IDFT →

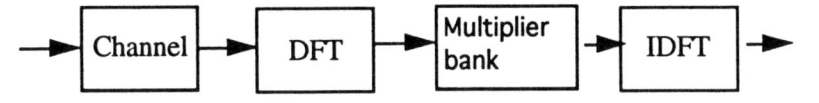

Fig. 2. Simplified block diagram of a frequency-domain equalizer

Denoting by $X(n) = (X_0(n), X_1(n), \ldots, X_{N-1}(n))^T$ the equalizer input during the nth DFT block and by $C(n) = (C_0(n), C_1(n), \ldots, C_{N-1}(n))^T$ its coefficient vector during the same block, a simple adaptation algorithm reads

$$C_k(n+1) = C_k(n) - \alpha . X_k^*(n) . E_k(n) \tag{9}$$

where α is a positive constant called the step-size parameter and the asterisk denotes complex conjugate. In this algorithm, (9) is an error signal given by

$$E_k(n) = Y_k(n) - A_k(n) \tag{10}$$

where $Y_k(n)$ designates the kth component of the equalizer output during the nth DFT block and $A_k(n)$ is the kth component of $A(n)$ which is obtained by taking the DFT of $a(n) = (a_0(n), a_1(n), \ldots, a_{N-1}(n))^T$ which is the symbol block corresponding to the nth DFT block.

In this algorithm, each coefficient minimizes the mean squared deviation of the equalizer output from the desired value at the corresponding frequency. Another adaptation technique consists of computing the equalizer coefficients using the equation

$$C_k(n) = A_k(n)/Y_k(n). \tag{11}$$

This block adaptation technique is applicable when some of the transmitted blocks consist of data symbols known from the receiver. It does not take into account additive noise and optimizes the equalizer according to the zero-forcing criterion.

Before closing this section, note that a linear frequency-domain equalizer performs circular convolution, and therefore a circular prefix must be inserted between successive N-point blocks at the transmitter so that the linear convolution of the channel becomes identical to circular convolution for the useful data symbols.

4. Walsh-Hadamard Transform

Replacement of the DFT in multicarrier transmission and frequency-domain channel equalization by a binary orthogonal transform would substantially simplify the receiver. One of the most well-known binary orthogonal transforms is the Walsh-Hadamard Transform (WHT) previously used in image processing [9]. The WHT is defined by a matrix transformation H_N given by the order-recursive relation

$$H_{2N} = \begin{pmatrix} H_N & H_N \\ H_N & -H_N \end{pmatrix} \tag{12}$$

and

$$H_2 = \begin{pmatrix} 1 & 1 \\ 1 & -1 \end{pmatrix} \qquad (13)$$

The WHT of a vector $x = (x_0, x_1, \ldots, x_{N-1})^T$ can also be written

$$X_n = \sum_{k=0}^{N-1} x_k (-1)^{\langle k,n \rangle} \qquad (14)$$

where $\langle .,. \rangle$ denotes the dot product in binary representation.

An equalizer structure in the WHT domain is depicted in Fig. 3. The equalizer comprises two forward WHTs, four multiplier banks and two inverse WHTs (IWHTs). Each set of forward WHT, multiplier bank, and inverse WHT in cascade attempts to synthesize a time-domain transversal filter. The block diagram of Fig. 3. clearly resembles the real filter representation of a complex transversal time-domain equalizer.

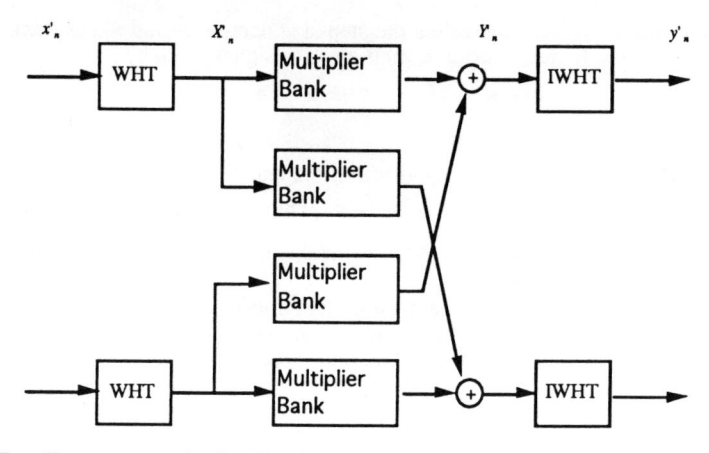

Fig. 3. Equalizer structure in the Walsh-Hadamard Transform domain

Focusing on the upper arm of the equalizer in Fig. 3., assume that $C = (C_0, C_1, \ldots, C_{N-1})^T$ denotes the coefficient vector in the multiplier bank, and that $c = (c_0, c_1, \ldots, c_{N-1})^T$ is the IWHT of C. Time-domain representation of the product in the WHT domain is [9]

$$y_n = \sum_{j=0}^{N-1} c_j x_{n \oplus j} \qquad (15)$$

where $\oplus$ denotes modulo-2 addition (or subtraction). This is known as dyadic convolution.

The problem is that the channel performs linear convolution, and using a circular prefix, this can be made equivalent to circular convolution which is the operation carried out by a frequency-domain equalizer based on the DFT. Unfortunately, no techniques are known which transform the linear convolution of the channel into dyadic convolution so that the WHT can be used for channel equalization. The WHT did not turn out to be appropriate for channel equalization (nor for multicarrier-like transmission where an IWHT is performed at transmitter and a WHT at receiver), and therefore our investigation shifted to number theoretic transforms (NTTs) which employ real, but nonbinary arithmetic.

5. Number Theoretic Transforms

We now examine the use of NTTs for transform-domain signal transmission or channel equalization. Let us denote by T the $n \times n$ matrix corresponding to an N-point transform. The transform X of an N-point vector x is written as

$$X = T(x) \qquad (16)$$

When the transform of an N-point vector obtained by circularly convolving two vectors x and y is the term-by-term product of their respective transforms X and Y, then transform T is said to verify the Circular Convolution Property (CCP). This can be written as

$$T(x * y) = T(x) \cdot T(y) \qquad (17)$$

where * stands for circular convolution.

As discussed in the previous section, the WHT does not have the CCP, and this has lead us to investigate a class of NTTs. The latter transforms use integer numbers, and all operations are carried out using modulo arithmetic. Note that arithmetic modulo an integer M does not make any distinction between two integers whose difference is a multiple of M. As an example, since the difference between 64 and 1 is a multiple of 9, 64 is equal to 1 in modulo-9 arithmetic. We write this as

$$2^6 = 64 = 1 \quad \langle 9 \rangle \qquad (18)$$

Where $\langle 9 \rangle$ stands for modulo 9. Consequently, modulo-M arithmetic only manipulates M elements, usually denoted

$$\overset{\bullet}{0}, \overset{\bullet}{1}, .., (\overset{\bullet}{M} - 1) \qquad (19)$$

but to simplify notation, the dots are often omitted. These M elements form a finite ring denoted Z_M.

Another important feature of an element of the finite ring Z_M is its order. The order of an element imposes the value of N, which is the number of points in the transform. The order of an element α, modulo M, is the smallest positive integer such that

$$\alpha^N = 1 \quad \langle M \rangle \qquad (20)$$

For example, according to (18), the order of the element 2, modulo 9, is 6.

The characteristic parameters of any NTT are thus M, α and N. M is the integer according to which all operations are computed in the modulo arithmetic, α is an integer that is often taken equal to 2 (or to a power of 2), and N is the order of α modulo M. As in the DFT, it is desirable to select a transform size N equal to a power of 2, so that fast algorithms can be used. As NTTs use integers exclusively, a scaling of the signals handled is necessary, but this does not involve any additional complexity, since signals are processed using digital circuits and interpreted as integer numbers anyway. The parameter M represents the dynamic range of the N-point transform. When scaling the signals, it is important to quantize them over the entire range of the elements of the ring Z_M. The lowest signal level is then mapped to an integer close to -M/2 and the maximum signal amplitude signal is mapped onto an integer close to +M/2.

It is easily verified that any NTT with a matrix T of the form

$$T = [\alpha^{nk}] \quad \langle M \rangle ; n = 0,..,N-1 ; k = 0,..,N-1 \qquad (21)$$

has the CCP given by (17). The DFT corresponds to $\alpha = W = e^{-i\frac{2\pi}{N}}$ and modulo arithmetic is automatically achieved because of the periodicity of the unit circle. Despite this, the DFT does not handle integers and is therefore not an NTT. Construction of NTTs and their properties are thoroughly exposed in [10].

In this work, we focused on two types of NTTs, namely the Fermat number theoretic transform and the Mersenne number theoretic transform, and we used them for transform-domain channel equalization. Fermat numbers are defined as

$$F_t = 2^{2^t} + 1 \tag{22}$$

where t is any integer.

Table 1 The first Fermat numbers F_t and the associated order N of $\alpha = 2$

t	0	1	2	3	4	5
F_t	3	5	17	257	665537	$\approx 4.10^9$
N	2	2	8	16	32	64
F_t dB	5	7	12	24	58	96

Fermat NTTs are suitable for data sequences with a length equal to a power of 2, but the integer M quickly becomes enormous. In practice, the dynamics of signals do not exceed a few dozen of dB, and this limits the transform size needed. Fig. 4. shows the frequency-domain and Fermat transform domain representations of a frequency-selective channel whose impulse response is given in Fig. 6.4.7.a of [7].

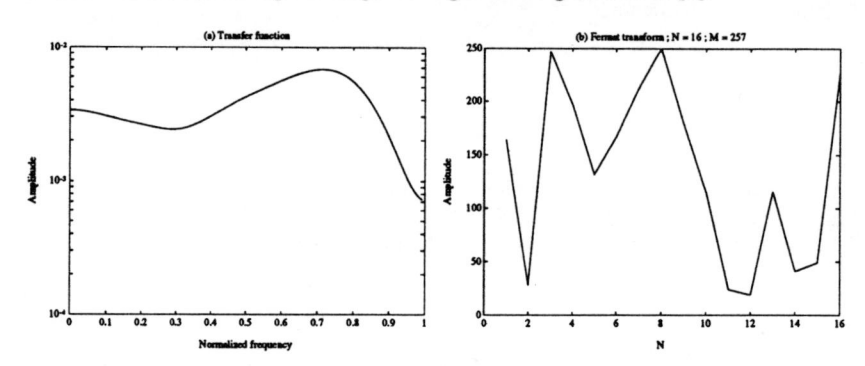

Fig. 4. Frequency domain (a) and Fermat transform domain (b) representations of the channel taken from [7].

Next, the Mersenne numbers are defined as

$$M_p = 2^p - 1 \tag{23}$$

where p is a prime number. In NTTs based on Mersenne numbers, the size of the sequences are therefore prime numbers, and this does not allow to use fast algorithms for computing the transform. Table 5.2 shows the first six Mersenne numbers as well as the associated order N of α and the dynamics M_p expressed in dB.

Table 2 The first Mersenne numbers M_p and the associated order N of $\alpha = 2$

p	2	3	5	7	11	13
M_p	3	7	131	127	2047	8191
N	2	3	5	7	11	13
M_p dB	4,8	8,4	14,9	21	33	39

This table shows that in the Mersenne transform, only a short sequence length (13 or 17) allows to handle a wide signal dynamics (39 dB or more).
 A simplified block diagram of an equalizer in the NTT domain is the same as that in Fig. 3. except that the WHT must be replaced by an NTT such as the Fermat transform or the Mersenne transform. Fig. 5. shows a typical performance curve corresponding to an equalizer in the Fermat transform domain.

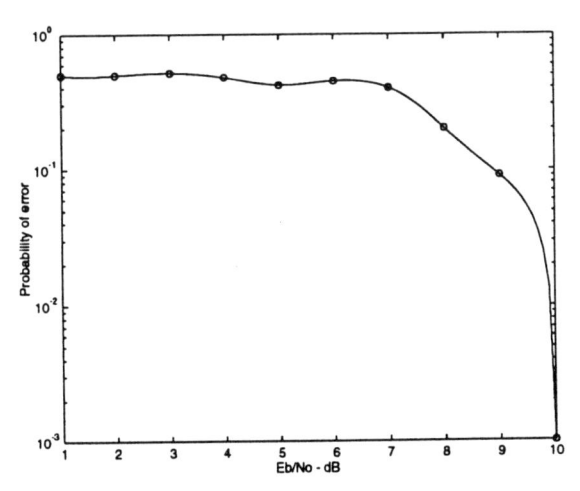

Fig. 5. Probability of error after equalization in the Fermat transform domain ($F_t = 257$)

This simulation corresponds to the channel whose transfer function is displayed in Fig. 4. This channel, taken from [7], corresponds to a telephone line with mild amplitude distortion. The modulation used is a quaternary phase shift keying (QPSK) signal format, and the F_t parameter used in the Fermat transform is $F_t = 257$. Note that for small E_b / N_0 values, the error probability is rather high, but the decrease is very steep for E_b / N_0 values larger than 7 dB.

6. Conclusions

In this paper, we first gave an overview of multicarrier transmission and highlighted its basic features for static as well as for nonstationary channels. We next presented frequency-domain channel equalization for single-carrier systems and emphasized its potential to cope with long impulse response channels.

Both of these techniques are based on the DFT. Multicarrier systems make use of one transform at transmitter and one at receiver, while frequency-domain equalization employs two transforms at the receiver. To reduce the receiver complexity, we have investigated the use of binary orthogonal transforms such as the WHT and of number theoretic transforms such as the Fermat transform and the Mersenne transform. It turned out that the WHT is not suitable, due to the fact that it does not translate circular convolution in the time domain into term-by-term product in the transform domain. In contrast, the Fermat transform and the Mersenne transform do possess this property and deserve further investigations for use in channel equalization.

REFERENCES

[1] J.S. Chow, J.-C. Tu, and J.M. Cioffi, "A discrete multitone transceiver system for HDSL applications," IEEE J. Select. Areas Commun., vol.9, pp.895-908, August 1991.

[2] M. Alard and R. Lassalle, "Principles of modulation and channel coding for digital broadcasting for mobile receivers," EBU Review, n° 224, pp. 3-25, August 1987.

[3] B. Le Floch, R. Halbert-Lassalle, and D. Castelain, "Digital sound broadcasting to mobile receivers," IEEE Trans. consumer Electronics, vol. 35, n°.3, August 1989.

[4] U. Reimers, "European perspectives in digital television broadcasting - Conclusions of the working group in digital television," EBU Technical Review, n°256, pp. 3-8, 1993.

[5] H. Sari, G. Karam, and I. Jeanclaude, "Channel equalization and carrier synchronization in OFDM systems," in Audio and Video Broadcasting Systems and Techniques, R. de Gaudenzi and M. Luise : Editors, pp. 191-202, Elsevier, Amsterdam, 1994.

[6] H. Sari, G. Karam, and I. Jeanclaude, "Transmission techniques for digital terrestrial TV broadcasting," IEEE Communications Magazine, vol.33, n°.2, pp.100-109, February 1995.

[7] J.G.Proakis , "Digital Communications," McGraw-Hill Book Company, New-York, 1989, 2nd ed.

[8] T. Walzman and M. Schwartz, "Automatic equalization using the discrete Fourier domain," IEEE Trans. Information Theory, vol. IT-19, pp. 59-68, January 1973.

[9] N. Ahmed and K.R. Rao, "Orthogonal transforms for digital signal processing," Springer-Verlag, Berlin - Heidelberg, 1975.

[10] D.F. Elliott and K.R. Rao, "Fast Transforms, Algorithms, Analyses, Applications," Academic Press inc, Orlando, 1982.

A Fuzzy-Rule-Based Phase Estimator Suited for Digital Implementation

Flavio Daffara

Laboratoires d'Electronique Philips S.A.S.
22 avenue Descartes, 94453 Limeil-Brévannes Cedex, France
e-mail: daffara@lep-philips.fr

This paper shows how fuzzy logic control techniques can be used to improve the behaviour of phase estimation in digital receivers. In particular, we have derived a fuzzy-rule-based phase estimator which permits a decrease in the steady-state phase variance to be achieved without increasing the convergence time.

1. INTRODUCTION

Digital demodulation has received an increasing interest during the last years because of its relatively low cost of implementation and good performance, which make it very attractive compared to the analog technique [1][2]. In addition, the digital technique can be employed in a larger and larger number of applications due to the continuous improvements in integrated circuit design and Digital Signal Processor (DSP) technologies.

One of the fundamental functions of a receiver is the synchronization, which consists in the recovery of the signal frequency, the clock timing and the carrier phase from the received signal and the use of them to achieve signal demodulation and data detection. In modern digital communications, efficient synchronization techniques are required to fully exploit the potentiality of modulation/coding schemes and not to cause a bottleneck in the overall system performance.

Synchronization in a digital demodulator leads to techniques which are different from the ones used in analog implementation. Such techniques have been thoroughly detailed in [3] and [4]. This paper deals with digital carrier phase recovery using a phase estimation approach, which is particularly indicated for burst mode transmissions because of its fast convergence properties. The aim of the paper is to exploit fuzzy logic control techniques in order to improve the performance of currently known phase estimators.

The paper is organized as follows: Section 2 describes the digital receiver structure we have considered. Section 3 gives an overview of how fuzzy control can be applied to phase estimation and more in general to synchronization. In Section 4 we derive a fuzzy-rule-based phase estimator and give its performance in terms of steady-state phase variance and acquisition time. Finally, Section 5 is for our conclusions.

2. SYSTEM DESCRIPTION

A typical block diagram of a fully-digital demodulator is shown on Figure 1. After

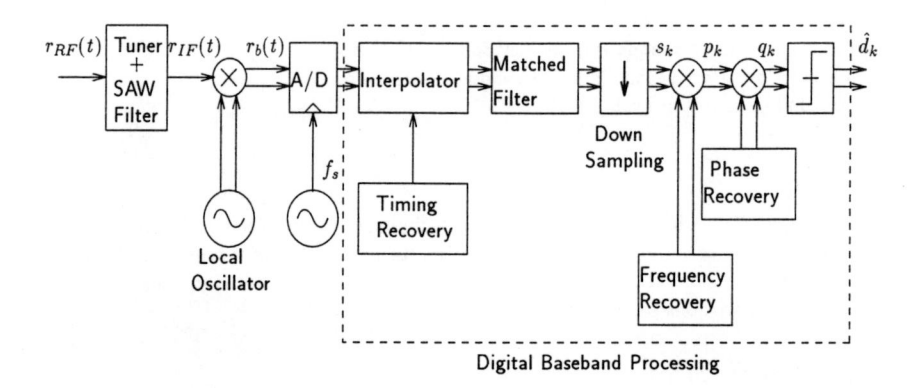

Figure 1. Fully-digital demodulator.

Radio Frequency (RF) to Intermediate Frequency (IF) conversion, the $r_{IF}(t)$ signal is down-converted to baseband by a free-running local oscillator. The baseband signal $r_b(t)$ is then sampled using a fixed and independent clock running at frequency $f_s = \frac{1}{T_s}$.

Using complex envelope notation, the samples of the received baseband signal can be written as

$$r_n = r_b(nT_s) = \sum_i d_i g(nT_s - iT - \epsilon T) e^{j(2\pi \Delta f nT_s + \phi)} + w(nT_s) \tag{1}$$

where:

- d_i is the sequence of the equiprobable and statistically independent random MxM-QAM symbols of variance σ_d^2, whose in-phase (I) and quadrature (Q) parts belong to the set $\{-(M-1)d, \ldots, -d, d, \ldots, (M-1)d\}$,

- $g(t)$ is the impulse response of the emission shaping filter convoluted with the channel impulse response,

- T_s (resp. T) is the sampling (resp. symbol) period,

- ϵ is the normalized error in the sampling instant,

- Δf and ϕ are the frequency and the phase error respectively,

- $w(nT_s) = w_I(nT_s) + jw_Q(nT_s)$ is a zero-meaned complex Gaussian random process whose real and imaginary components are statistically independent and have a variance of $\sigma_{w_I}^2 = \sigma_{w_Q}^2 = \frac{\sigma_d^2}{2\text{SNR}} \frac{T}{T_s}$ where SNR is the symbol-to-noise ratio.

Usually, sampling time is asynchronous with the symbol rate ($\frac{1}{T}$) and timing correction is accomplished by means of interpolation on the samples [3].

In the sequel we will assume perfect timing recovery so that the signal after downsampling can be expressed by

$$s_k = d_k h(0) e^{j(2\pi \Delta f kT + \phi)} + \underbrace{\sum_{i \neq k} d_i h((k-i)T) e^{j(2\pi \Delta f (k-i)T + \phi)}}_{\text{ISI}} + w_k \qquad (2)$$

where $h(t)$ is the impulse response of $g(t)$ convoluted with the impulse response of the matched filter. The frequency recovery unit corrects the frequency offset Δf present after baseband conversion, while the phase recovery unit is responsible for coping with the residual phase error.

3. FUZZY LOGIC CONTROL TECHNIQUES APPLIED TO PHASE ESTIMATION

The aim of this paper is to derive a fuzzy-rule-based Phase Estimator (PE) controller able to improve the performance of the phase recovery block. Figure 2 shows how such a controller should be interfaced with a PE. A set of rules describing the *desired* behaviour

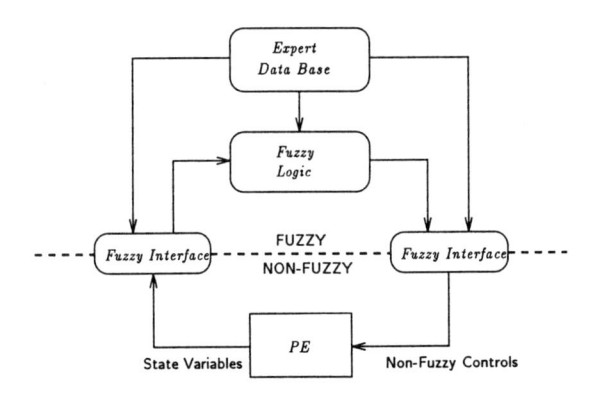

Figure 2. Fuzzy-rule-based phase estimator configuration.

of the PE constitutes the *Expert Data Base*. Such rules should be expressed in the form: if *condition* then *action*, where the *condition* and *action* parts are expressed in natural language and therefore contain subjective and not sharply defined classes of objects or some amount of uncertainty or inexactitude. For example the PE controller could make use of the information: "the PE has *almost converged*". Within fuzzy set theory [5] it is possible to give a meaning to such a sentence and a PE can have converged with a continuum of grades of membership (and not only converged or not converged).

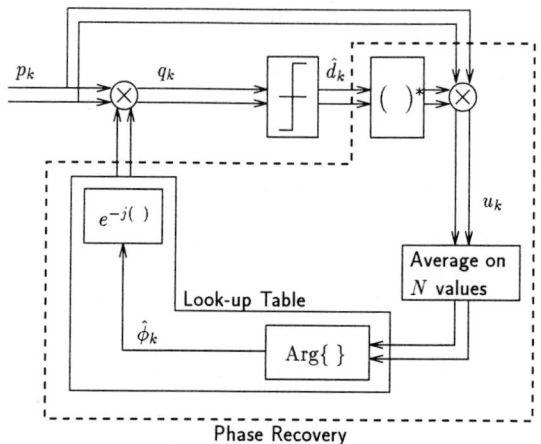

Figure 3. DD Arctan phase estimator.

Once the rules have been set, the PE state variables related to the *condition* part of the rules and the non-fuzzy controls dependent on the *action* part of the rules should be identified. Then the state variables (resp. non-fuzzy controls) should be mappped to ad-hoc fuzzy sets [5] describing the *conditions* (resp. *actions*). Every possible value of the state variables or controls should belong to at least one of the fuzzy sets. It is then possible to combine all the different rules (depending on the PE state variables) using a classical fuzzy logic method (Mamdani, Larsen, interpolation, etc. [6]) and obtain for each non-fuzzy control a value which is the most representative of the combination of all *actions*. To sum up, a fuzzy logic controller has to perform the three following steps:

1. *Fuzzification:* it is used to convert the input state variables to fuzzy values. Fuzzification determines membership degrees (from 0 to 1) of input state variables in fuzzy sets, which are denoted by input membership functions. The degree of membership (also called the rule's activation level) applies to the rule's output membership function to determine the degree to which the rule's specified output action occurs. The output membership function is most often determined by clipping or by scaling.

2. *Inferencing:* it determines the combined effect of all rules by combining different output fuzzy sets to create a resultant fuzzy set.

3. *Defuzzification:* it converts combined output membership functions to a single real value.

In the next section we shall apply the above mentioned method to improve the performance of the phase estimation block.

4. FUZZY-RULE-BASED PHASE ESTIMATOR

Figure 3 shows a phase estimator which is commonly known in literature as the Decision Directed (DD) Arctan PE [3].

The phase estimate $\hat{\phi}_k$ is given by:

$$\hat{\phi}_k = \text{Arg}\left\{\sum_{i=k-N}^{k-1} u_i\right\}$$

where Arg{} is the function giving the phase of a complex number, N is the length of the observation window and u_i is the complex signal given by

$$u_i = p_i \hat{d}_i^*$$

where p_i is the signal before the phase correction and $\hat{d}_i$ is the decision taken at time iT.

Such a phase estimator gives very good performance as far as the decision $\hat{d}_k$ is correct, i.e. as far as the SNR is high. However, if the number of errors in the decisions increases (at low SNR) the PE performance worsens.

The idea is to try to take into account only the decisions which are more likely to be correct and to discard the others. That is to say we try to use the following empiric rules:

- If the decision is correct then the PE should correct in the usual way, which gives good performance.

- If the decision does not seem to be correct then it is better not to correct at all (or at least less then usually) rather than correcting in the wrong direction.

In this case the *condition* part of the empiric rule can be identified as the "correctness of the decision" and the *action* part as the "strength of the u_k signal". We arbitrarily chose to relate the "correctness of the decision" to two state variables x_1 and x_2, which are the distances of the phase-corrected point q_k from the closest I and Q decision thresholds respectively. We also chose to relate the "strength of the u_k signal" to one control γ_k, which is a gain factor ranging from 0 to 1. The new PE is shown on Figure 4 where $u'_k = \gamma_k u_k$ is the new signal feeding the averaging circuit.

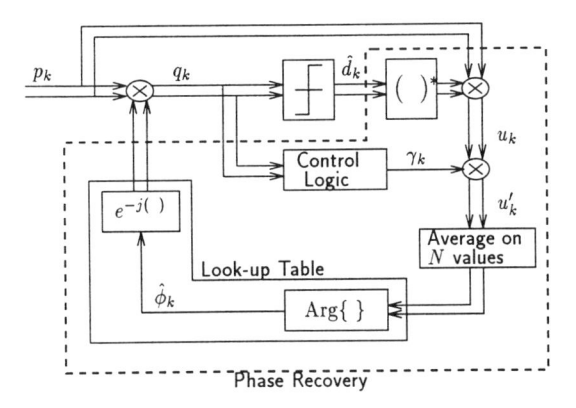

Figure 4. Fuzzy-rule-based DD Arctan phase estimator.

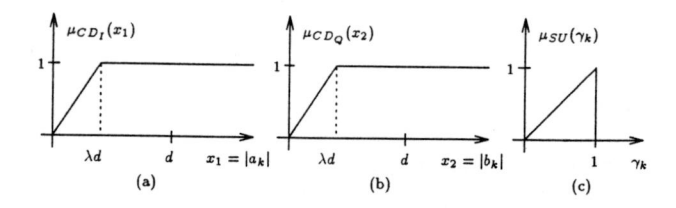

Figure 5. Membership functions of the fuzzy sets corresponding to the "correcteness of the decision" (a) and (b) and to the "strength of the u_k signal" (c) in the case of a QPSK constellation.

The membership functions [5] of the fuzzy sets corresponding to the "correctness of the decision" ($\mu_{CD_I}(x_1)$ and $\mu_{CD_Q}(x_2)$) and to the "strength of the u_k signal" ($\mu_{SU}(\gamma_k)$) are shown on Figure 5 as a function of x_1, x_2 and γ_k in the case of a QPSK constellation. The membership functions $\mu_{CD_I}(x_1)$ and $\mu_{CD_Q}(x_2)$ depend also on a parameter λ (see Figure 5), which has to be chosen carefully in order to optimize the fuzzy-controller.
Such a parameter defines some uncertainty regions in the complex plane as shown on Figure 6.

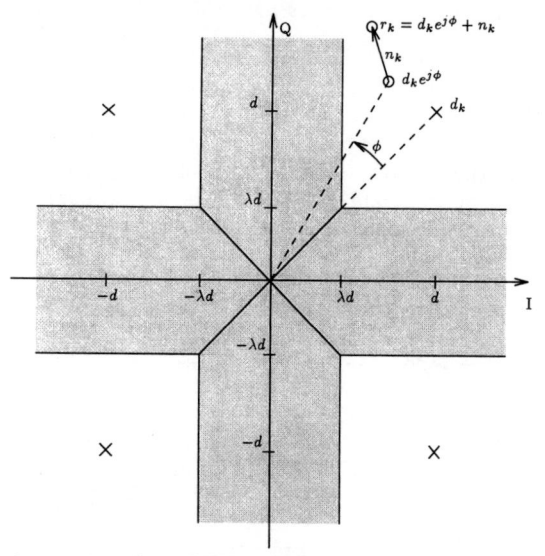

Figure 6. Uncertainty regions for a QPSK constellation.

Since the membership functions characterising the state variables are monothonic, we used the interpolation method [6] to derive the non-fuzzy control γ_k, i.e. $\gamma_k = \mu_{SU}^{-1}\{\min[\mu_{CD_I}(x_1), \mu_{CD_Q}(x_2)]\}$.

The performance of the so obtained fuzzy-controlled PE is given on Figure 7 (resp. Figure 8) for different values of SNR and λ in the case of a QPSK constellation with $N = 64$ (resp. $N = 256$).

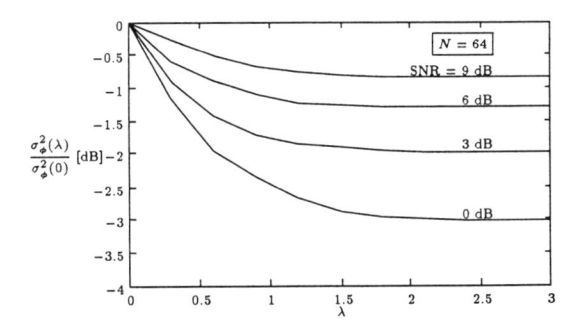

Figure 7. Gain in the phase estimate variance as a function of λ in the case of a QPSK constellation for $N = 64$.

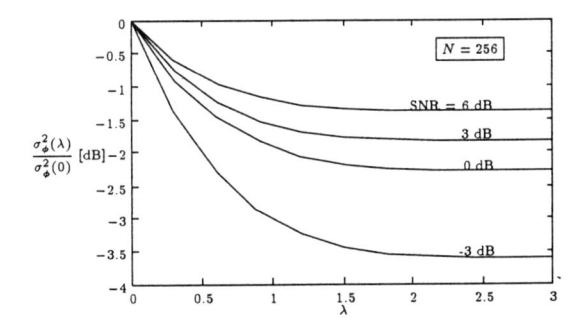

Figure 8. Gain in the phase estimate variance as a function of λ in the case of a QPSK constellation for $N = 256$.

The values for $\lambda = 0$ correspond to the classical DD Arctan PE and we can see that the fuzzy-controlled PE outperforms the classical one. Predictibly, the improvement is large at low SNR since the number of erroneous decisions is high, while it is small at high SNR because the decisions are almost completely correct. Figure 9 shows the acquisition time in the case $N = 64$ and SNR=3 dB. We can see that the fuzzy-controlled PE presents the same acquisition time as the classical one while having a lower phase variance. Conversely, we could decrease N for the fuzzy-rule-based PE (which would result in a faster convergence time) while keeping the same performance in terms of phase variance.

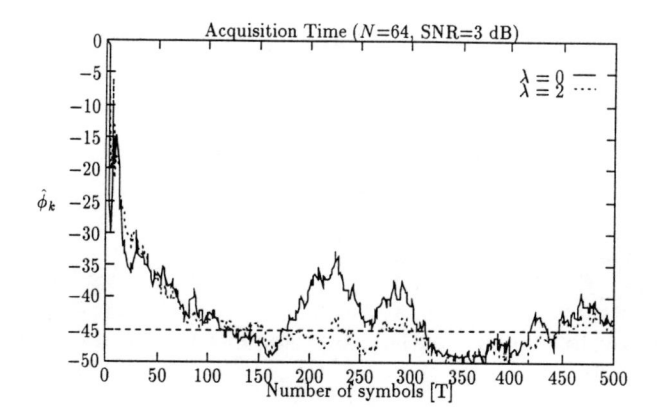

Figure 9. Convergence time in the case of a QPSK constellation with N=64 and SNR=3 dB. The initial phase error is -45 degrees.

5. CONCLUSIONS

In this paper we have shown the potentiality of using fuzzy logic control techniques to improve the behaviour of the phase estimation function in digital demodulators. In particular, we have derived a specific and simple example of a fuzzy-rule-based phase estimator. Even in this simple case (only 1 rule) an improvement in the steady-state phase variance can be achieved without increasing the convergence time. Modifications in the shape of the membership functions or the use of more sophisticated rules taking into account other conditions (SNR level, speed of convergence, value of N, etc...) are expected to give further improvement. Future work should also try to extend the method we have presented to other receiver functions such as frequency estimation, timing recovery and equalization.

REFERENCES

1. L.N. Lee, A. Shenoy and M.K. Eng, *Digital Signal Processor-Based Programmable BPSK/QPSK/offset-QPSK Modems*, Comsat Technical Review, Vol. 19, n. 2, pp. 195-234, Fall 1989.

2. J.J. Poklemba and F.R. Faris, *A Digitally Implemented Modem: Theory and Emulation Results*, Comsat Technical Review, Vol. 22, n. 1, pp. 149-195, Spring 1992.

3. F.M. Gardner, *Demodulator Reference Recovery Techniques Suited For Digital Implementation*, European Space Agency, ESTEC Contract n. 6847/86/NL/DG, August 1988.

4. T. Jesupret, M. Moeneclaey and G. Ascheid, *Digital Demodulator Synchronization - Performance Analysis*, European Space Agency, ESTEC Contract n. 8437/89/NL/RE, June 1991.

5. A. Kandel and W.J. Byatt, *Fuzzy Sets, Fuzzy Algebra, and Fuzzy Statics*, Proc. of the IEEE, Vol. 66, No. 12, December 1978, pp. 1619-1639.

6. D. Dubois and H. Prade, *Fuzzy Sets and Systems, Theory and Applicatons*, Academic Press, 1980.

Part 6

Recent Applications of DSP Techniques

Analog Envelope Constrained Filters for Channel Equalization

A. Cantoni, B. Vo, Z. Zang and K. L. Teo
Australian Telecommunications Research Institute,
Curtin University of Technology, Kent Street Bentley, W.A. 6102, Australia.

Abstract -The envelope constrained (EC) filtering problem is posed as the minimization of the noise gain of the filter, whilst satisfying the constraint that the filter's response to a specified input lies within a prescribed envelope. Two filter structures are considered. One is based on DSP and consists of an A/D converter, an FIR filter and an interpolator. The other is an all analog filter comprised of Laguerre networks. In both cases, it is shown that the EC filtering problem can be cast as a functional inequality constrained optimisation problem. A numerical method for solving the problem is developed. An example involving the application of EC filter to channel equalization is presented.

1. Introduction

Consider a time invariant filter used to process a given input signal $s(t)$ which is corrupted by white additive noise as shown in Fig. 1.1.(a). The noiseless output wave form $\psi(t)$ is required to fit into a *pulse shape envelope* defined by the upper and lower boundaries $\varepsilon^+(t)$ and $\varepsilon^-(t)$ as shown in Fig. 1.1.(b).

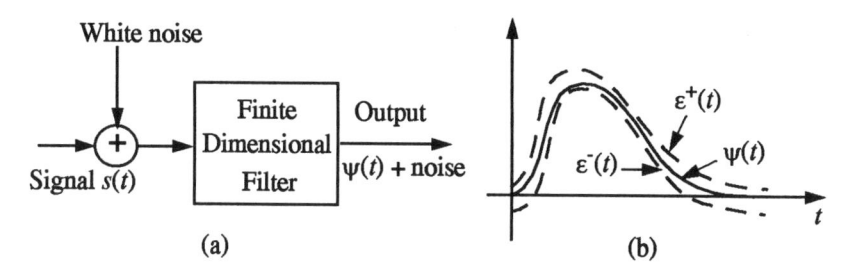

Fig. 1.1. Envelope Constrained filtering problem. (a) Block diagram. (b) Pulse shape envelope.

The *optimal envelope-constrained filter* is defined as that filter which minimizes the output noise power while satisfying the pulse shape constraints. To avoid the trivial solution $u(t) = 0$, it is assumed that there exists at least one point in the output mask at which the upper and lower boundaries have the same sign.

Envelope constrained (EC) filters, as given above, are more directly relevant than the commonly known least square filters in a variety of signal processing fields and in particular in communication channel equalization. In standards, the performance of digital links is often specified in terms of a mask applied to the received signal [1], [2]. The *envelope-constrained filter* design problem is directly applicable and the signal $s(t)$ would correspond to the test signal specified in the standard. In the literature, the problem has received little attention and has considered primarily a discretized version of the problem [3]-[6], except for [7] which considers piece-wise linear masks.

In this paper, the *envelope-constrained filter* design problem for two finite dimensional filter structures is formulated. One of the filter structures is realized by DSP techniques, while the other is realized by analog means. An efficient iterative algorithm for determining the parameters of the optimal filter is presented. We also present the mathematical foundation and numerical results for a digital transmission example to illustrate the performance of the proposed algorithm.

A note on notation; throughout this paper, ordering relations applies component-wise for vectors, and for any $a \in R^n$, $|a| = [|a_0|, |a_1|, ..., |a_{n-1}|]$ unless otherwise stated. The Hilbert space of all real-valued Lebesgue measurable and square integrable functions on the semi-infinite interval $[0, \infty)$ is denoted by $L^2[0, \infty)$.

2. DSP Based Filters

Consider the hybrid filter structure consisting of an A/D converter, an FIR filter and an interpolator as illustrated in Fig. 2.1.

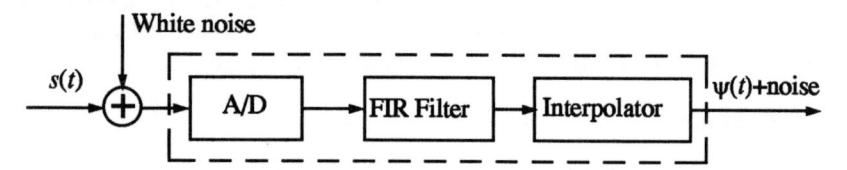

Fig. 2.1. Hybrid filter.

The analog input is sampled and processed by digital means and then interpolated, thus the filter has analog input and output. The resulting EC filtering problem is a finite dimensional optimization problem because the discrete-time filter component is an FIR filter. The interpolator could take on a number of different forms, for example a zero order hold which would correspond to a simple D/A converter. This paper considers linear interpolation since it offers better signal reconstructions.

Let $u = [u_0, ..., u_{n-1}]^T \in R^n$ be the vector of filter coefficients of the FIR filter. Suppose that without input noise, the A/D process produces a vector $s = [s_0, ..., s_{m-1}]^T \in R^m$, by sampling the continuous time input signal $s(t)$, at a period of T over the interval $[0, (m-1)T]$. Then the output of the FIR filter is $\psi = [\psi_0, ..., \psi_{N-1}]^T = Su \in R^N$, where $N = m + n - 1$ and

$$S = \begin{bmatrix} s_0 & 0 & \cdots & 0 \\ \vdots & s_0 & & \vdots \\ s_{m-1} & \vdots & \ddots & 0 \\ 0 & s_{m-1} & & s_0 \\ \vdots & & \ddots & \vdots \\ 0 & \cdots & 0 & s_{m-1} \end{bmatrix} \quad (1)$$

The output ψ of the FIR filter is then fed to an interpolator to produce the noiseless analog output $\psi(u, t)$ (this notation makes the dependence on the filter coefficients u more explicit).

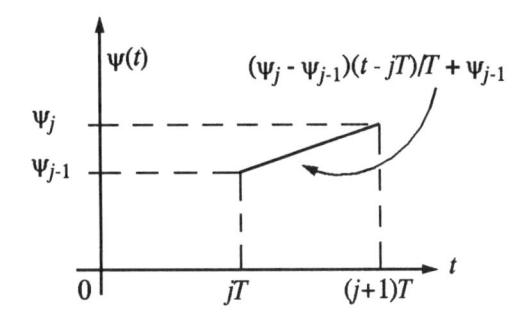

Fig. 2.2. Linear interpolation of two adjacent output points

Linear interpolation essentially fits a line between two adjacent points as depicted in Fig. 2.2. Note that a delay of T unit is incurred.

$$\psi(u, t) = \sum_{j=1}^{N-1} \left(\frac{(\psi_j - \psi_{j-1})}{T} (t - jT) + \psi_{j-1} \right) p(t - jT)$$

where $p(t)$ is a rectangular window with length T, given by

$$p(t) = \begin{cases} 1, t \in [0, T] \\ 0, \text{otherwise} \end{cases}$$

The envelope constraints required that the noiseless output $\psi(u, t)$ lies within the upper and lower boundaries $\varepsilon^+(t), \varepsilon^-(t)$ of the output mask for a specified time interval $[T, NT]$. The EC filtering problem is to determine an FIR filter u such that the output noise power is minimized.

It has been shown in [8] that:

(a) the noiseless output $\psi(u, t)$ can be written in the following form

$$\psi(u, t) = y^T(t) u$$

where

$$y^T(t) = \frac{1}{T}[(t-T)p(t-T), \ldots, (t-(N-1)T)p(t-(N-1)T)]\Xi S$$
$$+ [p(t-T), \ldots, p(t-(N-1)T)]\Gamma S$$

$$\Xi = \begin{bmatrix} -1 & 1 & 0 & \ldots & 0 \\ 0 & -1 & 1 & & \vdots \\ \vdots & & \ddots & \ddots & 0 \\ 0 & \ldots & 0 & -1 & 1 \end{bmatrix}_{(N-1) \times N}, \Gamma = \begin{bmatrix} 1 & 0 & 0 & \ldots & 0 \\ 0 & 1 & 0 & & \vdots \\ \vdots & & \ddots & \ddots & 0 \\ 0 & \ldots & 0 & 1 & 0 \end{bmatrix}_{(N-1) \times N},$$

(b) the output noise power is given by

$$u^T H u$$

where

$$H = \frac{\sigma^2}{6} \begin{bmatrix} 4 & 1 & 0 & \ldots & 0 \\ 1 & 4 & & & \vdots \\ 0 & & \ddots & & 0 \\ \vdots & & & 4 & 1 \\ 0 & \ldots & 0 & 1 & 4 \end{bmatrix}_{n \times n}$$

Thus the EC filtering problem for this filter structure can be posed as

$$\min u^T H u \text{ subject to } \varepsilon^-(t) \leq y^T(t) u \leq \varepsilon^+(t), \forall t \in [T, NT]$$

This is a finite dimensional optimization problem with a continuum of constraints. It is clear that the above problem has a convex cost and convex constraint set, hence there is a unique optimal solution.

3. Laguerre Based Filters

The use of Laguerre networks in filter realization problems have been well documented [9], [10]. The Laguerre methods provide concrete and novel techniques for approximating infinite dimensional systems by finite dimensional systems [9]. In this Section, we formulate the Envelope Constrained Filtering problem for the filter structure is realized using an analog Laguerre network as shown in Fig. 3.1.

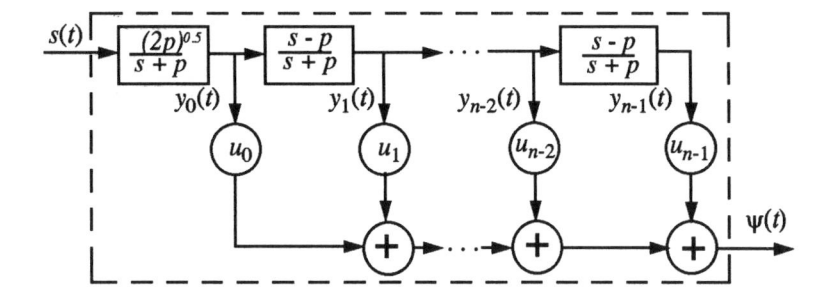

Fig. 3.1. A block diagram of a continuous-time Laguerre network.

The transfer function $U(s)$ for the Laguerre network above is easily shown to be

$$U(s) = \sum_{k=0}^{n-1} u_k \Theta_k^p(s),$$

where

$$\Theta_k^p(s) = \frac{\sqrt{2p}}{s+p}\left(\frac{s-p}{s+p}\right)^k, \quad p > 0, \quad k = 0, 1, 2, \ldots$$

The corresponding impulse response $u(t)$ is thus

$$u(t) = \sum_{k=0}^{n-1} u_k \theta_k(t)$$

where $\theta_k^p(t)$ denotes the inverse Laplace transform of $\Theta_k^p(s)$ and can be expressed in terms of the Laguerre polynomials (see [9], [10]) as

$$\theta_k^p(t) = \sqrt{2p} e^{-pt} L_k(2pt), \quad L_k(t) = \frac{e^t}{k!}\frac{d^k}{dt^k}(t^k e^{-t})$$

Let

$$\boldsymbol{u} = [u_0, u_1, \ldots, u_{n-1}]^T \in R^n$$

$$y_k(t) = \int_0^\infty s(\tau) \theta_k^p(t-\tau) d\tau, \quad k = 0, \ldots, n-1,$$

$$\boldsymbol{y}(t) = [y_0(t), \ldots, y_{n-1}(t)]^T \in R^n$$

Then, the noiseless output $\psi(t)$ of the Laguerre network invoked by the input signal $s(t)$ is given by (see also [11])

$$\psi(t) = \boldsymbol{y}^T(t)\boldsymbol{u}.$$

The envelope constraints requires the output $\psi(t)$ to lie within the upper and lower boundaries $\varepsilon^+(t)$, $\varepsilon^-(t)$ in the time interval $[0, \alpha]$.

It can be easily shown that for white noise, the output noise power is proportional to square of the $L^2[0, \infty)$ norm of the filter. Thus the gain to input noise power of the filter can be given by

$$\|u\|^2 = \int_0^\infty |u(t)|^2 dt = \sum_{j=0}^{n-1}\sum_{k=0}^{n-1} u_j u_k \langle \theta_j, \theta_k \rangle = u^T u$$

since the sequence $\{\theta_k^p(t)\}_{k=0}^\infty$ forms an orthonormal basis for the Hilbert space $L^2[0, \infty)$ [9], [10].

The EC filtering problem becomes,

$$\min \|u\|^2, u \in R^n \text{ subject to } \varepsilon^-(t) \le y^T(t) u \le \varepsilon^+(t), \forall t \in [0, \alpha]$$

This is a functional inequality constrained optimization problem with a convex cost and convex constraint set, hence there is a unique optimal solution.

4. Solution Method

It has been shown in the previous Sections that for the two types of filter structures considered, the EC filtering problem can be stated as problem (P) as follows

$$\min \ u^T H u \text{ subject to } \varepsilon^-(t) \le y^T(t) u \le \varepsilon^+(t), \forall t \in \Omega$$

where $u \in R^n$ is the parameter vector to be found, H is a positive definite matrix, $y(t) \in R^n$ is Lipschitz continuous on a compact interval $\Omega \subset R$.

In this Section, an algorithm for solving problem (P) is presented. The technique involves constructing an unconstrained problem to approximate problem (P) and then use a Newton-Raphson method with line search to determine the solution to the unconstrained problem.

For each $\upsilon, \gamma > 0$ an augmented cost function $f_{\upsilon, \gamma}$ is defined as:

$$f_{\upsilon,\gamma}(u) = u^T H u + \gamma \int_\Omega \left(g_\upsilon(y^T(t) u - \varepsilon^+(t)) + g_\upsilon(\varepsilon^-(t) - y^T(t) u) \right) dt \quad (2)$$

where the function g_υ, referred to as a penalty allocator, is any continuous function identically zero on $(-\infty, -\upsilon]$ but increasing otherwise as shown in Fig. 4.1.

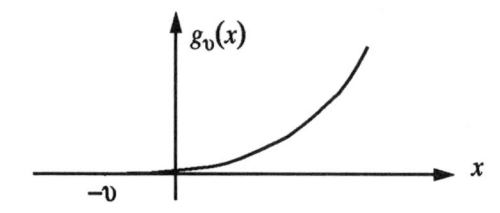

Fig. 4.1. Penalty allocator g_υ.

Problem (P) can be approximated by an unconstrained optimization problem $(P_{\upsilon, \gamma})$, defined for $\upsilon, \gamma > 0$ as:

$$\min f_{\upsilon, \gamma}(u). \qquad (3)$$

Since the original cost and the penalty are convex functions of u, problem $(P_{\upsilon, \gamma})$ has a unique solution [12]. It can also be shown that

(a) A sufficiently large penalty parameter γ forces the solution $u^*_{\upsilon, \gamma}$ of the approximate problem $(P_{\upsilon, \gamma})$ into the feasible region of the EC filtering problem.

(b) Provided that for each υ the penalty parameter γ is chosen according to part (a), the approximate solution $u^*_{\upsilon, \gamma}$ converges to the true solution u^* as the accuracy parameter υ tends to zero.

(c) If a feasible point $\bar{u}$ is known, then for any given error bound ε, the accuracy parameter υ can be calculated (without using any information on the solutions $u^*_{\upsilon, \gamma}$ and u^*) so that $\|u^*_{\upsilon, \gamma}\|^2 - \|u^*\|^2$, is less than the error bound ε.

Assuming that the feasible set $\mathcal{U} = \{u \in R^n : \phi(u, t) \leq 0\}$ of problem (P) has a non-empty interior and define for any $\upsilon > 0$,

$$\mathcal{U}_\upsilon = \{u : \varepsilon^-(t) + \upsilon \leq y^T(t) u \leq \varepsilon^+(t) - \upsilon, \forall t \in \Omega\}.$$

Then, the above results can be stated more concisely in the form of the following theorem, the proof of which can be found in [12].

Theorem 4.1. *Let u^*, $u^*_{\upsilon, \gamma}$ denote the solutions to problems (P) and $(P_{\upsilon, \gamma})$ respectively and let $\tau(\upsilon) = \sup_{\xi > 0} g_\upsilon(-\xi) \min(\alpha, \xi/\eta)$ where η is a number such that*

$$\eta |t_2 - t_1| [1, 1]^T \geq |\phi(u, t_2) - \phi(u, t_1)|, \forall t_2, t_1 \in \Omega.$$

(i) If $\gamma > \|u_\upsilon\|^2 / \tau(\upsilon)$ for some $u_\upsilon \in \mathcal{U}_\upsilon \neq \emptyset$, then, $u^*_{\upsilon, \gamma} \in \mathcal{U}$.

(ii) For each $\upsilon > 0$ such that $\mathcal{U}_\upsilon \neq \emptyset$, if γ is chosen so that $u^*_{\upsilon, \gamma} \in \mathcal{U}$, then

$$\lim_{\upsilon \to 0} u^*_{\upsilon, \gamma} = u^*.$$

(iii) For any $\bar{u} \in \text{int}(\mathcal{U})$, let

$$\delta(\bar{u}) = \min \{ \min_{t \in \Omega} \{ \frac{-\phi^+(\bar{u}, t)}{2\|\bar{u}\|^2} \}, \min_{t \in \Omega} \{ \frac{-\phi^-(\bar{u}, t)}{2\|\bar{u}\|^2} \} \}.$$

If $\upsilon > 0$ is such that $\mathcal{U}_\upsilon \neq \emptyset$, and γ is chosen so that $u^*_{\upsilon, \gamma} \in \mathcal{U}$, then,

$$0 < \upsilon \leq \delta(\bar{u}) \min \{\varepsilon, 2\|\bar{u}\|^2\} \Rightarrow 0 \leq \|u^*_{\upsilon, \gamma}\|^2 - \|u^*\|^2 \leq \varepsilon.$$

The minimum of the augmented cost is computed by the following update equation

$$u_{k+1} = u_k + \mu d_k,$$

where u_k is the filter's coefficient vector at the k th iteration, μ is the step size, and d_k satisfying $\nabla f_{v,\gamma}(u_k)^T d_k < 0$ is the descent direction.

The simplest of the descent directions is the steepest descent (SD) direction, $d_k = -\nabla f_{v,\gamma}(u_k)$. Another popular method is the Newton-Raphson (NR) algorithm, where the descent direction is given by $d_k = -(\nabla^2 f_{v,\gamma}(u_k))^{-1} \nabla f_{v,\gamma}(u_k)$. The step size μ can be obtained by performing some line search at each iteration.

Theorem 4.2. Suppose that g_v is twice continuously differentiable for each $v > 0$ with $|g''_v(x)| \le h(v)$. If $g''_v(x)$ is Lipschitz continuous, i.e. $|g''_v(x) - g''_v(y)| \le L|x - y|$, $\forall x, y \in R$, then, from an arbitrary starting point u_0, the sequence $\{u_k\}_0^\infty$ generated by the NR algorithm with step size μ chosen by the line search procedures of [12], converges to the solution $u^*_{v,\gamma}$ of $(P_{v,\gamma})$ at a quadratic rate.

5. Numerical example

To demonstrate the usefulness of the method developed in previous Sections, consider the equalization of a digital transmission channel consisting of a coaxial cable operating at the DX3 rate (45Mb/sec) [1]. An equalizing filter is required to take the impulse response of a coaxial cable with a loss of 30 dB at a normalized frequency of $1/\tau$ (where τ is the baud interval) as input and produce an output which lies within the envelope given by the DSX-3 pulse template (see Fig. 5.1.). At 20τ the input signal has decayed to a negligible level and thus can be taken as a time limited signal of duration 20τ.

The expression for the penalty allocator used in these example is given below,

$$g_v(x) = \begin{cases} 0, & x \le -v \\ -\dfrac{v}{\pi}\cos\left(\dfrac{\pi x}{2v}\right) + \dfrac{(x+v)}{2}, & -v \le x \le 0 \\ \dfrac{\pi x^2}{8v} + \dfrac{x}{2} + v\left(\dfrac{1}{2} - \dfrac{1}{\pi}\right), & x \ge 0 \end{cases} \quad (4)$$

It is easily verified that (4) is twice continuously differentiable with bounded and Lipschitz continuous second derivative. The convergence result of Theorem 7.4.2 applies.

For the hybrid filter structure, a 32 coefficients FIR filter is used and the input is sampled at every $T = \tau/8$. For the Laguerre network, 6 coefficients are used with a dominant pole value of $p = 8$. The solution is sought by applying the Newton-Ralphson algorithm with line search on the associated augmented cost.

Fig. 5.1. shows output of the hybrid equalizer with 32 coefficients on the interval [0, 8τ]. Note that the time axis are marked in units of τ. Fig. 5.2. plots the value of the augmented cost against the number of function evaluations used in computing the optimal filter coefficients.

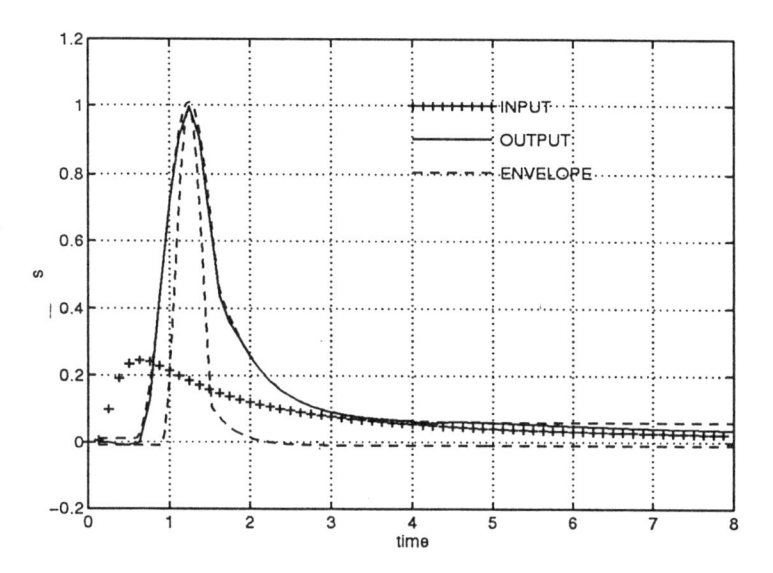

Fig. 5.1. DSX3 pulse template superimposed on coaxial cable response and hybrid filter output.

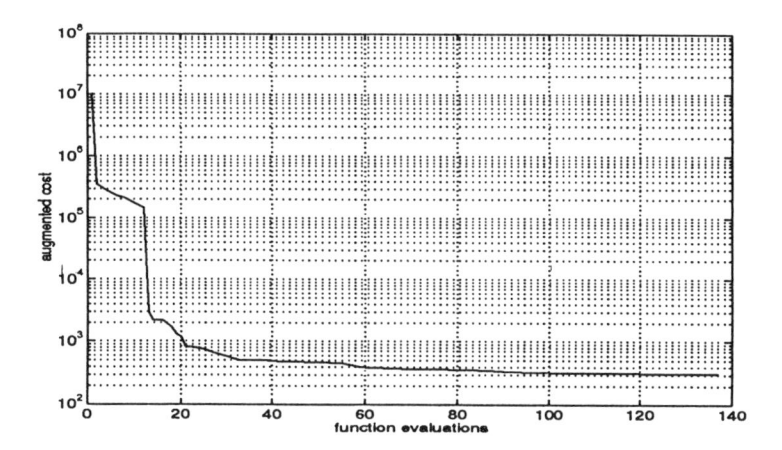

Fig. 5.2. Augmented cost for hybrid filter.

Fig. 5.3. shows the output of Laguerre equalizer on the interval $[0, 8\tau]$ and Fig. 5.4. shows the convergence behaviour of the augmented cost.

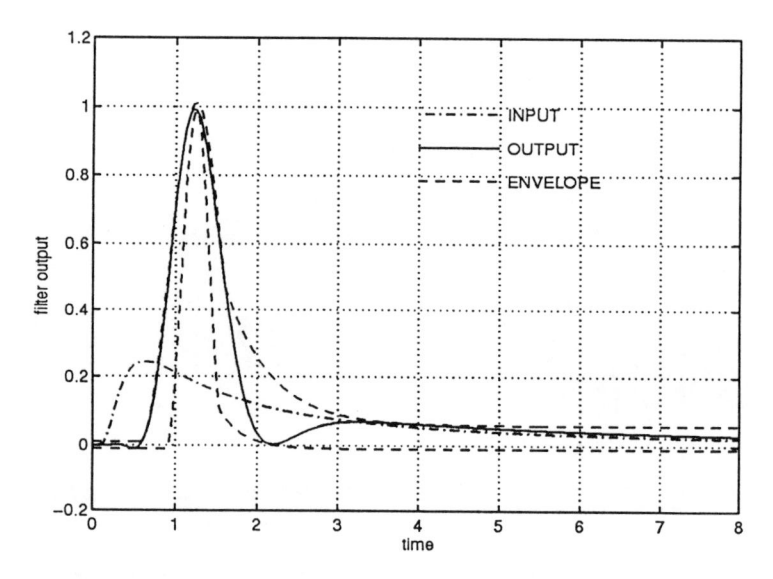

Fig. 5.3. DSX3 pulse template superimposed on coaxial cable response and Laguerre filter output

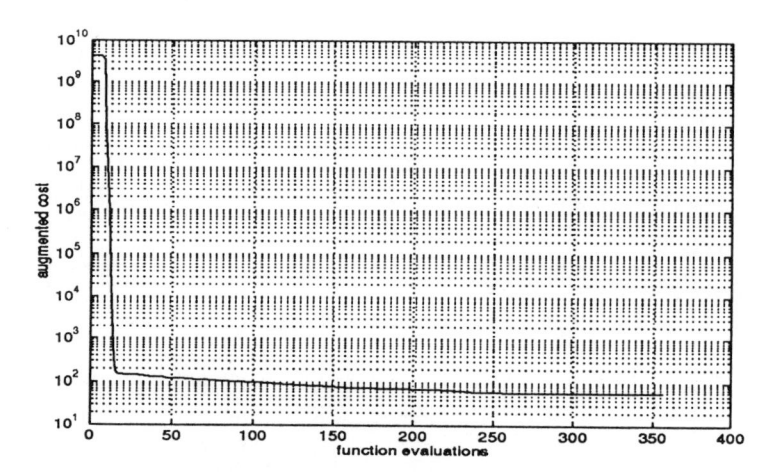

Fig. 5.4. Augmented cost function for Laguerre filter

Although the figures only show the output from 0 to 8τ, in both examples the outputs stayed within the envelope at all points on $[0, 20\tau]$.

6. Conclusions

It has been shown that with analog filters based on DSP technology or on Laguerre networks, the continuous-time EC filtering problem can be formulated as a finite dimensional optimization problem with a continuum of constraints. A solution to this problem was found by introducing an unconstrained problem via penalty techniques. It has been established that the approximate solution approaches the optimal EC filter in the limit. Numerical examples with channel equalization indicate satisfactory performance of the technique.

The iterative nature of the algorithm suggests that adaptive implementation can be readily realized. On-line or adaptive implementation of the algorithm is a prospective area of work. Other worthwhile areas of investigation include the generalization of linear interpolation to an arbitrary interpolation (for the hybrid filter structure) and the effect of component mismatch (for the Laguerre network).

References

[1] Bell Communications, "DSX-3 Isolated Pulse Template and Equations", *Technical Reference TR-TSY-000499*, pp 9-17, Issue 2, December 1988.

[2] CCITT, "Physical/Electrical characteristics of Hierarchical Digital Interfaces", *G.703*, Fascicle III, 1984.

[3] T. E. Fortmann, Athans, "Optimal filter design subject to output sidelobe constraints: Theoretical considerations", *Journal of Optimization Theory and Applications*, Vol. 14, No. 2, pp 179-197, 1974.

[4] R. J. Evans, T. E. Fortmann, and A. Cantoni, "Envelope-constrained filter, Part I, theory and applications", *IEEE Trans. Information Theory*, Vol. IT-23, pp. 421-434, 1977.

[5] R. J. Evans, A. Cantoni, and T. E. Fortmann, "Envelope-constrained filter, Part II, adaptive structures", *IEEE Trans. Information Theory*, Vol. IT-23, pp. 435-444, 1977.

[6] W. X. Zheng, A. Cantoni, B. Vo and K. L. Teo, "Recursive procedures for constrained optimization problems and its application in signal processing", *IEE Proc. - Vision, Image and Signal Processing*, Vol. 142, No. 3, pp. 161-168, 1995.

[7] J. W. Lechleider, "A new interpolation theorem with application to pulse transmission", *IEEE Trans. Communications*, Vol. 39, No. 10, pp 1438-1444, 1991.

[8] B. Vo, A. Cantoni and K. L. Teo, "Envelope constrained filter with linear interpolator", *Technical Report*, SPL-TM006, ATRI, Curtin University of Technology, Western Australia, 1995.

[9] P. M. Makila, "Laguerre series approximation of infinite dimensional systems", *Automatica*, Vol. 26, No. 6, pp. 885-995, 1990.

[10] G. Szego, *Orthogonal Polynomials*, American Mathematical Society Colloquium Publications Volume XXIII, American Mathematical Society Providence, Rhode Island. 1939.

[11] B. Vo, Z. Zang, A. Cantoni and K. L. Teo, "Continuous-time envelope constrained filter design via orthonormal filters", *Technical Report*, SPL-TM008, ATRI, Curtin University of Technology, Western Australia, 1995. (Accepted for publication in *IEE Proc. - Vision, Image and Signal Processing*).

[12] B. Vo, K. L. Teo and A. Cantoni, "Computational methods for a class of functional inequality constrained optimization problem", *World Scientific Series in Applicable Analysis - New trends in optimization*, R. P. Agarwal, World Scientific Publishing, Singapore, **Vol. 5**, pp. 447-465, 1995.

Design of All-Digital Wireless Spread-Spectrum Modems Using High-Level Synthesis

Ping Yeung[1], Ravi Subramanian[1], Marc Barberis[1], Michael Paff[2]

[1] Synopsys, Inc.
700 E. Middlefield Road
Mountain View, CA 94043-4033, USA

[2] Daewoo Semiconductor
675 Almanor Dr.
San Jose, CA 95134, USA

Abstract: The design of spread-spectrum receivers has been revolutionized over the last decade primarily due to the tremendous advances in VLSI technology together with the new capabilities provided by analog-to-digital conversion technologies. In this paper, we will demonstrate how it is now possible for communications engineers to actually control how they exploit the possibilities offered by VLSI technology. Specifically, we demonstrate a new design methodology where the communications system designer has the the ability to tie together algorithm exploration, architecture exploration, and VLSI implementation. A design example of a 15Kgate direct-sequence spread-spectrum cordless telephone modem ASIC is presented. Using this example, which contains a variety of signal processing operations occuring at a wide variety of sampling rates, we demonstrate algorithm exploration through parameterized dataflow simulations. Algorithm descriptions are then automatically translated into architecture-neutral or architecture-specific hardware descriptions. Architecture exploration is performed using behavioral synthesis. Finally, implementation onto digital VLSI is achieved using logic synthesis. For the spread-spectrum modem, the variety of algorithms (signal spreading, chip-filtering, code synchronization, and carrier synchronization) and the manner of specification makes architectural paritioning extremely important. We summarize our results for our spread-spectrum ASIC, and demonstrate the key benefits of this approach.

1 Introduction

The design of spread-spectrum receivers has been revolutionized over the last decade primarily due to the tremendous advances in VLSI digital signal processing technology together with the new capabilities provided by analog-to-digital conversion technologies. Digital VLSI has offered communications engineers the chance to realize algorithms that were often inconceivable in the analog domain.

Digital VLSI is also making the "parametric" receiver ubiquitous [Mey95]. The parametric receiver is one that estimates (using DSP techniques) the possibly time-varying characteristics from the received signals in a digital receiver, and uses these estimates to significantly improve the quality of the detection process. Examples of this include receivers that adapt the detection algorithms based on received signal strength, interfering signals, or particular types of IQ distortions on the received signal.

Although communications engineers can now realize algorithms using DSP that were inconceivable in the analog domain, they do not, to a large extent, control how they exploit various possible architectural strategies in implementing their algorithm. In fact, the ability for the communications engineer to actually explore the VLSI architecture space is typically very limited. In most design flows, algorithm developers innovate in their own world, exploring only the parameters that guarantee stability and good behavior of an algorithm or set of algorithms. This algorithmic specification is then typically manually translated into a hardware description language by an implementation team. This team then takes this new specification and uses traditional design automation tools to create silicon implementations. This gap leads to two key problems in the road to realization [Joh93]: (1) the problem of implementor preserving the algorithm designer's intent in the manual translation process, and (2) the problem of verifying the implementation at the system level. In this paper, we restrict our focus to the first problem.

In what follows, we will demonstrate how it is now possible for communications engineers to actually control how they exploit the possibilities offered by VLSI technology. Specifically, we demonstrate a new design methodology where the communications system designer has the the ability to tie together algorithm exploration, architecture exploration, and VLSI implementation. Central to this methodology is a new suite of tools which provides a direct path from high-level DSP system design, algorithm development, and simulation, through automated HDL code generation, to logic synthesis. This suite consists of Synopsys' COSSAP DSP system simulation tool [COS95], Behavioral Compiler [Beh94], and Design Compiler tools. The COSSAP tool allows high-level modeling of telecommunication and transmission systems, and creation and modeling of complex algorithms for signal processing. In addition, the tool provides automated behavioral and RTL VHDL and Verilog code generation. Design Compiler is used for logic synthesis from RTL VHDL and Verilog using a library based on a target technology. The Behavioral Compiler tool is used for architecture exploration for those signal processing algorithms that are characterized by low to medium throughput and a high cost of resources. Using COSSAP and Behavioral Compiler, one can perform synthesis in an exploratory manner, and arrive at an architectural solution with improved performance or gate count.

A design example of a spread-spectrum transceiver IC for portable wireless applications is described, and specific results for the design exploration through COSSAP, Behavioral Compiler, and Design Compiler are presented. Using this example, which contains a variety of signal processing operations occuring at a wide variety of sampling rates, we demonstrate algorithm exploration through parameterized dataflow simulations. Algorithm descriptions are then automatically translated into architecture-neutral or architecture-specific hardware descriptions. Architecture exploration is performed using behavioral synthesis. Finally, implementation onto digital VLSI is achieved using logic synthesis. For the spread-spectrum modem, the variety of algorithms and the manner of specification makes architectural paritioning extremely important. We summarize our results for our spread-spectrum ASIC, and demonstrate the key benefits of this approach.

2 Spread-Spectrum and DSP System Design

To date, most applications which require digital signal processing have typically been implemented using programmable general-purpose digital signal processors. These processors have a fixed architecture and an instruction set, and the signal processing operations are executed in a sequential manner utilizing shared resources. General-purpose DSPs have continued to advance, providing more and more computational power for new applications through new architectures and process technology advances.

Spread-spectrum applications, however, require signal processing capabilities well beyond those of even the most powerful new programmable DSP devices. These applications are characterized by signal processing rates well into the tens of megahertz. Examples of this are spread-spectrum-based cellular, PCS, and wireless Ethernet applications, as well as direct-broadcast satellite(DBS) and new broadband data services.

To see what paradigm shifts are taking place in the computational models required for DSP, consider that programmable general-purpose DSPs were typically used to process signals at bit rates up to hundreds of kilobits, using system clock rates of up to tens of megahertz. The new class of applications requires processing signals at bit rates up to tens of megabits, while still using system clock rates in the tens of megahertz. Thus, new computational architectures need to be created in order to handle the specific DSP algorithms required in the new applications. The integrated circuits required to process these new broadband signals must be implemented as an application-specific processor either through dedicated logic or through a customized programmable DSP.

Figure 1 shows the different regimes of signal processing in a direct-sequence spread-spectrum receiver [Mey95]. At the earliest point in signal digitization, the receiver samples the bandpass (IF signal) at anywhere from multiples of the IF frequency down to some multiple of the chip-signal bandwidth. Here, the receiver is focused on faithfully demodulating and digitizing the bandpass spectrum, capturing the chip energy. The next stage of processing typically involves sampling at a multiple of the chip rate, and focuses on the despreading operation. Here, the symbol energy is captured. This captured symbol energy is used to generate a detection statistic at the next stage of processing. The final stage of processing takes the narrowband signal and performs parameter estimation (eg. carrier phase and frequency offset estimation) and symbol detection.

Conventional wisdom says that the architecture for integrated circuits for high speed DSP applications like CDMA must be implemented by skilled logic designers utilizing schematic capture tools. Typically the system verification of these custom DSP chips is performed with a C-code program which mimics the target chip and models the external system parameters. This path to implementation and system verification has led to a design gap from system concept to implementation. We propose a new set of tools to close this gap in the process and provide a direct path from the high level system simulation, through automated HDL code generation, finally resulting in synthesized logic. Because the logic generation process is fully automated, a wide range of different logic configurations may be evaluated over the course of a few days. The solution which provides the lowest gate count, or the highest speed or the lowest power consumption may be selected from a wide range of different configurations.

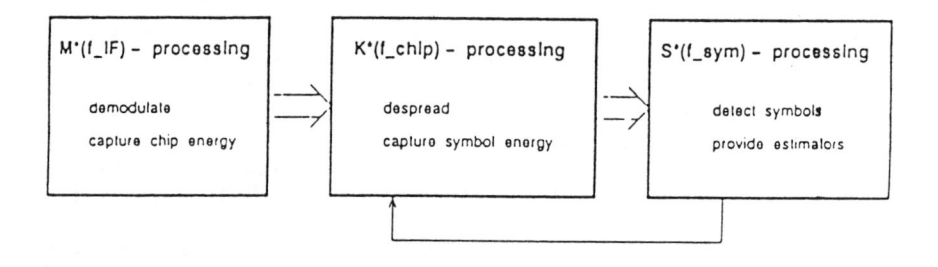

Fig. 1: Regimes of Digital Signal Processing In Spread-Spectrum Receivers

3 Design Methodology

The optimum design flow typically allows the engineer to rapidly explore the design space of algorithms, architectures, and implementation complexity. Design automation tools for analysis and modeling in the algorithm and architecture space are critical for efficient simulation and realization of DSP systems. Tight links from architecture to implementation are essential. Typically, because of the wide range of phenomenon to be captured, there exists no one tool to handle this herculean design task. Instead, several tools providing the appropriate level of modeling are used.

We illustrate and prove a new design flow for this process. This new design flow is shown in Figure 2. In this flow, we are concerned with system designers taking algorithms to architectures to implementation. These algorithms are comprehensively specified by algorithm developers using the COSSAP block-diagram editor. These block diagrams are used as input specifications for simulation and implementation. Extensive parameterized simulations are performed using the simulator. These features are used by algorithm developers to verify the performance and behavior of the algorithm(s). Implementation onto hardwired DSP architectures is done by designers first exploring the architecture space, and taking the optimum architecture to gates. This is done in COSSAP via Behavioral and RTL HDL code generation, and then using Synopsys' Behavioral Compiler and Design Compiler. Verification of these implementations is done at the system level by DSP system and hardware designers through co-simulation of the implementation-specific representation (HDL) with the system-level block-diagram specification.

The DSP system design tool COSSAP provides a block diagram editor which permits the designer to enter a behavioral description of his design in block diagram form which includes quantization effects, overflow considerations and all algorithmic processing. In addition to this high level model of the chip, the

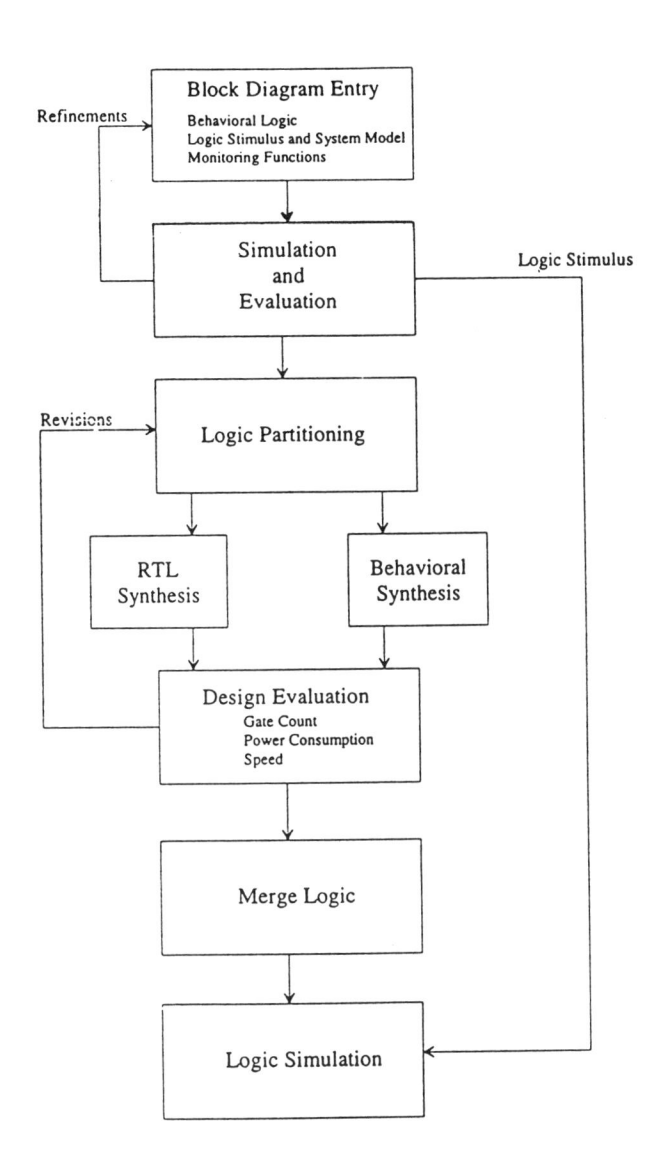

Fig. 2: Proposed Design Flow

designer may also include blocks which generate test signals, introduce noise and channel disturbances as well as model filters and other external system elements. Using this tool, the designer can then evaluate synchronization and detection algorithms in terms of acquisition and steady-state behavior, bit error rate performance, eye closure, evaluate signal statistics and optimize the behavioral logic configuration. The simulations performed at this level are purely functional and do not represent the effects of logic implementation or the target technology.

The next stage of design involves representing the algorithms in fixed-point, and using parameterized simulations to minimize the required wordlengths before embarking upon implementation.

Once the designer is satisfied with the functional performance of his fixed-point design, he can then concentrate on the logic implementation. The first task is to partition the design into a hierarchical structure which represents the physical partitioning of processing elements. Many different physical structures may be attempted, and all physical structures may appear different from the initial functional hierarchy. Each high level block of the physical hierarchy represents logic which will be grouped together for logic synthesis. In general, all the inputs to any given block should occur simultaneously and the throughput for all outputs should have the same requirements. High speed logic which requires only one clock for every input sample should be grouped together. Functions which have low throughput (many clock cycles per input sample) and require high cost resources such as multipliers and memory should be grouped together.

Once a group of blocks to be synthesized has been defined, there are two paths available. The most straightforward approach is to convert the block diagram description of each block into an RTL VHDL description. This description is then passed to the Synopsys Design Compiler which performs logic synthesis utilizing a library based on the target technology. From this level of synthesis, a total gate count can be established for that block as well as the estimated minimum clock period. This process should be repeated for each block to establish a baseline for the overall design.

For those blocks where operations corresponding to high-cost resources (eg. adders, multipliers) can be shared, the next step is to perform behavioral synthesis in an attempt to select the best architecture that will improve performance or reduce gate count. This is the most important step in moving from algorithm definition to architecture design. For example, if a design contained four multipliers, the Design Compiler will simply implement four multipliers in exactly the structure which was described at the block diagram level. Behavioral synthesis will attempt to implement one multiplier and timeshare resources in a manner which reduces the overall design cost. This process is initiated by translating the block diagram description of the block of interest into Behavioral VHDL. The Behavioral Compiler then produces a synthesized RTL VHDL result in the form of a .db file with directives [Beh94]. This is then applied to the Design Compiler. The Behavioral Compiler requires several inputs for each design iteration. The designer must specify the desired clock speed, the latency and the number of clocks per input sample. For faster clock conditions, the Behavioral Compiler will implement greater parallelism. It will schedule resources to achieve each set of design constraints. Thus, Behavioral Compiler creates an optimized design by performing architecture tradeoffs, creating the control logic, data path, and memory I/O, and automatically scheduling all operations in order to execute the DSP algorithms. Once the designer has investigated a particular block thoroughly, he can then determine which set of conditions produces the optimum result from the standpoint of design constraints such as area, speed, latency, and power consumption.

In certain situations, the designer may want to repartition the system and try different groupings of functions. This is typically done, for example, in the early stages of defining a chipset. Many different attempts can be made to optimize each specific logic function in different hierarchical configurations. In some designs it makes sense to build a prespecified architecture at the RTL level because the operations are very simple (eg. PN sequence generators, threshold comparators). For example, a design which contains a number of adders that cannot be shared, and other simple random logic to implement low-level control, may not be suited to behavioral synthesis even if many clocks per sample are available to share resources. In this case, the complexity of implementing the select functions and controller required to timeshare resources may outweigh the complexity of a fully parallel configuration.

The design flow presented provides the designer the freedom to investigate a wide range of possible solutions. He can break the design into many pieces and choose the best of many alternatives for each piece independently. In essence, the problem of algorithm exploration, architecture selection, and gate-level implementation can be partitioned into efficient design exploration stages using the tool suite described.

Once each logic block has been synthesized, the remaining effort is to merge the many pieces together. In the synthesis process, there is the potential for mismatch due to variations in delay. The final merging process requires that any delay mismatch be compensated to insure all the pieces work properly together. The final step is to simulate the synthesized logic. The same stimulus which was used to perform the system level simulations can be used to verify the logic performance. The system level tool thus serves as a system testbench for the complete ASIC.

4 Design Example: A Spread-Spectrum ASIC for Wireless Applications

As a method of evaluating the design tools, a spread-spectrum transceiver integrated circuit was designed which is suitable for portable wireless applications. Spread-spectrum techniques are increasingly being used in commercial wireless communication systems, such as cellular phones, GPS receivers, and wireless Ethernet links. A complete spread-spectrum transceiver is shown in Figure 3. As a first step prior to using the COSSAP tools, a model of the system was developed using C-code. This model provided a baseline description of the design to be implemented in COSSAP and effectively separated the issues related to system design from issues related to tool evaluation.

For this particular design example, the primary objectives are low power consumption and minimum gate count. This design provides three different processing speeds; one clock per sample, 10 clocks per sample and 128 clocks per sample. This variety of logic speed provides wide range of conditions for logic synthesis evaluation.

In the period of about 4 to 5 weeks, the complete receiver was entered into the COSSAP system and all modules were functionally demonstrated. These modules were the basic elements necessary for a spread-spectrum modem, including downconversion, demodulation, synchronization, detection, and control. Data was applied to the transmitter and received correctly by the receiver with no noise. System optimization with noise and interference to produce Bit-Error-Rate and acquisition performance curves would require a minimum of several more weeks. The initial 4 or 5 weeks also included time to learn the mechanics of the synthesis tools in addition to learning the basic COSSAP operation.

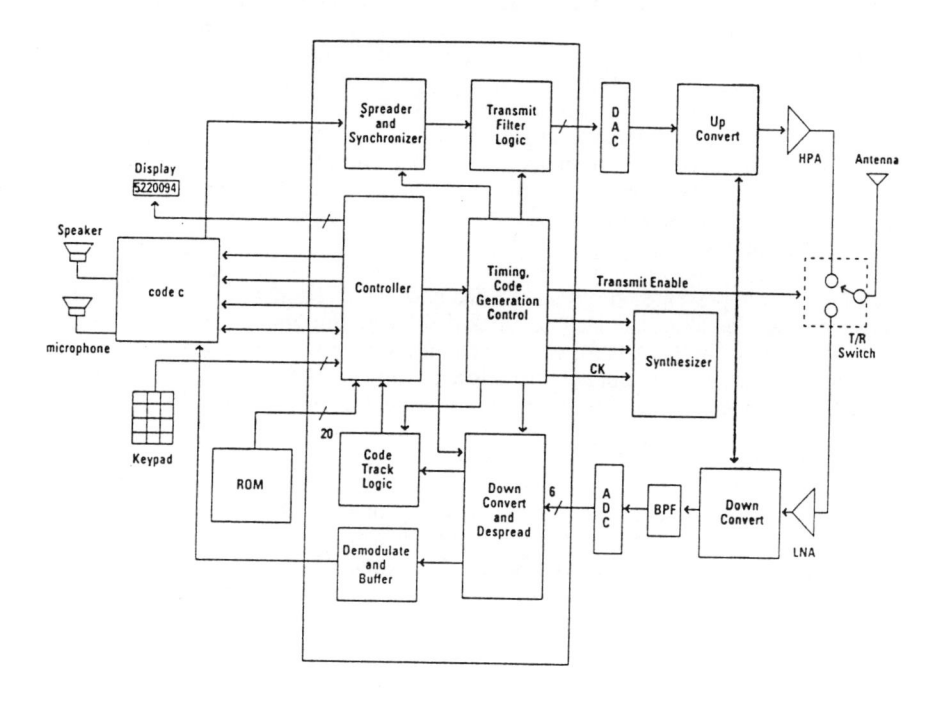

Fig. 3: Wireless Digital Spread-Spectrum Transceiver

Figure 4 illustrates the top level block COSSAP diagram of the spread-spectrum communication system to be evaluated. A transmitter receives external data and combines that information with the spreading sequence to produce a wideband, filtered spread spectrum signal. Noise is then added to this signal, followed by filtering, then frequency translation to an intermediate frequency. These last three steps are not blocks which would be implemented in logic, so they are implemented in the simulator in floating point logic. A model of an A/D converter is used to convert the floating point signal to a bit vector signal with a width of six bits. This word-length is parameterizable so that the optimum word length can be selected.

The clock for this circuit is limited to just over 12 MHz, which is not taxing for any current technology, but does provide many options for the synthesis programs. The logic identified as the Receiver Front End accepts samples at the clock rate and produces outputs at the clock rate. The timing circuits and the PN generator produce outputs at rates lower that the clock rate but must be treated as one-sample-per-clock circuits because of the rigid timing constraints on the outputs.

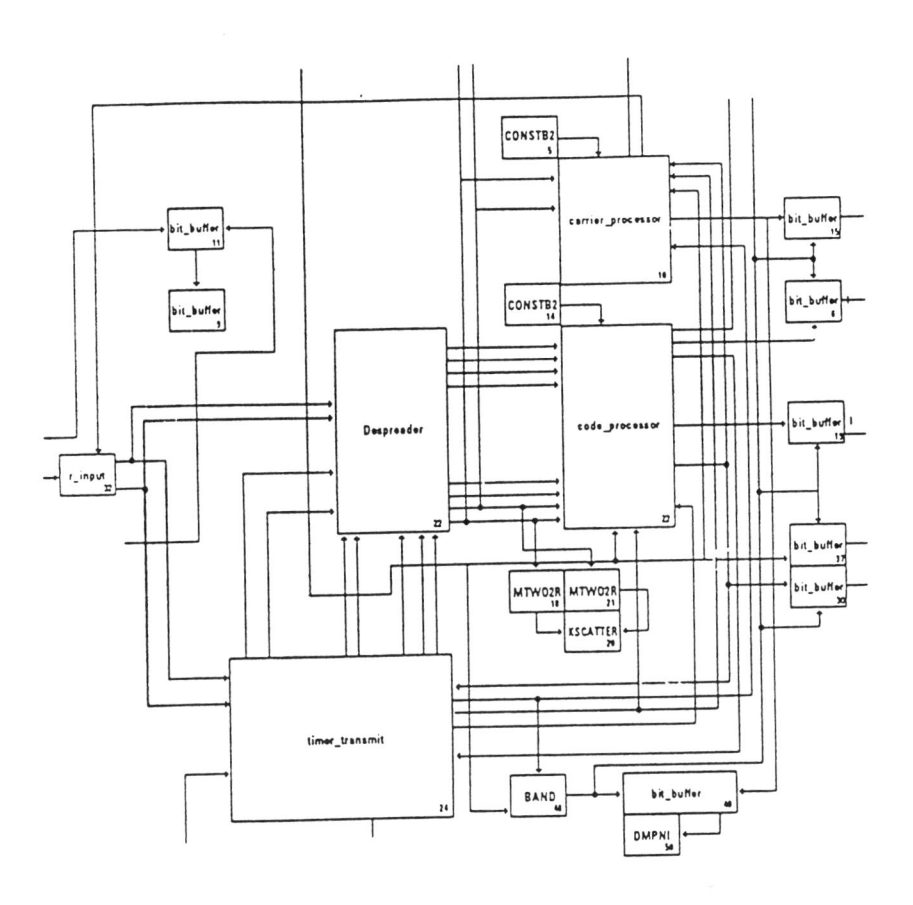

Fig. 4: COSSAP System Block Diagram

The circuit identified as despread logic accepts inputs at a rate of approximately one sample for every ten clocks. This circuit consists of eight accumulators with a one bit by 10 bit multiplier preceding each accumulator. Functionally, this block performs a correlation operation between the received noisy, digitized signal and a local reference signal.

The three blocks to the right identified as Data Demodulation, Carrier Tracking, and Code Tracking each receive one input sample for every 128 clock cycles.
This ratio of low data rate to clock rate is the first criteria used to identify possible candidates for gate-count reduction resource sharing. The designer must then check what types of operations actually constitute each of these algorithms. The criteria for identifying those algorithms where behavioral synthesis affords tremendous benefits are

- many (>>1) clock cycles available to process one data sample
- use of high-cost computational resources (eg. multipliers, adders)
- little use of random logic to construct algorithm (eg.bit manipulation,comparators)

In the data demodulation block, a PSK demodulator is realized. Data demodulation consists of computing a simple metric based on the received despread data and the most likely transmitted data symbols. This is relatively simple block that consists of comparators for the in-phase (I) and quadrature(Q) channels. There are no high-cost resources like multipliers, dividers, or adders. The most complicated element here is a shift-register used for bit-manipulation. This portion of the design will be taken straight to RTL and synthesized using Design Compiler.

Carrier tracking consists of estimating the carrier frequency and phase of the received signal so that the receiver can compensate for frequency and phase offsets. The carrier tracking algorithm used here is of the decision-aided(DA) type, where the I and Q samples of the received, despread signal are cross-multiplied and accumulated in cross-product streams. The extensive use of multipliers and accumulators makes resource sharing critical in this algorithm in order to ensure a design with reasonable area. Having 128 clock cycles to schedule this operation makes scheduling a shared resource a real possibility. Thus, this would be an ideal candidate for architectural exploration through behavioral synthesis.

Code tracking consists of a delay-locked loop which ensures that the local reference signal used in the despreading operation maintains extremely accurate timing so that the result of despreading is reliable data for detection. This operation consists of an accumulator, a multiplexer, a threshold comparator, and two PN shift-registers. Although there are 128 cycles to schedule this operation, there is little to be gained in trying to optimize this architecture consisting of an accumulator, and mostly random logic. This portion of the design will be taken straight to RTL with no architecture exploration as well.

The first step of the architecture definition process, once system simulation is complete, is to convert each of the five major blocks to RTL VHDL code, then directly synthesize the design using Design Compiler. For those blocks where architectural exploration is used, the process begins by first converting the design to Behavioral VHDL, followed by Behavioral synthesis to RTL VHDL then finally gate level synthesis utilizing the Design Compiler again. The results of this process are provided in Table 1.

Table 1: Gate Count Summary

	Candidate for Exploration	Behavioral Compiler	Design Compiler	Best of Both Worlds
Transmitter		n.a.	1385	1385
Rcvr. Front-End		n.a.	2619	2619
Despreader		n.a.	2675	2675
Carrier Processor	√	3572	4931	3572
Code Processor		n.a.	2811	2811
TOTAL		n.a.	14321	13062

As expected, the one clock per sample logic is best suited to the straightforward logic synthesis using the Design Compiler. Behavioral compiler yielded significant benefits in the case identified to meet the criteria for architectural exploration. The correlator logic and the code processor logic were produced using Design Compiler. This is because there are no high cost elements in either of these modules and the complexity required for sharing resources outweighs any benefit. The algorithms used for spread-spectrum code acquisition typically use accumulators with threshold-comparators, followed by control logic. These are not high-cost DSP operations. Figure 5 shows one such example, the code processor. More relevant for these blocks would be the word-length optimizations carried out in COSSAP to ensure only the minimum number of bits required are actually used in the correlator output and code-processor logic.

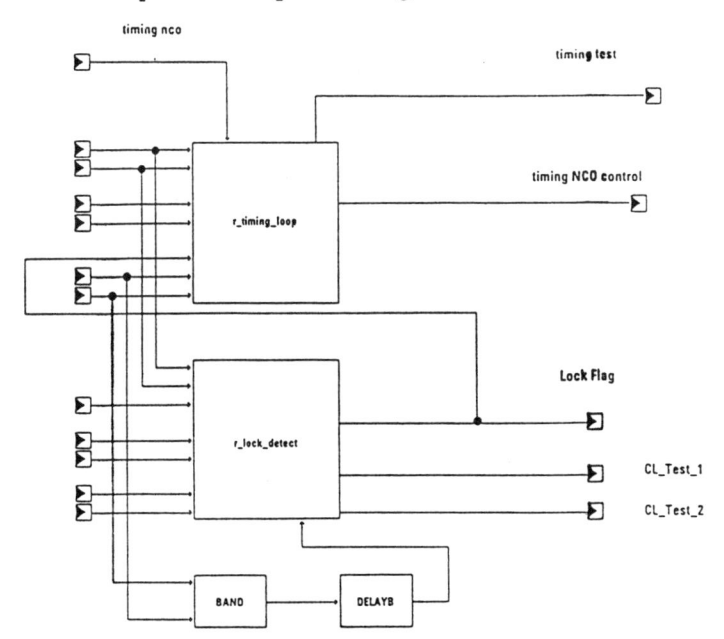

Fig. 5: Code Processor Block Diagram

The Carrier Processor, on the other hand, includes four 10 by 10 bit multipliers in order to execute a maximum-likelihood-based carrier estimation algorithm. The structure of the algorithm is shown in Figure 6. This algorithm is well conditioned and offers very reliable estimates of the carrier phase and frequency even in very low SNR environments. It did provide a significant improvement using Behavioral Compiler as compared with conventional logic synthesis by the Design Compiler. One multiplier costs about 900 gates, and the logic implemented by Design Compiler implements four separate devices. The Behavioral Compiler saved about 30% by implementing only one multiplier and using it sequentially to perform the four multiply operations.

In the course of performing architectural exploration using Behavioral Compiler, the designer has the power to control the exploration process through constraints. These constraints guide the exploration using criteria such as area, latency, and speed. The designer, for example, can control the clock speed, the I/O behavior, or how a certain operation is scheduled over several clock cycles. For the example at hand, we can keep the clock speed the same and control the number of clock cycles available per sample, thereby producing several different RTL architectural designs that all yield the same identical algorithmic result. These results are shown in Table 2. The five rows represent different results for five levels of throughput. At each throughput, two results are presented, one representing the estimate made by Behavioral Compiler as to the expected gate count and the second number is the actual gate count achieved by the Design Compiler. The estimated gate-counts derived from these results are well within ten percent of the actual gate-count results once the design is passed through Design Compiler.

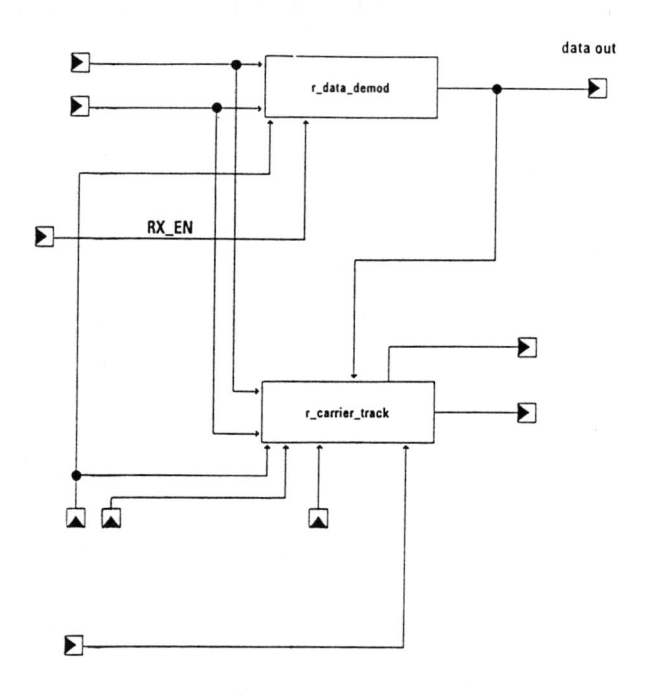

Fig. 6: Carrier Processor Block Diagram

Table 2: Behavioral Compiler Results for Carrier Processor

clocks/ sample	Gate Count (estimated)	Gate Count (actual)	Clock speed (ns) (estimated)	Clock speed (ns) (actual)
5 clocks/ sample	3860	3792	53	52.98
10 clocks/sample	3450	3572	53	52.23
20 clocks/sample	3424	3618	53	51.64
30 clocks/sample	3792	3891	53	53.13
40 clocks/sample	3744	3908	53	51.64

When the design compiler was used alone on the RTL version of this algorithm, the design required 4931 gates with a clock period of 43.8 ns and manual intervention. The library used for this analysis is for a commercially available, 1.5 micron CMOS gate array. All constraints used to control the architectural exploration of Behavioral Compiler resulted in required fewer gates with longer clock periods, as indicated in column 3. This is to be compared with the Design Compiler result of using the design compiler alone. Of the five trials, the best result occurred with 10 clocks per sample. More compilations could be performed to further optimize the results. At this point, one can see that the system designer can have knowledge about the implementation of a particular architecture very early in the design cycle.

5 Summary

COSSAP, Behavioral Compiler and Design Compiler have been demonstrated to be a powerful tool suite for developing custom integrated circuits to perform specific DSP algorithms. Unlike a fully sequential implementation in a standard or custom DSP, the design methodology made possible by these tools permits each specific logic operation to be optimized individually. This permits high throughput and low throughput operations to be implemented with the same design flow. This approach has the advantage of minimizing overall logic complexity and permitting the designer to more fully optimize his design for his specific application. Optimization here is through proper system partitioning, algorithm design, and architecture selection using the powerful link between system level (algorithm), behavioral level (architecture), and RTL level (implementation).

From the system-level block diagram specification, the process of generating RTL and Behavioral HDL and synthesizing logic is fully automated, thus avoiding potential problems with human translation. Indeed, the most significant advantage is that logic synthesis can be performed in days, sometimes hours instead of the many months required with conventional schematic capture tools. This methodology is unique in that it offers algorithm exploration and architectural exploration capabilities to system designers, so they can leverage their specialized knowledge about algorithms and particular sequence of computations in arriving at optimal architectures. It has been demonstrated that the three tools together offer significant productivity benefits and a powerful new methodology for developing custom ICs to perform specific DSP algorithms.

References

[Beh94] Behavioral Compiler Technology Backgrounder, Synopsys, Inc., 700 E. Middlefield Road, Mountain View, CA, 1994.

[COS95] COSSAP Overview and User Guide, Synopsys, Inc., 700 E. Middlefield Road, Mountain View, CA, 1995.

[deGau91] R. de Gaudenzi and M. Luise, "Decision-Directed Coherent Delay-Locked Loop for Direct-Sequence Spread-Spectrum," IEEE Transactions on Communication, COM-39, 1991, pp. 758-765.

[Fet88] G. Fettweis, R. Serra, and J. Stahl, "On the Interaction Between DSP Algorithms and VLSI Architecture," ESA Workshop on DSP Techniques Applied to Space Communications, ESTEC, Netherlands, Nov. 9-10, 1988.

[Hel68] C.W. Helstrom, "Statistical Theory Of Signal Detection," Pergamon Press, London, 1968.

[Jak74] W.C. Jakes, Jr. "Microwave Mobile Communications," J. Wiley and Sons, New York, 1974.

[Joh93] J. Johansson and J. Forskitt, "System Designs Into Silicon," IOP, Philadelphia, 1993.

[Lin73] W.C. Lindsey and M.K. Simon, "Telecommunications System Engineering," Prentice-Hall, Englewood Cliffs, NJ, 1973.

[Mey93] H. Meyr and R. Subramanian, "Advanced Digital Receiver Design," Proc. IEEE Zurich International Conference on Digital Communications, 1993.

[Mey95] H. Meyr and R. Subramanian, "Advanced Digital Receiver Principles and Technologies for PCS," IEEE Communications Magazine, January 1995, pp. 68-78.

[Pol84] A. Polydoros and M.K. Simon, "Generalized Serial Search Code Acquisition: The Equivalent Circular State Diagram Approach," IEEE Trans. Commun. COM-32, 1984, pp.1260-1268.

[Pro89] J. Proakis, "Digital Communications," 2nd Edition, McGraw-Hill, New York, 1989.

[Spi63] J.J. Spilker, "Delay-Lock tracking of Binary Signals," IEEE Trans. Space electr. and Telem., SET-9(1), 1963, pp.1-8.

Efficient VHDL Code Generation for Digital Receiver Design[1]

Thorsten Grötker, Uwe Lambrette and Heinrich Meyr

Lehrstuhl für Integrierte Systeme der Signalverarbeitung
RWTH Aachen

Abstract. The process of designing an ASIC implementation of a digital receiver is carried out on different levels of abstraction. This often involves the error-prone transition between different description styles which imposes obstacles on the joint optimization of algorithm and architecture.
In this contribution we present a design system (ADEN) that provides a link from system level development to VLSI implementation. It speeds up the design process of digital receiver ASIC implementations by providing an automated VHDL code generation based on a dynamic data-flow configuration that is used for system level simulation. The code generation is based on an extensible library (*ComBox*) which offers access to optimized architectures of communication system components. The library-based character of the ADEN system is seen not only as the key to enabling design reuse but also to speed up the design process by reducing the complexity of the verification task.
The design methodology together with the tool operation and the library concept will be explained. An actual design example is presented in form of a case study to demonstrate the effectiveness of this approach.

1 Introduction

Different levels of abstraction are suited for algorithm design and hardware architecture development when targeting towards a digital receiver. Using data-flow semantics for algorithm design has a major advantage: it does neither anticipate implementation specific details that do not influence the algorithmic performance nor implementation related signals like clock and reset. Hence, while providing a higher modeling and simulation efficiency ([1], [2]) as compared to a model on the register transfer level (RTL), it does not imply restrictions on the implementation.

On the other hand in a state of the art design process for fully synchronous digital hardware one will use hardware description languages (HDLs) to gain access to behavioral- and logic-synthesis. The RTL models will be simulated by means of an event-driven simulator.

To obtain efficient systems, algorithm and architecture have to be optimized jointly ([3]). Since a typical design process does not only involve a one way (top-down) transition from higher to lower levels of abstraction, this causes the troublesome transition between different models and description styles.

---
[1] This work was partly sponsored by the Deutsche Forschungsgemeinschaft under contract AZ Me 651/12-3.

Done manually this does not only slow down the design process (and make it unpredictable to a certain extent), but due to the ever growing system complexity verification becomes crucial.

We developed the ADEN system [4] to overcome these difficulties. ADEN takes a block-diagram used in the data-flow simulation system *COSSAP* as primary input specification and generates a synthesizable VHDL description of a synchronously clocked circuit. The implementations for each block of the data-flow graph can be selected from the *ComBox* implementation library (see fig. 1). The component characterization in the *ComBox* library provides

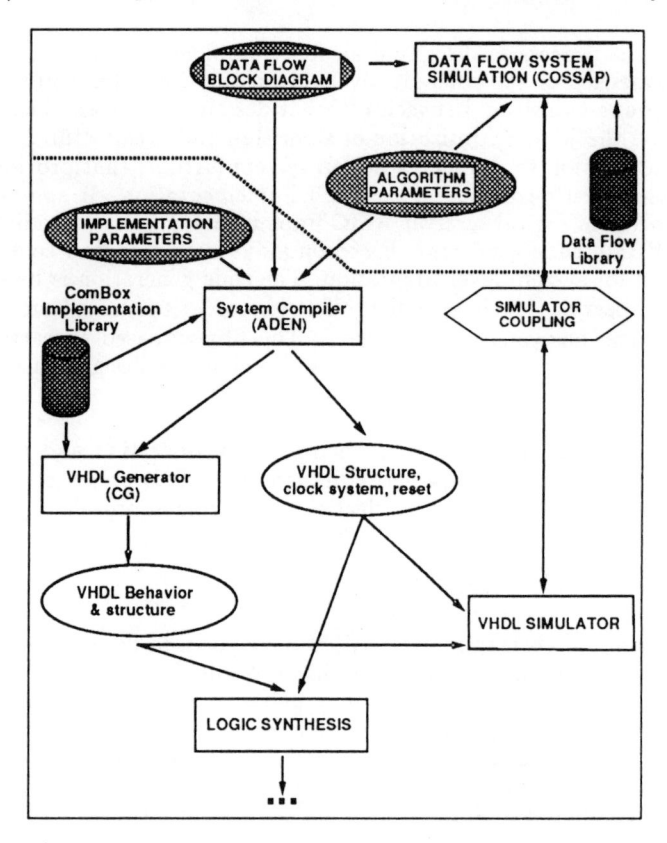

Figure 1 ADEN system overview

ADEN with information about data-flow and timing properties (in addition to the VHDL code itself) to take advantage of complex components like filters, interpolators or synchronizers when generating a multi-rate dynamic dataflow system. Related approaches ([5], [6]) do not provide this functionality. These require the specification of timing behavior and the configuration of timing and control related signals already on the system level and provide a very strict translation to a hardware description language.

In section 2 we take a closer look at the application domain. The *ComBox* library concept is explained in section 3. Section 4 contains a short description of the steps that are performed during VHDL code generation. The design methodology is described in section 5. In section 6 we present results from the design process of a DMSK transceiver for inter-vehicle communication that has been implemented using the ADEN software.

2 Application Domain

Medium to high throughput signal processing systems are modeled using blocks of comparingly coarse granularity such as filters or decoders. The number of different principal block functions ("classes") is relatively small.

The inter-block communication is typically simple with only little global control. The blocks work at different processing rates often employing dynamic data-flow. Thus, the entire system is of the multirate dynamic data-flow type.

Resource sharing among different blocks is not very common, since high data rates require the application of techniques such as parallel processing and pipelining. Additionally, wordlengths are very small (typically 3 to 10 bit for inner receiver components). Thus, the cost for control and multiplexing logic often overcompensates the area gain. Instead, regularity inherent to the underlying algorithms is used to find very efficient regular hardware structures.

Such optimized architectures are well known for a broad range of functional blocks. To take advantage of this knowledge and to enable design reuse a library-based approach has to be taken. A library concept has to cope with implementations that show multirate dynamic data-flow behavior as well as complex timing characteristics (e.g. due to pipelining or multi-cycle operations). Furthermore, the VHDL code generation has to deal with implementations that internally store algorithmic states (e.g. a filter).

Since there is a one-to-one mapping of operations in the block diagram onto resources in the hardware implementation, this approach is not suited for applications with low signal processing rates, when area costs can be reduced by the introduction of resource sharing. A complementary approach is to generate behavioral-oriented VHDL code [7] for such system parts and use *Behavioral Synthesis* tools [8] to generate an optimized RTL structure, that can be used for logic synthesis.

3 Library Concept

The clear separation of distinct levels in the *ComBox* library is the basis for the design methodology presented. The library embodies both data-flow and implementation level, which are separated by the dotted line in fig. 2. The library is organized in 4 layers. The topmost layer (*class*) contains different versions (*groups*) of the same basic algorithm (e.g. different types of equalizers), that show different algorithmic behavior (i.e. they do not operate identically *bit-true*). Hence, each *group* corresponds to a separate data-flow simulation model and may have different generic parameters. The implementation level consists of *primary* and *secondary* layer. On the *primary*

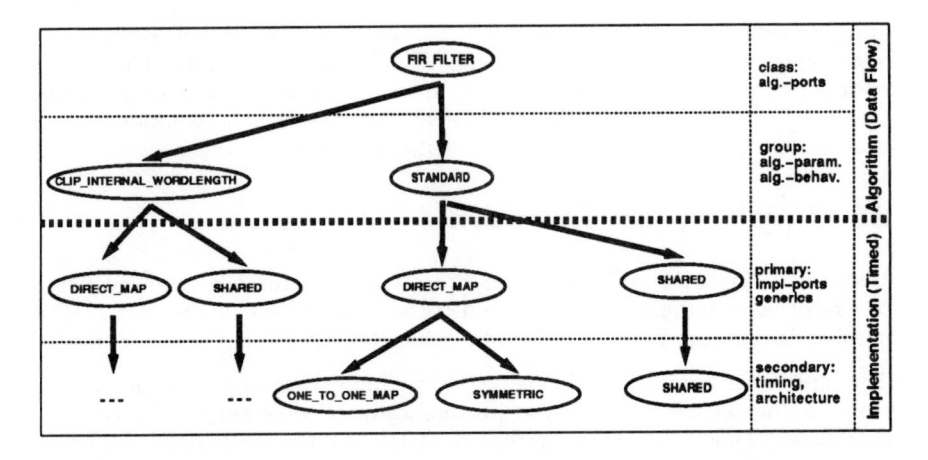

Figure 2 *ComBox* library concept

layer additional generic parameters and implementation ports (like `clock` and `reset`), that are not contained in the data-flow model, are declared. The timing (cp. Section 3.2 is defined on the *secondary* layer, which contains the actual architecture.

A functional verification of an implementation (as well as of the generated VHDL code) can be achieved using a simulator coupling (fig. 1) between the data-flow oriented system level simulator and an HDL-based event-driven simulator.

The designer can take great advantage of the library-based character of the HDL code generation. The library concept enhances the reliability and predictability of the design process. Due to the clear interface definitions and the availability of both data-flow model and implementation, design reuse is simplified. Since the use of complex optimized components is supported, the VHDL code generation may not only be used for rapid prototyping, but supports stepwise design refinement as well. This is stressed by the inherent technology-independence of the generated VHDL code.

3.1 Dataflow Model

Data-flow [9] systems are described as networks of blocks performing the signal processing and signals connecting the ports of those blocks. In static data-flow the blocks consume and produce a fixed number of samples at their ports during each activation. Although a broad range of algorithms can be described by means of static data-flow it imposes obstacles when modeling complete signal processing systems. Therefore ADEN makes use of a more general approach: dynamic data flow[9].

In order to obtain efficient implementations ADEN imposes some restrictions on the blocks' dynamic data-flow behavior:

- The data rates (the number of samples produced or consumed per activation; shown in circles at the blocks' ports in fig. 3) have to be specified for each port of a block.

- The data rates consist of two parts; a symbolic rate that is either 0 or 1 depending on a control condition and a static rate. The control condition has to depend solely on the current value of a so-called control port (e.g. `1 if (control is odd) else 0`).
- A control port has to be a static port with rate 1.

This data flow model allows to specify all configurations in the considered application domain without imposing unnecessary restrictions.

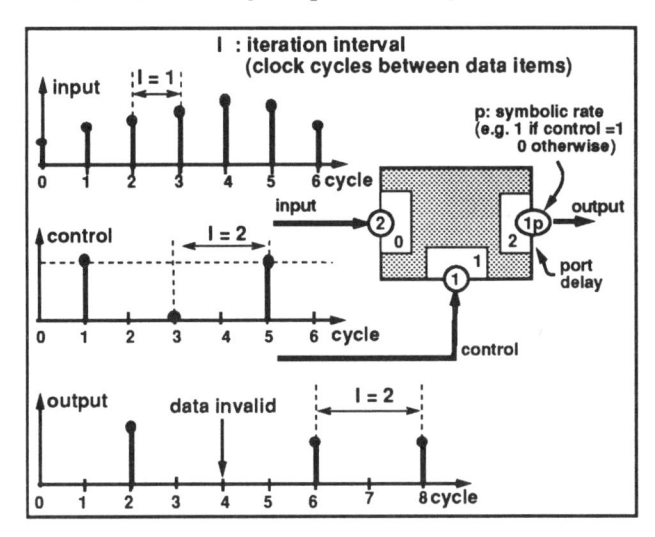

Figure 3 basic properties of a *ComBox* model

3.2 Timing Model

The target architecture is a fully parallel, synchronous circuit with central reset and clock generation subsystems. The blocks exchange data items at fixed time intervals called *iteration interval*. It is sufficient to specify one iteration interval for the block (this *intrinsic* iteration interval is 1 in the example shown in fig. 3), the ports' iteration intervals can be derived from it considering the static data rates. Furthermore the ports may have delays associated with them (*processing delays*, measured in multiples of clock cycles). In fig. 3 the port delays are indicated by numbers in rectangles at the blocks' ports. Note that these processing delays have to be fixed, i.e. independent of the current input values and the states of the blocks. Data-dependent processing delays can be modeled by means of dynamic data-flow.

4 VHDL Code Generation

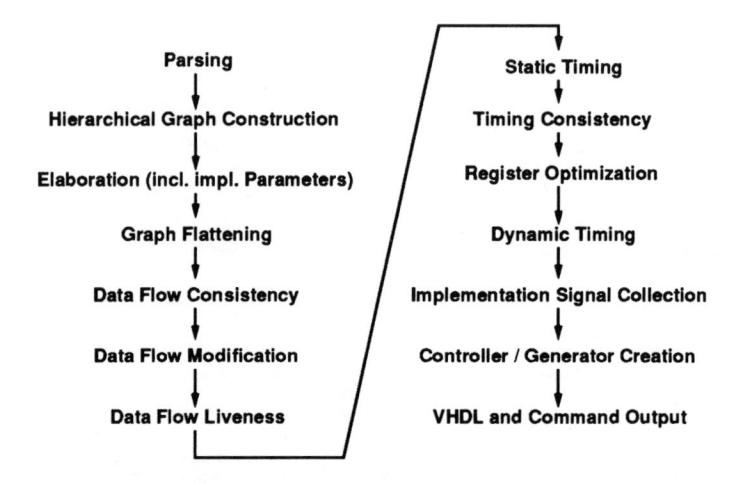

```
        Parsing                          Static Timing
           ↓                                   ↓
Hierarchical Graph Construction        Timing Consistency
           ↓                                   ↓
Elaboration (incl. impl. Parameters)   Register Optimization
           ↓                                   ↓
     Graph Flattening                   Dynamic Timing
           ↓                                   ↓
   Data Flow Consistency          Implementation Signal Collection
           ↓                                   ↓
   Data Flow Modification         Controller / Generator Creation
           ↓                                   ↓
     Data Flow Liveness            VHDL and Command Output
```

Figure 4   Code generation process

The main processing steps are depicted in fig. 4.

After parsing and elaboration ADEN checks whether the specified system has finite memory requirements, i.e. the graph is required to be *strongly consistent* [9,4]. Next, initial values (*algorithmic delays, separators*) are introduced as separate entities and the graph is tested for the absence of deadlocks (*liveness*).

The static timing yields the system's iteration interval and the number of registers (*shimming delays*) that have to be inserted to account for different processing delays in parallel branches and to implement initial values. After verifying that the implementation of each block is able to work at the newly computed iteration interval (*timing consistency*) the distribution of the registers that have to be inserted is optimized.

During dynamic timing ADEN modifies the circuit in a way that register loading can be suppressed for those blocks and signals within dynamic subgraphs that maintain algorithmic states. This is necessary to prevent corruption of internal states at points in time when no valid input data item is available. The implementation is based on gated clocks or load-enable registers.

Then the implementation signals (like `clock`, `reset` or `load-enable`) that are required by the different blocks are collected and subsystems are created by ADEN to generate these signals with respect to their special timing and control requirements.

Finally ADEN generates VHDL code for

- block instantiation and interconnection,
- hierarchical clock and reset generation subsystems,
- control signal generation,

- signal initialization and

- instantiation of registers (*shimming delays*) for timing synchronization and the implementation of algorithmic delays

as well as command scripts that can be used for (co-) simulation and logic synthesis of the generated VHDL code.

5 Design Methodology

The output of the data-flow level design phase is an arithmetic true description of the algorithm without implementation decisions made yet (the separation of data-flow and implementation level is indicated by a dotted line in fig. 1).

The amount of information gathered up to now marks the starting point of the implementation phase. Furthermore, at each subsequent step of the design process the data-flow model is a reference model for the functional verification of the implementation. The latter one can be achieved using a coupling between the *COSSAP* simulator and a VHDL simulator as shown in fig. 1 [10,5].

A user may now select implementations for each block in the data-flow graph from the *ComBox*-library and specify additional implementation parameters (e.g. the number of pipeline stages) if necessary. ADEN then generates VHDL code for the complete system based on the data-flow and timing characterization stored in the library (cp. section 3).

The code generation process can be controlled by various user options specifying e.g. the timing of reset and data signals, the generation of data-valid signals for external dynamic data-flow ports, the use of gated clocks or clocks with different phases and frequencies, and the minimum number of registers (*shimming delays*) to be placed on user-specified signals. The latter option can be used to influence the register distribution (e.g. to break long combinational paths)

The following logic synthesis process may unveil design bottlenecks not yet discovered. Measures to be taken include the use of different code generation options (ADEN), choosing different implementations for a specific block from the *ComBox*, selecting a different algorithm for a subsystem or re-designing a specific implementation. The structure of the *ComBox*-library (fig. 2) has been chosen to enable the designer to perform these actions with minimum effort.

The very advantage of using this approach is increased efficiency.

All information contained in the system level description can be reused, without the need for an error-prone manual transition so that the generated VHDL description will be "correct by construction" with respect to timing and interconnection. Additionally, a large number of pre-tested blocks from the *ComBox* library may be used. Thus, the reliability and predictability of the design is greatly increased.

It is the *locality of modifications*, which makes the modeling efficiency of dataflow oriented approaches superior compared to using less abstract models. *Locality of modifications* means, that changes of system parameters as

- number of algorithmic delays on a certain signal,
- control conditions or
- processing delays or interface behavior of a component

that may show global impact on the resulting circuit, can be modeled locally. Implementations kept in the *ComBox* can be developed independently; context-dependent modifications (e.g. suppression of register loading inside a conditional branch) are handled by the code generation. This decoupling increases the degree of concurrency in the design process.

6 Application Example

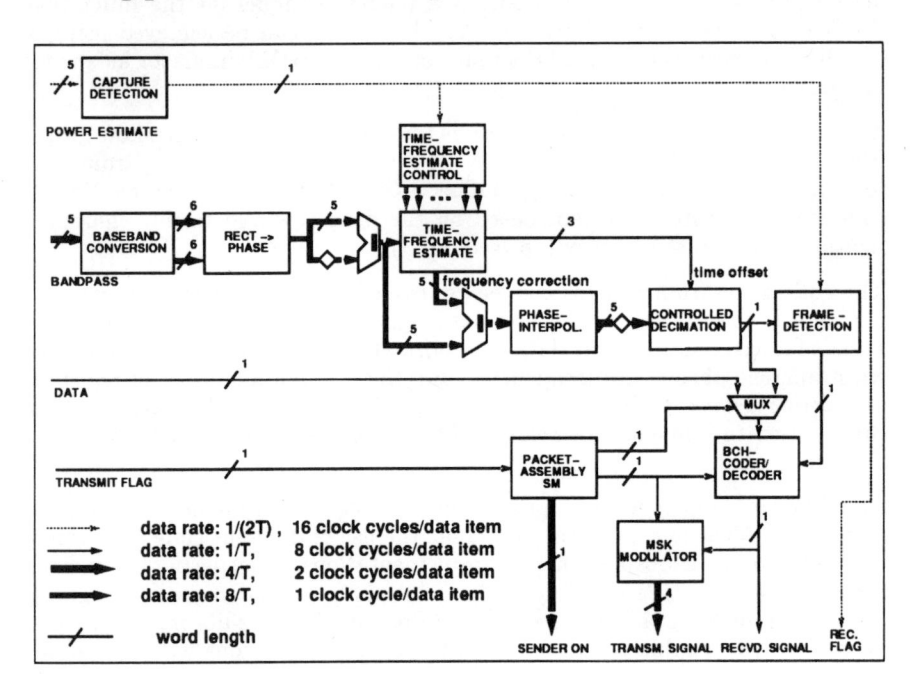

Figure 5 DMSK transceiver

ADEN has been used for the design of a DMSK transceiver for packet-based mobile communication networks shown in fig. 5. All participants in this scenario have equal priority. Transmission is not controlled by any global channel assignment unit - messages may be received as data bursts with multiple transmitters being active. Furthermore, ideal slot synchronization cannot be assumed. Thus, the receiver has to be able to recognize a capture event. Since the receiver has to cope with incoming packets independent of their timing, acquisition has to be performed once per single packet.

The channel can be assumed to be non time-selective due to the high data rate ($2.5 \frac{MBit}{sec}$) and transmission frequency. The slowly time-varying channel characteristics lead to error bursts exceeding the capability of simple FEC mechanisms. Thus, it was decided to rather implement an error-detection coding scheme and drop incorrect packets. The specification is summarized in table 1.

Table 1 DMSK transceiver specification

data rate (1/T)	2.5 MBit/sec
max. sampling rate	8/T
dynamic range	12-60dB
packet error rate	10^{-2}@14dB
frequency offset	±20%
coding	error detecting (511,493) BCH code
capture detection error	$\approx 10^{-3}$@$\Delta 10 dB$

The basic processing steps can be seen in fig. 5. A digital baseband conversion is is followed by a rectangular to phase conversion. The differential phase, which is computed next, is used to estimate timing and frequency offsets. The frequency correction is followed by an interpolation of the phase samples to provide 8 values per symbol, one of which is selected based on the estimated time offset. The sign of the interpolated phase values is used for detecting the respective data bit. The output is fed into the frame detector and the BCH codec.

28	16	493		18	1	length
010101...01	corr. word	data bits	...	parity	1	content

Figure 6 packet structure

The packet structure is depicted in fig. 6. It consists of an alternating sequence used for timing- and frequency-offset estimation, a 16-bit correlation word used by the frame synchronizer to detect the start of the payload, 493 data and 18 parity bits and an additional bit due to the differential modulation scheme.

The block diagram shows that the system uses multiple data rates; the necessary control flow can be modeled using static and dynamic data flow. The required data rate of 2.5MBit/s together with the eightfold oversampling at the input leads to a minimum clock frequency of 20MHz.

The resulting processing power is about 1.6GigaOps/Second.

The implementation phase started with a block diagram consisting of more than 600 primitive (i.e. non-hierarchical) blocks. The large number of blocks stems from a fine-grain specification of the transceiver which resulted from the fact that the *ComBox*-library did not yet contain many complex blocks such as complete decoders. This block diagram was used to assemble a first synthesizable VHDL description of the complete transceiver ($\approx$ 12000 lines of code). For a prototype version logic synthesis was carried out using an FPGA technology library, the results are displayed in fig. 7, where the dark bars represent the first version (which did not fulfill the timing requirements) and the white bars representing the final version.

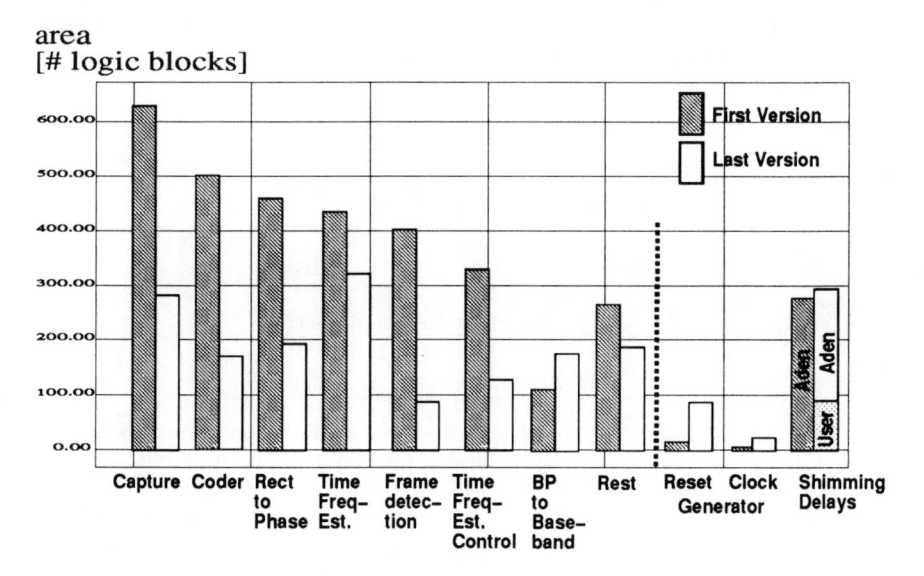

area
[# logic blocks]

Figure 7   logic synthesis results

The improvements that could be achieved (cp. fig. 7) were based on four major points; selection of different implementations for some data-flow blocks, choosing a different algorithm for the same principal functionality, assignment of registers to certain signals in order to break combinational paths (to fulfill the timing requirements) and replacement of complex blocks (e.g. the frame-synchronizer, cp. fig 5), that were specified in a fine-grain manner (i.e. composed of adders and XOR-gates) in the data-flow graph, by optimized architectures. These were then stored in the *ComBox* to allow for later reuse in other projects.

The actions that were taken exemplify the structure of the *ComBox*-library (cp. fig. 2 and section 3). Some examples are presented in the sequel.

For the subtractor used for the frequency offset correction (cp. fig. 5) a pipelined implementation was selected instead of a purely combinational. Since this implementation has and additional port (`clock`) an element from a different *primary* level of the *ComBox* had to be chosen.

Instead of the initial implementation of the argument function (quadrature components → phase, cp. fig. 7) we ended up with an area efficient version that operates correctly only if the absolute value of the input vector does not exceed certain boundaries. Since the algorithm was modified (both versions do not operate bit-true identically, hence they are stored in different *groups*) it was necessary to (re-) simulate the behavior to ensure that algorithmic requirements could still be met[11].

An example for a complex functional block for which the fine-grain specification has been replaced by an optimized architecture is the frame synchronizer. It compares the input data stream with the 16-bit correlation word

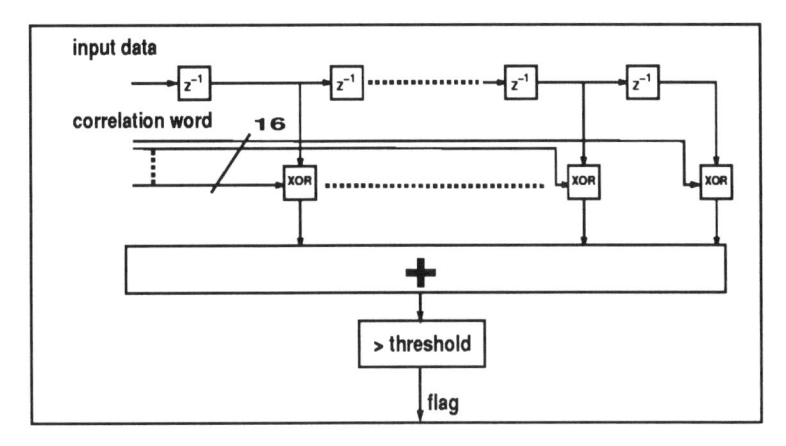

Figure 8 frame synchronizer

and signals if the number of matching bits is larger than a given threshold (fig. 8). The fine-grain model consisted mainly of XOR-blocks and an addertree. Since the data rate is $1/T$ there are 8 clock cycles to compute the result (cp. fig. 5). Hence, a version using resource sharing in the addertree is possible. Even better results - due to less multiplexing effort - could be achieved (see table 2) using a different architecture that stores the correlation instead of the input stream. This implementation has been placed as a primitive model in the *ComBox*.

Table 2 area results for different frame synchronizer implementations

adder tree	397 logic blocks
with resource sharing	116 logic blocks
correlation stored	69 logic blocks

The possibility to perform such optimizations within a single design environment turned out to be crucial for obtaining high-quality results in time. The automated VHDL generation, which allowed to reuse all of the information contained in the block diagram and freed the designer from error-prone tasks like controller generation, led to a very fast design process.

The design of the complete transceiver (starting from a fully specified dataflow block-diagram) consumed 4 person-months, which includes the exploration of algorithmic trade-offs as mentioned above.

7 Summary

We presented a tool for generating synchronous timed descriptions of digital receivers from dynamic data-flow system level configurations. It allows to make use of optimized architectures available for a broad range of communication system components. These architectures will be kept in the *ComBox* library which provides means to characterize their data-flow and timing properties. To provide maximum flexibility the approach is based on a strict separation between data-flow and implementation level. ADEN has proved effectiveness during the design of a DMSK transceiver for mobile communication networks.

References

[1] G. Jennings, "A case against event driven simulation of digital system design," in *The 24th Annual Simulation Symposium*, pp. 170–176, IEEE Computer Society Press, April 1991.

[2] P. Zepter and K. ten Hagen, "Using VHDL with stream driven simulators for digital signal processing applications," in *EURO-VHDL'91 Proceedings*, (Stockholm, Sweden), pp. 196–203, September 8-11 1991.

[3] O. J. Joeressen, G. Schneider, and H. Meyr, "Systematic Design Optimization of a Competitive Soft-Concatenated Decoding System," in *VLSI Signal Processing VI* (L. D. J. Eggermont, P. Dewilde, E. Deprettere, and J. van Meerbergen, eds.), pp. 105–113, IEEE, 1993.

[4] P. Zepter, T. Grötker, and H. Meyr, "Digital receiver design using VHDL generation from data flow graphs," in *Proc. 32nd Design Automation Conf.*, June 1995.

[5] Synopsys, Inc., 700 E. Middlefield Rd., Mountain View, CA 94043, USA, *COSSAP User's Manual: System Architecture*.

[6] P. Scheidt, "The DSP design link with Comdisco," *Synopsys Methodology Notes*, vol. 2, pp. 245–264, February 1992.

[7] Synopsys, Inc., 700 E. Middlefield Rd., Mountain View, CA 94043, USA, *COSSAP User's Manual: VHDL Code Generation*.

[8] Synopsys Inc., 700 E. Middlefield Rd., Mountain View, CA 94043, USA, *Behavioral Compiler User Guide*.

[9] E. A. Lee, "Consistency in dataflow graphs," *IEEE Trans. on Parallel and Distr. Systems*, vol. 2, pp. 223–235, Apr. 1991.

[10] P. Zepter, "Kopplung eines VHDL Simulators an einen Simulator für Signalverarbeitungsalgorithmen," in *GME Fachberichte 11 Mikroelektronik* (D. Seitzer, ed.), pp. 127–132, VDE Verlag, March 1993. in german.

[11] U. Lambrette, P. Zepter, R. Mehlan, and H. Meyr, "Rapid Prototyping of a DMSK Transceiver," in *Proceedings of the IEEE International Conference on Vehicular Technology*, June 1995.

[12] U. Lambrette: WWW page
http://www.informatik.rwth-aachen.de/ERT/Personen/lambrett.html.

[13] WWW home page containing information on the ADEN project:
http://www.informatik.rwth-aachen.de/ERT/Projekte/Tools/aden/aden.html.

Blind Joint Equalization of Multiple Synchronous Mobile Users for Spatial Division Multiple Access

Dirk T. M. Slock

Mobile Communications Department, Institut EURECOM
2229 route des Crêtes, B.P. 193
06904 Sophia Antipolis Cedex, FRANCE
slock@eurecom.fr

Abstract

We consider multiple (p) users that operate on the same carrier frequency and use the same linear digital modulation format. We consider $m > p$ antennas receiving mixtures of these signals through multipath propagation (equivalently, oversampling of the received signals of a smaller number of antenna signals could be used). We consider conditions on the matrix channel response for the existence of a Zero-Forcing Equalizer (ZFE) (which cancels inter-symbol and inter-user interference). In the noisefree case, we show how a ZFE can be obtained from linear prediction and the channel matrix itself can also be determined as a byproduct. The problem is one of signal and noise subspaces and we show a convenient way of solving the deterministic maximum likelihood problem using a minimal linear parameterization of the noise subspace. This parameterization is found as a byproduct in the linear prediction problem.

1 Matrix Channels

Consider linear digital modulation over a linear channel with additive Gaussian noise. Assume that we have p transmitters at the same carrier frequency and m antennas receiving mixtures of the signals. We shall assume throughout that $m > p$. For Spatial Division Multiple Access (SDMA), we shall indeed assume that multiple users operate simultaneously in the same frequency band, but are distinguishable in the spatial dimension. The received signals can be written in the baseband as

$$y_i(t) = \sum_{j=1}^{p}\sum_{k} a_j(k) h_{ij}(t-kT) + v_i(t) \quad (1)$$

where the $a_j(k)$ are the transmitted symbols from source j, T is the common symbol period, $h_{ij}(t)$ is the (overall) channel impulse response from transmitter j to receiver antenna i. We call these users synchronous because of the synchronicity of the symbol frequencies. The symbol phases (time delays) can be different for each user (unlike synchronous users in CDMA). Assuming the $\{a_j(k)\}$ and $\{v_i(t)\}$ to be jointly (wide-sense) stationary, the processes $\{y_i(t)\}$ are (wide-sense) cyclostationary with period T. If $\{y_i(t)\}$ is sampled with period T, the sampled process is (wide-sense) stationary. Sampling in this way leads to an equivalent discrete-time representation. We could also obtain multiple channels in the discrete time domain by oversampling the continuous-time received signals, see [1],[2],[3]. In fact, since the timing of the various users will not be aligned, it is probably a good idea to use oversampling.

We assume the channels to be FIR. In particular, after sampling we assume the (vector) impulse response from source j to be of length N_j. Without loss of generality, we assume the first non-zero vector impulse response sample to occur at discrete time zero, and we can assume the sources to be ordered so that $N_1 \geq N_2 \geq \cdots \geq N_p$. Let $N = \sum_{j=1}^{p} N_j$. The discrete-time received signal can be represented in vector form as

$$\begin{aligned}
\mathbf{y}(k) &= \sum_{j=1}^{p} \sum_{i=0}^{N_j-1} \mathbf{h}_j(i) a_j(k-i) + \mathbf{v}(k) \\
&= \sum_{i=0}^{N_1-1} \mathbf{h}(i) \mathbf{a}(k-i) + \mathbf{v}(k) \\
&= \sum_{j=1}^{p} \mathbf{H}_{j,N_j} A_{j,N_j}(k) + \mathbf{v}(k) = \mathbf{H}_N \mathbf{A}_N(k) + \mathbf{v}(k),
\end{aligned}$$

$$\mathbf{y}(k) = \begin{bmatrix} y_1(k) \\ \vdots \\ y_m(k) \end{bmatrix}, \mathbf{v}(k) = \begin{bmatrix} v_1(k) \\ \vdots \\ v_m(k) \end{bmatrix}, \mathbf{h}_j(k) = \begin{bmatrix} h_{1j}(k) \\ \vdots \\ h_{mj}(k) \end{bmatrix} \quad (2)$$

$$\mathbf{H}_{j,N_j} = [\mathbf{h}_j(N_j-1) \cdots \mathbf{h}_j(0)], \mathbf{H}_N = [\mathbf{H}_{1,N_1} \cdots \mathbf{H}_{p,N_p}]$$
$$\mathbf{h}(k) = [\mathbf{h}_1(k) \cdots \mathbf{h}_p(k)], \mathbf{a}(k) = \left[a_1^H(k) \cdots a_p^H(k)\right]^H$$
$$A_{j,N_j}(k) = \left[a_j^H(k-N_j+1) \cdots a_j^H(k)\right]^H$$
$$\mathbf{A}_N(k) = \left[A_{1,N_1}^H(k) \cdots A_{p,N_p}^H(k)\right]^H$$

where superscript ^{H} denotes Hermitian transpose.

2 FIR Zero-Forcing (ZF) Equalization

We consider a structure of equalizers as in Fig. 1 to not only cancel the intersymbol interference for every source separately, but also cancel the interference between different sources. We assume the equalizer filters to be FIR of length L: $F_{ji}(z) = \sum_{k=0}^{L-1} f_{ji}(k) z^{-k}$, $j = 1,\ldots,p, i = 1,\ldots,m$. We introduce $\mathbf{f}_j(k) = [f_{j1}(k) \cdots f_{jm}(k)]$, $\mathbf{f}(k) = \left[\mathbf{f}_1^H(k) \cdots \mathbf{f}_p^H(k)\right]^H$, $\mathbf{F}_{j,L} = [\mathbf{f}_j(L-1) \cdots \mathbf{f}_j(0)]$, $\mathbf{F}_L = [\mathbf{F}_{1,L}^H \cdots \mathbf{F}_{p,L}^H]^H$, $\mathbf{H}(z) = \sum_{k=0}^{N_1-1} \mathbf{h}(k) z^{-k}$ and $\mathbf{F}(z) = \sum_{k=0}^{L-1} \mathbf{f}(k) z^{-k}$. The condition for the equalizer to be ZF is $\mathbf{F}(z)\mathbf{H}(z) = \text{diag}\{z^{-n_1} \cdots z^{-n_p}\}$ where $n_j \in \{0,1,\ldots,N_j+L-2\}$. The ZF condition in the multi-user context means that both Inter-Symbol Interference (ISI) and Inter-User Interference (IUI) should be zero. The ZF condition can be written in the time-domain as

$$\mathbf{F}_L \, \mathcal{T}_{L,p}(\mathbf{H}_N) = \begin{bmatrix} 0\cdots 0 \; 1 \; 0\cdots 0 & \cdots & 0\cdots 0 \\ \vdots & \ddots & \vdots \\ 0\cdots 0 & \cdots & 0\cdots 0 \; 1 \; 0\cdots 0 \end{bmatrix} \quad (3)$$

where $\mathcal{T}_{M,p}(\mathbf{H}_N) = [\mathcal{T}_M(\mathbf{H}_{1,N_1}) \cdots \mathcal{T}_M(\mathbf{H}_{p,N_p})]$ and $\mathcal{T}_M(\mathbf{x})$ is a banded block Toeplitz matrix with M block rows and $[\mathbf{x} \; 0_{n \times (M-1)}]$ as first block row (n is the number of rows in $\mathbf{x}$). (3) is a system of $p(N+p(L-1))$ equations in Lmp unknowns. To be able to equalize, we need to choose the equalizer length L such that the system of equations (3) is exactly or underdetermined. Hence

$$L \geq \underline{L} = \left\lceil \frac{N-p}{m-p} \right\rceil . \quad (4)$$

We assume that $\mathbf{H}_N$ has full rank if $N \geq m$. If not, it is still possible to go through the developments we consider below. But lots of singularities will appear and the non-singular part will behave in the same way as if we had a reduced number of channels, equal to the row rank of $\mathbf{H}_N$. Reduced rank in $\mathbf{H}_N$ can be detected by inspecting the rank of (noise-free) $\mathbf{E}\mathbf{y}(k)\mathbf{y}^H(k)$. If a reduced rank in $\mathbf{H}_N$ is detected, the best way to proceed (also when quantities are estimated from data) is to preprocess the data $\mathbf{y}(k)$ by transforming them into new data of dimension equal to the row rank of $\mathbf{H}_N$.

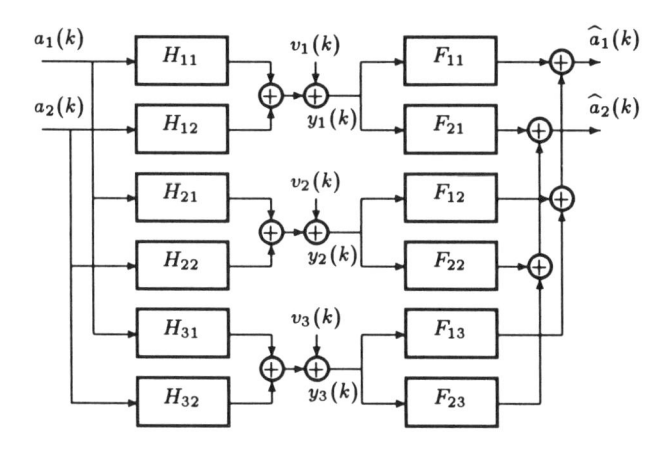

Figure 1: Channel and linear equalizer for $m = 3$ channels and $p = 2$ sources.

The matrix $\mathcal{T}_{L,p}(\mathbf{H}_N)$ is a block Toeplitz block matrix. It can be shown that for $L \geq \underline{L}$ it has full column rank if the following assumptions are satisfied

(A1) $\text{rank}(\mathbf{H}(z)) = p$, $\forall z$ and $\text{rank}(\mathbf{h}(0)) = p$. In this case, $\mathbf{H}(z)$ is called irreducible in systems theory,

(A2) $\text{rank}([\mathbf{h}_1(N_1-1) \cdots \mathbf{h}_p(N_p-1)]) = p$, in which case $\mathbf{H}(z)$ is called column reduced, see [4].

Assuming $\mathcal{T}_{L,p}(\mathbf{H}_N)$ to have full column rank, the nullspace of $\mathcal{T}_{L,p}^H(\mathbf{H}_N)$ has dimension $L(m-p)-N+p$. If we take the entries of any vector in this nullspace as equalizer coefficients, then the equalizer output is zero, regardless of the transmitted symbols.

To find a ZF equalizer (corresponding to some delays n_j), it suffices to take an equalizer length equal to $\underline{L}$. We can arbitrarily fix $\underline{m} = \underline{L}(m-p)-N+p$ equalizer coefficients (e.g. take $\underline{m}$ equalizer filters of length $\underline{L}-1$ only). The remaining $p(p(\underline{L}-1)+N)$ coefficients can be found from (3). This shows that for $m > p$, a FIR equalizer suffices for ZF equalization (and interference cancellation)!

3 FIR Zero-Forcing Transmission Filters for SDMA

For a mobile communications system, the multi-user receiver considerations of the previous section apply mostly to the base station (BS) since it is not unrealistic to have multiple antennas at the BS and to deploy the necessary processing power required for multi-user detection. So the receiver considerations for the BS apply to the uplink. Dual considerations apply to the downlink. Indeed, in order to implement SDMA at the BS, the BS not only needs to unravel the mixture of uplink signals it receives from the simultaneous users, it also needs to transmit the simultaneous downlink signals so that each mobile user receives its signal with low IUI. To this end, the BS can employ a transmission filter matrix linking the p signals to be transmitted to its m antennas so that the signals will arrive at the p mobile users without ISI or IUI! Indeed, $H_{ij}(z)$ is the sampled impulse response from user j to BS antenna i in the uplink. Assuming we have channel reciprocity (if not, the channel impulse responses will simply be different between uplink and downlink), the channel impulse response from antenna i to mobile j is also $H_{ij}(z)$. Hence, the transfer matrix on the downlink from the m antennas to the p mobile users will be $\mathbf{H}^T(z)$ ($p \times m$), where superscript ^{T} denotes transpose. Let $G_{ij}(z)$ be the transmitter filter linking the signal to be tansmitted to mobile j to antenna i, then $\mathbf{G}(z)$ is the overall ($m \times p$) transmitter filter to be used on the downlink. Using this transmitter filter, the signals intended for each mobile user will arrive at the respective users without ISI or IUI if

$$\mathbf{H}^T(z)\mathbf{G}(z) = \mathrm{diag}\{z^{-n_1} \cdots z^{-n_p}\} \tag{5}$$

for certain delays n_i. As in the ZFE case, an FIR filter matrix $\mathbf{G}(z)$ can be found if the channel satisfies the conditions mentioned before.

The channel impulse response may satisfy the reciprocity property if Time Division Duplex (TDD) is used. In that case, uplink and downlink occur at the same carrier frequency. For reciprocity, we also need to assume that the channel will remain constant during the time it takes to switch from uplink to downlink, and that the same analog filters are used in the BS and in the mobile sets. Assuming reciprocity to hold, the channel impulse response matrix $\mathbf{H}(z)$ can be estimated from the uplink and then the transmission filter $\mathbf{G}(z)$ for the downlink can be determined from that channel estimate. If channel reciprocity does not hold (because Frequency Division Duplex is used or because the channel varies too rapidly), then the channel impulse response for the downlink should be estimated using feedback from the users. Or other approaches for the downlink, based on direction-of-arrival (DOA) information (that can be estimated from the uplink), have to be deployed [5].

4 Channel Identification from a Training Sequence

A vector of L consecutive received samples of the vector channel can be expressed as

$$\mathbf{Y}_L(k) = \mathcal{T}_{L,p}(\mathbf{H}_N)\, \mathbf{A}_{N+p(L-1)}(k) + \mathbf{V}_L(k) \tag{6}$$

where $\mathbf{Y}_L(k) = \left[\mathbf{y}^H(k-L+1) \cdots \mathbf{y}^H(k)\right]^H$ and $\mathbf{V}_L(k)$ is defined similarly, and $\mathbf{A}_{N+pL}(k) = \left[A^H_{1,N_1+L}(k) \cdots A^H_{p,N_p+L}(k)\right]^H$. Consider now $k = L-1$ and let $\mathbf{A}_{N+p(L-1)}(L-1)$ contain training sequence data. Due to the commutativity of convolution, we can also write

$$\mathcal{T}_{L,p}(\mathbf{H}_N)\, \mathbf{A}_{N+p(L-1)}(L-1) = \mathcal{A}\, \mathbf{H}_N^{tT} \tag{7}$$

where superscript ^{t} denotes transposition of the $(m \times 1)$ block elements and $\mathcal{A}$ is a structured block matrix filled with the elements of $\mathbf{A}_{N+p(L-1)}(L-1)$. For additive white Gaussian noise, the maximum likelihood channel estimate becomes simply the least-squares estimate

$$\widehat{\mathbf{H}}_N^{tT} = \left(\mathcal{A}^H \mathcal{A}\right)^{-1} \mathcal{A}^H \mathbf{Y}_L(L-1)\,. \tag{8}$$

The variance of this estimate is proportional to $\dfrac{mN}{mL} = \dfrac{N}{L}$, which is proportional to p if the N_i are approximately equal. Hence, to achieve a certain estimation quality, L increases proportionally to p, the number of users. The training sequence length for user i is $N_i + L - 1$. Hence, the training sequence overhead becomes prohibitive if we want to allow a considerable number of simultaneous users. This motivates us to research blind channel estimation techniques, that do not require training sequences.

5 Channel Identification from Second-order Statistics: Frequency Domain Approach

Consider the noise-free case and let the sources be temporally white but possibly correlated among themselves with $p \times p$ covariance matrix $R_\mathbf{a}$. Then the power spectral density matrix of the stationary vector process $\mathbf{y}(k) = \mathbf{H}(z)\mathbf{a}(k)$ is

$$S_{\mathbf{yy}}(z) = \mathbf{H}(z)\, R_\mathbf{a} \mathbf{H}^H(z^{-*})\,. \tag{9}$$

The following spectral factorization result has been brought to our attention by Loubaton [6]. Let $\mathbf{K}(z)$ be a $m \times p$ rational transfer function that is causal and stable. Then $\mathbf{K}(z)$ is called minimum-phase if $\mathbf{K}(z) \neq 0$, $|z| > 1$. Let $S_{\mathbf{yy}}(z)$ be a rational $m \times m$ spectral density matrix of rank p. Then there exists a rational $m \times p$ transfer matrix $\mathbf{K}(z)$ that is causal, stable, minimum-phase, unique up to a unitary $p \times p$ constant matrix, of (minimal) McMillan degree $\deg(\mathbf{K}) = \frac{1}{2}\deg(S_{\mathbf{yy}})$ such that

$$S_{\mathbf{yy}}(z) = \mathbf{K}(z)\, \mathbf{K}^H(z^{-*})\,. \tag{10}$$

In our case, $S_{\mathbf{yy}}$ is polynomial (FIR channel) and $\mathbf{H}(z)$ is minimum-phase since we assume rank $(\mathbf{H}(z)) = p$, $\forall z$. Hence, the spectral factor $\mathbf{K}(z)$ identifies the channel

$$\mathbf{K}(z) = \mathbf{H}(z) R_{\mathbf{a}}^{1/2} \Phi \qquad (11)$$

where $R_{\mathbf{a}}^{1/2}$ is any particular (e.g. triangular) matrix square-root of $R_{\mathbf{a}}$ and Φ is a $p \times p$ unitary matrix. So the channel identification from second-order statistics is simply a multivariate MA spectral factorization problem. The remaining factors $R_{\mathbf{a}}^{1/2}$ and Φ can be identified by exploiting higher-order moments (see [7] and references therein) or the discrete distribution nature of the sources [8].

6 Gram-Schmidt Orthogonalization, Triangular Factorization and Linear Prediction

6.1 UDL Factorization of the Inverse Covariance Matrix

Consider a vector of zero mean random variables $Y = \begin{bmatrix} y_1^H & y_2^H & \cdots & y_M^H \end{bmatrix}^H$. We shall introduce the notation $y_{1:M} = Y$. Consider Gram-Schmidt orthogonalization of the components of Y. We can determine the linear least-squares (lls) estimate $\widehat{y}_i$ of y_i given $y_{1:i-1}$ and the associated estimation error $\widetilde{y}_i$ as

$$\begin{aligned} \widehat{y}_i &= \widehat{y}_i|_{y_{1:i-1}} = R_{y_i y_{1:i-1}} R_{y_{1:i-1} y_{1:i-1}}^{-1} y_{1:i-1}, \\ \widetilde{y}_i &= \widetilde{y}_i|_{y_{1:i-1}} = y_i - \widehat{y}_i \end{aligned} \qquad (12)$$

where $R_{ab} = \mathrm{E} ab^H$ for two random column vectors a and b. The Gram-Schmidt orthogonalization process consists of generating consecutively $\widetilde{Y} = \begin{bmatrix} \widetilde{y}_1^H & \widetilde{y}_2^H & \cdots & \widetilde{y}_M^H \end{bmatrix}^H$ starting with $\widetilde{y}_1 = y_1$. We can write the relation

$$L Y = \widetilde{Y} \qquad (13)$$

where L is a unit-diagonal lower triangular matrix. The first $i-1$ elements in row i of L are $-R_{y_i y_{1:i-1}} R_{y_{1:i-1} y_{1:i-1}}^{-1}$. From (13), we obtain

$$\mathrm{E}(LY)(LY)^H = \mathrm{E}\widetilde{Y}\widetilde{Y}^H \Rightarrow L R_{YY} L^H = D = R_{\widetilde{Y}\widetilde{Y}}. \qquad (14)$$

D is indeed a diagonal matrix since the $\widetilde{y}_i$ are decorrelated. Equation (14) can be rewritten as the UDL triangular factorization of R_{YY}^{-1}

$$R_{YY}^{-1} = L^H D^{-1} L. \qquad (15)$$

If Y is filled up with consecutive samples of a random process, $Y = [y^H(k)\ y^H(k-1)\cdots y^H(k-M+1)]^H$, then the $\widetilde{y}_i$ become backward prediction errors of order $i-1$, the corresponding rows in L are backward prediction filters and the corresponding diagonal elements in D are backward prediction error variances. If the process is stationary, then R_{YY} is Toeplitz and the backward prediction errors filters and variances (and hence the UDL factorization of

R_{YY}^{-1}) can be determined using a fast algorithm, the Levinson algorithm. If Y is filled up in a different order, *i.e.* $Y = \left[y^H(k)\ y^H(k+1)\cdots y^H(k+M-1)\right]^H$, then the backward prediction quantities become forward prediction quantities, which for the the prediction error filters and variances are the same as the backward quantities if the process $y(.)$ is scalar valued.

If the process $\mathbf{y}(.)$ is vector valued, we shall still carry out the Gram-Schmidt orthogonalization scalar component by scalar component. In the time-series case, this is multichannel linear prediction with sequential processing of the channels. If the matrix $R_{\mathbf{YY}}$ is singular, then there exist linear relationships between certain components of $\mathbf{Y}$. As a result, certain components y_i will be perfectly predictible from the previous components and their resulting orthogonalized version $\tilde{y}_i$ will be zero. The corresponding diagonal entry in D will hence be zero also. For the orthogonalization of the following components, we don't need this y_i. As a result, the entries under the diagonal in the corresponding column of L can be taken to be zero (minimum-norm choice for the prediction filters in those rows). The (linearly independent) row vectors in L that correspond to zeros in D are vectors that span the null space of $R_{\mathbf{YY}}$. The number of non-zero elements in D equals the rank of $R_{\mathbf{YY}}$.

6.2 LDU Factorization of a Covariance Matrix

Assume at first that $R_{\mathbf{YY}}$ is nonsingular. Since the $\tilde{y}_i$ form just an orthogonal basis in the space spanned by the y_i, $\mathbf{Y}$ can be perfectly estimated from $\tilde{\mathbf{Y}}$. Expressing that the covariance matrix of the error in estimating $\mathbf{Y}$ from $\tilde{\mathbf{Y}}$ is zero leads to

$$0 = R_{\mathbf{YY}} - R_{\mathbf{Y}\tilde{\mathbf{Y}}} R_{\tilde{\mathbf{Y}}\tilde{\mathbf{Y}}}^{-1} R_{\tilde{\mathbf{Y}}\mathbf{Y}} \Rightarrow$$
$$R_{\mathbf{YY}} = \left(R_{\mathbf{Y}\tilde{\mathbf{Y}}} R_{\tilde{\mathbf{Y}}\tilde{\mathbf{Y}}}^{-1}\right) R_{\tilde{\mathbf{Y}}\tilde{\mathbf{Y}}} \left(R_{\tilde{\mathbf{Y}}\tilde{\mathbf{Y}}}^{-1} R_{\tilde{\mathbf{Y}}\mathbf{Y}}\right) = U^H D U \tag{16}$$

where D is the same diagonal matrix as in (14) and $U = L^{-H}$ is a unit-diagonal upper triangular matrix. (16) is the LDU triangular factorization of $R_{\mathbf{YY}}$. In the stationary multichannel time-series case, $R_{\mathbf{YY}}$ is block Toeplitz and the rows of U and the diagonal elements of D can be computed in a fast way using a sequential processing version of the multichannel Schur algorithm.

When $R_{\mathbf{YY}}$ is singular, then D will contain a number of zeros, equal to the dimension of the nullspace of $R_{\mathbf{YY}}$. Let J be a selection matrix (the rows of J are rows of the identity matrix) that selects the nonzero elements of D so that JDJ^H is a diagonal matrix that contains the consecutive non-zero diagonal elements of D. Then we can write

$$R_{\mathbf{YY}} = (JU)^H (JDJ^H)(JU) \tag{17}$$

which is a modified LDU triangular factorization of the singular $R_{\mathbf{YY}}$. $(JU)^H$ is a modified lower triangular matrix, its columns being a subset of the columns of the lower triangular matrix U^H. A modified version of the Schur algorithm to compute the generalized LDU factorization of a singular block Toeplitz matrix $R_{\mathbf{YY}}$ has been recently proposed in [9].

7 Signal and Noise Subspaces

Consider now the measured data with additive independent white noise $\mathbf{v}(k)$ with zero mean and assume $E\mathbf{v}(k)\mathbf{v}^H(k) = \sigma_v^2 I_m$ with unknown variance σ_v^2 (in the complex case, real and imaginary parts are assumed to be uncorrelated; colored noise with known correlation structure but unknown variance could equally well be handled). From (5), we get for the structure of the covariance matrix of the received signal $\mathbf{y}(k)$

$$R_L^{\mathbf{y}} = \mathcal{T}_{L,p}(\mathbf{H}_N) R_{N+p(L-1)}^{\mathbf{a}} \mathcal{T}_{L,p}^H(\mathbf{H}_N) + \sigma_v^2 I_{mL} \tag{18}$$

where $R_{N+p(L-1)}^{\mathbf{a}} = E\mathbf{A}_{N+p(L-1)}(k)\mathbf{A}_{N+p(L-1)}^H(k)$. We assume $R_M^{\mathbf{a}}$ to be nonsingular for any M. For $L \geq \underline{L}$, and assuming (A1), (A2), $\mathcal{T}_{L,p}(\mathbf{H}_N)$ has full column rank and σ_v^2 can be identified as the smallest eigenvalue of $R_L^{\mathbf{y}}$. Replacing $R_L^{\mathbf{y}}$ by $R_L^{\mathbf{y}} - \sigma_v^2 I_{mL}$ gives us the covariance matrix for noise-free data. Given the structure of $R_L^{\mathbf{y}}$ in (18), the column space of $\mathcal{T}_{L,p}(\mathbf{H}_N)$ is called the signal subspace and its orthogonal complement the noise subspace.

Consider the eigendecomposition of $R_L^{\mathbf{y}}$ of which the real positive eigenvalues are ordered in descending order:

$$\begin{aligned} R_L^{\mathbf{y}} &= \sum_{i=1}^{N+p(L-1)} \lambda_i V_i V_i^H + \sum_{i=N+p(L-1)+1}^{mL} \lambda_i V_i V_i^H \\ &= V_S \Lambda_S V_S^H + V_\mathcal{N} \Lambda_\mathcal{N} V_\mathcal{N}^H \end{aligned} \tag{19}$$

where $\Lambda_\mathcal{N} = \sigma_v^2 I_{(m-p)L-N+p}$ (see (18)). The sets of eigenvectors V_S and $V_\mathcal{N}$ are orthogonal: $V_S^H V_\mathcal{N} = 0$, and $\lambda_i > \sigma_v^2$, $i = 1,\ldots,N+p(L-1)$. We then have the following equivalent descriptions of the signal and noise subspaces

$$Range\{V_S\} = Range\{\mathcal{T}_{L,p}(\mathbf{H}_N)\}, V_\mathcal{N}^H \mathcal{T}_{L,p}(\mathbf{H}_N) = 0. \tag{20}$$

8 The Instantaneous Mixture Case

We shall consider the noiseless case and we can assume w.l.o.g. that the first p rows of $\mathbf{h}(0)$ are linearly independent (the ordering of the channels can always be permuted to achieve this). The covariance matrix of $\mathbf{y}(k) = \mathbf{h}(0)\mathbf{a}(k)$ is $R_1^{\mathbf{y}} = \mathbf{h}(0) R_\mathbf{a} \mathbf{h}^H(0)$. By carrying out the Gram-Schmidt orthogonalization of the components of $\mathbf{y}(k)$, we obtain the triangular factorizations we discussed above. In particular

$$\begin{aligned} LR_1^{\mathbf{y}} L^H &= D = \text{blockdiag}\{D_p, 0_{(m-p)\times(m-p)}\} \\ \Rightarrow R_1^{\mathbf{y}} &= U_p^H D_p U_p \end{aligned} \tag{21}$$

where U_p^H is a $m \times p$ matrix of the generalized lower triangular form we discussed above. Taking $R_\mathbf{a}^{1/2}$ to be triangular, we arrive at

$$\mathbf{h}(0) = U_p^H D_p^{1/2} \Phi R_\mathbf{a}^{-1/2} \tag{22}$$

where Φ is a $p\times p$ unitary matrix. Φ and $R_{\mathbf{a}}^{1/2}$ represent $\frac{1}{2}p(p-1)$ and $\frac{1}{2}p(p+1)$ degrees of freedom respectively. If we don't know $R_{\mathbf{a}}$, we can determine $\mathbf{h}(0)$, using the LDU factorization of $R_1^{\mathbf{y}}$, as $U_p^H D_p^{1/2}$, up to p^2 degrees of freedom. If $R_{\mathbf{a}}$ is known, e.g. $R_{\mathbf{a}} = \sigma_a^2 I_p$, then $U_p^H D_p^{1/2}$ determines $\mathbf{h}(0)$ up to only Φ, i.e. up to only $\frac{1}{2}p(p-1)$ degrees of freedom.

In general, if $\mathbf{h}(0)$ is determined using subspace techniques from U_p^H, then the only part of $\mathbf{h}(0)$ that can be determined uniquely from $R^{\mathbf{y}}$ is $\mathbf{h}(0)T = \mathbf{h}'(0) = [I_p \; *]^H$ which is related to $\mathbf{h}(0)$ by a non-singular $p \times p$ matrix T, representing p^2 degrees of freedom. Note also that $LU_p^H = [I_p \; 0]^H$. Hence

$$\widetilde{\mathbf{y}}(k) = L\mathbf{y}(k) = L\mathbf{h}(0)\mathbf{a}(k) = \begin{bmatrix} I_p \\ 0 \end{bmatrix} D_p^{1/2} \Phi R_{\mathbf{a}}^{-1/2} \mathbf{a}(k) \qquad (23)$$

or $\widetilde{y}_{1:p}(k)$ is just a linear transformation of $\mathbf{a}(k)$.

9 Blind Equalization and Channel Identification from Second-order Statistics by Multichannel Linear Prediction

9.1 ZF Equalizer and Noise Subspace Determination

We consider again the noiseless covariance matrix or equivalently assume noisefree data: $v(t) \equiv 0$. We shall also assume the transmitted symbols to be uncorrelated, $R_M^{\mathbf{a}} = R_{\mathbf{a}} \otimes I_M$, though the noise subspace parameterization we shall obtain also holds when the transmitted symbols are correlated.

Consider now the Gram-Schmidt orthogonalization of the consecutive (scalar) elements in the vector $\mathbf{Y}_L(k)$. We start building the UDL factorization of $R_L^{\mathbf{y}}$ and obtain the consecutive prediction error filters and variances. No singularities are encountered until we arrive at block row $\underline{L}$ in which we treat the elements of $\mathbf{y}(k+\underline{L}-1)$. From the full column rank of $\mathcal{T}_{\underline{L},p}(\mathbf{H}_N)$, we infer that we will get $\underline{m} \in \{0, 1, \ldots, m-p-1\}$ singularities. If $\underline{m} > 0$, then the following scalar components of $\mathbf{Y}$ become zero after orthogonalization: $\widetilde{y}_i(k+\underline{L}-1) = 0$, $i = m+1-\underline{m}, \ldots, m$. So the corresponding elements in the diagonal factor D are also zero. We shall call the corresponding rows in the triangular factor L singular prediction filters.

For $L = \underline{L}+1$, $\mathcal{T}_{\underline{L}+1,p}(\mathbf{H}_N)$ has m more rows than $\mathcal{T}_{\underline{L},p}(\mathbf{H}_N)$ but only p more columns. Hence the (column) rank increases by p. As a result $\widetilde{y}_i(k+\underline{L})$, $i = 1, \ldots, p$ are not zero in general while $\widetilde{y}_i(k+\underline{L}) = 0$, $i = p+1, \ldots, m$. Furthermore, since $\mathcal{T}_{\underline{L},p}(\mathbf{H}_N)$ has full column rank, the orthogonalization of $y_{1:p}(k+\underline{L})$ w.r.t. $\mathbf{Y}_{\underline{L},p}(k)$ is the same as the orthogonalization of $y_{1:p}(k+\underline{L})$ w.r.t. $A_{N+p(\underline{L}-1)}(k+\underline{L}-1)$. Hence, since the $\mathbf{a}(k)$ are assumed to be uncorrelated, only the components of $y_{1:p}(k+\underline{L})$ along $\mathbf{a}(k+\underline{L})$ remain: $\widetilde{\mathbf{y}}(k+\underline{L})|_{\mathbf{Y}_{L,p}(k)} = \mathbf{h}(0)\mathbf{a}(k+\underline{L})$ and for the rest of the details of the orthogonalization of the components of $\mathbf{y}(k+\underline{L})$, we can refer to section 8. In particular, $\widetilde{y}_{1:p}(k+\underline{L})$ are just a linear transformation of

$\mathbf{a}(k+\underline{L})$. This means that the corresponding (p outputs) prediction filter is (proportional to) a ZF equalizer! Since the prediction error is white, a further increase in the length of the prediction span will not improve the prediction. Hence $\widetilde{\mathbf{y}}(k+L) = \mathbf{h}(0)\mathbf{a}(k+L)$, $L \geq \underline{L}$ and the (block of m) prediction filters in the corresponding block row $L+1$ will be appropriately shifted versions of the (block) prediction filter in (block) row $\underline{L}+1$. In particular also for the prediction errors that are zero, a further increase of the length of the prediction span cannot possibly improve the prediction. Hence $\widetilde{y}_i(k+L) = 0$, $i = p+1,\ldots,m$, $L \geq \underline{L}$. The singular prediction filters further down in the triangular factor L are appropriately shifted versions of the first $m-p$ singular prediction filters. Furthermore, the entries in these first $m-p$ singular prediction filters that appear under the 1's ("diagonal" elements) are zero for reasons we explained before in the general orthogonalization context. So we get a (rank p) white prediction error with a finite prediction order. Hence the channel ouput process $\mathbf{y}(k)$ is *autoregressive*. Due to the structure of the remaining rows in L being shifted versions of the first ZF equalizer and the first $m-p$ singular prediction filters, after a finite "transient", L becomes a banded lower triangular block Toeplitz matrix.

Consider now $L > \underline{L}$ and let us collect all consecutive singular prediction filters in the triangular factor L into a $((m-p)(L-\underline{L})+\underline{m}) \times (mL)$ matrix $\mathcal{G}_L$. The row space of $\mathcal{G}_L$ is the (transpose of) the noise subspace. Indeed, every singular prediction filter belongs to the noise subspace since $\mathcal{G}_L \mathcal{T}_{L,p}(\mathbf{H}_N) = 0$, all rows in $\mathcal{G}_L$ are linearly independent since they are a subset of the rows of a unit-diagonal triangular matrix, and the number of rows in $\mathcal{G}_L$ equals the noise subspace dimension. $\mathcal{G}_L$ is a banded block Toeplitz matrix of which the first $m-p-\underline{m}$ rows have been omitted. $\mathcal{G}_L$ is in fact parameterized by the first $m-p$ singular prediction filters. Let us collect the nontrivial entries in these $m-1$ singular prediction filters into a column vector G_N. So we can write $\mathcal{G}_L(G_N)$. The length of G_N can be calculated to be $Nm - p^2$ which equals the number of degrees of freedom in $\mathbf{H}_N$ for identification with a subspace technique (in which case we can only identify $\mathbf{h}(k)T = \mathbf{h}'(k)$ where T is such that $\mathbf{h}'(0) = [I_p \; *]^H$). So $\mathcal{G}_L(G_N)$ represents a minimal linear parameterization of the noise subspace.

9.2 Channel Identification

From the discussion above, it is now not difficult to see that in the LDU factorization of $R^{\mathbf{y}}$, the lower triangular factor $(JU)^H$ is banded and becomes block Toeplitz after a finite transient. Indeed, for $L \geq \underline{L}$, the $L+1^{\text{st}}$ block column of $(JU)^H$ is $E\mathbf{y}(k:\infty)\widetilde{y}_{1:p}^H(k+L) =$

$$\left[0_{mL\times p}^H \; \mathbf{h}^H(0) \cdots \mathbf{h}^H(N_1-1) \; 0 \cdots \right]^H E\mathbf{a}(k+L)\widetilde{y}_{1:p}^H(k+L) \qquad (24)$$

which hence contains the channel impulse response, apart from a $p \times p$ multiplicative factor.

9.3 Channel Estimation from Data using Deterministic ML

See [3] for channel estimation from an estimated covariance sequence by subspace fitting for $p = 1$. That approach can straightforwardly be extended to the case of general p. The details for deterministic maximum likelihood have been worked out in [10] for $p = 1$. Basically, we use

$$P^\perp_{\mathcal{T}_{M,p}(\mathbf{H}_N)} = P_{\mathcal{G}_M^H(G_N)} .$$

The essential number of degrees of freedom in $\mathbf{H}_N$ and G_N is $mN-p^2$ for both. So $\mathbf{H}_N$ can be uniquely determined from G_N and vice versa. Due to the (almost) block Toeplitz character of $\mathcal{G}_M$, the product $\mathcal{G}_M \mathbf{Y}_M(k)$ represents a convolution. Due to the commutativity of convolution, we can write $\mathcal{G}_M(G_N)\mathbf{Y}_M(k) = \mathcal{Y}_N(\mathbf{Y}_M(k))[1\ G_N^H]^H$ for some properly structured $\mathcal{Y}_N(\mathbf{Y}_M(k))$. This leads us to formulate the DML problem as

$$\min_{G_N} \begin{bmatrix} 1 \\ G_N \end{bmatrix}^H \mathcal{Y}_N^H(\mathbf{Y}_M(k)) \left(\mathcal{G}_M(G_N)\mathcal{G}_M^H(G_N)\right)^{-1} \mathcal{Y}_N(\mathbf{Y}_M(k)) \begin{bmatrix} 1 \\ G_N \end{bmatrix} \quad (25)$$

which can be solved iteratively in the IQML fashion. In [11], a robustified version of this IQML procedure was presented which can be straightforwardly extended to the multi-source case.

References

[1] L. Tong, G. Xu, and T. Kailath. "A New Approach to Blind Identification and Equalization of Multipath Channels". In *Proc. of the 25th Asilomar Conference on Signals, Systems & Computers*, pages 856–860, Pacific Grove, CA, Nov. 1991.

[2] D.T.M. Slock. "Blind Fractionally-Spaced Equalization, Perfect-Reconstruction Filter Banks and Multichannel Linear Prediction". In *Proc. ICASSP 94 Conf.*, Adelaide, Australia, April 1994.

[3] D.T.M. Slock and C.B. Papadias. "Blind Fractionally-Spaced Equalization Based on Cyclostationarity". In *Proc. Vehicular Technology Conf.*, pages 1286–1290, Stockholm, Sweden, June 1994.

[4] T. Kailath. *Linear Systems*. Prentice-Hall, Englewood Cliffs, NJ, 1980.

[5] B.E. Ottersten. "Base-Station Antenna Arrays in Mobile Communications". In E. Biglieri and M. Luise, editors, *Signal Processing in Telecommunications, Proc. of the 7th Tyrrhenian Int. Workshop on Dig. Communications*. Springer-Verlag, 1995.

[6] Ph. Loubaton. "Egalisation autodidacte multi-capteurs et systèmes multivariables". GDR 134 (Signal Processing) working document, february 1994, France.

[7] B. Laheld and J.-F. Cardoso. "Adaptive Source Separation with Uniform Performance". In M. Holt, C. Cowan, P. Grant, and W. Sandham, editors, *Proc. EUSIPCO-94, SIGNAL PROCESSING VII: Theories and Applications*, pages 183–186, Edinburgh, Scotland, Sept. 13-16 1994.

[8] S. Talwar, M. Viberg, and A. Paulraj. "Blind Estimation of Multiple Co-Channel Digital Signals Using an Antenna Array". *IEEE Signal Processing Letters*, SPL-1(2):29–31, Feb. 1994.

[9] K. Gallivan, S. Thirumalai, and P. Van Dooren. "A Block Toeplitz Look-Ahead Schur Algorithm". In *Proc. 3rd International Workshop on SVD and Signal Processing*, Leuven, Belgium, Aug. 22-25 1994.

[10] D.T.M. Slock. "Subspace Techniques in Blind Mobile Radio Channel Identification and Equalization using Fractional Spacing and/or Multiple Antennas". In *Proc. 3rd International Workshop on SVD and Signal Processing*, Leuven, Belgium, Aug. 22-25 1994.

[11] D.T.M. Slock and C.B. Papadias. "Further Results on Blind Identification and Equalization of Multiple FIR Channels". In *Proc. ICASSP 95 Conf.*, Detroit, Michigan, May 1995.

Base-Station Antenna Arrays in Mobile Communications

BJÖRN OTTERSTEN, PER ZETTERBERG

SIGNAL PROCESSING
Royal Institute of Technology
100 44 STOCKHOLM

Abstract

This paper describes the utilization of antenna arrays at the base stations of mobile communication systems. Multiple antennas can provide a processing gain to increase the base station range and improve coverage. Also, by exploiting the angular selectivity of an antenna array at the base stations of a wireless system, users may be spatially multiplexed to increase system capacity. We address several aspects of the reception and transmission problems that arise when the spatial dimension is considered. Both the forward and reverse channels are discussed and an overview of some interesting research topics in the area is presented.

1 Introduction

The dramatic expansion of mobile communications over the last years has emphasized the importance of efficient use of frequency bandwidth. There is an increasing demand for capacity in wireless systems which traditionally directly translates into a demand for more bandwidth which is quite limited. Also, the infrastructure investment costs are often a limiting factor when deploying a new system that must have wide area coverage. It is therefore of great interest to increase the range by employing antenna arrays.

Traditional telecommunication schemes multiplex channels in frequency and/or time. However, the spatial dimension is in general used in a very rudimentary fashion by, for example, using certain channels in certain geographical areas (cell planning). By employing an array of antennas, it is possible to multiplex channels in the spatial dimension just as in the frequency and time dimensions. Recently, a more efficient use of the spatial dimension has appeared as a means of increasing the capacity in wireless communication systems without exploiting additional bandwidth [1, 2]. By employing an array of antennas at the base stations of a cellular system when receiving and transmitting over the communication channel, the spatial dimension may be used to separate several users operating in the same channel. This more efficient use of the spatially dimension will be termed Spatial Division Multiple Access (SDMA).

Using an array of antennas is also a way of increasing the gain of the system thereby extending range and improving coverage. Of course, the hardware requirements are more demanding but this permits a sparser infrastructure and will often be more cost effective. In general, increasing the range of cellular systems is of great interest initially, for example, when deploying the new PCS system in the US. However, we may expect that demand for increased system capacity will follow shortly after adequate coverage is achieved.

2 Exploiting the Spatial Dimension

To achieve increased range in a cellular systems, it may be argued that the mobile to base communication (up link) is the critical link. It is desirable that the mobiles operate at low powers and thus, for acquisition, the base stations must be able to detect weak signals of short duration in a noisy and possibly interfering environment. In the down link (base to mobile communication), increased range may be achieved by for example increasing the transmit power.

When receiving communication signals at an antenna array, the proposed signal processing methods for distinguishing different messages, can be grouped in two main categories; those that exploit array response information and those that do not. Since in general, the array configuration is known, the array response is known (or may be calibrated) for an incident wave from a given location [3]. This assumes that the scenario is well behaved in the sense that the propagation may be modeled by a single, or small number of paths. These methods will be referred to as using *directional information* and include techniques proposed in e.g. [1, 4, 5]. The other class of methods, make little or no assumptions on the array response but rely on other properties for separating the signals. In [4, 6], a reference signal is assumed available which may be correlated with the array output to achieve signal separation. This reference signal may be a known training sequence, a known code sequence [7], or may be generated by feeding back decisions [8]. There are a number of methods that make use of the constant modulus property or finite alphabet of communication signals to separate them [9, 10].

To achieve increased system capacity by employing an array of antennas at the base stations, the frequency reuse distance may be decreased [11, 12] or the frequency channels may be reused with in a cell [1] (or a combination thereof). In both cases, the interference in the system induced by other users is of course increased. In the up-link, this is manifested by the cross-talk problem. Mobiles operating on the same channel (frequency/time slot) with dramatically different signal amplitudes caused by, for example, fading are difficult to separate. It is difficult to adequately suppress the stronger signal when estimating the weaker signal resulting in cross-talk. In some sense the down link problem may be even more sever, especially in frequency division duplex (FDD) systems [1]. The fading caused by local scattering around the mobile (or the base station) is observable in the up-link but unobservable in the down link due to the uncorrelatedness of the fading processes at the different frequencies. The up and down link channels are not reciprocal. The down link problem has received limited attention. In [11] a method is proposed which does not exploit directional information whereas in [1] a model based approach using this information is proposed.

3 A Spatial Channel Model

In [2, 13] a model of the flat fading due to local scattering is developed taking the spatial dimension into account. The propagation between the mobile and the array is modeled as a superposition of a large number of rays originating from local scatterers in the vicinity of the mobile. We assume independent scattering, an angular distribution of the scatterers which is Gaussian (as seen from the array), and that the relative time delays for different propagation paths are small compared to the inverse of the bandwidth of the communication signal.

Assuming a uniform linear array with element spacing Δ in wavelengths, the signal received at the array may then be modeled as

$$\mathbf{x}(t) = \mathbf{v}s(t) + \mathbf{n}(t) \tag{1}$$

$$\mathbf{x} \in N(0, \mathbf{R}(\theta, \sigma, S, \sigma_n)) \tag{2}$$

$$\mathbf{R}(\theta, \sigma, S, \sigma_n) \approx S\, \mathbf{a}(\theta)\mathbf{a}^*(\theta) \odot \mathbf{B}(\theta, \sigma) + \sigma_n^2 \mathbf{I} \tag{3}$$

$$\mathbf{a}(\theta) = [1, e^{j2\pi\Delta \sin\theta}, \ldots, e^{j2\pi\Delta(m-1)\sin\theta}]^T \tag{4}$$

$$\{\mathbf{B}(\theta, \sigma)\}_{kl} = e^{-2[\pi\Delta(k-l)]^2 \sigma^2 \cos^2\theta} \tag{5}$$

where, $\mathbf{x}(t)$, is a complex valued ($m \times 1$) vector, $s(t)$ is the complex envelop of the transmitted signal, $S = |s(t)|^2$ is a function of the constant amplitude of $s(t)$, $\mathbf{n}(t)$ is the additive spatially white noise, $\odot$ denotes element-wise multiplication, and $\mathbf{v}$ is a complex, Gaussian random vector with a distribution function parameterized by the nominal direction to the mobile, θ, and the angular spread (standard deviation), σ, see Figure 1.

Equations (1-5) model the Rayleigh fading of the channel. The vector $\mathbf{a}(\theta)$ is often termed the array response vector and represents the array output to a point source from direction θ. Frequency selective fading may be incorporated in the model by adding time delayed versions of the signal with different spatial characteristics. Also, interfering sources on the same frequency channel may easily be added to the model.

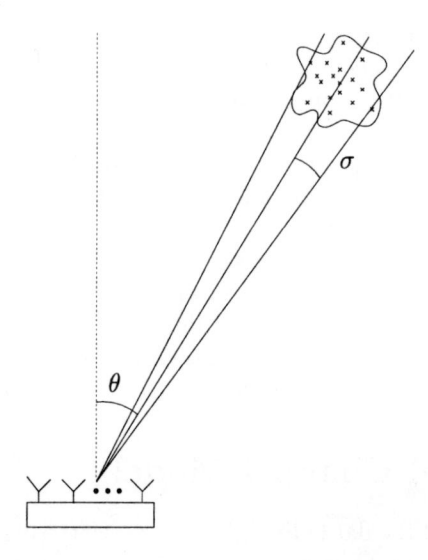

Figure 1: Geometry of the model characterizing the local scattering in the vicinity of the mobile.

Propagation Modeling and Data Experiments

The spatial channel model described above has been validated against experimental data collected by Ericsson Radio Systems. In the field experiments, a transmitter has been placed in urban areas approximately 1km from the receiving array [12]. The data has been processed to gain insight into propagation effects as well as into the behavior of some receiving algorithms. The standard deviation, σ, of the angular distribution is a critical parameter for SDMA systems, [1]. In [2, 13] the angular spread is found to be between two and six degrees in the experiments when the transmitter is placed 1km from the receiving array.

4 SDMA in the Up Link

Here, we will discuss some observations related to the model presented above. Due to the local scattering, the wavefront at the array represented by $\mathbf{v}$ is not planar, i.e.,

$$\mathbf{v} \neq \mathbf{a}(\theta), \quad \text{for any } \theta . \tag{6}$$

This may be interpreted as spatial diversity, i.e., the correlation between antenna elements decreases with distance, this is seen in the structure of the second moment of $\mathbf{v}$ in (3). The flat fading becomes less sever at the array as the diversity increases, i.e., σ increases. Techniques that make no use of directional information, e.g., [6] efficiently exploit this fact and perform better as the angular spread increases. Methods that are based on directional information, $\mathbf{a}(\theta)$, for estimating the signals [4, 14] will in general deteriorate as the angular spread becomes larger. These methods which are related to traditional *beamforming* techniques, are derived from a point source model. This behavior is not surprising since $\mathbf{v}$ will not correspond to an array response vector for any θ.

If the goal is to increase the range of a cellular system, this model error is not critical, however, the situation is quite different when attempting to host multiple mobiles in the same cell on the same frequency channel. This is most noticeable in situations where multiple mobiles are present and the signal amplitudes differ significantly. When estimating the weaker signal, it will be difficult to impose orthogonality to the stronger signal when constrained to the point source model resulting in cross talk between the spatial channels.

Note that the signal at the array (neglecting noise) is still a low rank process lending itself to subspace based estimation techniques. However, the point source array response model is not appropriate.

5 SDMA in the Down Link

Note that we may model the down link spatial channel statistics as in (1) however, in most current FDD systems the up and down link flat fading may be considered independent. If the main objective is increased range, this does not pose a major problem. However, the unobservable down link channel this one of the main obstacles if the intention is to also increase system capacity. An array could be employed at the mobile site as well, but in many applications this is not considered a feasible solution. Another alternative is to attempt to estimate the channel by employing feedback [15]. This requires a complete redesign of protocols and signaling and is probably only possible in environments which are very slowly varying. This technique may be feasible for *movable* (rather than mobile) systems such as indoor wireless local area networks.

If we are attempting to increase capacity in current FDD systems in the down link, the information gained from the signal separation techniques in the up link, can not be used directly. Since the channels are not reciprocal, it is not possible to reuse an "optimal" weight vector obtained from receive data, in the transmit mode. One must at least attempt to "transform" the weights to the transmit frequency. However, this is not a well conditioned problem unless an array model is introduce. When using an array model to transform weight vectors, directional information is exploited. It should be noted that in [11], a transmit scheme is proposed which does not use directional information. The down link scheme is based on statistical information estimated in the up link to take into account the unobservable fading. However, the frequency duplex distance is not compensated for causing the

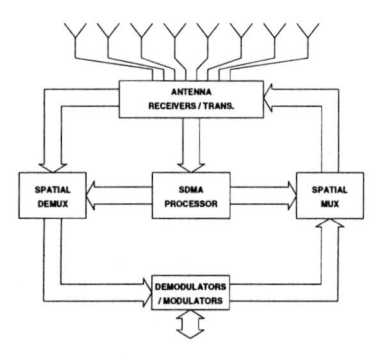

Figure 2: Possible configuration of an SDMA system.

system to degrade in the presence of strong direct paths.

In time division duplex (TDD) systems, the up and down link channels can be considered reciprocal if there is limited movement between receive and transmit. Up link channel information may then be used to achieve spatially selective transmission and thus increasing capacity [16]. However, efficient use of the spatial dimension in current FDD cellular systems requires the use of directional information. Array response modeling is feasible for medium to large size cells with high placement of the base station antennas avoiding near field scattering.

6 SDMA System

There are two main approaches for increasing capacity with antenna arrays. The frequency reuse distance may be decreased or multiple mobiles may be allocated to the same cell (or some combination of the above). In [1] it is shown that when directional information can be used, multiple mobiles per cell is a more efficient way of increasing capacity. Figure 2 displays the general structure of a SDMA system. There are several advantages with this approach to increasing capacity. To fully exploit decreased reuse distance, one must suppress signals to mobiles in other cells by forming nulls in the down link transmit pattern. This is very difficult even in a synchronous TDMA system because of propagation delays. With a small reuse distance, the desired and interfering signals fade independently at the mobile causing problems. Also, allocating mobiles to a frequency/spatial channel is easier when treated locally within a cell. By reducing the reuse distance, capacity is maximized when all frequencies are used in all cells whereas in the other scheme, at least in theory, capacity is hardware limited.

7 Summary

Providing adequate coverage and sufficient capacity are two challenging problems for wireless communication systems. Antenna arrays at the base stations of cellular systems can increase range compared to current systems. The capacity problem can be significantly mitigated by *spatial division multiple access (SDMA)* techniques. SDMA supports multiple connections on a single conventional channel, based on user localization steered reception and transmission schemes, and therefore offers capacity increases over current wireless system implementations.

References

[1] Per Zetterberg and Björn Ottersten, The spectrum efficiency of a basestation antenna array system for spatially selective transmission, *IEEE Transactions on Vehicular Technology*, **44**(3):651–660, August 1995.

[2] Per Zetterberg, Mobile communication with base station antenna arrays: Propagation modeling and system capacity, Licentiate thesis TRITA-SB-9502, Signals, Sensors & Systems, February 1995.

[3] D.H. Johnson and D.E. Dudgeon, *Array Signal Processing – Concepts and Techniques*, Prentice-Hall, Englewood Cliffs, NJ, 1993.

[4] S. Andersson, M. Millnert, M. Viberg, and B. Wahlberg, "An Adaptive Array for Mobile Communication Systems", *IEEE Trans. on Veh. Tec.*, **40**(1):230–236, 1991.

[5] S.C. Swales, M.A. Beach, D.J. Edwards, and J.P. McGeehan, The performance enhancement of multibeam adaptive base-station antennas for cellular land mobile radio systems, *IEEE Trans. Vehicular Technology*, **39**:56–67, Feb. 1990.

[6] J.H. Winters, Optimum combining in digital mobile radio with cochannel interference, *IEEE Trans. Vehicular Technology*, **33**(3):144–155, August 1984.

[7] A.F. Naguib, A. Paulraj, and T. Kailath, "Capacity Improvement with Base-Station Antenna Arrays in Cellular CDMA", *IEEE Trans. Vehicular Technology*, **43**(3):691–698, Aug. 1994.

[8] E. Lindskog, A. Ahlén, and M. Sternad, "Combined spatial and temporal equalization using an adaptive antenna array and a decision feedback equalization scheme", In *IEEE International Conference on Acoustics, Speech and Signal Processing*, Detroit, USA, May 1995.

[9] S. Talwar, M. Viberg, and A. Paulraj, "Blind Estimation of Multiple Co-Channel Digital Signals Using an Antenna Array", *IEEE SP Letters*, 1:29–31, Feb. 1994.

[10] B.G. Agee, A.V. Schell, and W.A. Gardner, "Spectral Self-Coherence Restoral: A New Approach to Blind Adaptive Signal Extraction Using Antenna Arrays", *Proc. IEEE*, **78**:753–767, Apr. 1990.

[11] G. Raleigh, S.N. Diggavi, V.K. Jones, and A. Paulraj, A blind adaptive transmit antenna algorithm for wireless communication, In *Proc. ICC*, 1995.

[12] U. Forssén, J. Karlsson, B. Johannisson, M. Almgren, F. Lotse, and F. Kronestedt, "Adaptive Antenna Arrays for GSM900/DCS1800", In *Proc. IEEE Veh. Technol. Conf.*, pages 605–609, 1994.

[13] Tõnu Trump and Björn Ottersten, Estimation of nominal direction of arrival and angular spread using an array of sensors, To appear in Signal Processing, Elsevier, 1995.

[14] B. Ottersten, R. Roy, and T. Kailath, "Signal Waveform Estimation in Sensor Array Processing", In *Proc. 23rd Asilomar Conf. Sig., Syst.,Comput.*, pages 787–791, Nov. 1989.

[15] D. Gerlach and A. Paulraj, Adaptive transmitting antenna arrays with feedback, *IEEE SP Letters*, **1**(10):150–152, October 1994.

[16] P. Mogensen, F. Frederiksen, J. Wigard, and S. Petersen, "A Research Study of Antenna Diversity and Data Receivers for DECT", In *Proc. Nordic Radio Symposium*, Saltsjöbaden, Sweden, April 1995.